CREATIVE ENGINEERING DESIGN

FOURTH EDITION

Brian S. Thompson

Professor of Mechanical Engineering
Michigan State University

Okemos Press

Okemos Michigan

Creative Engineering Design

Fourth Edition
Second Printing July 1997

Printed in the United States of America

Okemos Press
Scientific and Technical Books

Okemos MI 48805-0085

ISBN 0-9630471-8-3

The logo of the Okemos Press is derived from alchemical symbols for lead ♄ and gold ☉

To Anne and our creative engineering designs:
Ashley, Christopher, and Andrew.

Contents

Preface for the Student

Our twentieth century lifestyles expose us to a wide variety of products. They include tennis rackets and toothbrushes, light bulbs and lasers, dog collars and distensible catheters. All of these products are the result of a product realization process in which artifacts are conceived, manufactured, brought to market, supported during their service life, and finally disposed of. This process is characterized by the repeated confrontation of problems which have no obvious single solution and which are solved using an iterative cycling of creativity and analysis. This iterative cycling is the very heart of design.

Design is not a new process; it has been practiced for millennia as evidenced by the very first creations of *Homo habilis:* flint hand tools with special shapes, bone needles, and the spear. Thus design has been practiced for at least two million years—long before the advent of the scientific method, mathematics, or computer technologies. While these relatively modern methodologies and technologies are highly developed, the design process is not well understood and it is often classified as an art rather than a science because of its complexity. Indeed we have not yet developed an algorithm that definitively creates an optimal solution for an open-ended design problem. The wide range of family automobiles on our streets attests to this assertion.

In spite of this limitation, the implementation of good design methodologies is crucial for the success of any company because the design process is largely responsible for the quality of its products. Studies have shown that over 70 percent of the life-cycle cost of products is determined during the design phase and therefore this process is responsible for the success of the company and ultimately the success of any industrialized nation. The challenges of contemporary engineering practice and the associated societal responsibilities demand a high level of creativity in the engineering design process. Indeed it is of paramount importance in the development of any product.

Creative Engineering Design presents the essential ingredients of the design process from the recognition and definition of the problem, through conceptualization, to the transformation of that vision into plans for manufacture of a product and its support during service. As a student, you have taken numerous classes devoted to analysis (mathematics, machinery dynamics, heat transfer) but you have probably not been exposed to an environment involving creativity in engineering design. Without some knowledge of this discipline you will find it difficult to design anything. Unlike the purely analytical single-solution problems encountered in your classes in the engineering sciences, your design classes expose you to methodologies for solving open-ended problems. These problems—for which there is a multitude of solutions—are extremely important because they are common in all walks of life. Examples include deciding where to go for a vacation, what type of car to purchase, or which job to take! Therefore it is of great personal benefit to become proficient at addressing them.

Solutions to most engineering design problems can only be developed by adopting an interdisciplinary approach that is not constrained by the artificial barriers imposed upon you in your more formal classes such as chemistry, entomology, or the differential calculus. An integrated approach must be employed to synthesize a viable solution. This solution will be achieved by adopting a methodology in which you must visualize new configurations and concepts, in which you must develop appropriate mathematical models, and in which you must select real materials and manufacturing processes subject to economic and many other constraints.

This book has been written in an attempt to help transform you, today's student, into tomorrow's professional engineer by integrating generic design rules with engineering reality.

The book's design has the following attributes:

1. The text is formatted with a relatively narrow column width, just like the format of a newspaper, to facilitate rapid reading. Each page contains wide margins that hold two types of notes. The first category of marginal notes collectively comprise a summary of the book. Each note provides a synopsis of one or more important ideas presented on that page. They can be of great use when you are doing last minute revision before a quiz. Furthermore they can help you locate a section of text on a specific topic at a level below that of the subheadings. A second kind of marginal note presents thought-provoking quotations that provide additional perspectives to the main theme.

   You should use the margins to write your own personal notes. These notes could include your own ideas on the important features of a section. You could write of your own experiences relative to the theme of the page. However, the important thing is to realize that this book is one of *your* own personal tools in *your* design education. Please use it as such.

2. Sets of design rules are highlighted at appropriate locations throughout the book. They serve as a source of easy reference when you work on design projects.

3. Each chapter concludes with a summary of important material covered in the chapter and how it relates to other chapters. This can be an invaluable tool when you review for exams or work on design projects. Each chapter also contains a *Key Concepts* section of the ideas presented in that chapter.

4. *Review Questions* are presented at the end of each chapter that cover the important topics and concepts discussed in the chapter. The answers to these questions can be found in the text. These review questions will assist in your own personal evaluation of how well you have understood the chapter.

5. A list of more challenging *Problems* follows the review questions in each chapter. Although these problems are based upon the topics covered in the chapter, they often require some additional research along with an ability to think and reason. They assume various forms but are invariably problems that have several possible solutions. These problems may serve as the basis for class assignments.

6. Each chapter concludes with a section entitled *Further Reading*. The books and papers contained in this list provide you with the resources for developing an in-depth appreciation of the subject matter covered in the chapter. This can be particularly valuable when you are required to undertake project work.

7. The *Glossary* provides an easily located source of the definitions of all the major terms used in the book. This is a useful resource if you are studying the chapters nonsequentially.

I hope that this student-orientated format will help you learn the rudiments of a systematic approach to creative engineering design. It is also my wish that in some small way this book will ignite a desire within you to become proficient at solving design problems and through this you will develop a quest for knowledge about engineering design and its methodologies. This mind-set will be of great use in your future, whether you plan to become an engineer, a manager, or a beach bum. A knowledge of how to solve life's never-ending sequence of open-ended ill-defined problems creatively will always be a great asset.

Brian S. Thompson
East Lansing, Michigan

Preface for the Instructor

Design is the very heart of engineering. Therefore instructors have a responsibility to the student, the profession, and ultimately the nation to provide a stimulating environment where design skills can be acquired and honed prior to graduation. The methodology for effectively teaching the design process and for the solution of ill-defined open-ended problems is still a topic of considerable debate. Indeed it is an open-ended problem in itself! Viable solutions to this educational problem require much innovation in the already crowded undergraduate curriculum, the current reward system in academia, and the faculty-intensive nature of teaching design.

The author is convinced, however, that innovative solutions must include design-build-test-report projects that involve groups of students. These projects mimic industrial design practices and help the student evolve from an academic environment into industrial practice. However, we must be sensitive to the significant challenges that confront our students studying the field of engineering design. These challenges occur for several reasons, and especially so when students participate in group projects. They include the following:

- Students must plan and then execute a project often lasting several months. This mandates project planning and project management.
- Group projects require students to cooperate harmoniously on a long-term task. This motivates the student to develop and refine interpersonal skills.
- Engineering design involves both creativity and analysis. They are the two extremes of a range of cognitive processes. While students have experience with analysis, they are invariably weak in creativity.
- Students are required to clarify and define the design problem before beginning to solve it. Again this contrasts with their experiences in the engineering sciences where instructors assign carefully selected, carefully formulated, single-discipline problems.
- Students are required to make oral presentations to the class and to write a series of design reports while they maintain a design notebook. These are new challenges but they facilitate the development of the vital communication skills needed by practicing professional engineers.
- Students must be able to develop mathematical models for predicting the behavior of a system or a part. This involves making appropriate assumptions, creating computer models, and comparing the results with experimental test data. This iterative process is a new experience.

- Students must undertake economic analyses while they work at the boundaries of engineering and management, and they must establish a financially acceptable domain within which to work on their project.
- Students must be able to transform sketches into working physical entities. This requires a good knowledge of the engineering sciences, manufacturing processes, and design-for-manufacture principles. Again, this is a new experience for many students.
- Engineering design is a boundless subject requiring the application of knowledge from many fields, including the sciences, economics, engineering, business practices, marketing, the arts, and psychology. This is quite a challenge. However this challenge must be met if we are to educate students who are qualified to enter professional practice as engineering designers.

The author is convinced that the design experience should incorporate group projects that mimic the complete product realization process. Design methodologies cannot be learned and understood by simply attending lectures or by only completing "paper studies." The student should be challenged in design courses by the necessity to complete projects involving the rigorous definition of problems, the creation of conceptual designs, the evaluation and transformation of these concepts into plans for manufacture, the fabrication of parts, and the testing of the product. Such experiences provide the basis for learning with excitement, frustration, satisfaction, and fun; they require active learning; they mimic the product realization process.

Naturally these projects take different forms depending upon the academic maturity of the students. Thus in the capstone senior-level design class, for example, students should be exposed to *bone fide* industrial projects. These projects will involve regular interactions between the students and a representative of the industrial sponsor, and they should include visits to the sponsor's facilities. This is a very motivational environment in which students operate at the dusk of their formal educational experience. Before this, student design-build-test competitions involving small-scale projects, such as shoe-box-sized vehicles powered by birthday candles, can be employed effectively to teach several important aspects of creative engineering design. These class projects can create great excitement. Furthermore, at all levels students should be encouraged to develop interdisciplinary mathematical models, to recognize their mistakes, and then to rectify them by employing an iterative procedure. This procedure involves predicting the behavior of their proposed design and then testing that prediction.

Creative Engineering Design has evolved from class notes used to teach undergraduate senior-level design classes during the past 20 years. However, the basic ideas can be taught at the sophomore level perhaps in conjunction with a class that introduces basic manufacturing processes. This book presents a systematic approach to engineering design that is predicated upon the philosophy that with all branches of human activity—whether an intellectual activity or a manual activity—a systematic approach will guarantee the greatest progress in the minimal time. The book is written primarily for undergraduate mechanical engineering design courses focused upon the teaching of generic methodologies. This is accomplished in courses where students are exposed to a series of lectures on the multiple facets of the design process before these concepts are cemented together by a project activity involving groups of students. The principles and strategies presented in these lectures can be significantly enhanced using illustrative examples from engineering practice. Hardware should be brought to class for discussion and *in vogue* items from newspapers and periodicals also provide design concepts to which students can relate. The class notes embedded in this book have proven to be of great utility in the teaching of these courses which attempt to rectify the imbalance in the undergraduate curriculum caused by the bias towards the analytical techniques of the engineering sciences.

It is quite impossible to include all of the disciplines relevant to engineering design in a single text. This text is not directed at the markets targeted by handbooks or reference texts. Rather, it is directed at the undergraduate student population in mechanical engineering with the objective of helping the student develop a systematic approach to solving open-ended design problems.

The early chapters of the book offer a description of design, its history, morphology, and role in the generation of products. Subsequent chapters focus on the various phases of the design process from design specification through conceptual design to product design. Chapter 1 provides an introduction to mechanical engineering design in which the interdisciplinary nature of the field is emphasized. Chapter 2 focuses upon the history of design. The impact of design activities upon the evolution of civilizations is discussed. Subsequently Chapter 3 presents a case-study of the *Voyager* aircraft that flew around the world without refueling. The objective of this chapter is to expose the student to an open-ended design problem and the task of creating a solution to this challenge.

Chapter 4 presents a procedure for solving open-ended design problems. The primary phases are discussed and this chapter provides a global road map for the student in the solution of design projects. The product realization process is presented in Chapter 5. This chapter is concerned with the influence of design methodologies on industrial practices. The numerous facets of the product realization process are highlighted. Sequential and concurrent engineering are defined and contrasted. This chapter concludes the introduction to the broad field of engineering design. Attention in Chapter 6 focuses on environmental issues throughout the product realization process. This topic is becoming increasingly important and the pivotal role of the design community is discussed. Design principles are presented for materials management, energy utilization, and extending the service life of products.

Chapter 7 considers the first stage of the design process: the establishment of a design specification. This specification is formatted using quality function deployment, design parameters, and determination of the relative weight of these parameters. Creativity is an essential component in design and this topic is the subject of Chapter 8. A number of psychological impediments to creativity are discussed. Chapter 9 presents strategies for generating conceptual designs where creativity is at a premium. Chapter 10 concludes Part One of the book by presenting methods for evaluating the different concepts generated during the conceptual design phase.

Part Two of the book is dedicated to product design with an emphasis on design-for-manufacturability. Chapter 11 discusses product design in which the vision of the final product is transformed into hardware. Chapter 12 is devoted to the general rules for design-for-manufacture. They are quite independent of the materials or the manufacturing processes under consideration. This is followed by a discussion of design for assembly in Chapter 13. Chapters 14 through 17 are concerned with design-for-manufacture of monolithic materials by sand casting, forging, welding, and machining. The impact of each manufacturing process on the design of the product is considered in detail. The focus of attention in Chapter 18 is the commercial fibrous polymeric composite materials. The attributes of fibrous composites are discussed and common processes are described before a methodology is introduced for designing parts in these materials. This is an increasingly important field.

The first section of the Appendix contains a discussion of the importance of communication skills. Guidelines are presented for the written, spoken, and the unspoken word. This is a particularly important part of the engineering profession that is frequently overlooked in a crowded undergraduate curriculum. The second part of the Appendix describes a variety of student design-build-test projects, many of which have been used in American Society of Mechanical Engineers (ASME) competitions. The third section of the Appendix contains a glossary of important terms and concepts. This is followed by a bibliography.

The author is conscious of human frailties and limitations. Consequently, he realizes that errors must have inevitably escaped his scrutiny as he toiled to complete this book. He would be grateful if readers would advise him of any errors and he welcomes suggestions for improvements.

The author hopes that this book will stimulate instructors to further cogitate upon the role of creative engineering design activities in the shaping of societal values at the dawn of the 21st century. The solution of open-ended problems, of which design is one set, is of great utility in numerous walks of life. It is important that our students develop a good grounding in solving this class of problems. Design is one of the oldest pursuits of human civilizations, yet it is often practiced as an art with little scientific basis. This realization, coupled with a profound appreciation that engineering and design are synonymous, has been responsible for increased scrutiny of the design process. Indeed the current consensus is that design should be emphasized more in university curricula in order to better educate the next generation of engineers. It should be woven throughout the curriculum from freshman classes to the senior level classes. This book represents the author's attempt to embrace and further this notion, albeit acknowledging the fact that, while *Homo sapiens sapiens* has amassed mountains of knowledge in many fields, the knowledge associated with design and the effective teaching of this subject is more akin to a flat plain. Creative engineering design remains one of the remaining intellectual challenges of the 20th century.

Brian S. Thompson
East Lansing, Michigan

Homines dum docent discunt.

Even when they teach, men learn.

Seneca (4 B.C.–A.D. 65)

Acknowledgments

On February 5, 1676, Isaac Newton wrote a letter to Robert Hooke, the famous physicist, in which he commented on his own achievements in the following way. "If I have seen a little further [than others], it is by standing on the shoulders of giants." If I have achieved anything in this printed work it is because of this very reason. It is based upon the accomplishments of current and previous generations of gifted individuals.

The author wishes to formally acknowledge with thanks the numerous individuals who have directly contributed to this book, both physically and through mental stimulation. The author's interest in solving open-ended design problems was first stimulated during his undergraduate education by the teaching of Mr. A. A. Fogarasy at the University of Newcastle-upon-Tyne in England. Lectures by Professor L. Maunder further reinforced this interest. These gifted individuals were responsible for formulating a crucial script that was indelibly written.

The author also wishes to acknowledge with thanks the numerous insightful comments and suggestions on aerodynamic principles offered by his young soccer-playing colleague, Professor M. M. Koochesfahani, during the preparation of the section of the book devoted to aerodynamic biomimetics. These stimulating interactions on aerodynamic principles have been of great value to the author, both in the class room and on the soccer field where the control of a single synthetic spherical body subjected to unsteady aerodynamical environmental conditions assumes considerable significance.

The author would also like to thank the following reviewers for their helpful comments: Professor A. Ward, University of Michigan, C. Cloet, University of Leuven, Belgium, and Bruce K. Bright of DeWitt High School. Some of the chapters were reviewed by Julia Damore and Craig Gunn, Director of Communications in the Department of Mechanical Engineering at Michigan State University. This book would not have been completed without the intellectual stimuli, discussions, guidance, and comradeship provided by Dr. Robert N. Hammer and Dr. Paul W. W. Hunter of the Okemos Press.

This book has taken several years to complete and it has included a number of memorable experiences. I think that these are eloquently captured by Winston Churchill in the following quotation: "Writing a book is an adventure. To begin with, it is a toy and amusement; then it becomes a mistress, and then it becomes a master, and a tyrant. The last phase is that just as you are about to be reconciled to your servitude, you kill the monster, and fling it out to the public." I can readily relate to these sentences.

Brian S. Thompson
East Lansing, Michigan

Sources of Illustrations

The author wishes to thank the following individuals and organizations for permission to reproduce their illustrations and photographs.

Pergamon Press for illustrations from M. F. Ashby's *Materials Selection in Mechanical Design*

A. Simpson of Airlife Publishing Ltd.

George Mochnal of the American Forging Association

American Society of Mechanical Engineers

D. Seim of Arden Fasteners

P. Nestor of W. M. Berg

R. L. Saccone II of DFI Pultruded Composites

DRM Associates

S. Nichols of Ford Motor Company

Gene Wistehuff of G. M. Powertrain, Inc.

Gordon L. Hall

Institute of Scrap Recycling Industries

McGraw-Hill, Inc.

Medisense, Inc.

Munro & Associates

B. Fryday of NASA

Society of Manufacturing Engineers

Society of Plastics Industry

Dean Johnson of Suminoto Machinery Corporation of America

M. H. Evans of Rolls Royce plc

Texas Instruments, Inc

L. E. Repaci of WD-40 Company

J. S. Szpondowski of Wyandotte Industries, Inc.

Part One

Prologue: Engineering Design

1

A scientist discovers that which exists. An engineer creates that which never was.

Theodore von Kármán
(1881–1963)

Objectives

- To illustrate the multidisciplinary character of engineering design.
- To discuss different models of the design process.
- To emphasize the importance of design in the engineering curriculum.
- To present a methodology for solving open-ended design problems.
- To introduce project planning for a student design assignment.

Contents

What if...

You're an engineer and your boss has a problem. Now *you* have the problem—but no ideas for its solution. What a situation!

Remember that engineering design course you had in college? Could it help?

You're more likely to create an idea if you know what the problem *really* is. So define it—clearly—in your own mind. That gives you a few ideas, but they aren't very good. You know that fresh ideas often arise if you juxtapose seemingly unrelated ideas. After all, engineers can use ideas from chemistry, or history, or today's newspaper. Creativity is a complicated, interdisciplinary process.

And it's iterative. So around the circle you go, each time rethinking everything. Each time the idea gets better developed, its reduction to engineering practice improved.

Then explain the ideas to others. Surprising how that clears your head, even if the other person doesn't know much!

1.0 Introduction

Engineering is defined in the *Encyclopaedia Britannica* as "the application of scientific principles to the optimal conversion of natural resources into structures, machines, products, systems, and processes for the benefit of mankind."

Engineering design is at the very heart of this activity. It is the global process associated with creation of a product to satisfy a need in the marketplace. In this chapter you will see how this complex process commences with the identification of that need and culminates in specific detailed plans for manufacturing and supporting a product that satisfies the need.

Engineering design is the process associated with development of a product to satisfy a need in the marketplace.

A product is an artifact that is the result of a design process.

Engineering design has the following attributes:

- It's an interdisciplinary process
- It's complicated
- It involves both creativity and analysis
- It's an interactive process
- It's an iterative process

You will see how each of these characteristics occurs in the design process in the subsequent chapters, but first let's examine design and the design process and how it fits into the university engineering curriculum.

1.1 Design

Design activities are essential to creative problem solving and the design of products. The design methodology presented here and illustrated in Figure 1.1 is equally applicable to designing an engineering system, planning landscaping for a home, or shaping clay on a potter's wheel. It requires the creation of ideas, their analysis and evaluation, and the selection of the most promising idea for future development and refinement.

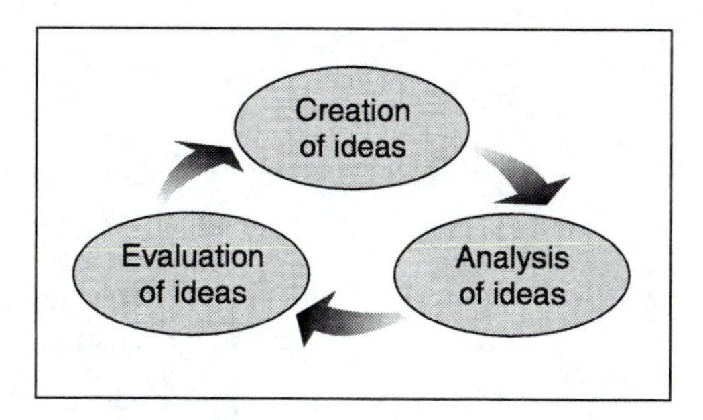

Figure 1.1

A methodology for creative problem solving

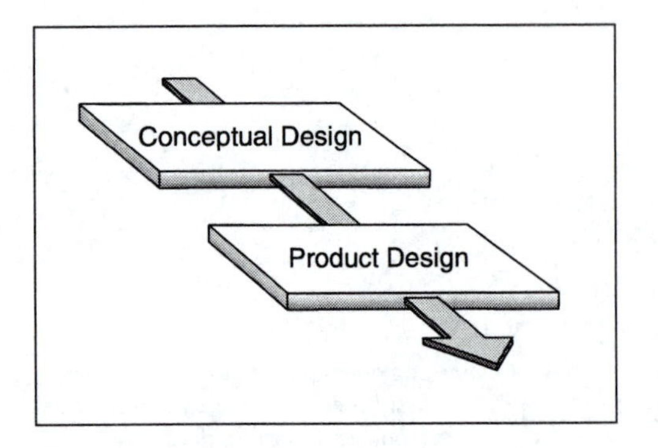

Figure 1.2

The two phases of engineering design

Engineering design takes place in two principal phases: the *conceptual design phase* and the *product design phase* (Figure 1.2). During the conceptual design phase, function is the primary concern. Different concepts are created that could potentially satisfy the requirements for the product. This creative activity often involves recording ideas as rough sketches or written explanations in notebooks. Generally, the broader the set of conceptual designs, the greater the probability of creating a viable product.

The most promising concept provides a basis for the product design phase. This phase transforms the winning concept into engineering drawings and detailed plans for manufacturing the product. Materials, shapes, and other manufacturing attributes of all parts and subassemblies are specified.

The task of defining and implementing the design process varies, and several different formulations have been proposed. In the principal methodologies proposed in the literature, various terms and subtasks are emphasized. To illustrate these differences, four primary design models are summarized below and their terminologies are briefly defined. They are illustrated in Figure 1.3.

Model 1. In the United States, the design process is modeled as the two stages mentioned above—the conceptual design phase and the product design phase.

Model 2. In Germany, considerable research has been devoted to the development of systematic models for the design process. For example, Verein Deutscher Ingeniere (VDI), the German society of professional engineers, has developed guideline VDI 2221 entitled *Systematic Approach to the Design of Technical Systems and Products.* This approach is based upon several distinct stages listed in Figure 1.3 that are each characterized by an iterative process involving creativity, analysis, and evaluation.

Model 3. One of the popular German models was proposed by Pahl and Beitz in their text *Engineering Design.* It comprises four stages: *task clarification, conceptual design, embodiment design* and *detail design. Embodiment design* involves the creation of preliminary layouts and shapes of the individual components as the technical product is developed from the original concept. During this phase the original qualitative concept is subjected to mathematical analysis for the first time and quantitative results are generated. The best preliminary layouts are selected for further refinement. *Detail design* is the last stage of the traditional design process in which the function, materials, shape, and manufacturing details are put into final form. Detail drawings are made for each part, assembly drawings are completed, and production documentation is developed.

Define the task			
Model 1	Model 2	Model 3	Model 4
Conceptual design	Determine functions and structures Search for solution principles	Conceptual design	Basic design
Product design	Divide into realizable modules Complete the overall layout	Embodiment design	Form design
	Prepare manufacturing documentation	Detail design	Detail design

Figure 1.3

Four primary design methodologies

Model 4. In other parts of Europe, the design process is often considered to consist of *basic design, form design,* and *detail design.* The *basic design* stage is the same as the conceptual design phase while *form design* is the same as the embodiment design phase or the early portion of the product design phase. The alternative term form design is motivated by an emphasis on developing an appropriate form or shape for each part. The detail design phase is the same as that described earlier.

All four models describe engineering design in slightly different ways. The design process is essentially the same although the terminologies differ.

1.2 Why Design is Important in the Undergraduate Engineering Curriculum

In a good undergraduate engineering curriculum, a warp and weft of design protocols are woven through the fabric of all courses. Exposure to design should begin in your freshman year and design competence should gradually increase until, in the senior year, it attains maturity in a capstone design projects class. This completes your formal undergraduate engineering education. One can argue that an engineering design class focused on creative thinking is one of the most important educational experiences of your degree program. This statement is especially true if the projects are generated by industrial sponsors and you interact regularly with a representative from that sponsor.

Mechanical engineering design classes provide an educational experience in which you use your intellectual ability to apply scientific knowledge and knowledge from other fields to solve open-ended engineering design problems.

To attain these goals, you must do the following:

- Define the problem
- Develop a plan
- Be creative
- Integrate disciplines
- Develop mathematical models
- Perform an economic analysis
- Communicate with others
- Transform academic experiences into engineering practice

Planning the Project

Design class projects may take several weeks or a whole semester to complete. On the completion date a final report, possibly an oral presentation, and perhaps a working model of your design will be required. Furthermore, there may be interim reports and presentations to be made on specified dates. To achieve these objectives in a timely manner you should develop a plan for accomplishing the project (Table 1.1). In this project—as in many aspects of life—you must manage your time carefully.

Remember, time is money.

Benjamin Franklin
(1706–1790)

To develop a plan for a project you must first determine:

1. The objective of the project.
2. The beginning and the ending dates as well as other important interim dates for the project.
3. The tasks that must be accomplished to achieve the final objective.
4. What must be accomplished in each task. The tasks should be precisely defined.
5. The estimated duration of each task.
6. The sequencing of the tasks.

Table 1.1

Things you must know to plan a project.

Now you have a plan for your project that permits you to accomplish the assigned task in the allotted time. If you do not plan your work carefully, you will undoubtedly become overburdened immediately before project deadlines and will have difficulty completing the assignment.

The successful execution of a project requires planning and scheduling. *Planning* is the identification and sequencing of the tasks in a project. *Scheduling* is the assignment of the tasks to specific calendar dates.

Planning is the identification and sequencing of the tasks in a project.

Scheduling is the assignment of the tasks to specific calendar dates.

Several methods have been developed to help you plan and schedule projects. The simplest is the *Gantt chart*—a kind of horizontal bar chart. Figure 1.4 (a greatly simplified Gantt chart) suggests how to make one. The horizontal axis displays

In studies, whatsoever a man commandeth upon himself, let him set hours for it.

Francis Bacon (1561–1626)

the passage of time—best represented by the actual dates on the calendar rather than "Week Six," for example. (Actual dates help you quickly see when a project deadline occurs.) The vertical axis represents the tasks to be accomplished. These are laid out in horizontal blocks. Notice that some tasks can be worked on concurrently whereas others require prior completion of certain tasks. Gantt charts are adequate for simple projects (your design class project, for example) but they are inadequate for complex projects in which activities are interrelated.

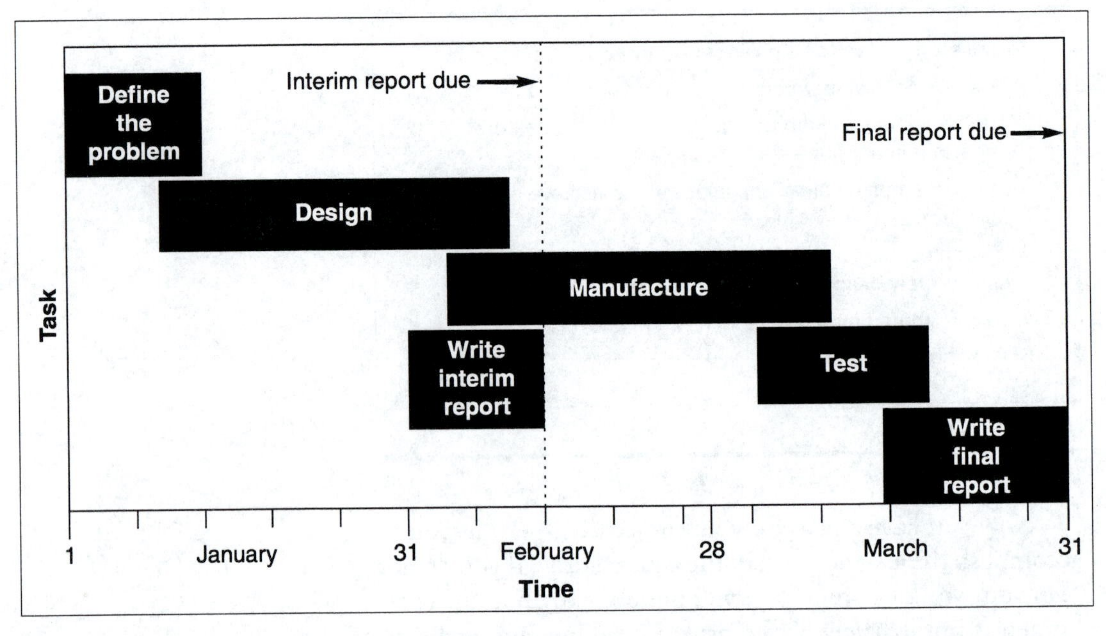

Figure 1.4

A Gantt chart. The horizontal axis displays time while the vertical axis lists tasks to be accomplished.

Once you have developed a Gantt chart for one of your own projects, you should display it in a prominent place where you can see it daily and keep focused.

CPA is sometimes called critical path method *(CPM)*.

Complex projects are best scheduled using logic network methods such as *critical path analysis* (CPA) and the *program evaluation and review technique* (PERT). While these methods differ in how they estimate the duration of individual activities, both use a network of interconnecting arrows and circular nodes to represent the duration, direction of flow, and connectivity of activities and events of a project. An *activity* is a time-consuming task that must be accomplished as part of a project. An *event* is the end of an activity and the beginning of another activity. It is a time for making a decision or it is the termination of a task. Figure 1.5 shows a network logic diagram.

An *activity* is a time-consuming task that must be accomplished as part of a project.

An *event* is the end of an activity and the beginning of another activity.

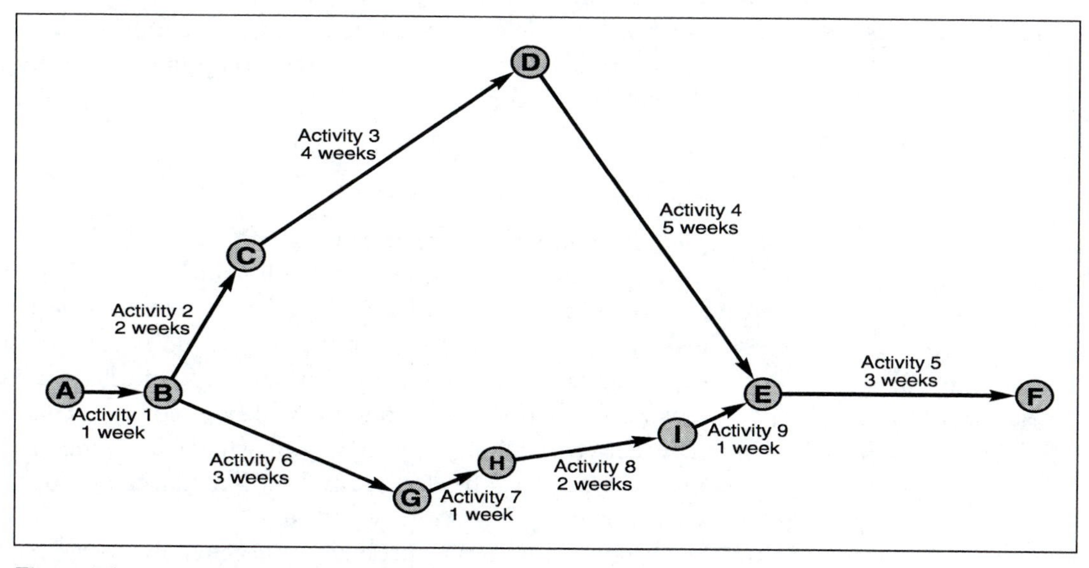

Figure 1.5

A network logic diagram.

In order to construct a CPA or PERT diagram, a list of all the tasks to be accomplished must be developed. Then the connectivity of these tasks must be determined by deciding what must be done before each activity can be started. Finally the event that signals successful completion of each task must be identified. This information provides the basis for a graphical presentation.

CPA or PERT techniques illustrate which tasks must be completed before subsequent tasks can be started. Thus, referring to Figure 1.5, Activity 1 must be done before

Activity 2 and Activity 3 can begin. Similarly, Activity 5 cannot begin until Activity 4 and Activity 9 have both been completed.

The *critical path* is the longest path through the network.

The *critical path* is the longest path through the network. It determines the fastest time in which the project can be completed. Any delay on this path affects the completion date.

In Figure 1.5 this path is traced out by events A, B, C, D, E, and F. The other path denoted by A, B, G, H, I, and E contains slack time. Thus if Activity 9 was started on time but there is a delay of one week in completing it, then this delay would not compromise the completion of the overall project on schedule.

Definition of Problems

The ability to define any problem is an invaluable asset in life, whether you are purchasing insurance, selecting an employer, or being confronted by an engineering design task. The correct problem formulation is the essential first step toward a practical solution. If the problem is defined *incorrectly,* then the solution also will be incorrect. This is especially true in engineering design where the requests of the customer must be clarified and translated into appropriate engineering terminology.

If a problem is defined *incorrectly,* then the solution will be incorrect.

Engineering design problems are generally *open-ended or ill-defined problems.* They are quite different from conventional analytical problems and are often vague, not clearly understood, and are devoid of explicit constraints. Moreover they have no definitive solution—there are many potential solutions.

An *open-ended* or *ill-defined problem* is one that is often vague, not clearly understood, devoid of explicit constraints, and has many solutions.

Your mechanical engineering design projects class will provide an opportunity for you and your student group to address and define problems that are typical of those you will face as a professional engineer. Upon being presented with the skeleton of a problem, you (or your student group) must first define the problem in the context of a number of disciplines such as manufacturing, service, and economics.

Engineering problems are defined in the context of disciplines such as manufacturing, service, and economics.

The experience of addressing a problem for which there is no single solution contrasts with other classes in the undergraduate curriculum in which you are confronted by many problems that rarely occur in the real world. These well-defined, closed-end problems have an obvious goal and clearly prescribed methods for achieving the solution. Such problems often involve idealized conditions characterized by terms such as rigid bodies, isothermal processes, inviscid flow, and spring-mass-dashpot models. These terms are associated with classes in the engineering sciences that focus upon distinct disciplines in isolation from others. Classes of this kind are necessary in the education of design engineers because analysis is an essential ingredient of the design process, but you must learn to consider carefully the assumptions embedded in these mathematical models, their bounds of applicability, and their consequences in design situations. Real-world problems are almost always interdisciplinary.

Synthesis and Creativity

A mechanical engineering design projects class at most universities exposes you for the first time to techniques for enhancing creativity and innovation through a series of lectures and complementary projects. You start the project section of the course with an outline of a problem.

You will create a number of potential solutions. The creation of these concepts is an extremely important phase of the whole design process. In the intensely competitive world of engineering, it might be argued that skill in creating concepts—which is generally responsible for the creation of innovative, cost-effective products—may indeed be more important than analytical skills. Certainly the creative process has a major impact on the success of a product and its final cost.

Integration of Disciplines

A mechanical engineering design projects class involves integration of concepts from many disciplines—those you are studying concurrently and those you have studied in

Real-world problems are not constrained by artificial academic barriers.

previous years at the university, both inside and outside of engineering classes. Artificial barriers created by addressing engineering problems through courses such as heat transfer, vibrations, and solid mechanics no longer apply in the design experience. *You* decide what is important and what the appropriate mathematical assumptions are. Think of yourself as an emancipated student!

You must decide just how sophisticated the mathematical model must be to analyze a given situation. Should you adopt a simple formulation where the theoretical predictions may be quite meaningless in practice, or should you develop a more sophisticated model that may become mathematically intractable? For example, could heat be transferred only by forced convection or will all three primary modes of heat transfer be needed? Would a linear theory be adequate for modeling a phenomenon or must more sophisticated nonlinear models be employed?

A design-build-test project will require you to develop your own mathematical models. Indeed, you will need to consider several disciplines simultaneously and make appropriate assumptions.

The design project experience challenges you to integrate concepts from all of the engineering sciences, the pure sciences, and also the arts. This stimulating environment permits you to gain experience with tools for creative thinking in a multidisciplinary real-world environment.

Economic Analysis

The final cost of a product generally has a major influence upon the success of that product in the marketplace.

A design project usually requires you to undertake an economic analysis because the final cost of a product has a major influence upon the success of the product. This activity often does not appear in other engineering courses. An economic analysis is an important activity in any industrial organization, and the experience of costing at the boundaries of the technical and commercial domains will afford you valuable experience for your professional life.

A cost analysis provides a measure of the success of the design process employed by an organization. The final cost of a product depends upon four major factors which are themselves interdependent because of the iterative, multidisciplinary aspect of the design process (Figure 1.6).

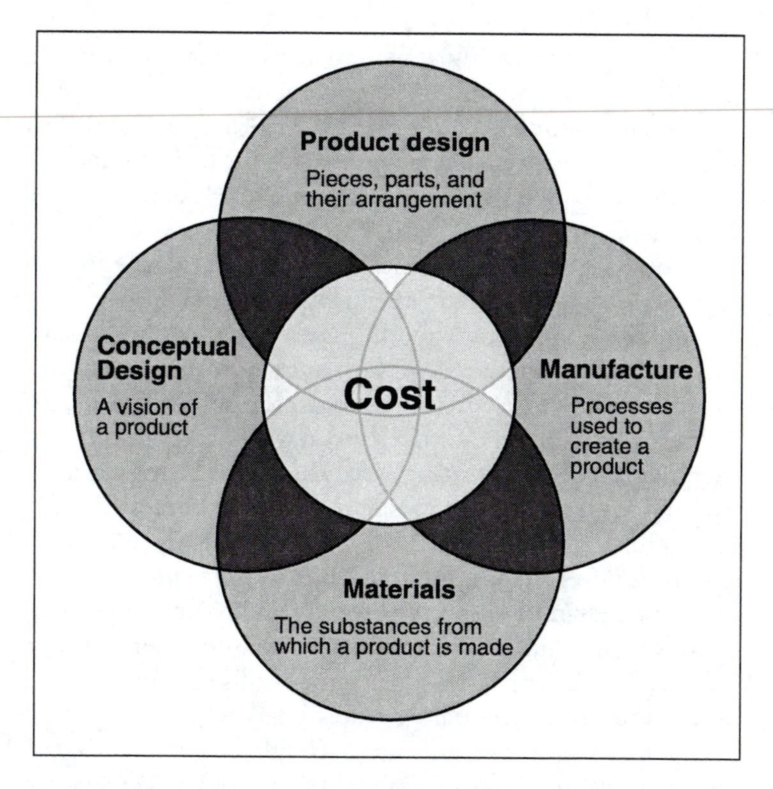

Figure 1.6

The interrelationships between cost, conceptual design, product design, materials, and manufacture.

Communication

While analytical and creative skills are necessary in engineering, you will soon discover the importance of communication skills. They are an essential part of engineering activities in any industrial enterprise. Communication includes reading, technical writing, speaking, listening, personal interaction, and the preparation of drawings and other graphics.

Communication is an essential ingredient of the engineering activities in any industry.

For an industrial organization to function effectively, ideas and information must flow easily among individuals and departments. This is especially important during the engineering design process.

Design projects classes enable you to develop various communication skills: reading, technical writing, speaking, listening, personal interaction, and the preparation of drawings and other graphics.

Your design class allows you to develop both writing and speaking skills. Not only must you analyze information and formulate responses, you must also write accurate and clear class project reports. These parallel the professional technical reports that you will someday write for an employer or a customer.

Your reports are themselves the result of solving an open-ended problem and they can therefore be developed using the method presented in Figure 1.1. You should start with an appropriate report format and an outline of the major topics. Write a first draft and, after setting it aside for some time, analyze, evaluate, and rewrite it. Continue this iterative process of writing, analyzing, evaluating, and rewriting until you achieve a document of acceptable quality.

There are only two powers in the world, the sword and the pen; and in the end the former is always conquered by the latter.

Napoleon I (1769–1821)

Your design projects class provides you with the opportunity to develop speaking skills. End-of-term presentations on your design project may be delivered to an audience of students and industrial sponsors. Student designers who are unable to develop communication skills are unlikely to attain their full potential in professional practice.

Communication skills are discussed in more detail in Part A of the Appendix.

Communication skills are particularly relevant to *program reviews* or *design reviews* which are business meetings held periodically during the evolution of a product. At these meetings designers explain to senior management what has been accomplished during the previous reporting period. Management can then make decisions about the program. Meetings of this kind involve identification of perceived weaknesses in a proposed solution, and generally the design team will be required to present a detailed plan of work for the next phase of the program. These presentations are important because no matter how innovative and creative the design team has been, its efforts will have been largely wasted unless the engineers can communicate with management, explain their progress, and convince decision-makers of the worth of a proposed solution.

Design reviews are scheduled team meetings for assessing progress and reviewing the evolution of the product.

This situation is made worse by the Not-Invented-Here (NIH) syndrome which prevails in many industrial and government facilities. This attitude represents the reluctance of many people to accept other people's ideas, concepts, and philosophies. Thus design presentations must be planned very carefully. Indeed, the NIH syndrome has been a characteristic of humanity throughout civilization, as evidenced by a statement made in 1513 by Niccolò Machiavelli:

> "It must be remembered that there is nothing more difficult to plan, more doubtful of success, nor more dangerous to manage, than the creation of a new system. For the initiator has the enmity of all who would profit by the preservation of the old institutions and merely lukewarm defenders in those who would gain by the new ones."

Transition to Engineering Practice

Because groups of students work together on a project in your design class, you have an opportunity to participate in a team effort that mimics professional practice. Teams are common in industry and your ability to work with other team members is an important skill for your professional growth.

You can develop tools for investigating new subjects that are unfamiliar to you. These tools undoubtedly will include the ability to search for, locate, and retrieve information as well as the knack of sifting through it, assimilating it, and ultimately employing it for analysis and synthesis. If the group project is sponsored by an industrial organization, then you must achieve prescribed goals in a specified time and operate within a budget in a pseudoindustrial setting. In order to attain these goals, you will need to develop some management skills that ultimately will be of great use to you.

> Libraries are the wardrobes of literature, whence men, properly informed, may bring forth something for ornament, much for curiosity, and more for use.
>
> John Dyer (1700–1758)

Furthermore, you must establish a work schedule that incorporates both short-term and long-term goals as well as dates by which these goals must be accomplished. Time management is particularly important for projects lasting several weeks.

Times for weekly project review meetings should be set up by your student group. The agenda for these meetings should be based upon the "3P Principle", namely, progress,

The *3P Principle:*

- progress
- problems
- proposals

problems, and proposals. Therefore, during each meeting, the progress made toward achieving the goals established at the last meeting should be reviewed; the problems encountered should be discussed and analyzed; and proposals should be developed for addressing these problems. Finally, it should be clear to each group member what he or she must do before the next meeting.

The design projects class can provide excitement, responsibility, industrial experience, and an opportunity to develop people-skills, conflict-resolution skills, and communication skills while you develop a broader understanding of engineering. Such characteristics are commendable, especially in the context of the old Chinese proverb:

> Tell me, I forget,
> Show me, I remember,
> Involve me, I understand.

1.3 The Student Design-Build-Test Project

At the end of this chapter and throughout the first part of this book we shall chronicle the evolution of a small shoebox size design-build-test project from one of the author's design classes. Projects of this kind are frequently used in the undergraduate curriculum to encourage students to develop skills for solving open-ended design problems. We conclude each chapter with a discussion of how the principles presented in the chapter apply to the student project.

The objective of this project was for each student group to design, build, and test a vehicle powered only by ten burning birthday candles. A contest was held at the end of the semester to determine which vehicle travelled the farthest along a smooth corridor. The following rules applied:

- Candles were nominally 5 mm in diameter and 50 mm long.
- The motion of the vehicle had to be self-initiating.

- Vehicle construction material could cost no more than $50.
- Tracks and guides were prohibited.
- The change in total mechanical potential energy between the initiation of motion and the final resting position of the vehicle had to be zero.
- There were no restrictions on the size, shape, or principle of operation of the vehicle.

At the first meeting of the class, the project was introduced and the rules described. Students organized themselves into groups of four, established a Gantt chart to plan and execute the project, and assigned tasks and duties to individual group members.

Many students immediately embarked upon the conceptual design phase (Chapter 9) and discussed possible ways in which burning candles could power a traditional four-wheeled vehicle. Some students began to think of ways to measure how much energy they could obtain from a burning candle. Others were stunned by the assignment and uncertain how to begin. To them, the assignment appeared to be quite impossible. Other students realized that the problem should be defined before they attempted to create solutions. This aspect of the design process is discussed in Chapter 7.

Summary

The heart of all engineering activities is design: an interdisciplinary process involving iterative protocols, creativity, analysis, problem definition, and communication.

Design engineers should possess the following attributes:

- Creative skills
- Analytical skills
- Communication skills
- Integration skills
- Managerial ability

Design classes, such as a senior-level design projects class, expose you to open-ended problems that require a precise definition, after which you can seek solutions through a blend of both creative and analytical thinking. In a design class the traditional boundaries defining academic disciplines dissolve as you move from an educational to an industrial environment.

Design classes include exposure to defining complex problems, creative thinking strategies, multidisciplinary engineering problems outside the constraints imposed by classes in the engineering sciences, the development of viable models for engineering situations by making appropriate assumptions, and situations where communication skills must be employed.

Chapter One has focused upon the characteristics of undergraduate courses dedicated to the teaching of the generic design process. Chapter Two continues this introductory theme by discussing the importance of design in society, with particular emphasis on the role of engineering design throughout history.

Key Concepts

Analysis. A separation or decomposition of a whole entity into its individual parts for examination. This examination of the parts typically focuses upon their characteristics, function, interrelationships, and behavior.

Basic design. The same as conceptual design. This phrase pertains to a phase of the design process and also to a vision for a product.

Communication. The act of transmitting information orally, by writing, or by other means.

Conceptual design. A concept proposed as the basis for a product in response to a design specification. Usually many concepts are proposed.

Cost analysis. An economic analysis undertaken to determine the cost of an entity. Economic analyses typically comprise two types: the cost of a project in which the focus is the resources required to undertake a design activity, and the product cost, in which the total cost of materials and the manufacturing of the product is ascertained.

Creativity. An inventive characteristic associated with randomness, imagination, synthesis, irregularity, sensuality, nonjudgmental scrutiny, and visualization.

Critical Path. The critical path is the longest path through the network of tasks.

Critical Path Analysis. A method used to represent the interrelationship between tasks making up a project. Arrows are used to represent activities or tasks; events are the points at which arrows meet.

Detail design. The same as the final stage of the product design phase. During this phase final decisions are made about manufacturing details, materials, shapes, and functions of parts.

Definition of problems. All problems must be defined properly before solutions are sought.

Design review. A design review is a meeting held periodically during the evolution of a product to assess progress since the previous meeting.

Integration of disciplines. The integration of knowledge from several disciplines is invariably required to generate solutions to a design problem.

Economic analysis. The study of the cost of developing and manufacturing a product. The cost of a product has a major impact upon its success in the marketplace.

Embodiment design. The process of transforming a conceptual design into the preliminary layouts and shapes of the individual components of the embryonic product. Quantitative data are developed for the first time through mathematical analysis.

Engineering. The application of scientific principles to practical ends.

Engineering Design. The technical element of the product realization process in which specifications for products and their manufacture are established.

Form design. This is the same as the early phase of the product design stage or the same as embodiment design.

Gantt chart. A horizontal bar chart illustrating tasks to be completed *vs.* time. The chart indicates those tasks that are prerequisites for subsequent tasks and those tasks that can be worked on concurrently.

Management. The art, act, or manner of directing or controlling the activities of a group of individuals undertaking a task.

Manufacturing. The process of producing component parts, assemblies, and complete products.

Multidisciplinary activities. The generation of viable solutions to engineering design problems generally requiring the integration and utilization of diverse disciplines in both the arts and the sciences.

Open-ended problems. A class of ill-defined problems that are not clearly understood, are vague, and that have no unique answers. Numerous solutions may exist.

PERT diagram. Program evaluation and review technique. A network of interconnecting lines (activities) and nodes (events) representing the connectivity of tasks making up a project.

Planning. The identification and scheduling of tasks in a project.

Product. The result or artifact of a design process.

Product design. The design details for a product that is manufacturable. Materials, tolerances, surface finishes, dimensions, heat treatment and other relevant data as well as bought-out parts are specified.

Scheduling. The assignment of tasks to a specific time table.

Team. A group of individuals who work together toward some objective.

Review Questions

When you have finished this chapter, you should be able to do the following:

1. Describe how the problems associated with a design class differ from those of a class in the engineering sciences.
2. Discuss the primary ingredients of a cost analysis.
3. List some communication skills required in engineering practice and explain (with examples) why they are important.
4. Explain why engineering design problems are interdisciplinary.
5. List six attributes of a senior-level design projects class.
6. Discuss the task of defining design problems.
7. Explain the role of creativity in the context of engineering design.
8. Describe the role of the evaluation process in design.

Problems

1. Develop a device to catch mice in a more humane way than the traditional method. Discuss the design process in the conceptual design and product design phases.

2. Suppose that you were assigned a position as leader of a group of engineering students confronted with the following design problem: Design a machine that will pick up as quickly as possible all the black wooden balls from a mixture of 50 white and black balls 1 cm in diameter held within a 1 m × 1 m enclosure bounded by walls 10 cm high. Describe what you would do as leader of the group.

3. Imagine that you are required to develop an alternative cooling system for household refrigerators that would avoid the use of CFC's. Describe, without necessarily developing a solution, how you would approach the problem.

Further Reading

Cross, N. *Engineering Design Methods.* New York: John Wiley & Sons, 1989.

Dieter, G. *Engineering Design.* New York: McGraw-Hill, 1983.

Fleming, Q. W., J. W. Bronn, and G. C. Humphreys. *Project and Production Scheduling.* Chicago: Probus Publishing Company, 1987.

Kerzner, H. *Product Management: A Systems Approach to Planning, Scheduling, and Controlling.* New York: Van Nostrand Reinhold Company, 1979.

Pahl, G., and W. Beitz. *Engineering Design.* English edition edited by K. Wallace. London: The Design Council, 1984.

Pugh, S. *Total Design: Integrated Methods for Successful Product Engineering.* Reading, MA: Addison-Wesley, 1991.

Suh, N. P. *The Principles of Design.* Oxford: Oxford University Press, 1990.

Whitehouse, G. E. *Systems Analysis and Design Using Network Techniques.* Englewood Cliffs, NJ: Prentice Hall, 1973.

Ullman, D. G. *The Mechanical Design Process.* New York: McGraw-Hill, 1992.

Design in History

2

If no use is made of the labors of past ages, the world must remain always in the infancy of knowledge.

Marcus Tullius Cicero
(106–46 B.C.)

Objectives

- To emphasize that design philosophies permeate all aspects of society.
- To provide an historical perspective of design creativity.
- To explain the importance of design in engineering.
- To illustrate the role of engineering design in the evolution of civilizations.
- To discuss engineering design education.

Contents

What if...

You're a Cro-Magnon engineer. The date is 10,000 B.C. You've been killing your dinner—and your enemy—with sticks and stones. It's not very effective. What to do? Develop a better weapon!

Even Cro-Magnon people practiced engineering design, and for them it must have been pretty amazing, sophisticated stuff. How easily could *you* design a bow and arrow if you'd never seen one before?

The slow storage of energy in the deformed bow. The sudden release of that energy to launch—what? A lightweight stick with a hard, sharp, lethal tip. Send it on a long trajectory.

Indeed, the design of the most sophisticated weapon system of its time. Could you have done it? A Cro-Magnon engineer did!

2.0 Introduction

Design, which is derived from the Latin root *designare*, is defined in Webster's *New World Dictionary* as

- a plan; scheme; project; purpose; aim; intention
- a thing planned for or outcome aimed at
- a working out by plan or development according to a plan
- a secret, usually dishonest, or selfish scheme
- a plan or sketch to work from; pattern
- the art of making designs or patterns
- the arrangement of parts, details, form, color, etc., so as to produce a complete artistic unit; artistic or skillful invention
- a finished artistic work or decoration

The creative activities associated with the word *design* manifest themselves in many different fields, such as:

fine arts	typography	clothing
landscape architecture	engineering	tapestry
basketry	language arts	mosaic art
furniture	music	pottery
flower arrangement	jewelry	carpets
sculpture	architecture	

However, the skills necessary to create a stunning oil painting are quite different from those required to design a space shuttle, for example. Artistic painting involves the brain's right side which is largely responsible for the recognition of complicated visual patterns, aesthetic endeavors, and the expression of emotion. The design of a spacecraft involves not only the right side of the brain but also the left side which controls the ability to utilize mathematics, logic, and language.

Engineering design integrates the sciences with the arts, as evidenced by the beauty and functionality of the Concorde airplane, the Golden Gate suspension bridge, and the Sydney Opera House (Figure 2.1). Different facets of design require different skills and capabilities, and while these have evolved throughout time, they still are not well understood.

Figure 2.1

The Sydney Opera House.

c. 1,750,000 B.C.
Flint tools
c. 10,000 B.C.
Bow and arrow
c. 4000 B.C.
Plow
c. 3000 B.C.
Wheel
c. A.D. 105
Paper
c. 1100
Magnetic compass
c. 1440
Movable type
c. 1593
Thermometer
1690–1769
Steam engine
1826
Photography
1837
Telegraph
1860
Internal combustion engine
1879
Incandescent light
1903
Airplane
1930
Analog computer
1942
Nuclear reactor
1947
Transistor
1957
Sputnik
1971
Microprocessor

Figure 2.2

Inventions of man though the ages.

2.1 Historical Perspective

The history of design and its associated technologies is as old as humanity itself (Figure 2.2). The first evidence of design in the life styles of hominids dates from the beginning of the Paleolithic period—the Old Stone Age—some 1,750,000 years ago when primitive human beings, *Homo habilis,* fabricated simple shelters to provide protection against the elements. In addition, they fashioned chopping tools and hand axes—and in doing so distinguished themselves from other creatures.

Most creatures possess creative instincts: beavers build dams and birds weave nests. But instinctive behavior is not easily modified in response to a changing environment. Beavers have always built dams and birds have always built nests; their behavior changes very little. In contrast, humans can think logically and develop new ways to address new problems. We have the innate ability to respond quickly and creatively to a changing environment. That ability is an essential part of engineering design.

During the Old Stone Age prehistoric people learned to create fire by striking pieces of iron pyrite and flint together to generate sparks. The exploitation and control of fire had tremendous cultural consequences. It provided for two activities requiring design skills: the pottery industry in the Mesolithic period, and the first smelting of metals and the beginnings of foundry practices in the Bronze Age. Near the end of the Paleolithic period, Neanderthal man began to form more advanced hand tools that required the assembly of a head and a handle from materials with quite different properties. This activity was important from an engineering perspective because it required the selection of appropriate materials, the processing of these materials, and the creative assembly of individual parts into a new artifact.

About 10,000 B.C., an early form of modern man—Cro-Magnon man—designed the bow-and-arrow. This innovation demonstrated a maturing degree of design creativity. These people invented the crafts of sewing and weaving that provided the basis for the design of clothing, tapestry, and carpets. Cave paintings made during this period were the first recorded art works (Figure 2.3). The oldest, in caves along the border between France and Spain, date from approximately 20,000 B.C.. Communicating ideas by creating forms and shapes on a flat surface has, of course, considerable significance for engineers because they frequently draw parts and assemblies to transmit ideas during the design and manufacturing phases of the creation of a product.

Figure 2.3

Prehistoric cave painting, Lascaux, France.

When fire began to be used for cooking, it triggered the evolution of pottery for pots and pans. These artifacts were the result of a creative engineering process involving the shaping of soft, pliable clay into objects that could be fire-hardening into rigid ceramic vessels.

Plants and animals were domesticated in the late Neolithic period. The development of agriculture involved a significant engineering component as agricultural tools and implements were created from wood and stone to till the earth, to thresh grain, and to grind grain into flour. At about the same time, an embryonic textile industry began to manufacture cloth from flax, reeds, and animal hair. This industry stimulated engineering design activities that were responsible for the development of primitive machinery.

Writing is the great invention of the world.

Abraham Lincoln (1809–1865)

Around 3,000 B.C., writing appeared in both Egypt and what is now Iraq. While this mode of communication was developed primarily to record business transactions, it also led to a new outlet for the creativity of *Homo sapiens sapiens* through poetry and book writing. It is of interest to note how these two Middle Eastern countries created quite different solutions to the challenge of recording business transactions on a transportable medium (Figure 2.4). The Sumerians in the Tigris-Euphrates valley developed cuneiform writing to record facts and ideas on clay tablets using sharp, needle-like instruments. The tablets were then dried in the sun to create rigid, durable blocks. In contrast, the Egyptians living on the Nile delta developed hieroglyphic writing with brushes and ink to record ideas on papyrus, a flexible sheet manufactured from a reed abundant in the delta region.

Figure 2.4

Cuneiform writing.

These two quite different solutions to the same problem is typical of the design process. The different ways to record business transactions which developed in Egypt and Mesopotamia demonstrate that there is no "correct" solution to a design problem.

At the dusk of the Neolithic period and the dawn of the Bronze Age several events occurred which have considerable engineering significance. These included the invention of the wheel and the development of a high-temperature kiln that further stimulated metal smelting. Smelting triggered the evolution of the casting process that permitted the fabrication of tools and weapons from an alloy of tin and copper. This earliest metallurgical feat some 5000 years ago defined the era as the Bronze Age. Typical artifacts include urns, helmets, swords, shields, and drinking cups.

The stepped pyramid at Saqqarah in Egypt, built around 2630 B.C., is the first product to be designed by an engineer whose name (Imhotep) was recorded.

Around 2630 B.C. during the Bronze Age, the stepped pyramid at Saqqarah near Memphis in Egypt was designed by an engineer named Imhotep. His work represented a new solution to the problems of protecting a corpse from desecration and of providing a dwelling for the deceased in the afterlife. This ancient wonder of the world, which was constructed to protect the Egyptian king Djoser, is the first product to be designed by an engineer whose name is archivally recorded.

Humans possess the ability to think logically and to find solutions to new classes of problems. This ability is an essential part of engineering design.

Subsequently, the successors of Imhotep elevated the civil engineering profession to new heights by utilizing analytical techniques to design the Great Pyramid of Cheops (2550 B.C.) at Giza (Figure 2.5). These techniques involved precise arithmetic and geometry. The largest deviation from a right angle at any of the corners of the Great Pyramid is approximately 0.05 degree, and the maximum difference in length of a 755 foot (230 meters) side is approximately 0.65 foot (20 cm). These engineering accomplishments are astonishing!

The rise of Greek civilization in approximately 1000 B.C. witnessed the dawn of a new era of creativity in the Mediterranean region. These people employed the most advanced material of that period: iron. Exploitation of the superior properties of this hard, strong, and abundant material was responsible for further advances in the field of engineering design. Perhaps the greatest individual of this era was Archimedes (287–212 B.C.) who was both a scientist and an engineer. He is generally credited with inventing a variety of military weapons in addition to the compound pulley and the Archimedean screw-pump (Figure 2.6).

Archimedes was a scientist and engineer who lived during a period of creative design in the eastern Mediterranean region.

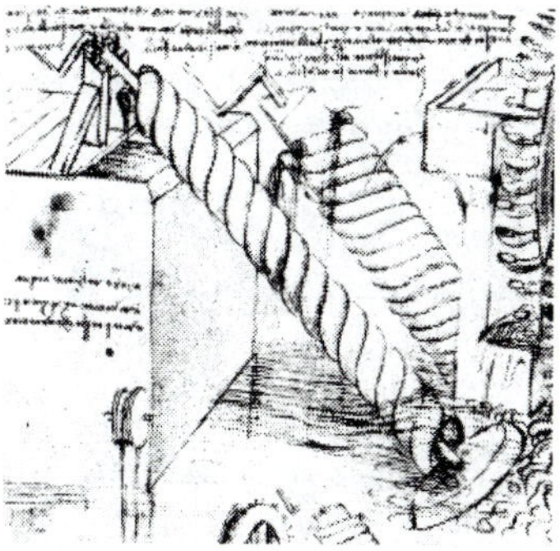

Figure 2.6

An Archimedian screw-pump. (From a sketch of a water-lifting device by Leonardo da Vinci.)

Figure 2.5

The Great Pyramid of Cheops (Photo courtesy of Gordon L. Hall).

The creation of the kiln to smelt metals, the development of the associated metallurgical processes, and the use of arithmetic and geometry in technology heralded the emergence of *Homo sapiens* as the dominant species on earth. For the first time these hominids began to employ complex engineering principles.

During the Iron Age, *Homo sapiens* began to employ complex engineering principles.

The Emergence of Engineering

Engineering was destined to have a significant role in the advancement of civilization. Its emerging principles can be found in the definition of engineering formulated by the Engineers Council for Professional Development in the United States (see box).

> *Engineering* is the application of "scientific principles to design or develop structures, machines, apparatus or manufacturing processes, or works utilizing them singly or in combination; or to construct or operate the same with full cognizance of their design; or to forecast their behavior under specific operating conditions; all as respects an intended function, economics of operation and safety to life and property."
>
> Engineers Council for Professional Development in the United States

Perhaps it is worthy of note that the words *ingenious* and *engineer* both derive from Medieval Latin *ingeniator* or *ingeniare* (to design or contrive). These in turn come from the same Latin root *ingenium,* meaning an inborn talent or skill. Linguistically, therefore, design and engineering are one and the same.

An engineer is a creative individual whose role is to *do*.

The role of the engineer is to *do*, while in contrast, the role of the scientist is to *acquire knowledge*. Moreover, engineering practice involves a confluence of ideas from both the arts and the sciences because diverse classes of knowledge are applied to solve practical problems. These activities may require the conception of original solutions to challenging societal, technological, and environmental problems as well as the prediction of the performance and cost of new processes and inventions.

Like its Latin root, the Middle English word *engine* means "to contrive." The individual associated with this design activity was called the "engine-er," a name originally applied to persons serving in the military. The primary work of these military engineers was to design engines of war such as catapults and assault towers. Julius Caesar employed his

Figure 2.7

Part of the Bayeux Tapestry that was embroidered in the eleventh century A.D. It depicts instruments of war and the construction of a ship engineered around 1066.

praefectus fabrum to protect military engineers and their machines of destruction. According to the Doomsday Book (1086), William the Conqueror employed an engineer, Waldivus Ingeniator, to oversee his activities. Thus it is evident (Figure 2.7) that at the time of the Battle of Hastings 900 years ago, military, civil, and mechanical engineers were already assuming a primary role in shaping societal values. This role has continued unabated as the time-line approaches the dawn of the twenty-first century.

A machine-based civilization began to emerge in Europe during the Middle Ages.

The evolution of engineering design during the Middle Ages was relatively slow and sustained. During this time a machine-based civilization began to appear in Europe using a confluence of indigenous ideas with those from the more mature civilizations in the Middle East and China. This led to the design and manufacture of the windmill and the water wheel for generation of power, to the development of new plows for more efficient cultivation, to the invention of the lathe, the invention of the printing press incorporating movable metal type, the creation of mechanical clocks, and the devising of linkage mechanisms for converting circular motion into reciprocating motion. Innovative designs of this period were to provide a foundation for the Renaissance, a period of great learning and change.

Scientific and artistic thinking bloomed during the Renaissance. Leonardo da Vinci, a superbly innovative designer, lived during this era.

The Renaissance in Europe during the fourteenth, fifteenth, and sixteenth centuries was responsible for the transition from the medieval world to the modern world. This period of revival involved both the arts and the sciences.

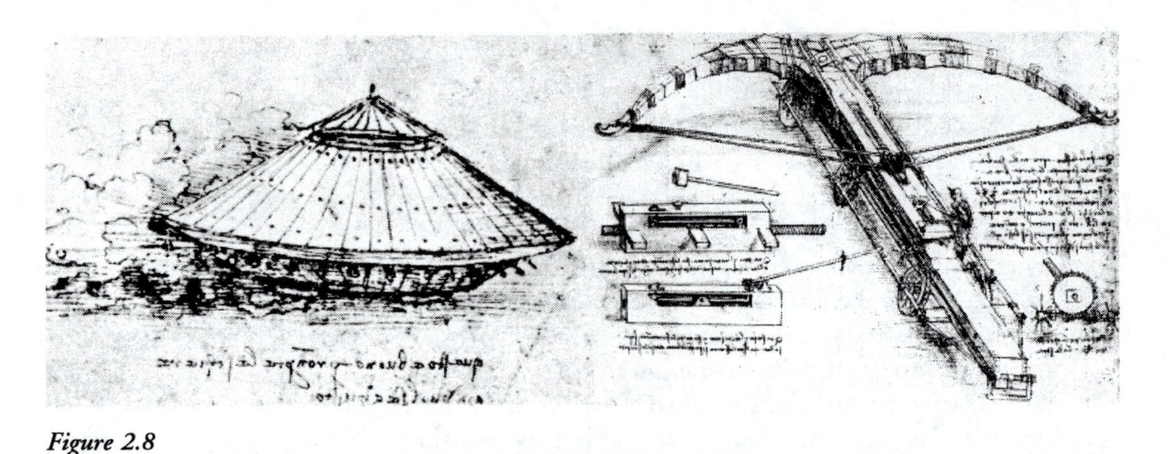

Figure 2.8

Designs for a military tank and a huge crossbow from Leonardo da Vinci's notebooks.

Indeed, it witnessed the first revolution of scientific thinking since the Greeks established the basis for statics some 1500 years earlier. A consequence of this revolution was a vigorous period of invention and creative design led by a true Renaissance man, Leonardo da Vinci (1452–1519). This genius had enormous talent in both the arts and the sciences as evidenced by his paintings, such as *Mona Lisa* and *The Last Supper*, and his notebooks which include over 4000 pages of diagrams and text in bioengineering, optics, fluid mechanics, physiology, and machinery. The notebooks contain designs for thrust ball bearings, a parachute, submarines, airplanes, moveable bridges, complex gear trains, and many other devices (Figure 2.8).

Thus the sixteenth and seventeenth centuries witnessed the dawn of a new era of engineering design as traditional approaches were gradually superseded by analytical approaches involving rudiments of the engineering sciences. The *art* of engineering, involving a trial-and-error approach or empirical methods that were communicated verbally, was augmented by scientific methods based upon theories initiated by a number of intellectuals. Leonardo da Vinci was probably the first to recognize the significance of relating theory to practice. Galileo was the father of astronomy, and his observations were a logical precursor to the development of modern mathematics. Kepler established laws of planetary motion and concepts of geometry, Newton formulated the laws of motion and the infinitesimal calculus, and Napier developed logarithms. At about the same time Boyle studied the behavior of gases and Hooke established laws to describe the behavior of materials. There were many others.

The radically new and revolutionary philosophies proposed by these and other gifted individuals coincided with a period of significant social change that triggered the Industrial Revolution in Great Britain during the period 1750 to 1900. Numerous innovative engineering systems were designed, and these products were partially responsible for catapulting nations with agrarian economies into industrialized nations. Steam power was exploited in foundries, textile mills, ships, railroad locomotives, and coal mines. Distinct branches of engineering began to be recognized as the field of engineering began to mature and diversify. Furthermore, learned societies formed in the fields of civil, mechanical, electrical, and chemical engineering, and these societies stimulated the practice of design and the growth of the engineering profession.

The Industrial Revolution in Great Britain stimulated the design of numerous engineering masterpieces.

Engineering Design Education

In the early eighteenth century, universities began to play a pivotal role in the evolution of design practices and the dissemination of engineering knowledge. Indeed, these institutions had been in existence for several centuries before this. For example the University of Bologna was founded in the eleventh century and Oxford University in the twelfth. However, the primary objective of these early universities was to educate scholars for a career in the church.

Education is not the filling of a pail, but the lighting of a fire.

William Butler Yeats
(1865–1939)

A college for teaching technical subjects was established in Prague in 1707 and another was founded in Paris in 1716. Soon afterwards similar technical universities were founded in Austria and Germany. Individuals at these institutions were responsible for the development of design theories and practices. In spite of the crucial role of design in engineering,

Education is not preparation for life; education is life itself.

John Dewey
(1859–1952)

Design education at German universities has been responsible for the strong design skills of their graduates and ultimately for the strength of German industries.

the first chair of engineering design at a German university was not established until 1965 at the Technical University of Munich. This visionary action was quickly replicated at the primary Technische Hochschulen and the teaching of design at these and other German universities has been responsible for the strong design skills of their graduates and ultimately for the strength of German industries.

Early universities were free to develop curricula without interference from outside governing bodies; there was considerable academic freedom. However this gradually changed and now most educational programs in institutions of higher education are subject to the scrutiny of professional review boards. The review board responsible for accrediting engineering curricula in the United States is the Accreditation Board for Engineering and Technology. Individuals representing this review board often criticize the teaching of engineering design and also design curricula when they evaluate engineering colleges at universities. These criticisms are motivated by the crucial importance of good design practices to the prosperity of the industrial sector and ultimately of the nation.

Other federal organizations perceive that the overall quality of design in the United States is inferior to that in other nations, and universities have been targeted as one of the culprits for this situation. A report[1] of the National Research Council states that

> "...engineering design education in the United States is poor and strengthening engineering design education is critical to the long-term development of engineers who are prepared to become good designers and leaders who will provide a lasting foundation for U.S. industry's international competitiveness."

Numerous reports recommend closer ties between government, industries, and universities. Furthermore, a revision of the definition of the university design experience has been proposed to the Accreditation Board for Engineering and Technology[1]:

---

[1] *Improving Engineering Design,* National Research Council, National Academy Press, Washington D.C., 1991

> "The engineering design component of a curriculum must include nearly all of the following features: development of student creativity, use of open-ended problems, development and use of design methodology, formulation of design problem requirements and specifications, in-depth consideration of alternative solutions, feasibility considerations, production processes, advanced design methodologies, simultaneous engineering design, life cycle considerations, detailed system descriptions, and participation in an interdisciplinary group on a design project. Further, it is essential to include a variety of realistic constraints, such as economic factors, safety, reliability, aesthetics, ethics, and social impact. Finally, design courses should be integrated into the curriculum so as to provide a continual increase in the design competence of a student progressing through an engineering program."

This proposal provides the basis for an educational experience that is superior to many current engineering curricula and it exposes the student to the very best industrial design practices.

2.2 The Student Design-Build-Test Project.

You will recall that we shall end each chapter of Part One of this text with a review of the design-build-test student project. The assignment described at the end of Chapter 1 involves the design, building, and testing of a vehicle to be powered by ten burning birthday candles. Some of the inventions and scientific discoveries described below and a review of the evolution of both science and engineering influenced the final design of the vehicles constructed and tested by some student groups.

The burning of birthday candles involves a number of phenomena including the generation of light, gaseous combustion products, and heat. Methods to convert heat into mechanical work have been investigated for centuries. These methods were researched by the students and reported in their end-of-semester written reports. They included the following:

- A.D. 100 Hero's aeolipile (the first heat engine—a steam turbine)
- A.D. 1629 Branca's impulse turbine
- A.D. 1690 Papin's heat engine
- A.D. 1698 Savery's pulsometer pump
- A.D. 1706 Newcomen's steam engine
- A.D. 1769 Watt's steam engine
- A.D. 1803 Trevithick's steam engine
- A.D. 1807 Cayley's hot-air engine
- A.D. 1816 Stirling's engine
- A.D. 1870 Ericsson's solar-energy engine
- A.D. 1884 Parson's steam turbine
- A.D. 1937 Whittle's jet engine

If men could learn from history, what lessons it might teach us!
Samuel Taylor Coleridge (1772–1834)

Ancillary technologies and methodologies relevant to the project—feedback control, for example—were also reviewed. Such reviews included the works of Ktesibios of Alexandria who invented a float valve to regulate water flow in a water clock about 300 B.C., as well as James Watt's feedback control schemes on steam engines which included the "flyball" governor and boiler draft regulation to control steam pressure.

Students did not restrict their literature review to the applied sciences. They also reviewed some science to obtain a deeper understanding of the fundamental phenomena relevant to the design problem that confronted them. These searches revealed some of the following accomplishments:

- 200 B.C. Archimedes' solar-powered furnace
- A.D. 1824 Carnot's heat-engine cycle
- A.D. 1840 Joule's establishment of the foundations of thermodynamics

- A.D. 1850 Clausius' postulation of the first and second laws of thermodynamics
- A.D. 1954 Chain, Fuller, and Parson's solar cell

Other studies included reviews of the gas laws of Boyle and Charles, pyroelectricity, shape-memory alloys, and research on the physical and chemical aspects of combustion such as those studied by Bacon in 1620, Lavoisier in 1772, Priestley in 1774, and Davy in 1815, as well as Faraday's famous lectures on the burning of a candle.

This historical review of ways in which heat can be transformed into mechanical work provided the basis for new conceptual designs. It was not permitted to stifle imagination or suppress creativity; rather it prevented the students from making the same mistakes that others made in the past. It is not necessary to reinvent the wheel.

Summary

Engineering design is a creative activity whose goal is to conceive and plan the form, structure, manufacture, and operation of an object, a machine, or other product. It involves artistic as well as engineering, scientific, and mathematical skills.

Design is as old as humanity itself. It began in the Stone Age when primitive humans first fashioned tools and structures. The design process evolved as humans began to distinguish themselves from creatures of instinct. About 10,000 to 20,000 B.C. an early human designed the bow and arrow; the crafts of sewing and weaving developed, and cave paintings became the first known works of art. Shortly afterward, agricultural tools were created and primitive industries began. Writing appeared around 3000 B.C. and rapidly became important as a communication and recording tool.

During the Bronze Age about 5000 years ago, the Egyptians elevated the civil engineering profession to new heights when they introduced arithmetic and geometry into the construction of the pyramids. As Greek civilization emerged in about 1000 B.C., iron began to be used, the compound pulley and the screw were invented, and people began to employ complex engineering principles.

A machine-based civilization evolved in Europe during the Middle Ages with the invention of devices such as the windmill, water wheel, lathe, printing press, and mechanical clock.

The Renaissance in Europe some 500 years ago marked a period of vigorous creation and invention in the arts, sciences, and engineering. Trial-and-error approaches began to be replaced by analysis and the application of scientific theory. Universities started teaching technical subjects and they soon played an important role in the development of engineering.

Chapter Two has chronicled the evolution of engineering design. Chapter Three builds upon the first two chapters by discussing an industrial case study of a successful design activity. Chapter Four then presents a systematic procedure for undertaking design projects.

Key Concepts

Engineering design. The technical element in the creation of products that involves the application of knowledge and techniques from engineering, science, aesthetics, economics, and psychology in establishing specifications for products and their associated production processes. It is the technical process by which engineering descriptions and specifications are formulated to insure that a product will possess the desired behavior, performance, quality, and cost.

Review Questions

1. Define in your own words what you believe the word *engineering* means.
2. Define the word "design."
3. Develop a list of disciplines that involve design activities.
4. How do the thought processes associated with the design of a car differ from those associated with the design of a floral display?
5 How do the design activities of humans differ from those of animals?
6. List some of the early products designed by humans.
7. How did the domestication of animals motivate design activities?
8. Contrast cuneiform writing and hieroglyphic writing.
9. What analytical techniques were employed to construct the Great Pyramid of Cheops?
10. During what era did engineering principles first begin to be employed in practice?
11. How does the role of an engineer differ from the classical definition of a scientist?
12. How are *design, creativity*, and *engineering* related linguistically?
13. In which field was the word engineer first used? What did these individuals do?
14. List several artifacts that were designed in the Middle Ages.
15. During which era did Leonardo da Vinci live and what did he accomplish in his lifetime?
16. When were the rudiments of trial-and-error superseded by analytical methods in the engineering design process?
17. The analytical methods of engineering design juxtaposed upon a ferment of social change during the eighteenth and nineteenth centuries were responsible for a revolution in Great Britain. What were the consequences of this?

Problems

1. Keyboards are used in many products ranging from pocket calculators to cash registers at the check-outs in stores and from telephones to personal computers. Develop a report that chronicles the evolution of this technology from the mechanical keyboards used for typewriters in early Victorian times through the following generation of electromechanical keyboards to modern electronic keyboards. The most common keyboard uses the QWERTY pattern of keys, named from the first six characters on the left side in the top row of letter keys. Is this a good format?
2. The automobile is the dominant mode of transportation in the United States. Develop a report on the evolution of the car. Pay attention to the effect of new technologies on automobile design, highway systems, traffic control in cities, the design of cities, and finally upon the environment.
3. Write an essay on the role of engineering design during the Industrial Revolution.
4. Discuss the various aspects of design evident in an airport terminal.
5. Write a report describing the design and construction of the Great Pyramid of Cheops.
6. Develop a biography of Leonardo da Vinci with particular emphasis on his creativity.
7. Write a report chronicling the evolution of time-keeping devices throughout the ages. Emphasize the different methods employed for this task, such as water-based devices, mechanical systems, sundials, flammable ropes with knots, electronic devices, and atomic clocks.
8. Write a report on the human quest to fly. Begin with Greek mythology involving Daedalus and Icarus, the balloons of the Montgolfier brothers, the pioneering work of Sir George Cayley, the Wright Brothers, and beyond.

9. Develop a report on the life work of Thomas A. Edison, a man often described as the greatest inventor of all time. He is credited with 1,093 patents. How did Edison's work affect society? What do you consider his most important invention and why?
10. Write a report on the evolution of weapons systems throughout the ages from a design perspective. Provide illustrative examples of how new generations of weapons were responsible for the modification of tactical procedures. For example, what happened at the Battle of Crécy in 1346? What was the crucial weapon in the battle and how did it assure victory?
11. Write an essay describing the accomplishments and methods employed by one of the following eminent engineers: Sir Barnes Wallace, Rudolf Diesel, Charles Kettering, or Isombard Kingdom Brunel.
12. Develop a report discussing the different devices employed to generate artificial light throughout the millennia.

Further Reading

Burstal, A. F. *A History of Mechanical Engineering.* London: Faber and Faber, 1963.

Burke, James. *Connections.* Boston: Little, Brown & Company, 1978.

Eder, W. E. "A European Outlook on Engineering Design in Education," *Journal of Engineering Education*, **82**, (2), 118–122.

Galluzzi, P., ed. *Leonardo da Vinci: Engineer and Architect.* Montreal: The Montreal Museum of Fine Art, 1987.

Gibbs-Smith, C., and G. Rees. *The Inventions of Leonardo da Vinci.* Oxford: Phaidon Press Limited, 1978.

Gombrich, E. H. *The Story of Art.* London: Phidon Press, 1967.

Mumford, L. *Technics and Civilizations.* New York: Harcourt, 1934.

Pacey, A. *Technology in World Civilization—A Thousand Year History.* Cambridge, MA: MIT Press, 1990.

Parsons, W. B. *Engineers and Engineering in the Renaissance.* Baltimore: Williams and Wilkins, 1939.

Reuleaux, F., and C. L. Moll. *Construktionlehre für den Maschinenbau.* Braunschweig: Vieweg, 1862.

Reuleaux, F. *Der Construkteur.* Braunschweig: Vieweg, 1882.

Singer, C. and T. I. Williams, eds. *A History of Technology.* Vols. I, II, III, IV, V. New York: Oxford University Press, 1954–1984.

The Encyclopaedia Britannica, 15th ed. Chicago: Encyclopaedia Britannica, Inc., 1992.

A Case Study in Design

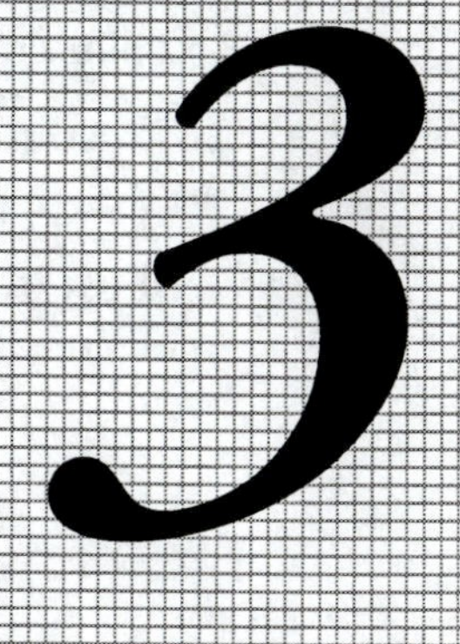

*Nobody will fly for a thousand years!**

Wilbur Wright
(1867–1912)

*In a fit of frustration, Wilbur Wright uttered these words in 1901. Two years later his brother Orville made the first powered flight in a heavier-than-air craft of their own invention and construction.

Objectives

- To illustrate the design process by considering an open-ended aeronautical design problem.
- To illustrate the multidisciplinary character of the design process.
- To illustrate cascading effects of decisions made in the design process.

Contents

What if...

You are the pilot of an experimental aircraft taking off from Edwards Air Force Base in California in an attempt to set a new long distance flight record. A wing tip scrapes along the runway and a drag-reducing winglet is knocked off. How do you feel? What should you do?

Landing to repair the damage is not an option because the aircraft was not designed to land with the tanks full of fuel! Flying the aircraft around the world may now be impossible because of the increased drag. Are the fuel consumption estimates correct? Are all other calculations accurate?

This situation occurred on December 14, 1986. The pilot, Dick Rutan, improved the performance of the plane by violently maneuvering the aircraft to break off the other winglet. The now-symmetrical wing configuration allowed him and his copilot Jeana Yeager to set a new world endurance record for aircraft by circumnavigating the earth without stopping or refueling. Whew! What a design!

3.0 Introduction

This chapter presents an interdisciplinary case study of an open-ended aeronautical design problem. A superficial overview of some typical questions that must be answered in this undertaking are presented in an attempt to stimulate your thinking about the design task. The need to integrate concepts associated with numerous engineering science disciplines is emphasized. Finally you will learn how decisions made about one aspect of the problem have a cascading effect upon other facets of the design process.

3.1 The Design of the Rutan *Voyager*

This case study illustrates the development of a strategy to fly around the world without stopping and without refueling. The goal and various strategies to analyze the problem and to investigate possible solutions might be stated as follows:

Goal: To fly nonstop around the world in an aircraft without refueling.

Strategy 1: Telephone the ticket agents at one of the airlines, such as *TWA* or *United*, for a seat reservation on one of their scheduled flights.

Result: No scheduled flight exists. The flight envelopes of 727s, DC-10s, and 747s prevent these subsonic aircraft from flying nonstop around the world without refueling.

Strategy 2: Telephone *Air France* to charter a supersonic Concorde.

Result: A Concorde cannot fly nonstop around the world without refueling. There is no commercial aircraft capable of flying nonstop around the world without refueling.

Strategy 3: Contact the U.S. Air Force and the air forces of other countries for information about military aircraft.

Result: There are no military aircraft that can fly around the world without refueling.

Figure 3.1

The Rutan Voyager *aircraft which, in December 1986, was the first plane to fly nonstop around the world without refueling. (Source: Visions, Inc., New York, NY).*

Strategy 4: Since there are no commercial or military aircraft that can fly round the world without refueling, design and build a plane that can do so.

Result: This challenge was addressed by a team in California, beginning in 1982. After defining this multidisciplinary open-ended design problem, several aircraft concepts were generated by the team. The potentially best concept was selected and subjected to further synthesis, analysis, and refinement. This design-build-test program created the Rutan *Voyager* aircraft pictured in Figure 3.1. It flew nonstop around the world without refueling in December, 1986. The two-person crew traveled 26,678 miles in nine days.

The task of designing and manufacturing this new aircraft demanded the integration and use of diverse information from the undergraduate curriculum in mechanical engineering, from other fields of engineering, and from several nontechnical fields. Creation of this plane was a response to an open-ended problem with no unique solution and there was no single correct answer. Indeed, several types of planes could have accomplished this task. Furthermore, an iterative solution had to be adopted in order to develop the aircraft successfully. The task mandated a combination of creative thinking and analytical thinking. The same approach also had to be adopted for creating the flight plan and for planning communications, provisions, safety equipment, navigation aids, autopilots, etc.

The Rutan *Voyager* aircraft was created using a design-build-test program

Voyager was designed by employing creativity, analysis, and evaluation.

A number of questions pertaining to the creation of this aircraft can be formulated in response to the single-sentence problem statement at the beginning of this section. These questions relate to development of a design specification or a problem definition for the plane. Some of these questions are:

Range?	Over 24,855 miles, the circumference of the earth
Type of power plant?	Turboprop? Gas turbine? Ramjet? Piston engine? Turbojet? Rocket? Turbofan?
Propeller configuration?	2 blades? 3 blades? 4 blades? 5 blades? Double contrarotating?
Propeller location?	Tractor? Pusher?
Number of power plants?	One? Two? Three? More?
Location of power plants?	Within the fuselage? Slung below a wing? Supported above a wing?
Fuel consumption?	
Volume of fuel tanks?	
Location of fuel tanks?	Fuselage? Wings? Drop-off tanks?
Aircraft takeoff weight?	
Wing configuration?	Triplane? Low-wing monoplane? Biplane? High-wing monoplane? Mid-wing monoplane?
Wing geometry?	Straight? Swept-back? Swept forward? Delta?

Wing loading?	
Landing gear?	Fixed? Retractable? Tail wheel? Tricycle? Bicycle? Multibogey?
Tail configuration?	Tailless? Lifting canard? Boom-mounted? Cruciform? Butterfly? Right-angle vertical? High-T horizontal? V-tail? Swept-back vertical? Inverted-V? Twin-vertical? Triple-tail?
Airfoil geometry?	
Thrust–weight ratio?	
Lift-drag ratio?	
Crew station?	
Number of pilots?	One? Two? Three? More?
Cruising speed?	
Cruising altitude?	
Maximum altitude?	
Survival equipment?	

These questions provide a flavor of the task confronting a group of individuals wishing to design an aircraft capable of flying nonstop around the world without refueling. The design of the *Voyager* was governed by the Breguet theorem which determines the range of an aircraft. To maximize the range, the plane required a high ratio of thrust efficiency to specific fuel consumption, a high lift-to-drag ratio, and a high ratio of fuel weight to gross aircraft weight. A group of design characteristics had to be maximized: the lift of the wings, the engine efficiency, the reliability of the autopilot, and the crew safety. At the same time a group of characteristics had to be minimized: fuel consumption, the aircraft weight, and the aerodynamic drag. The creation of such an aircraft required considerable ingenuity, and the final product was a combination of custom built parts and standard vendor-supplied products such as adhesives, engines, fuel, nuts, and bolts.

The iterative process of designing the *Voyager* implicitly began with a market analysis described earlier in strategies 1, 2, 3, and 4. An economic analysis was undertaken and industrial sponsors were sought to help finance the venture because the *Voyager* would ultimately cost two million U.S. dollars to develop.

A definition of the problem was developed which included specifications for the aircraft performance and the principal systems such as the engines, airframe, and fuel supply system. The plane had to fly three and a half times longer than the longest nonstop flight by a commercial aircraft and had to fly twice as far as the then-current world long-distance record.

Fuel storage and fuel consumption were major challenges in the design of the aircraft—especially its airframe—because these parameters dictated the configuration of the plane. Critical design parameters were the weight of the fully-fueled aircraft, the location and geometry of the fuel tanks, and the efficiency of the engines. Engines consume fuel, and fuel has significant weight; thus it takes more fuel to carry more fuel. This design cycle continues since each additional liter of fuel demands more tankage. This tankage requires more material, which imposes additional weight on the airframe, and this additional weight requires additional strength in the wings and supporting structure. Indeed, each extra kilogram of weight of *Voyager* at takeoff required the aircraft to carry an additional six kilograms of fuel. Weight savings were of crucial importance. The extent to which this philosophy was pursued can be illustrated by noting that only the upper wing surfaces were painted!

The critical design considerations were the fuel tanks, the efficiency of the engines, and the weight of the aircraft.

Thus the optimization problem confronting the design team was to accommodate the necessary fuel to propel the plane on its flight around the world yet to minimize the aircraft's takeoff weight. Clearly this problem can only be solved with knowledge from many different fields such as:

- **aircraft engines**—combustion phenomena, the chemistry of aircraft fuels, heat transfer, lubrication, machinery dynamics, and control theory
- **aerodynamics**—lift and drag characteristics of a structure that must accommodate a prescribed volume of fuel—as well as the engines, pilots, food, navigation equipment, etc.
- **lightweight aerospace structures**—strength of materials, ma-

terials science, vibrations, and advanced manufacturing technologies for aerospace materials

Creative features of *Voyager* included honeycomb structures and drag-reducing winglets that are analogous to naturally occurring features.

The *Voyager* weight was minimized by exploiting lightweight polymeric fibrous composite materials (discussed in Chapter 18). These materials permit the designer to tailor the material characteristics specifically for each application. This is typically accomplished by selecting appropriate fibers, matrices, fiber orientation, and other laminate characteristics. The macroscopic design of these composite materials requires the engineer to undertake both creative thinking and analytical thinking in order to predict the stresses and failure modes of structures. Composite laminated skins are very strong, low-density materials and, when they are used in conjunction with honeycomb paper cores, they create innovative lightweight high-stiffness structures. The honeycomb structure of core material is a biomimetic idea derived from studies of bees and beehives (Chapter 9).

The *Voyager* airframe was fabricated from composite materials with continuous carbon fibers embedded in an epoxy matrix. This technology permitted *Voyager* to weigh only 939 pounds (426 kg) without engines. Installation of the engines doubled that weight, and the addition of fuel resulted in a takeoff weight of over 7,000 pounds (~3200 kg)!

It is a remarkable engineering achievement when an aircraft carries over seven times its own weight in fuel at takeoff. Seventeen tanks dispersed throughout the aircraft were used to hold 1,200 gallons (4500 L) of fuel. These tanks had to be filled in a predetermined sequence in order to distribute the loading imposed on the fragile craft. Indeed the loading on the main wing during takeoff bent it so much that it scraped the runway and the wing tips were damaged. These drag-reducing winglets, incidentally, were a biomimetic feature borrowed from the hawk family.

The volume of fuel needed for such a flight required large tanks which dominated the early conceptual designs proposed for the airframe. This particularly creative phase of the design process produced several different aircraft concepts that had to be evaluated. The *Voyager* airframe design that emerged had two long fuel tanks in separate fuselages connected by long slender primary wings which also stored fuel, as did two small canard wings at the front of the aircraft. *Voyager* was primarily a flying fuel tank!

Calculation of the airflow round this novel design was accomplished with small conventional computers using simplified vortex-lattice mathematical models and engineering intuition. This contrasts with the sophisticated fluid dynamics calculations of the aircraft industry where supercomputers are used for numerical calculations involving Navier Stokes equations.

Voyager's wing span was 111 feet (34 meters); the width at the roots was 2 feet (0.6 meter), while the width at the tips was only one foot (0.3 meter). In rough weather these wing tips flexed and vibrated with an amplitude as great as nine feet (2.7 meters)! The central fuselage housed the two pilots in cramped quarters that mandated noise attenuation and the application of ergonomic principles. Noise-control measures included active noise-cancelling earphones.

The overriding concern with weight reduction during the design process favored an aircraft with small wings called canards just like the Wright brothers' airplane (Figure 3.2). These small wings were located ahead of the main wings, as can be seen in Figure 3.1, and they operate in unison with the vertical fins to provide flight stability. In most aircraft the center of mass is forward of the aerodynamic center of the plane to ensure stability, and the airplane has a tendency to fly with a nose-down attitude. This is usually counteracted by horizontal stabilizers at the rear of the aircraft to generate downward forces that raise the nose. Under these conditions the main wings of the aircraft must withstand the loading caused by the aircraft weight plus the additional loading caused by these stabilizers. Thus there is a weight penalty associated with this class of designs. Such a penalty was deemed unacceptable for an aircraft which had to fly around the earth without refueling; consequently the canard design was adopted.

The necessity to minimize the aircraft weight motivated the lightweight airframe with canards.

Figure 3.2

The Wright brothers' airplane, now in the Smithsonian Museum.

The canards at the front of the *Voyager* generate lift, or an upward loading on the aircraft, to raise its nose and ensure stable flight. Consequently the main wings could be reduced in size and weight because both the main wings and the canards developed lift. This was an important feature. Furthermore this airframe is inherently safer than a conventional design with rear stabilizers. If an aircraft loses airspeed, its front wings stall first. With a conventional design, a plane has a tendency to fall out of the sky when it loses power because the main wings, which provide the primary lift, are the first set of wings to lose lift. With a canard design it is only the canards that lose lift. The nose drops automatically; the aircraft dives and recovers airspeed and lift. This characteristic is a positive attribute when pilots become tired and inattentive during long record-breaking flights.

Pilot fatigue was also alleviated by an autopilot to relieve the pilot of continuously monitoring altitude, speed, and direction. *Voyager's* autopilot was an off-the-shelf item which had to be modified by the vendor to accommodate the unique flight characteristics of the plane and to take into account the large change in weight—and hence response—of the aircraft during flight.

The challenge of finding the best power plant for the *Voyager* involved consideration of energy lost to aerodynamic drag, the change in weight of the aircraft during flight (*Voyager* weighed five times more at takeoff than at landing), and the fuel efficiency of the engine.

To minimize aerodynamic drag, *Voyager* flew at a nominal altitude of 8,000 feet (even though it was capable of attaining 20,000 feet) and at the relatively low speed of 110 miles per hour. Acceptable engine efficiency at low speeds was attained by specifying two piston engines located on the center line of the plane, one pushing and the other pulling the aircraft. The strategy was to use both engines for takeoff, but as the weight of the aircraft progressively declined, one engine would be throttled back. Eventually the aircraft would be flown on only one engine. The second engine could then be used for emergency situations.

The task of achieving an acceptable engine efficiency at all aircraft weights was solved by using two engines.

If a single engine had been used to propel *Voyager,* it would have been necessary to continually reduce its power to maintain the optimal speed of 110 miles per hour as the aircraft weight was reduced by the consumption of fuel. Because aircraft engines are not efficient over a broad range of throttle conditions, that approach would have lowered engine efficiency.

Another subtle feature of *Voyager's* engine configuration was that both engines developed thrust along the center line of the craft, with this thrust vector passing through the plane's center of mass. Conventional twin engine designs have thrust vectors separated from each other and the operation of only one engine makes the plane more difficult to fly. Because *Voyager's* engine configuration made it relatively easy to fly, this design provided a less stressful environment for the pilot.

The two engines of the *Voyager* were quite different in their design characteristics. The front engine was a well-proven standard air-cooled machine capable of developing 130 horsepower (97 kw), and it was assigned the role of auxiliary power plant. It provided power during takeoff and the first 80 hours of the flight. In addition, it was used when *Voyager* had to climb above thunderstorms.

The main engine was a pusher engine at the rear of the fuselage. It was liquid-cooled, developed 110 horsepower (82 kw), and operated successfully for the entire record-breaking flight except for a five minute period when, inadvertently, it was not supplied with fuel.

The selection of a heavy liquid-cooled main engine involved several design trade-offs.

While a liquid-cooled engine is somewhat heavier than an air-cooled one, it offers advantages in superior power output and efficiency. The decision to use this kind of engine involved a number of design trade-offs which are quite common in solving open-ended problems of this kind. A liquid-cooled engine's superior efficiency and power are achieved primarily because a constant engine temperature is maintained. This permits use of closer engine tolerances. An air-cooled engine must accommodate the variable temperature of the cooling air. Metal parts contract and expand in response to temperature changes, and consequently there are changes in the clearances between moving parts. This translates into either increased engine wear as clearances are reduced and the flow of lubricants is restricted or a reduction in engine efficiency is reduced as clearances rise and gases leak past the piston rings.

The rear liquid-cooled engine of *Voyager* was able to operate with a compression ratio of 11-to-1—high compared to the equivalent air-cooled engine that typically operates with a compression ratio of approximately 7-to-1. Furthermore, the liquid-cooled engine achieved a 20 percent superiority in fuel consumption, and the overall efficiency of the engine was 40 percent superior to the air-cooled design. These were significant advantages in the design of the *Voyager*.

Following a series of iterations involving creative thinking as well as analysis and testing, *Voyager* was completed and, in July 1986, was test flown over California for four and a half days without landing or refueling. This major test of the aircraft and its pilots led to several modifications before the record-breaking flight in December 1986. This process is typical of the development of a new engineering system.

The creation of an airplane requires a delicate balance between the aerodynamic characteristics governing the primary features of the airplane, such as the wings and fuselage, the propulsion system, flight control systems, and the undercarriage. All of these primary subsystems dictate the performance of the plane. Clearly the design of an aircraft is a complex interdisciplinary problem. A decision concerning one aspect of the machine often has a cascading effect that immediately changes another facet of the aircraft. Compromises and numerous iterations are, therefore, a characteristic of the design process. Nevertheless this process was completed and *Voyager* was successfully flown nonstop around the world without refueling.

Compromises and numerous design iterations were characteristic of the *Voyager* program.

Disciplines Relevant to the Design of *Voyager*

Some of the undergraduate courses and disciplines relevant to the design and manufacture of an aircraft with the ability to fly nonstop around the world without refueling are listed below:

acoustics	economics	materials science
advanced manufacturing	ergonomics	medicine
aerodynamics	finance	meteorology
aerospace structures	geography	navigation aids
avionics	heat transfer	numerical techniques
CAD techniques	hydraulics	optimization
chemistry	ichthyology	ornithology
composite materials	instrumentation	thermodynamics
controls	kinematics	tribology
design strategies	law	vibrations
dynamics		

3.2 The Student Design-Build-Test Project.

In this chapter we examined a "real life" example of how to undertake a design-build-test project. The problem was defined before the design team entered the iterative cycling of creativity, analysis, and evaluation. The problem that students encountered in their design-build-test project was defined by the objectives and rules listed in Chapter 1.

The initial deliberations of the student groups paralleled those of the *Voyager* designers. A series of rhetorical questions helped to establish the design parameters and explore different possibilities. Questions students asked themselves included the following:

Power plant?
Wheel configuration?
Steering mechanism?
Efficiency of the heat engine?
How much energy from a burning candle?
Weight of the vehicle?
Self-starting capability?
How to reduce inefficiencies of the vehicle.
How to transform thermal energy into kinetic energy.
Facilities for manufacturing the desired parts?
Adequate knowledge to analyze the vehicle's subsystems?
How to develop interdisciplinary mathematical models for predicting vehicle performance?

Summary

The Rutan *Voyager* set a new endurance record for long distance flight in December 1986 by flying nonstop around the world without refueling.

The design process was initiated by posing a series of questions governing the characteristics of the experimental aircraft. Its range was governed by the ratio of thrust efficiency to fuel consumption, the ratio of lift to drag, and the ratio of fuel weight to gross aircraft weight. These aircraft attributes dominated the conceptual design of the airframe.

An iterative cycling of creativity, analysis, and evaluation—with numerous trade-offs and compromises—was employed to arrive at the final airframe design. This design embodied a number of innovative features including lightweight honeycomb structures, strong lightweight composite materials, drag-reducing winglets, a canard wing design, two engines on the primary center-line axis of the aircraft, and two long, slender boom fuel tanks connecting both sets of wings.

Voyager's design employed techniques from a broad range of engineering and scientific disciplines. Many procedures embedded in this real-life design program will be examined in Chapter Four which presents a systematic design algorithm. It will enable you to address the design-build-test projects of your undergraduate studies and also the design problems that will challenge you in professional practice.

Key Concepts

Analytical thinking. Evaluative thought processes which utilize reason, logic, facts, mathematics, and so on to make value judgements.

Creative thinking. Inventive thought processes intended to produce new ideas.

Design activities involve many disciplines. The design and building of an aircraft not only involves a range of engineering subdisciplines but also non-engineering disciplines such as economics, medicine, and ergonomics.

Design trade-offs and cascading consequences. The design process involves the resolution of conflicting and compromizing situations. Decisions made about one problem influence others.

Iterative process. An action that is done repeatedly, often many times over, with modification of the initial conditions between each iteration in a quest for a better or different outcome.

Problem analysis. The investigation of possible solutions to a problem, usually involving identification of the goal, development of possible strategies or ideas for its attainment, and sufficient data collection to decide on the probable result of each strategy.

Review Questions

1. List some of the questions concerning the characteristics of aircraft that need to be answered in the design of an aircraft to circumnavigate the world without refueling.
2. List the disciplines that need to be involved in the design of *Voyager* and describe why they are needed.
3. What is the theorem that governs the endurance of an aircraft. How does this theorem influence the design of the airplane? Discuss these design activities and how they conflict with each other.
4. Discuss the problem definition for *Voyager* and also the primary components of the plane.
5. What were the major challenges in the design of the *Voyager* airframe?
6. What was the primary optimization problem confronting the *Voyager's* design team? List the three primary fields of the engineering sciences that dominated the solution process.
7. How were lightweight, stiff structures fabricated for *Voyager*?
8. What are the advantages of canard wings for *Voyager*?
9. List the primary decisions to be made in the selection of an aircraft power plant.
10. Why did *Voyager* have two engines configured on the center line of the aircraft? Discuss the trade-offs associated with the selection of the engines for *Voyager*.

11. Why did *Voyager* fly at a nominal altitude of 8,000 feet at a speed of 110 mph?
12. How was the idea for drag-reducing winglets at the tips of *Voyager's* main wings conceived?

Problems

1. Suppose you are responsible for designing a system for transporting an astronaut to Mars and successfully bringing him or her back to earth. Devise various concepts and methods to evaluate these concepts.
2. It was said that every additional kilogram of mass of *Voyager* required an extra six kilograms of fuel. Since the volume of a fuel tank varies as the cube of its dimension, it could be argued that the *Voyager* should have been built twice as big. What assumptions are made in the 1 kg mass to 6 kg fuel rule?
3. The *Voyager* aircraft is not the only design solution for flying around the world nonstop without refuelling. Develop and discuss an alternative design.
4. The intercontinental transportation of commercial goods could be made far more effective if ships were able to move at greater speeds. Discuss means of transporting material across the oceans at greater speeds than is possible by present-day surface vessels.

Further Reading

Yaeger, J., D. Rutan, and P. Patton. *Voyager*. New York: Alfred A. Knopf, 1987.

A Systematic Design Procedure

4

Method is like packing things in a box; a good packer will get in half as much again as a bad one.

Robert Cecil
(1830-1903)

Objectives

- **To present a methodology for systematically solving open-ended design problems.**
- **To summarize the primary ingredients in each phase of the design process.**

Contents

What if...

You have to create something—a new design, a product. The creative process may be hard or easy, depending upon how you approach it. Worry, which can take more out of you than work, usually results when you try to see the completed project in finished form.

There is no mystery about creating something. You follow a series of logical steps, each a manageable task with characteristic problems that can be overcome. The secret of success is seeing each step clearly and completing each task in an orderly fashion. Viewed as a finished product, a design project can seem overwhelming; as a series of manageable steps, it becomes possible.

At any stage of creativity, anticipate bad ideas; they are, after all, better than no ideas and may contain the germ of a gem! Great designs don't spring forth in full-blown beauty. If you follow a plan, they emerge slowly through many revisions and after much reflection, modification, and reconstruction.

4.0 Introduction

Chapter 4 introduces the systematic design process that underlies the philosophy of this book. This approach is so basic and has such general validity that it can be used to solve open-ended problems in all walks of life. As you study this chapter, you may wish to consider how you could apply this procedure to your own personal projects and problems.

4.1 A Systematic Design Procedure

Every useful product satisfies a need. It follows, therefore, that the design of a product begins with a statement of the problem that defines the need. Potential solutions are created, analyzed, and refined before the final solution is attained. This is accomplished by employing a *design methodology*—a complex procedure consisting of two principal stages, the *conceptual design* and the *product design*. (Recall Figure 1.2. in Chapter One.) Conceptual design is strongly dependent upon creative thinking. It is characterized by the creation of numerous potential solutions to a problem—solutions that may begin as no more than brief notes and rough sketches. The analysis and evaluation of these concepts permit the best potential solution to be identified.

Every product derives from a need.

A *methodology* is a set of rules for solving a problem.

There is no approved solution to any tactical situation.

General George S. Patton, Jr. (1885-1945)

The best potential solution provides input to the product design phase of the design process. In this phase, sketches and drawings of the best concept are refined to transform the concept into plans for a real product. This transformation is accomplished by an iterative combination of creativity and analysis often involving mathematical models and computational simulations. The product probably consists of parts and assemblies that may include castings, forged and welded structures, belts, bearings, valves, and other off-the-shelf items.

A good conceptual design requires good creative thinking.

A systematic design procedure is proposed that includes the following steps:

- Definition of the problem
- Creation of potential solutions
- Analysis and evaluation of these concepts
- Selection of the best concept
- Iterative refinement of the best concept
- Transformation of this concept into plans for a product

This sequence of events is illustrated in Figure 4.1 which shows the development of a paper airplane. The first three cartoons at the left of the figure describe the conceptual design phase and the three on the right represent the product design phase. The problem is first defined in order to determine what the airplane must do. Then many different potential solutions are created in response.

This book is intended to be used, not just read or looked at. Its wide margins are for your notes; write in them as you think about how the text relates to your design problems.

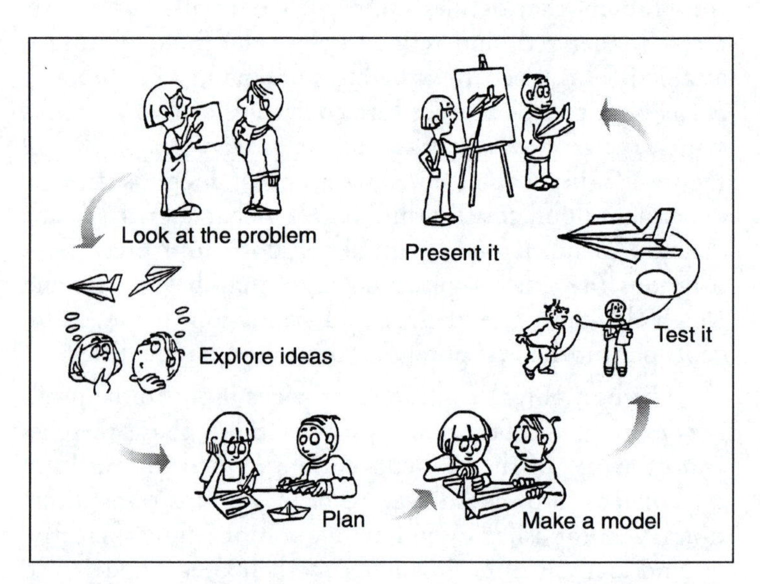

Figure 4.1

The design of a paper airplane.

These various concepts are analyzed and evaluated in order to select the best one for further development. The development process is called *product design;* it transforms the concept into detailed plans for the manufacture of a product.

The *product design process* transforms a conceptual design into plans for manufacturing the product.

Near the end of the product design phase, before the design is completed, a physical model of the airplane—a *prototype*—is generally built and tested. This prototype permits the design to be thoroughly evaluated. Performance characteristics are experimentally determined and compared with those stipulated in the design specification before the product enters production. This is an iterative activity that generally involves modification of the product.

A *prototype* is a physical model of the product used during the development phase to determine experimentally whether it satisfies the specification.

These steps in the design process are presented in Figure 4.2 which shows its principal features as a flow diagram. Note that the design process is divided into distinct tasks. Each task has a feedback loop coupling it to all previous tasks.

Each stage of Figure 4.2 involves the cycling of cognitive processes: creative thought, analytical thought, and decision making. This cycling process, which was presented in Figure 1.1, involves abrupt switching between creative and analytical thinking.

Invention is little more than a new combination of those images which have been previously gathered and deposited in the memory. Nothing can be made of nothing; he who has laid up no materials can produce no combinations.

Sir Joshua Reynolds
(1723-1792)

The characteristics of the design process are reasonably clear but the precise sequencing is uncertain and problem-dependent. Continual feedback, updating, and refinement are essential ingredients of all creative design activities.

Usually the design process is initiated by a market analysis or a request from a customer and it evolves from the definition of the problem to the final product design. Important as the design process is, it represents only part of bringing a product to market. The entire process by which products are conceived, designed, manufactured, brought to the marketplace, and subsequently supported is often called the *product realization process.*

The *product realization process* encompasses the conception, design, manufacture, and support of products.

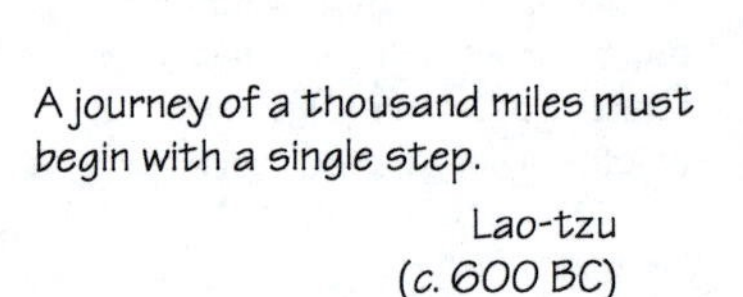

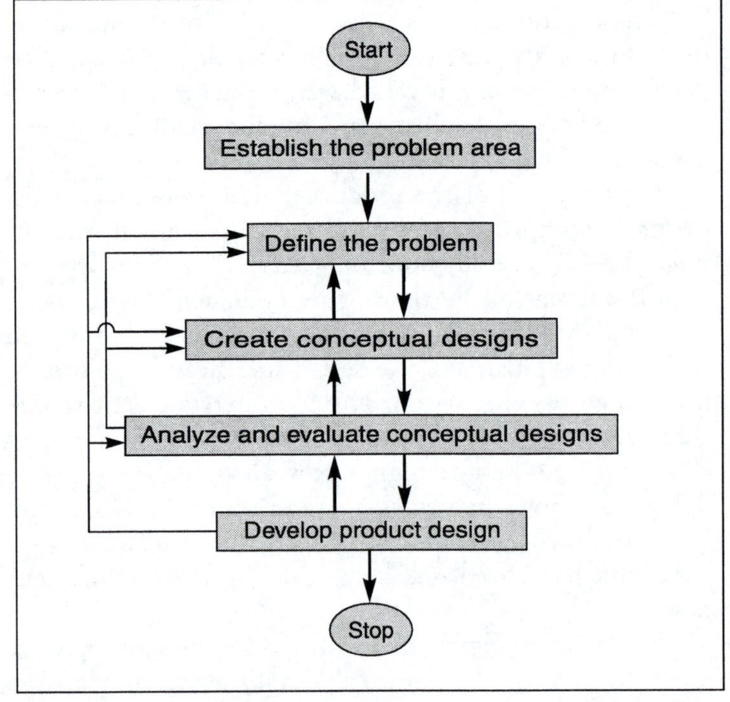

Figure 4.2

Essential ingredients of the design process.

The design process—a part of the total product realization process—will be examined further in the next chapter. Here we shall simply outline the essential features of the five steps shown in Figure 4.2.

1. **Establish the problem area**

   - Investigate the marketplace and identify customer needs. The ways in which customer requirements can be translated into technical requirements are described in more detail in Section 7.2. Ascertain the characteristics of the product that is to be created. Benchmark the products of the principal competitors (Chapter 7).

- Consider constraints that will affect the product, such as legislation, industrial standards, and environmental restrictions (Chapter 6).

2. **Define the problem**

- Become intimately familiar with the problem.
- Collect and analyze all relevant data that will enhance the solution process.
- Definition of the problem is a crucial step in the design process because if the problem is not defined correctly, the solution will be inappropriate and resources will be wasted.
- Develop a design specification for the product. This specification should include both technical and nontechnical requirements. Technical requirements include such things as power consumption and maintenance, whereas nontechnical requirements often include economic considerations and aesthetics. The development of a design specification is discussed in Chapter 7.
- Determine the relative importance of the parameters defining the product. This is discussed in Section 7.3.

It is a profound mistake to think that everything has been discovered; as well think the horizon the boundary of the world.

Lemierre

3. **Create conceptual designs**

- Conceptualization is primarily concerned with creativity, which is the subject of Chapters 8 and 9. Various conceptual designs are created to provide the basis of a product that will satisfy the design specification. These potential solutions are often freehand sketches that show only essential features and principles of operation. The primary concern is functionality.
- Conceptual designs can be generated by various strategies such as brainstorming, reviewing the patent literature, consulting books containing standard elements of construction, and considering biomimetic approaches. These strategies are discussed in Chapter 9.

Things seen are mightier than things heard.

Alfred, Lord Tennyson (1809-1892)

Evaluation is always a threat, always creates a need for defensiveness...

Carl Rogers (1902-1987)

A poor conceptual design cannot be significantly enhanced by superb product design or manufacturing prowess.

4. **Analyze and evaluate conceptual designs**

- In analyzing and evaluating conceptual designs, you often must use mathematical models to develop order-of-magnitude estimates of both cost and performance.
- Compare each of the conceptual designs with the criteria in the design specification. Often this is done by assigning numerical values to the various attributes (Chapter 10). Select the most promising conceptual design for further scrutiny, refinement, analysis, and development during the subsequent product design phase.
- Generally no single concept will be so outstanding that it is clearly the best one when the conceptual designs are first evaluated. Consequently several different conceptual designs often must be further refined until the best is clearly evident. Evaluation is important because a poor concept cannot be rectified by superb product design or manufacturing.

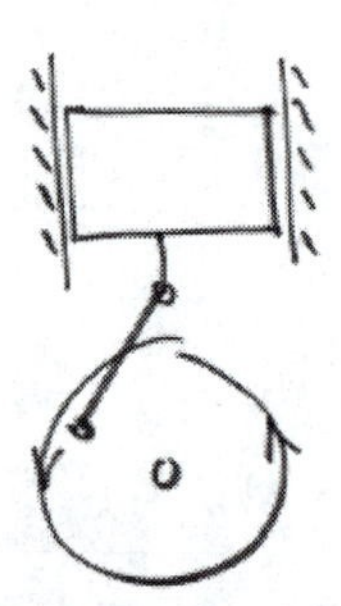

Figure 4.3

A single-cylinder engine: a conceptual sketch and the final product (cutaway view).

5. **Develop the product design**

- The conceptual design (the embryonic product) is gradually transformed into engineering hardware by the design of assemblies and individual parts (Figure 4.3). This activity, called the product design phase, involves both analysis and synthesis (Chapter 11).
- Design specifications for major assemblies, subassemblies, and individual parts must be developed (Section 7.3).
- Specifications are developed for each part: material, shape, heat treatment, surface finish, tolerances, and the manufacturing processes to be employed.
- Sophisticated calculations may be needed to ensure that each part can be manufactured economically. This topic is discussed in Chapter 12.

- Decide which parts and assemblies should be purchased from vendors.
- To facilitate manufacture and inspection, detailed drawings of all components, assemblies, and subassemblies are made. CAD/CAM software is frequently used.
- An inappropriate product design can ruin an excellent conceptual design.
- Product design provides the basis for the manufacture of the product. With most products, prototypes are made and tested before production begins.
- Plan for the efficient manufacture of the product by meticulously considering the interactions between product attributes and manufacturing process attributes. This is called *design-for-manufacture* and is discussed in Part Two of this book (particularly in Chapter 12).

Design for manufacture is the practice of designing a product with an emphasis on issues pertaining to ease of manufacture.

- The product is then launched into the marketplace. To measure the success of the product and provide relevant feedback to the design team, the response of the customer should be solicited. This can be done by monitoring parts replacement, service calls, service life, etc., as well as by soliciting customer opinions. Subsequently, the design is modified and the product improved.

4.2 The Student Design-Build-Test Project

This chapter has described the essential methodology for designing, building, and testing a birthday-candle-powered vehicle. The approach adopted by some student groups is described below; this was implemented in accordance with the Gantt chart established at the initial planning meeting described in Chapter 1:

- Definition of the problem. (Chapters 1, 3, and 7)
- Establish protocols for the design process. (Chapters 2, 5, and 6)
- Create potential solutions. (Chapters 8 and 9)
- Evaluate conceptual designs. (Chapter 10)
- Make a working model. (Chapters 11, 12, and 13 and later chapters)
- Test the model and then refine the design.

This philosophy is captured by the following paragraphs in a student report from one of the author's classes:

> The design group started by focusing on the motive source. Fifteen conceptual designs were generated. After evaluation, two solution candidates were identified. Some of the most important criteria for a motive source are reliability, ability to start without mechanical energy input, and high conversion of thermal power to vehicle motion. The two solution candidates were a hybrid thermoelectric-photovoltaic motive source, and an open Rankine cycle.
>
> In order to make a decision between the two motive sources, a detailed analysis was undertaken for each one. Most of the calculations on the open Rankine cycle focussed on the heat exchanger and on maximizing the outlet steam temperature. It was assumed that the more the steam was superheated, the more it would expand the bellows-crank mechanism and produce more work. The hybrid motive source calculations focussed on the emission spectrum and radiative power output of the candles, the number of thermocouples required to power a 1.5 volt DC motor, and the heat transfer from the flue gas to the measuring junctions of the thermocouples. Prototypes of both systems were built and tested, and refinements in the analyses were made accordingly. The chassis was also analyzed and a three wheel design was settled upon. Also, a steering mechanism was designed to prevent the vehicle from crashing into the walls of the corridor during the competition.
>
> The results of the analyses and the prototype testing indicated that the best design for the motive source was the hybrid motive source. The analysis of the open Rankine cycle was lengthy and time consuming. Also, the Rankine cycle prototype was not reliable; it did not always produce

sufficient power to move the vehicle. The hybrid motive source was chosen because of the shortcomings of the Rankine cycle but also because it was reliable. It always provided enough power to propel the vehicle. The hybrid motive source takes advantage of two modes of conversion of thermal energy to mechanical work; thus it captured more of the energy content of the candles. The two converters were linked in parallel so that if one failed, the other would provide enough energy to keep the car moving. Also, although each source separately is inherently inefficient, their combination increases the overall efficiency of the motive source.

Given more time, it is possible that a reliable working model of an open Rankine cycle engine could have been produced. Also, a better thermodynamic model needs to be developed to accurately describe the power output of the bellows-crank mechanism.

Summary

In this chapter a general procedure for addressing engineering design problems has been presented. This procedure involves recognition of the problem, formulation of the design specification, creation of several concepts as potential solutions to the problem, evaluation of these concepts, and the transformation of the best concept into detailed plans for the manufacture of the product. The outline in this chapter contains references to subsequent chapters that describe the process in detail.

In the next chapter we shall discuss the design processes used by industrial organizations as they develop products. That chapter will end our broad introduction to engineering design. The remainder of the book will discuss details of the design process.

Key Concepts

Benchmarking. Establishment of the characteristics of competing products.

Conceptual design. The creation of designs satisfying the product specification

Design methodology. A procedure for solving a design problem in which the creative and analytical aspects of the design process are defined as a series of steps involving feedback, updating, and refinement.

Design specification. A comprehensive statement of the requirements to be fulfilled by a product.

Evaluation. The determination of the relative worth of various conceptual designs proposed as the basis for a product.

Market analysis. The development of a comprehensive understanding of the market for a proposed product. Projected sales, advertising, pricing, product life, disposal, etc., must be considered.

Mathematical model. A mathematical representation of a physical system used to predict behavior.

Product realization process. How a product is conceived, designed, manufactured, brought to market, and subsequently supported.

Prototype. The first product created from a conceptual design. It is tested and evaluated carefully to determine if it satisfies the design specification.

Systematic design procedure. The development of plans to manufacture a product requires the definition of the problem, the creation of numerous potential solutions, the evaluation of these concepts, and the transformation of the most promising concept into the product.

Vendors. Companies selling parts or assemblies used in a product. Sometimes called suppliers.

Review Questions

1. What is the role of creativity in engineering design?
2. Cite a specific example of the use of mathematical modelling in the design of a particular product.
3. How does conceptual design differ from product design?
4. What is the role of evaluation in design?
5. What are the two types of thought processes employed in design activities? How do they differ?
6. List the primary steps in the systematic design procedure presented in this chapter.
7. What is prototyping and what purpose does it serve?
8. What is the generic design cycle that is embodied in the systematic design procedure described in this chapter?
9. How is the design process initiated?
10. What is a design specification?
11. What is benchmarking?
12. In which phase of the design process are computer-aided design and drafting packages used?
13. The methodology presented in this chapter can be used to solve open-ended problems. List various kinds of open-ended problems.
14. What form do the designs that emerge from the conceptual design phase often take?

Problems

1. Develop in outline a systematic procedure for the design of a machine that would sort mail by size, arrange face-up, frank the stamp by date, read the address, stamp with a bar code, and then sort by zip code into bins.
2. Develop a systematic procedure for the design and testing of a display on the dashboard of an automobile that would inform the driver of the current fuel efficiency (mpg) of the car.

3. One problem facing the developers of ski resorts is how to transport skiers back to the top of the mountain as quickly and efficiently as possible. Develop a systematic procedure for the design of an appropriate transport system.

Further Reading

Cross, N. *Engineering Design Methods.* New York: John Wiley, 1989.

Pahl, G. and W. Beitz. *Engineering Design.* English edition edited by K. Wallace. London: The Design Council, 1984.

Pugh, S. *Total Design: Integrated Methods for Successful Product Engineering.* New York: Addison-Wesley, 1991.

Ullman, D. G. *The Mechanical Design Process.* New York: McGraw-Hill, 1992.

How Products Are Created

5

Where tumbling waters turn enormous wheels
And hammers, rising and descending, learn
To imitate the industry of man.

John Dyer
(1700-1758)

Objectives

- To define the product realization process
- To discuss the influence of design on industrial practice
- To characterize and contrast sequential and concurrent product development
- To summarize the benefits of implementing concurrent engineering in several industries

Contents

What If...

You are responsible for the creation of new products for your company. You have been losing market share to foreign competition which creates better products at lower prices. This situation is alarming.

One day you meet a university co-op student. She works part-time in industry and studies part-time at her university. The unfortunate status of your company becomes a topic of conversation. The student advises you that she has recently taken a course in concurrent engineering in which products are developed by multidisciplinary teams involving individuals from different company departments. She claims that concurrent product development offers many advantages over traditional sequential engineering. It sounds interesting. If you were to implement these ideas, it would be a major change in the way your company operates.

Would it work for you and increase your profits? Would your employees accept it? Should you take the risk?

5.0 Introduction

In the last chapter we presented a systematic design procedure. Now we will examine how this procedure fits into the process used to create products. The theme for this chapter, therefore, is the creation of products in industry. We will explain how the design process introduced in Chapter 4 influences the manufacture and support of new products.

A *product* is an artifact that is the outcome of a design and manufacturing process.

We will discuss two primary forms of product realization—sequential product development and concurrent product development—before concluding with performance data comparing concurrent engineering with traditional sequential engineering practices in several companies.

5.1 The Product Realization Process

Every product is the result of a *product realization process.* This process begins with identification of the type of product needed in the market place, continues through design and manufacturing, and culminates with customer purchases and ultimately disposal of the product. The intervening time is spent making difficult decisions involving complicated trade-offs between conflicting requirements. Analysis, synthesis, and evaluation need to be undertaken. These procedures are completed by diverse groups throughout the organization. Thus product development involves significant risk and may involve conflict.

Product realization process: the conception, design, manufacture, and support of new products.

The primary steps of the product realization process (PRP) are presented in Figure 5.1. However, the precise protocols will vary from company to company and the process employed by each company will not be carved in stone. Instead it is a dynamic process that evolves in response to new technologies, new methodologies, and feedback from individuals throughout the company and outside.

A company's product realization process should be continually reviewed and refined.

The implementation of a product realization process requires a corporate decision because it affects everyone in the company. Furthermore, because the process involves

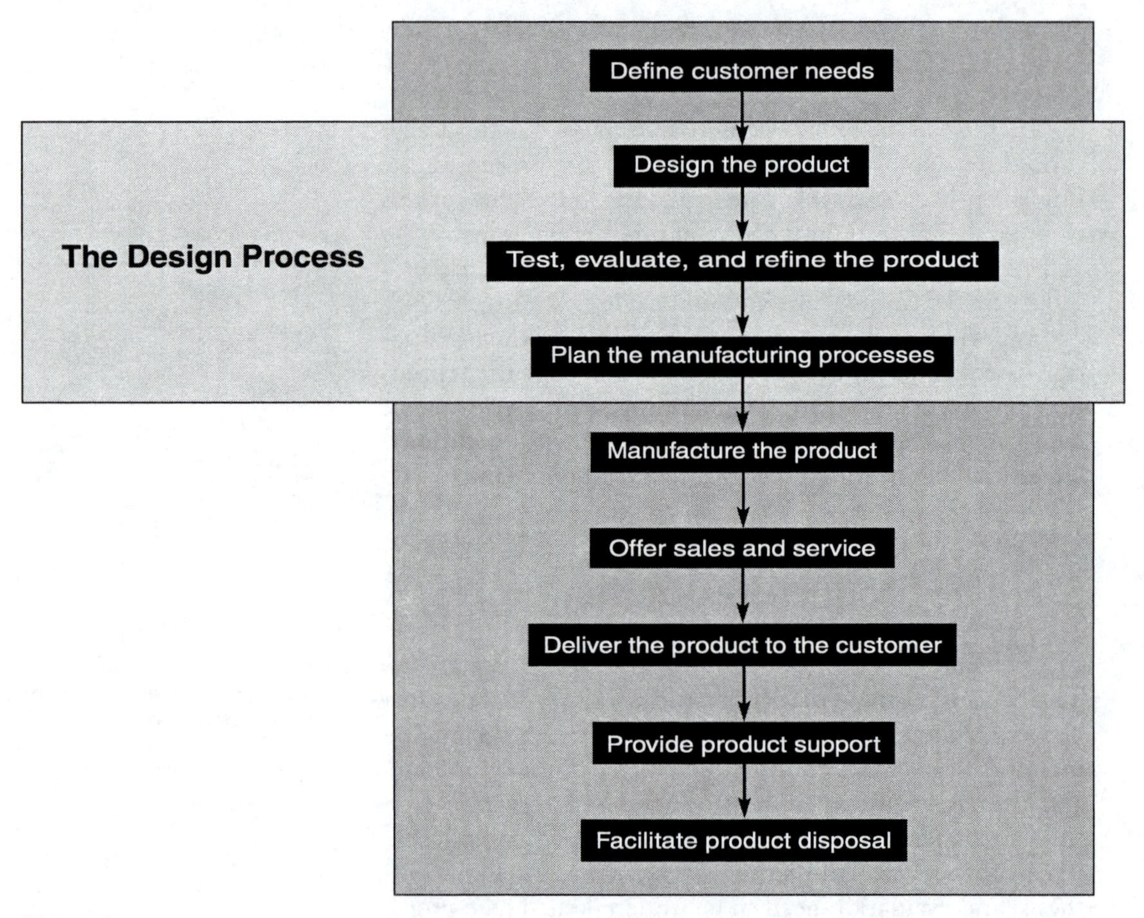

Figure 5.1

The primary steps of the product realization process.

individuals from many different disciplines, there must naturally be extensive cooperation among company departments such as finance, marketing, research and development, design, manufacturing, and sales.

The PRP is initiated by the needs of the customer or the perceived needs of the marketplace.

The product realization process is initiated by the definition of the customer's needs and the desired performance of the product. This activity may involve a review of technical, marketing, and business opportunities. Competitive products are analyzed and benchmarks established for the new product. Ultimately a comprehensive specification will be developed by considering many parameters such as aesthetics, reliability, quality, and cost.

Develop a comprehensive design specification.

Having defined the problem, some companies undertake a major review of the situation to plan for the evolution of the product through the design and manufacturing phases of the product realization process. These activities focus on identification of areas where performance can be improved or costs reduced because of technological advances. Input from diverse groups is sought to identify technologies for future generations of products. Planning and scheduling documents are created for the development of design, manufacturing, and production systems.

As described in the previous chapter, the design process begins in response to the product specification. A number of different conceptual designs are created and evaluated in order to select the most promising concept. This concept is then analyzed further to transform it into a product through the specification of materials, processes, and shape. The product is designed to minimize performance variation despite wide variation in product parameters and operating conditions. Furthermore, it is designed for ease of manufacture.

Prototypes of the product are manufactured in order to evaluate the design and ascertain whether it complies with the design specification. Thus the product is subjected to rigorous testing, evaluation, and refinement. This activity is an important procedure for reducing time to market and for providing a basis for evaluation of quality and producibility.

Manufacture prototypes, or physical models, to test the product.

Multifunctional teams should be employed to plan the manufacturing protocols and processes to manufacture the product to the required specifications at minimal cost. This phase of the product realization process is initiated by undertaking cost analyses, identifying the most appropriate manufacturing methods, establishing quality metrics, designing factory layouts, and determining the skill levels and training of personnel. These activities should incorporate input from parts suppliers and from the manufacturers of any new equipment to be used.

Manufacturing parts requires controls to ensure that they comply with specifications.

The production phase of the product realization process is dedicated to the manufacture of the product and to ensuring that the parts, subassemblies, and the ultimate product comply with the specifications. These activities involve in-process quality control and statistical process control which provide a basis for continuously improving the design and manufacturing processes. This assists the development of future products.

Having designed and manufactured a product, the company must interact with customers in order to sell and subsequently service the product throughout its life. These activities require technical and business decisions. The marketing effort for a consumer product usually involves the news media, while a product for an industrial customer with technical requirements may mandate a technical sales approach involving trade publications, specialized brochures containing performance data, etc.

One product realization process—a list of the steps describing the product realization process employed by the Polaroid Company—is illustrated below. This sequence assures that essential business and technical considerations related to product evolution are considered, evaluated, and understood by the whole corporation.

1. Explore the business, marketing, and technical fields, and prospect for opportunities.
2. Define the needs of the customer and attempt to improve the product realization process continuously.
3. Define a long-range, customer-focused product line and define the architecture of a family of future products.
4. Develop a comprehensive product performance specification. This should be undertaken by the product development team comprising members from engineering, finance, manufacturing, and marketing.
5. Insure that the product specification requires no inventions.
6. Changes should not subsequently be made to the product specification since all groups should agree to this specification before the process evolves further. The only exception to this rule is a change made by the customer.
7. Establish a benchmark process which contains goals and is motivated by the necessity for continuous improvement.

8. Simultaneously develop core technology building blocks for future products.
9. Initiate the design process using appropriate CAE/CAD/CAM tools. Employ multidisciplinary professional engineers and designers.
10. Concurrent with the design process, design the manufacturing process too.
11. Establish a reusable mathematical model database for the product technology and employ it for the simulation, analysis, and modeling of all product designs during the iterative phases.
12. Develop an information process for tracking world-class engineering design practices and share successful generic design processes with other U.S. companies and the universities.

Each of the above steps requires action and commitment at all levels in the company. It is important to observe that the specification for the product is defined and also agreed upon by all groups before initiation of the design phase. This minimizes unexpected scheduling problems and cost overruns.

A recent report published by the American Society of Mechanical Engineers based upon research funded by the National Science Foundation discusses the product realization process. The study attempted to identify the "best practices" of progressive companies within the United States and how these practices are related to undergraduate teaching. Five categories of best practices were identified:

Knowledge of product realization processes
Product realization process team skills
Design skills
Analysis and testing skills
Manufacturing skills

These primary categories included 56 subcategories of skills and knowledge that are presented in Table 5.1. This long list of best practices provides a measure of just how complex the product realization process is. While this complexity might be somewhat intimidating for you, this research concluded that, as an entry level engineer with a B.S.

An entry-level engineer with a BS degree should have skills in the following disciplines:

- Teams
- Communication
- Design-for-manufacture
- CAD systems
- Ethics
- Creative thinking
- Design-for-performance
- Design-for-reliability
- Design-for-safety
- Concurrent engineering
- Sketching and drawing
- Design for cost
- Statistics
- Reliability
- Geometric tolerancing
- Value engineering
- Design reviews
- Manufacturing processes
- Systems perspective
- Design-for-assembly

Knowledge of PRP	PRP Team Skills	Design Skills	Analysis and Testing Skills	Manufacturing Skills
Knowledge of Product Realization Process Bench marking **Concurrent Engineering** Corporate vision and product fit Business functions Industrial design	Project management tools Budgeting Project risk analysis **Design reviews** Information processing **Communication** **Sketching and drawing** Leadership Conflict management **Professional ethics** **Teams and teamwork**	Competitive analysis **Creative thinking** Tools for "customer centered" design Solid modeling/ rapid prototyping systems **Systems perspective** **Design for assembly** Design for commonality platform **Design for cost** Design for disassembly Design for environment Design for ergonomics (human factors) **Design for manufacture** **Design for performance** **Design for reliability** **Design for safety** Design for service/ repair **CAD systems** **Geometric tolerancing**	Finite element analysis Design of experiments **Value engineering** Mechatronics (Mechanisms and controls) Process improvement tools Statistical process control Design standards (e.g., UL, ASME) Testing standards (e.g., ASTM) Process standards (e.g. ISO 9000) Product testing Physical testing Test equipment **Application of statistics** **Reliability**	Materials planning and inventory Total quality management **Manufacturing processes** Manufacturing floor/worker cell layout Robotics and automated assembly Computer integrated manufacturing Electromechanical packaging

Table 5.1

The 56 best practices in the product realization process (Source: ASME Book MP3695, *December 1995).*

degree, you should have skills in the top twenty "best practices" shown in bold in Table 5.1.

5.2 Influence of the Design Process on Industrial Practice

Poor design probably causes 85 percent of the problems with new products: consumers complain that products do not work properly or are unreliable, and manufacturers complain that the product costs too much and takes too long to bring to market. Both manufacturers and customers want cheaper and better products. It follows, then, that the design process influences the profitability and prestige of any com-

Eighty-five percent of the problems with new products are the result of poor design procedures. These problems typically include

- products that cost too much
- products that do not work
- products that are unreliable
- products that take too long to bring to market

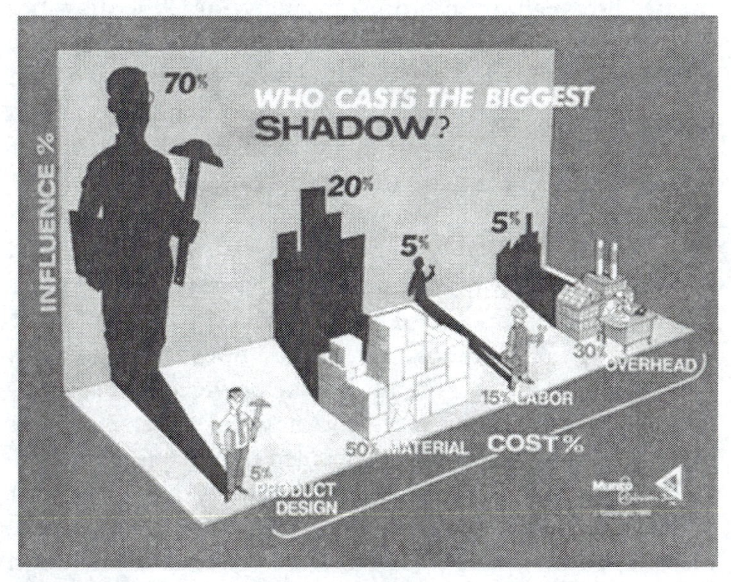

Figure 5.2

Product costs: a comparison of traditional accounting practice and the influence of these costing components on the total cost of a product. (Courtesy of Munro & Associates, Troy, Michigan)

pany trying to market new, cost-effective, high-quality products.

The cost of the design process and its influence on the total cost of a product are illustrated in Figure 5.2. This diagram shows the percentage of the total cost of a product associated with design activities, materials, labor, and the financial burden according to traditional accounting practice. Approximately 5 percent is attributed to design, about 50 percent to materials, 15 percent to labor, and about 30 percent to fixed overhead. These data are representative of information published by numerous companies such as the Big Three auto makers in the United States and the aeroengine operations of Rolls Royce in England. Notice, for example, that although design *costs* are only 5 percent of the total cost, design *decisions* are responsible for 70 percent of the total cost.

Design decisions have the greatest impact on the cost of a product.

Rolls Royce undertook a study of the costs associated with manufacturing operations during the design phase of the product realization process by analyzing 2000 drawings describing conceptual designs and product designs. The study concluded that the greatest potential for savings in manufacturing costs occurs during the conceptual design phase, although significant savings can be achieved in the product design phase and the process planning phase.

The greatest potential for savings in manufacturing costs occurs during the conceptual design phase.

During the conceptual design phase a number of alternative solutions should be generated, evaluated, and refined because different working principles largely dictate manufacturing costs. For example, should a family automobile be powered by a reciprocating internal-combustion engine or a gas turbine? If the automobile is to be powered by a conventional power source with reciprocating parts, then engineers must, for example, generate a detailed manufacturing drawing of a connecting rod during the product design phase or component design phase. Decisions must be made about the material from which these parts are to be fabricated and decisions must be made about the shape, tolerances, surface finish, etc. Subsequently, during the process planning phase, the most appropriate manufacturing processes must be as-

No amount of skilful invention can replace the essential element of imagination.

Edward Hopper (1882-1967)

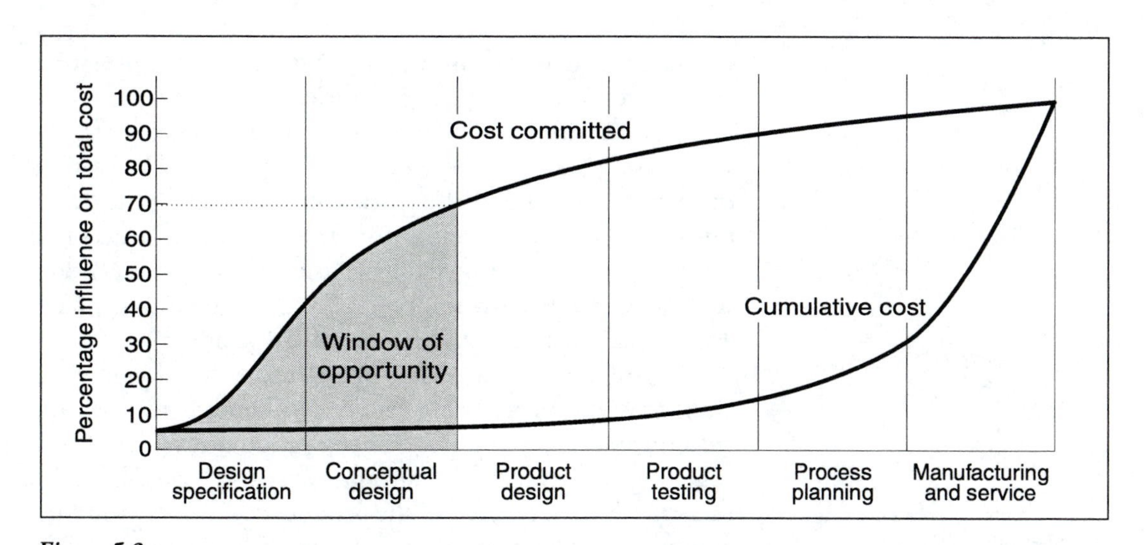

Figure 5.3

Cost profiles relative to the product realization process.

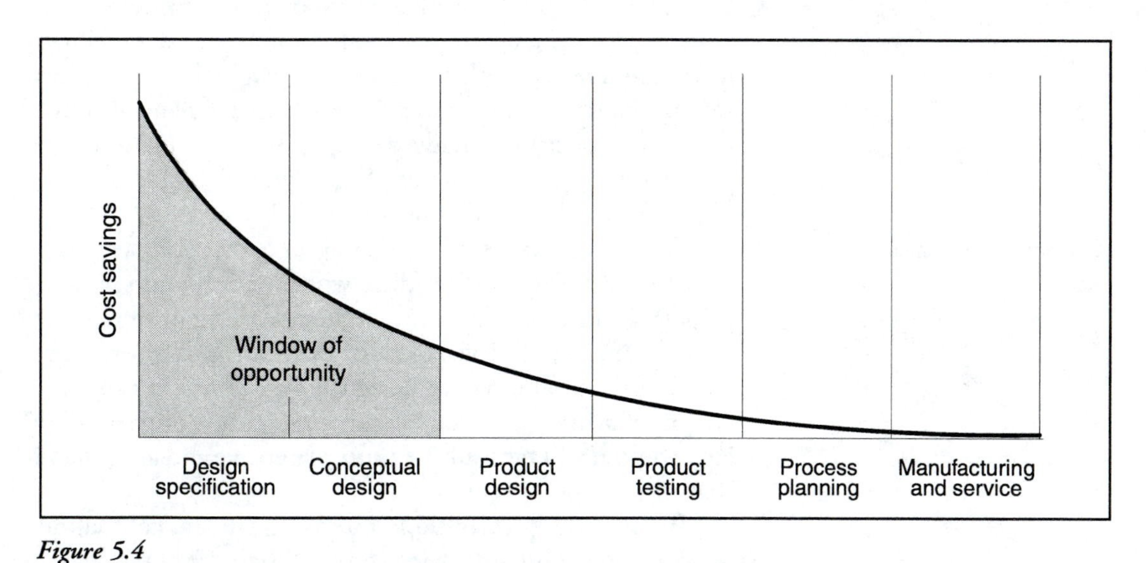

Figure 5.4

Potential for financial savings during design and manufacture.

certained and decisions concerning quantity, time scale, tooling, and manufacturing facilities must be made.

Figures 5.3 and 5.4 present costs associated with the various stages of the product realization process. The abscissa of each figure presents the passage of time during the development of a design specification, the creation of conceptual designs, the development of product designs, the product testing phase, the manufacturing process planning phase, and the manufacture and service of the product.

Figure 5.3 reinforces the observation that design decisions have far-reaching consequences. Upon completion of the conceptual design phase of the product realization process, only 5 percent of the cumulative cost typically has been incurred, but 70 percent of the product cost has already been committed. Therefore there is very little opportunity for potential cost savings during the remainder of the product realization process as the cumulative costs steadily increase. It is evident that the crucial phases of product realization are the design specification phase—involving both design and manufacturing parameters—and the conceptual design phase in which creative solutions to the original problem are sought. Figure 5.4 illustrates the substantial financial consequences of decisions made during the design phase of the product realization process. Subsequently there are few opportunities for significant financial savings.

Approximately 70 percent of the manufacturing cost is committed by the end of the conceptual phase of the design process.

During the conceptual design phase, decisions are made about how a proposed product will function and about its principal features. For example, the operating principle of a product may exploit viscid fluid flow and utilize only two parts while an alternative design may utilize a mechanical system of gears and linkages with a multitude of parts. This latter design will typically be more expensive and less reliable than the former.

KISS is an acronym meaning

Keep
It
Simple,
Stupid

Product design decisions following the conceptual design decisions only influence about 30 percent of the manufacturing cost of a product. The product design phase progresses through the specification of materials, geometries, tolerances, surface finishes, dimensions, heat treatment processes, bought-out parts, and other relevant manufacturing data. It terminates with the development of detailed plans for manufacturing a product.

Throughout the design phase, decisions are made that affect the quality of a product. High quality does not arise by accident. It can only arise by design; it cannot be *manufactured* into a product. A quality product cannot be manufactured unless it derives from a quality design. This is because the design should specify in detail the attributes of each part. These attributes are in turn generated by a series of manufacturing operations and manufacturing parameters that dictate the quality of the manufacturing procedures and hence the quality of the product. Consequently the sooner the issue of product quality is addressed in the product realization process, the bigger the return on the investment. The payoff is measured in financial savings and reduced engineering time. This observation is symbolized in Figure 5.5. Thus as a rule-of-thumb there is a 100:1 saving in product quality at the conceptual design stage but only a 1:1 saving when the product is at the manufacturing phase.

Quality can only be *designed* into a product; it cannot be *manufactured* into a product.

Product quality issues should be incorporated early in the product realization process for maximum effectiveness.

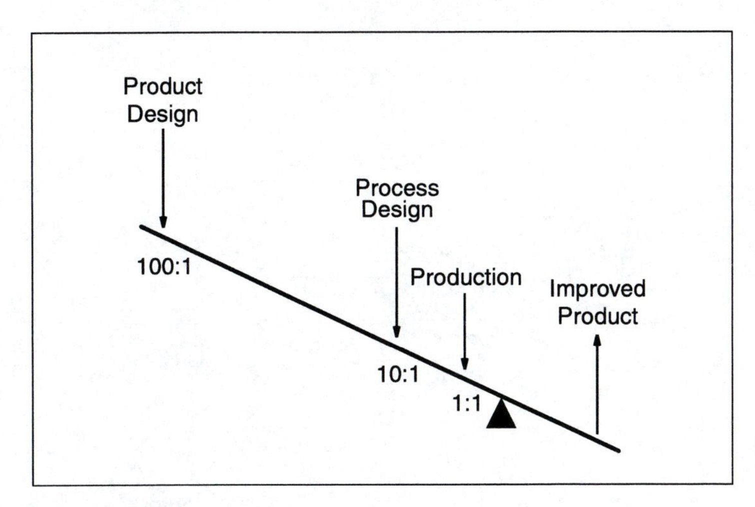

Figure 5.5

Quality-lever analogy quantifying the consequences of introducing quality considerations during the different stages of the product realization process.

Techniques employed in the early phases of the design process preceding the transition from the conceptual design phase to product design is a principal focus of this book. The motivation for this special emphasis is evident from Figure 5.5 which shows the significant influence of early design decisions on the total cost of a product. This observation has stimulated a rethinking of industrial practices governing the development of products. Consequently more progressive companies have shifted from traditional sequential product development to concurrent product development.

Sequential Product Development

Sequential engineering is characterized by design engineers creating products in isolation from other professional groups such as manufacturing engineers.

Sequential product development, or *sequential engineering,* is the traditional approach to the development of products. It is typically undertaken by companies organized with isolated departments in which product data is transmitted sequentially from department to department. The product is developed chronologically from the initial market re-

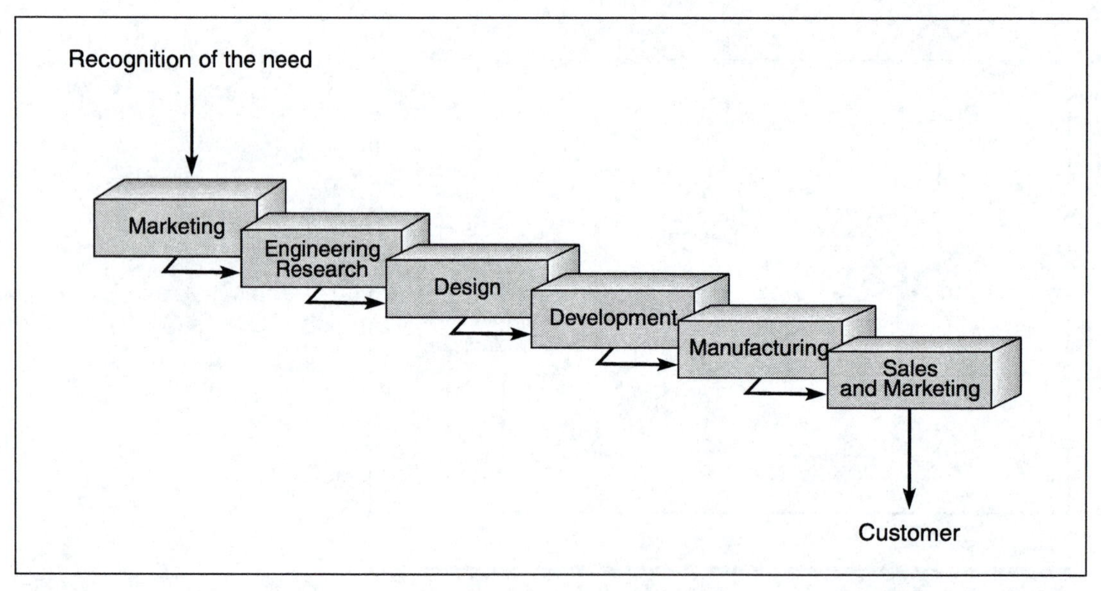

Figure 5.6

Sequential product development.

search by the marketing department through engineering research, design, development, and manufacturing to sales and marketing. This process is represented in Figure 5.6.

Sequential engineering is very inefficient.

Sequential engineering generally involves numerous costly engineering changes during product realization and the resulting lengthy concept-to-customer phase impedes sales. Furthermore, classical sequential product development fosters rivalry and isolation among departments because of poor information flow and impaired team spirit.

Under the sequential product development philosophy, design engineers create products in isolation from other professional groups in a company. Indeed the company operates as a set of discrete and isolated departments. Engineers in the design department submit drawings and specifications to engineers in the manufacturing department who are then required to manufacture the product. This practice is frequently termed a "throw it over the wall" approach or "throw it over the fence" engineering (Figure 5.7). It is an undesirable practice because design engineers frequently

Figure 5.7

Sequential product development uses an inefficient "throw it over the wall" strategy.

specify parts with shapes or other characteristics that are difficult or expensive to manufacture. Alternatively, manufacturing engineers redesign parts for ease of manufacture, but the consequences of the redesign process on the performance of the product may not be taken into account.

However, the ultimate objective of any company is to develop and sell a product that is desired by the customer. This is a challenging activity in a competitive marketplace and consequently all departments in a company must remain harmoniously focused upon that objective. Sequential product development does not foster such an environment.

In a well-organized system, all of the components work together to support each other.

W. E. Deming (1900-1993)

Concurrent Product Development

The collaborative approach in which interdisciplinary teams concurrently develop products is called *concurrent* or *simultaneous engineering.*

The major shortcomings of sequential product development in the product realization process are presently being addressed in some progressive companies by teams of designers, manufacturing engineers, and support services personnel who *concurrently* develop products. This approach is illustrated in Figure 5.8. Interdisciplinary groups create cost-effective products by combining the interests and con-

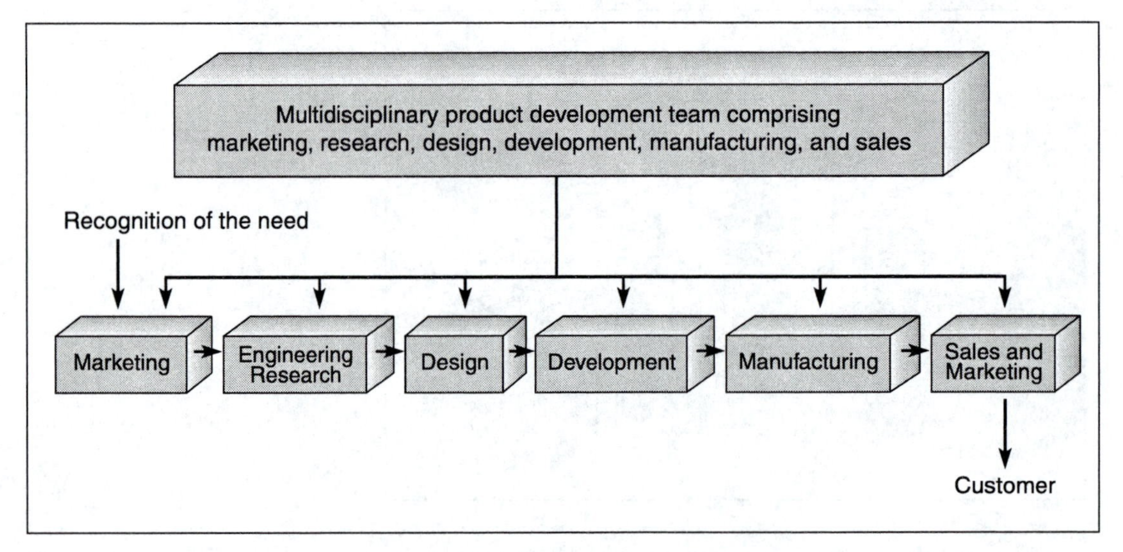

Figure 5.8

Concurrent product development.

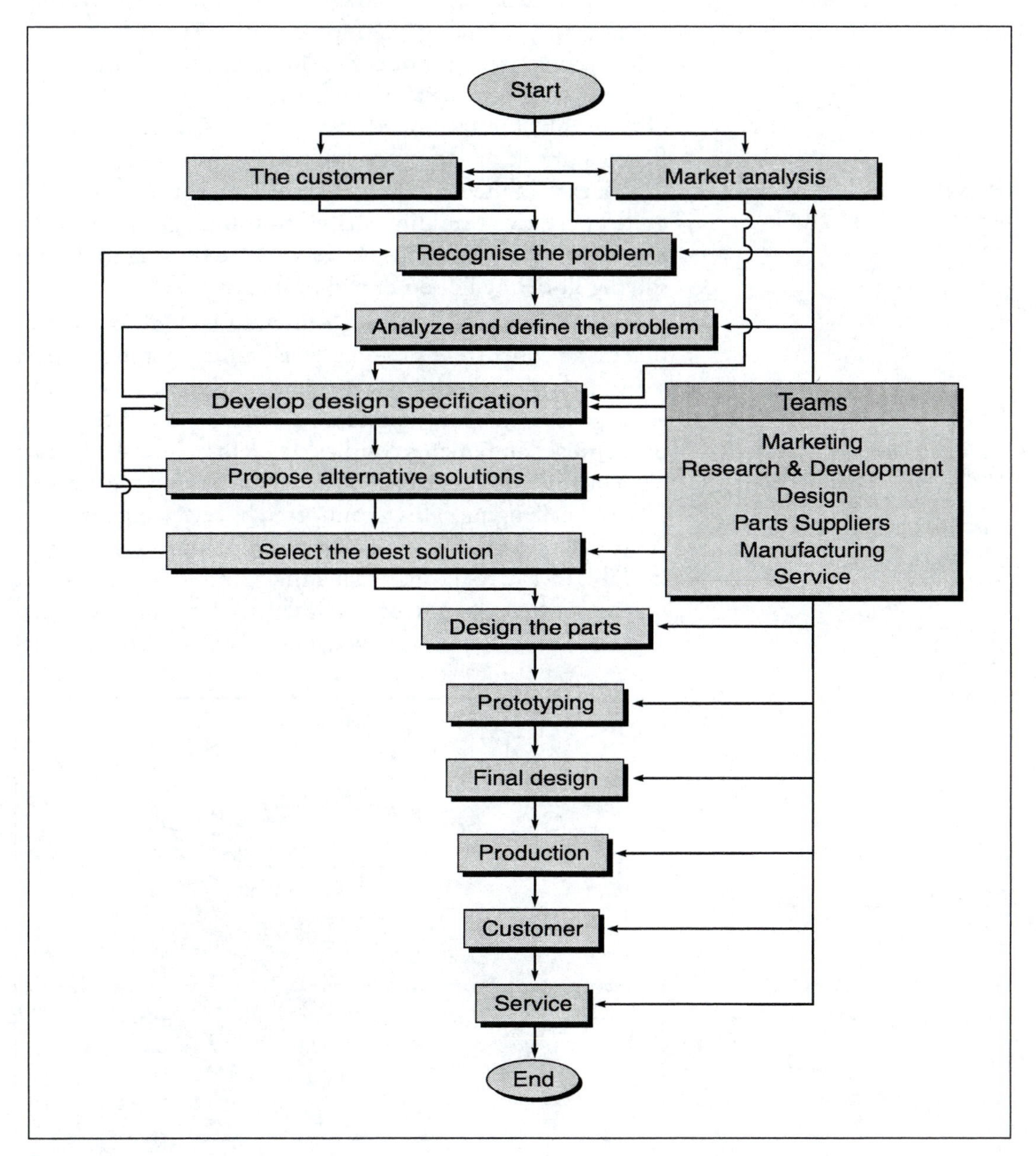

Figure 5.9

Concurrent product development.

straints of such areas as marketing, design, manufacturing, service, and recycling. Compare Figure 5.8 with Figure 5.6. In concurrent product development, the entire multidisciplinary group makes the relevant decisions at every stage of the product realization process. A more detailed algorithm for concurrent product development is presented in Figure 5.9. In each of the major stages of this procedure, an iterative cycle of creative thinking, analytical thinking, and decision making occurs. Recall the process presented in Figure 1.1 for solving ill-defined open-ended problems.

Always use concurrent engineering practices. They are the most effective.

The desires and concerns expressed by a diverse group practicing concurrent product development imply that they are functioning in a very large domain. Numerous parameters govern the solution to the design specification and numerous requirements conflict. Therefore the task of creating a quality product to satisfy a specific need in the marketplace is challenging. This is reinforced by reviewing the list of "best practices" in the product realization process presented in Table 5.1. Traditional sequential product development practices often do not permit the problem to be defined correctly, and consequently its solutions are not optimal

The task of producing a quality product is challenging because of the many concerns and constraints influencing the product realization process.

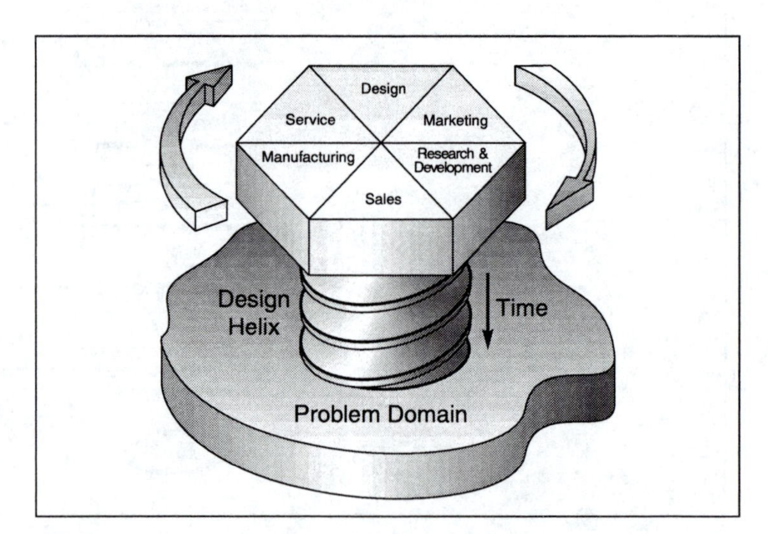

Figure 5.10

The concurrent design process.

because the global problem, with all of the inherent issues, has not been correctly defined or addressed coherently. Only subproblems are addressed.

The concurrent design philosophy of requiring each disciplinary group to be actively involved with all of the principal design decisions is illustrated in Figure 5.10.

The collaborative approach of concurrently or simultaneously developing products by interdisciplinary teams is called *concurrent engineering* (CE), *simultaneous engineering* (SE), concurrent product development (CPD), integrated product development (IPD), or unified life-cycle engineering (ULE).

Each of these groups will impose constraints on the design process and each will develop a list of attributes that the product should contain. These activities are summarized by design-for-*X* rules, where *X* represents the particular group activity. Thus, for example, these teams employ *design-for-manufacture* (DFM) strategies as a subset of this multidisciplinary product realization process. When these strategies

People must be involved in the changes that affect their lives.

Joseph Damore (1952–)

Some synonyms:

CE	concurrent engineering
SE	simultaneous engineering
CPD	concurrent product development
IPD	integrated product development
ULE	unified life cycle engineering

Some design-for-X rules are:

- Design-for-serviceability
- Design-for-assembly
- Design-for-reliability
- Design-for-quality
- Design-for-disassembly
- Design-for-manufacturing

- Reduced number of parts in a product
- Faster product assembly
- Simpler and cheaper manufacture
- Fewer last-minute modifications
- Improved product reliability
- Lower costs
- Products precisely satisfy the customer's requests
- Shorter concept-to-customer phase
- Earlier financial break-even point
- Shorter time to market
- Fewer subsequent changes to the product

Table 5.2

Some of the significant benefits of concurrent engineering.

When these design-for-manufacturing strategies are implemented, they ensure that all manufacturing constraints are embodied in the design of the product.

are implemented, they ensure that all manufacturing constraints are embodied in the design of the product.

The significant benefits of concurrent engineering are summarized in Table 5.2.

The kernel of concurrent engineering is seen in the following *Ten Commandments of Concurrent Engineering* formulated by DRM Associates of Palos Verdes Estates, California.

• Understand your customer

Create a multidisciplinary group to find out exactly what the customer wants.

The first step in concurrent engineering is to understand what the customer wants. Only then can the design problem be correctly defined and a design specification established. This is the essential first step in creating a solution to the problem and a successful product. Groups such as marketing and product management should be involved in the data-gathering process because it increases the probability that the product will satisfy the customer's needs. A variety of techniques can be used to translate the customer's requests into a good product made under appropriate manufacturing and quality specifications.

Departments represented in a product development team include:

- marketing
- finance
- sales
- design
- manufacturing
- research & development
- product support

• Deploy product development teams

Profound knowledge comes from outside the organization.

W. E. Deming
(1900-1993)

The early involvement of a diverse multidisciplinary product development team comprised of individuals from marketing, finance, product support, design, and manufacturing in the product realization process provides the necessary understanding of product and manufacturing considerations for ensuring the evolution of a viable product for the marketplace. The goal of the team is to reduce iterations in the design process, reduce the time to market, and to reduce the problems associated with the manufacturing phase. Upon attaining these goals, the product typically will be cheaper to produce, more robust, more reliable, and easier to support during its life cycle.

The goal of the multidisciplinary team is a product of high quality at low cost. Use design-for-manufacture strategies.

• **Employ design-for-manufacture strategies**

In order to optimize the product performance and life cycle cost, it is essential to adopt strategies which integrate both design considerations and manufacturing process constraints. Typical approaches include design-for-assembly, design-for-cost, and design-for-quality.

• **Seek early counsel from suppliers**

Seek input from vendors early in the product realization process.

Suppliers and subcontractors are intimately familiar with their products, their product applications, and their inherent manufacturing constraints. Consequently these resources should be exploited early in the product realization process in order to fully explore the prospects for developing superior products at lower costs.

• **Employ digital product models**

Use a single computer database in the product realization process that is accessible to all.

A single source of data describing the product should be employed during the product realization process and this source should be accessible to all members of the development team. This information should be stored in a computer so it can be updated and refined continually as the product design evolves. Thus this set of digital data describing the current status of the product is a dynamic data set, not a static one. Such a single source of product information provides consistent information to the diverse groups of individuals in the product development team. This strategy minimizes data errors and minimizes data handling by the different groups.

• **Integrate CAD/CAM tools**

Computer storage of product data is the first step in exploiting the capabilities of computers in the product realization process. Subsequently a set of integrated CAD/CAM tools should be used to represent, analyze, and iteratively refine the product and to process data. This strategy should minimize errors, enhance product quality, and reduce the time from concept to customer.

Integrate computational tools for design and manufacture.

• **Simulation of product and process**

Computer simulations of the product performance should be undertaken in order to develop a more mature design that will reduce the finances and time invested in prototype design-build-test cycles. Typical tools include finite element analysis (FEA) techniques and the verification of the paths of numerically-controlled (NC) machine tools.

• **Quality engineering and reliability**

Use a Failure Mode and Effects Analysis (FMEA) to determine potential failure modes and rank their importance.

The product realization process should incorporate periodic design review sessions to ensure that all risks and design concerns have been adequately addressed. Reliability techniques and quality engineering principles should be employed to develop a good appreciation for the relationships between product-and-process parameters as well as characteristics defining quality, reliability, and performance.

• **Create an efficient work environment**

Design teams should not be hindered by artificial bureaucratic barriers.

Teams involved with concurrent engineering practices should contain a blend of highly motivated and experienced personnel who are freed from bureaucratic constraints. Furthermore, they should have access to all available productivity enhancing facilities. Thus teams should operate in an effective and efficient work environment that encourages development of quality products in minimal time.

• **Continually enhance the design process**

Continually monitor product quality relative to design and manufacturing parameters.

The design process employed by an industrial organization should be the subject of continual scrutiny in order to enhance the process of concurrent engineering. The future competitiveness of the organization depends upon it. Technical tools, methodologies, and practices should be reviewed to remove activities that do not add value to the product. Product quality should be the criterion for these decisions. Opportunities for enhanced market share can be identified by benchmark comparisons of the company products with those of competitors.

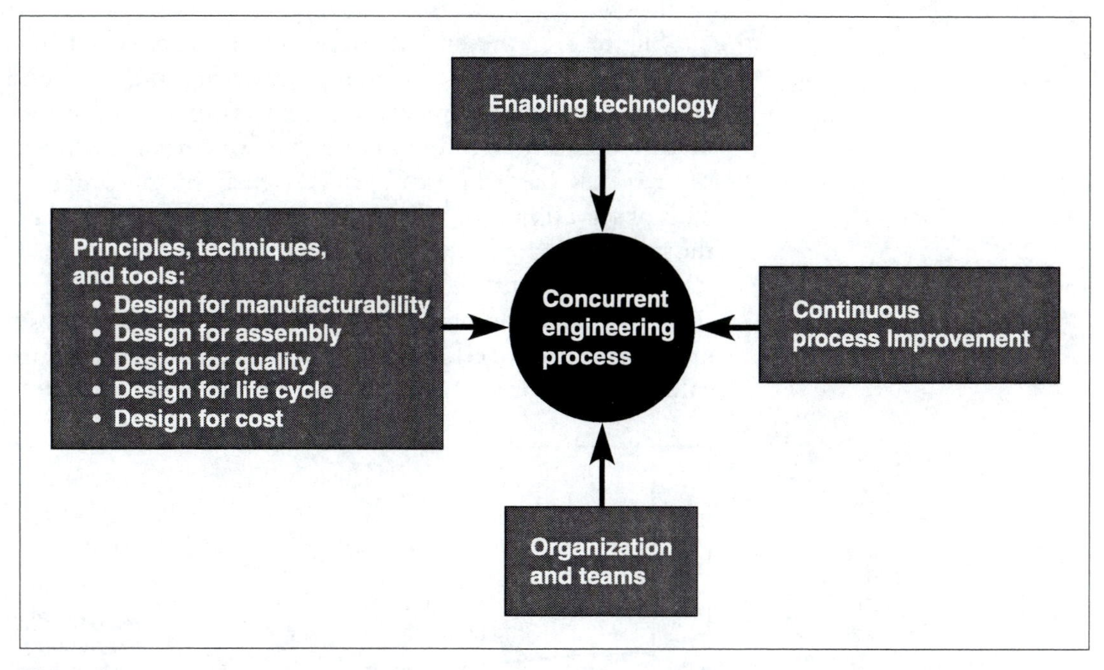

Figure 5.11

The principal components of concurrent engineering (Courtesy DRM Associates).

These guidelines are presented in Figure 5.11 which illustrates the global attributes of the field of concurrent engineering.

Comparison of Sequential and Concurrent Engineering

Concurrent engineering has long been practiced in Japan where there is a more uniform culture within a company than within a typical United States company. Furthermore, many European companies, especially in eastern Europe, have practiced design-for-manufacturability approaches for decades. These approaches belong to a subset of concurrent engineering principles. However, there is a major shift in many progressive United States companies from sequential engineering to concurrent engineering.

Figure 5.12 presents the level of effort expended by a company practicing sequential engineering and a second company practicing concurrent engineering in the development of the same product. Note that concurrent engineering requires the greatest effort very early in the program. This observation is to be anticipated from Figure 5.4 where the greatest opportunities for savings occur in the very early stages of product development. Thus the key to cost-effective manufacturing practice is good design practice. Subsequently the level of effort decreases and assumes a minimum value after the onset of manufacturing the product.

The key to cost-effective manufacturing is good design.

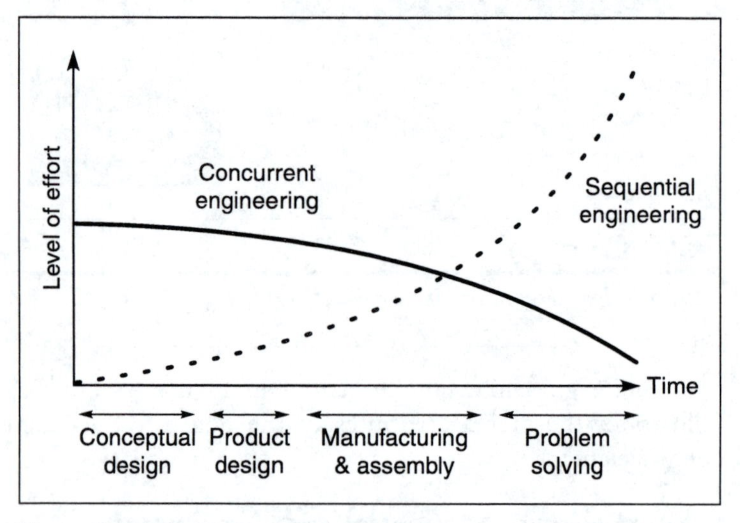

Figure 5.12

A comparison of the level of effort expended during the product realization process when companies practice concurrent engineering and sequential engineering.

The reverse is true when a sequential engineering philosophy is adopted. The level of effort gradually increases during the design and development phase and it is highest if major problems are being addressed after manufacturing begins. Because problem-solving activities during this latter phase often involve changes to tooling and production facilities, these changes will be very expensive.

Figure 5.13 presents a comparison of the number of engineering product changes made by two companies creating the same product by two different approaches involving concurrent engineering and sequential engineering. Several observations can be drawn from this graphical presentation:

Characteristics of concurrent engineering:

- large number of early design changes
- minimal changes during manufacture
- faster concept-to-customer phase

- The company practicing concurrent engineering made a large number of design changes early in the design phase, while the company practicing sequential engineering made most of the engineering changes at the end of the development phase and just prior to initiating production.
- The company practicing concurrent engineering had a minimal number of product changes as production began but the company practicing sequential engineering made many changes after production began. These changes were being made when the product was being sold to customers. Consequently some customers purchased a product of inferior quality.

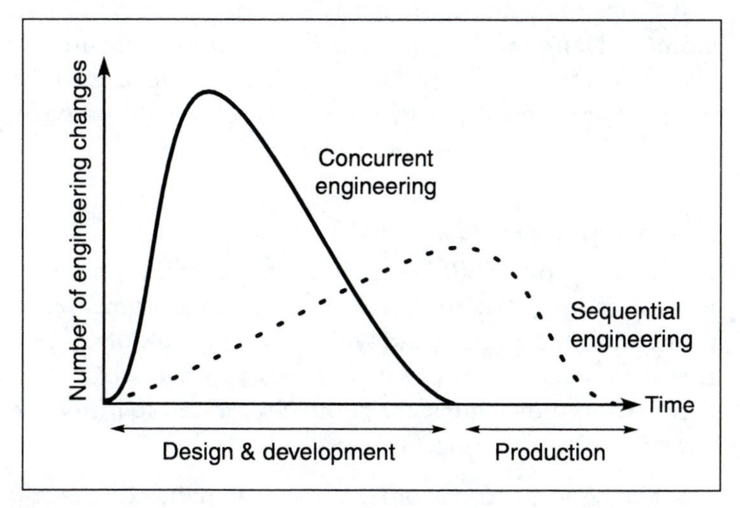

Figure 5.13

The number of product changes as a function of time for sequential versus concurrent engineering in two companies developing the same product.

- The company practicing concurrent engineering developed the product from concept-to-customer faster than the company practicing sequential engineering. Thus a car designed in the United States in the early 1980s using sequential engineering took over five years to develop from the original concept to delivery to the customer. Using concurrent engineering, the same activities took the Japanese 3.5 years. As a result, Japanese vehicles were being sold for 18 months before domestic vehicles with similar features appeared in the marketplace to disturb the monopoly enjoyed by these imports.

Concurrent engineering benefits all industrial sectors:

- automotive
- defense
- electronics
- furniture
- domestic appliances
- industrial equipment
- sporting goods

Concurrent engineering can benefit the entire industrial sector. The Big Three auto makers introduced concurrent engineering to their operations in the mid 1980s. Major U.S. machine tool builders such as Cross, Ingersoll Milling Machine, and Lamb Technicon also have implemented concurrent engineering practices. These strategies have been introduced into the defense industry (Northrop, etc.) and the U.S. electronics industry (DEC, Xerox, etc.). In Europe, concurrent engineering is employed by Bosch, Comau, Daimler-Benz, Dowty, Fiat, Krupp, Rolls-Royce, and Volkswagen, among others. Benefits to companies which have implemented this teamwork approach are illustrated by the following industrial case studies.

Digital Equipment Company

In the early 1980s. the Digital Equipment Company took 30 months to complete the concept-to-customer phase for new products that would typically become obsolete in 24 months. Because this phase took much longer than their competitors' procedures, the company decided to introduce concurrent engineering. As a result:

- The concept-to-customer phase was reduced from 30 months to 18 months.
- The break-even point was reduced to 12 months.
- The lifetime cost of the product was reduced by $75 million.

- Assembly costs of magnetic drives were reduced by 55 percent.
- The number of parts for a magnetic drive unit were reduced by 52 percent.
- Sales increased by 100 percent.

Northrop Corporation

The company spent six years developing a program of concurrent engineering and reaped the following benefits in the production of fighter aircraft:

- The number of product changes was reduced by 75 percent.
- The volume of work-in-process was reduced by 50 percent.
- The duration of the manufacturing phase was shortened by 33 percent.

General Motors Corporation

The power train division employed concurrent engineering to redesign the housing for an automatic transmission. The original housing was a complex single part that was replaced by three housings with the following advantages:

- The number of tools required to machine the housing was reduced by 30 percent.
- The cost of the material was reduced by 50 percent.
- The volume of scrap material was reduced by 60 percent.

These three industrial examples involving the introduction of concurrent engineering to the electronic equipment industry, the defense industry, and the automotive industry serve to illustrate the significant advantages of concurrent engineering in product development.

5.3 A Student Design-Build-Test Project.

At this stage of the project, the design specification has been established and teams of students are working concurrently on all of the tasks associated with the design, building, and testing of their vehicles. Team members have assumed individual responsibility for different aspects of the project, and communication between the team members has been established by exchanging e-mail addresses, telephone numbers, and agreeing on the time for group meetings. Hobby shops and children's toy stores have been searched for affordable items that can be used to build the vehicle within the financial constraints imposed upon this project. The product realization process is being addressed by adopting the philosophy presented in Figures 5.8 and 5.9.

One student group's report discussed the issues of concurrent engineering as follows:

> The group decided to use a concurrent engineering approach because, according to the lectures, it offers us some distinct advantages over other methods. For example it should permit us to do this project in the shortest time, and with our class load that's a big plus. It also offers the chance to keep the cost of the car less than the financial constraint of $50.
>
> One of the big concerns was how to manufacture the pieces for the car. Clearly because we are only making one car, mass production methods cannot be used. This is not good because a casting facility would be useful to make a particular part for one of the concepts. This concept is based on the Christmas tree ornament idea where hot air rises and rotates a turbine wheel to turn the wheels of the car through a gear box. Preliminary tests have shown that there is not enough power by this means to rotate the wheels of the car. If the blades of the turbine could be made with a more airfoil-like shape instead of the bent sheet metal blades of our car, then perhaps more energy could be extracted from the moving air.

Summary

This chapter has focused upon the important role of design in industrial product development. Design methods have a significant impact upon industrial procedures and profitability since design is the heart of engineering. While design procedures permeate all aspects of the product realization process, they dominate during the conceptualization phase where there is the greatest potential for reaping the largest financial savings.

Products should be developed by practicing concurrent engineering rather than sequential engineering because multidisciplinary teams are more able to effectively address the challenges of creating high-quality products. The rationale for this assertion was discussed and industrial data presented to substantiate this thesis. The introduction to engineering design and the crucial role it plays in any industrial organization is now complete.

The remainder of this book is dedicated to describing the details of the different phases of the design process in mechanical engineering—a field that will undoubtedly continue to play a major role in the prosperity of every industrialized nation for the foreseeable future.

Key Concepts

Conceptual design. A concept proposed as the basis for a product in response to a design specification.

Concurrent design. An engineering design practice that combines into one integrated procedure the concerns of marketing, functional product and process design, production, field service, recycling, and disposal.

Concurrent engineering. An integrated approach to the development of a new product in which multidisciplinary teams work in parallel from conceptualization through to the introduction of the product in the marketplace. The team typically contains individuals from marketing, research, design, development, manufacturing, purchasing, and service. Concurrent engineering is sometimes called unified life-cycle engineering, or integrated product development.

Concurrent engineering is better than sequential engineering. Interdisciplinary groups practicing concurrent engineering have several advantages over the more traditional approach of sequential engineering.

Concurrent product development. The same as concurrent engineering. It is sometimes called integrated product development.

Design-for-manufacturability. Design for manufacturability (or design for manufacture) is the practice of designing a product with an emphasis on ease of manufacture. The implementation of this philosophy reduces the development costs, reduces the time for product development and hence the time-to-market of a product, and it also insures cost-effective manufacture.

Design-for-*X* (DFX). A collection of techniques emphasizing different aspects of design such as manufacturability, assembly, service, etc. These individual topics are the *X's*. These techniques are different from the traditional design focus of product function.

Design phase provides the biggest savings. The biggest savings can be accrued during the design phase of the product realization process.

Design specification. A comprehensive, detailed statement of the requirements to be fulfilled by a product.

Diverse skills of the product realization process. The skills needed to effectively undertake a product realization process are very diverse and require a knowledge of teams, design, analysis, testing, and manufacturing.

How design influences industrial practice. Poor design decisions are responsible for 85% of the problems with new products and are responsible for 70% of the lifetime costs of a product.

Manufacturing. The process of producing component parts, assemblies, and complete products. This process includes fabrication, assembly, testing, storage, and distribution.

Process planning. The selection of the most appropriate manufacturing processes for producing a product from data emulating from the design process. Decisions must typically be made concerning quantity, tooling, production rates, tolerances, surface finishes, time scales, and manufacturing facilities.

Product. An artifact that is manufactured as the outcome of a design process. This term describes artifacts produced in all segments of the manufacturing sector of the economy such as electrical, mechanical, chemical, electronic, etc., and furthermore it applies to large plant installations, complex machinery, small devices, and also individual parts.

Product design phase. This phase of the design process is initiated by a concept and terminates with the design of a product that is manufacturable through the specification of materials, tolerances, surface finishes, dimensions, heat treatment, and other relevant data, and also bought-out parts.

Product realization process. The process describes how new and improved products are conceived, designed, manufactured, brought to market, and subsequently supported. This process includes determining customers' needs, translating these needs into engineering specification, designing the product, determining the appropriate production processes, deciding upon the appropriate support processes prior to operating and coordinating all of these activities.

Product testing. The procedure for experimentally determining the performance of the product relative to the performance written in the design specification. This activity is typically undertaken in a laboratory using prototypes.

Quality. The totality of features and characteristics of a product or service that bear on its ability to satisfy stated or implied needs. Quality is closely related to cost; consequently quality involves the ability of a product to satisfy the customer's requirements at a price that is acceptable to the customer.

Quality engineering. An engineering philosophy which primarily increases the competitiveness of new products while concurrently increasing their quality and decreasing their cost.

Sequential engineering. The classical method of developing products involving a number of discrete isolated departments in a company in which product data are transmitted from department to department in a serial algorithm. Thus the product is developed in a chronological sequence of events from the initial market research through to the final sale to the customer. This typically involves the sequential evolution of the product from the market research group through engineering research, design, development, and manufacturing to sales. This philosophy is also termed sequential product development.

Sequential product development. The same as sequential engineering.

Terms in Table 5.1 are included in the Glossary in Appendix C at the end of the text.

Review Questions

1. List the advantages of concurrent engineering.
2. What are some of the disadvantages of sequential engineering?
3. Which phase of the design process offers the greatest potential for cost savings? Explain the reasoning behind your decision.
4. What are the principal features of concurrent engineering?
5. Compare concurrent engineering and sequential engineering in the context of the number of changes made to the product during the product realization process. What are the consequences of these profiles?

6. How significant were the enhancements developed by Northrop, General Motors, and Digital Equipment Company when each company introduced concurrent engineering to the product realization process? Discuss the consequences of implementing this new paradigm.
7. What is the purpose of prototyping?
8. List some of the advantages of developing products by employing multidisciplinary teams.
9. In manufacturing operations, what are the magnitudes of typical savings that arise from concurrent design methods and where do they occur?
10. Upon completion of the conceptual design phase, what percentage of the product cost has been committed?
11. What is the quality lever analogy and what are the typical data associated with this mechanical principle?
12. Describe over-the-wall engineering and its major disadvantages.
13. Describe how the design process fits into the overall product realization process. Define each term and explain how each relates to the others.

Problems

1. Concurrent engineering methodologies propose the establishment of cohesive relationships with suppliers. Why is this important? Discuss the different types of partnerships.
2. Discuss the different ways in which a company can maximize its concurrent engineering activities.
3. Discuss whether or not concurrent product development facilitates innovation and creativity.
4. Describe how you approach the task of designing a device to power an oceanfront home by generating electricity from the motion of ocean waves.
5. What mistakes can occur if there is inadequate independence in decision making on a concurrent product development team? Illustrate your answer with a variety of situations involving different tasks.

6. A company practicing concurrent engineering can minimize costs by creating a seamless domain without interfaces between information sources. These sources pertain to both the processes and the product. Contrast the minimization of process interfaces and the minimization of product interfaces.
7. Can concurrent engineering be described by a Gantt chart or PERT? Discuss.
8. Complex concurrent design situations can be simplified by dividing the global task into smaller tasks. What are the advantages of dividing the product realization process into smaller tasks before they are reassemble?
9. No process works perfectly in every situation. Develop a list of situations where the potential benefits of concurrent engineering could be thwarted. How can these situations be identified and resolved?
10. Teamwork and synergy are two essential attributes of a concurrent product development team. Discuss other attributes you regard as essential.

Further Reading

Allen, C. W., ed. *Simultaneous Engineering*. Dearborn, MI: Society of Manufacturing Engineers, 1990.

"The application of Design-to-Cost at Rolls Royce," NATO/AGARD Lecture Series 107, May 1980.

Dertouzos, M. L., R. K. Lester, and R. M. Solow. *Made in America*. Cambridge MA: MIT Press, 1989.

Grayson, C. J. and C. O'Dell. *American Business a Two-minute Warning: Ten Changes Managers Must Make to Survive Into the 21st Century*. New York: Free Press, Macmillan, 1988.

Hartley, J. R. *Concurrent Engineering*. Cambridge, MA: Productivity Press, 1992.

Miller, L. C. G. *Concurrent Engineering Design.* Dearborn, MI: Society of Manufacturing Engineers, 1993.

Womack, J. P., D. T. Jones, and D. Roos. *The Machine That Changed the World.* New York: Harper Collins, 1991.

Green Design

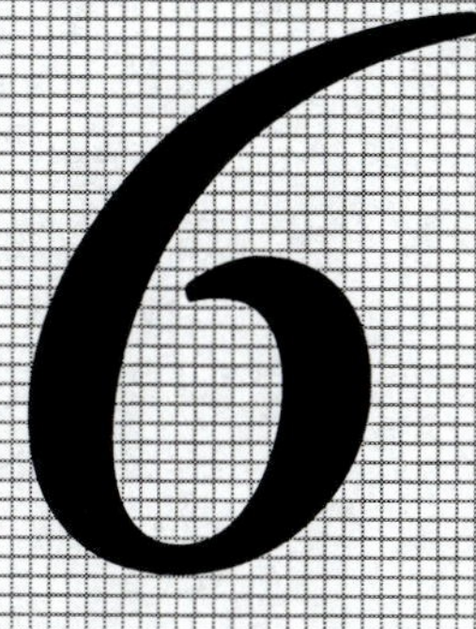

Its not easy bein' green.

Kermit the Frog

Objectives

- To emphasize the crucial role of the design community in reducing pollution.
- To present design rules for managing materials, for efficient energy utilization, and for extending the service life of a product.
- To describe the principles of reducing, reusing and recycling.

Contents

What If...

You are invited by The White House to coordinate the efforts of the Big Three U.S. auto makers in their development of a new generation of environmentally-friendly family cars. These cars should be capable of travelling 100 miles per gallon of gasoline while significantly reducing exhaust pollutants. Furthermore you are required to achieve this goal with no reduction in performance, no loss of passenger carrying capacity, and no reduction in safety.

How is this to be accomplished? Can new lightweight materials reduce the body weight? Can further aerodynamic styling significantly reduce energy losses? Can the vehicles be recycled and the service life of some parts be extended? Should you consider programs to develop hybrid vehicles with fuel cells, new types of batteries, flywheels, and other classes of energy-storage devices to power them?

6.0 Introduction

The theme of this chapter is the adverse impact of products upon the planet and how the design community can introduce methods for reducing pollution. This theme is relevant in the manufacture of products, their service life, and their final disposal. In this chapter you will learn rules to help you make environmentally-friendly design decisions by using principles of *green design*. The cornerstones of green design are:

Save energy
Use less material
Lengthen service life
Recycle more

Green design is an approach that considers the complete life cycle of a product from extraction of raw materials through manufacture and use to final disposal.

6.1 Global Pollution

The United States contains only five percent of the world's population but consumes twenty-five percent of the energy used in the world. Americans enjoy one of the highest standards of living on the planet, and consume more and more energy in doing so. For example, the average household consumption of electricity* in 1971 was 7,380 kilowatt hours; in 1981 it was 8,825 and in 1991 it was 9,738 kilowatt hours.

The United States contains only five percent of the world's population but it is responsible for twenty-five percent of the energy used in the world.

In addition to the consumption of the largest quantity of energy per capita of any nation on the planet, the population in the United States generates the largest amount of waste per capita. Moreover, the amount of waste is at least twice as great as that generated in the western European nations. Municipal solid waste generated by commercial and domestic activities amounts to approximately two kilograms for each person each day in the United States. The waste dumped into landfills each week includes for example approximately six billion pens and forty-four million newspapers—equivalent to 500,000 trees.

Waste can be defined as those products or residuals which are discarded because they cannot be used economically.

*Edison Electric Institute

The packaging of products accounts for one-third of municipal solid waste.

Approximately one-third of the weight of municipal solid waste is product packaging. This high percentage has been responsible for aggressively targeted legislation worldwide to encourage the packaging industry to become more efficient and reduce this waste. The German packaging law, for example, mandates that manufacturers and distributors recover and subsequently recycle their own packaging waste.

In 1988, landfills received 180 million metric tons of municipal solid waste. This figure is responsible for the fact that approximately one in every six trucks in the United States is a garbage truck. The Environmental Protection Agency estimates that only 11 percent of the waste stream is recycled, 13 percent is incinerated, and the remainder—the vast majority—is disposed of in landfills. Unfortunately, landfills in many states in the U.S. are near capacity and if the current conditions prevail, then one half of the currently available landfills will be filled to capacity before the year 2000. This situation is further exacerbated by public resistance to the siting of waste management facilities.

Only 11 percent of the waste stream is recycled, 13 percent is incinerated, and the remainder is disposed of in landfills.

Public resistance to landfills has led to the word *NIMBY*–Not In My Back Yard!

The major disadvantage of landfills is that they have no positive attributes except providing for the disposal of waste. Additional disadvantages include their cost and the potential for environmental damage. Better evaluation of the properties of this "waste" might reveal a basis for generating revenue through the reuse of products and the recycling of materials.

Some communities incinerate waste to recover some of its energy and reduce its bulk before it is dumped into landfills. However, incineration has the disadvantage that hazardous emissions from smokestacks can include toxic ash.

Industrial wastes typically have a much greater environmental impact than municipal waste.

While municipal waste is cause for concern, it is almost insignificant compared to industrial waste which typically has a much greater environmental impact. The industrial sector consumes 38 percent of the energy generated in the United States, and 90 percent of that energy is from the combustion of fossil fuels such as coal, natural gas, and petroleum.

Municipal solid waste, which is currently generated at an annual rate of 160 million metric tons, only accounts for about 2 percent of the total solid waste as defined by the U.S.

Resource Conservation and Recovery Act. Industrial waste comprises approximately 700 million metric tons of hazardous waste and 11 billion metric tons of nonhazardous waste. It is therefore clear that industrialists, and the design community in particular, must assume responsibility for a large proportion of the pollution of the planet. Consequently they have a pivotal role to play in the reduction of environmental pollution.

Industrialists, and the design community in particular, must assume responsibility for a large proportion of pollution.

Environmental pollution—the generation of waste products that are discharged into the environment—has been associated with *Homo sapiens sapiens* throughout history. This has been of little consequence until the past two or three hundred years because individuals traditionally lived in sparsely populated rural areas and the pollutants were widely scattered and readily dispersed. With the transition to more industrial rather than agrarian economies during the industrial revolution, the situation changed. People and factories began to generate relatively large amounts of pollutants in small areas and the delicate natural equilibrium between the flora and fauna of the local and global ecosystems was upset.

Environmental pollution is the discharge of waste products into the environment.

Pollution at the dawn of the twenty-first century assumes several forms. These include:

- *air pollution* by gases and particulates
- *water pollution* by contaminants from sewerage systems, industrial effluent, and agricultural runoff; and from solid wastes in domestic landfills and the spoil heaps associated with mining operations
- *soil pollution* from the inappropriate use of fertilizers and pesticides
- *noise pollution* from transportation systems and industrial complexes
- *radiation pollution* from electronic devices and radioactive materials

Wild beasts and birds are by right not property merely of the people alive today, but the property of the unborn generations whose belongings we have no right to squander.

President Theodore Roosevelt (1858–1919)

Since environmental pollution affects the three ingredients essential for life—our air, water, and soil—pollution control is clearly one of the serious problems confronting us today.

Environmental pollution affects the air, water, and soil.

Pollution reduction will involve cooperation between engineers, scientists, and legislators.

The attenuation of pollution will undoubtedly involve a concerted approach by scientists, engineers, and legislators. But the solution to this challenge is complex because of depleted natural resources, third-world poverty, a rapidly increasing world population, and economics. Especially in the western world the environmental situation is complicated by the fact that pollution is caused by facilities that are highly desired by society. Factories are responsible for much air and water pollution, but they provide employment and they manufacture products that are desired by people.

Consider the automobile. It pollutes the air but in the western world it is considered essential because of the freedom and convenience it affords. The design of cars sold in the United States is subject to such government regulations as the Clean Air Act imposed on automobile emissions and the Corporate Average Fuel Economy (CAFE) standards imposed on gasoline consumption. Nevertheless, automobiles are a major source of pollution in cities where individuals prefer to drive cars rather than ride less-polluting and more energy-efficient buses or trains.

Science is always wrong. It never solves a problem without creating ten more.

George Bernard Shaw
(1856-1950)

Some environmental pollution is caused for no apparently good reason. Consider the popular high-tech sneakers called L. A. Gear Lights that were designed in 1992. Their popularity is attributed to the colored lights in the soles that flash brightly as the heels strike the ground. In the original design, the effect was caused by movement of a mercury slug that completed an electric circuit when it bridged electrical contacts. While no health problem exists for wearers of these shoes, there is a potential health problem when the shoes are discarded in landfills from which the mercury can leach into the water table.

Mercury is a liquid metal at room temperature. Its vapor is extremely poisonous. A few drops of mercury spilled on the floor of a room can introduce enough mercury vapor into the air to exceed the maximum safe level by a factor of 200.

Legislation in states such as Minnesota has caused L. A. Gear to redesign the switch in the shoes to use a metal ball instead of a mercury slug to complete the electrical circuit. The manufacturer has provided a toll-free telephone number for customers to request prepaid packing with which to return shoes containing mercury to a federal recycling facility in Texas.

Other social causes of pollution include synthetic materials that have been developed to save people money or time. The McDonalds fast-food chain was the focus of consumer criticism for pollution caused by its synthetic packaging materials. The company responded by initiating the recycling of polystyrene packaging, they introduced napkins manufactured from recycled paper, and they changed the design specification for their drinking straws so that they could be manufactured with a thinner wall. This action reduced their solid waste by approximately 500,000 kilograms each year. Other fast-food chains implemented similar initiatives.

Fast-food chains have become more conscientious about their packaging materials.

Furthermore, all beverages sold in Denmark must be retailed in returnable bottles, and since 1995 all plastic containers in Italy must be biodegradable. In Great Britain, the government is contemplating the introduction of legislation restricting the excessive packaging of products, legislation compelling retailers to recycle more materials, and legislation banning the manufacture of beverage bottles from combinations of different plastics because mixed plastics are harder to recycle. Thus there are numerous governmental and industrial initiatives to address the impact of products on the environment.

One of the major global environmental concerns is global warming. This phenomenon involves the emission of gaseous pollutants which permit sunlight to enter the atmosphere and heat the earth by radiation but prevent the heat from returning to the earth's atmosphere and ultimately into space. The ambient temperature increases and climates change because of this "greenhouse effect." The primary culprit is carbon dioxide, the volume of which must be reduced to a level so that vegetation and other agents can absorb enough of the gas to restore ecological equilibrium.

A major global environmental concern is global warming.

Gases causing global warming are associated with the burning of fossil fuels—particularly coal in power stations and gasoline in automobiles. To reduce the emission of these gases, the consumption of fossil fuels must be reduced and more energy-efficient products must be developed by the engineering community.

Most of these initiatives will be focused on utility companies and automobile manufacturers but undoubtedly everyone will be affected because the use of energy is essential to the industrial, commercial, and residential sectors of the economy. Designers will place greater emphasis on the size and weight of products in order to reduce fuel consumption and hence the cost of shipping products from the factory to the customer. Efficient energy utilization will mandate the evolution of more fuel-efficient vehicles, manufacturing plants, domestic appliances, and commercial equipment.

The Environmental Protection Agency establishes and enforces antipollution laws.

These environmental concerns have been embraced by citizens, and governments worldwide have created agencies and legislation to address them. In the United States, the Environmental Protection Agency (EPA) was founded in 1970 to establish and enforce pollution standards. Meanwhile in western Europe, the European Community has established a number of laws controlling emissions from automobiles and power stations.

The challenge that confronts the design community is to create products that impose minimal damage on the environment while they enhance the quality of human life.

Having painted the global picture of pollution and its impact upon the design of products, our next logical step is to question the effect of these environmental concerns upon the design process. The challenge that confronts the design community is to create products that impose minimal damage on the environment while they enhance the quality of life. To attain this goal, an accounting practice must be implemented to ensure that the inventory of renewable resources is maintained, with nonrenewable resources depleted at the minimal rate. Furthermore, pollution must be kept low enough for the environment to assimilate.

6.2 Green Design

The impact of a product upon the environment is a direct consequence of design decisions made during the product realization process.

The numerous products that characterize twentieth century lifestyles are largely responsible for global pollution. The impact of a product upon the environment is a direct consequence of design decisions made during the product realization process. These decisions include the specification of materials, the length of the service life, and the energy efficiency of the product. The challenge confronting the international community is how to address these environmental concerns.

The task is to establish rules that promote products that impose a reduced burden upon the environment

- during the extraction of raw materials from which the product will be made
- during the manufacture of the product
- during the distribution to the consumer
- during the use of the product by the consumer
- during the disposal of the product

While the designer has traditionally been concerned only with a product until it was launched into the marketplace, the designer who is sensitive to the environment must not only be concerned with a product's complete life cycle from the cradle to the grave but also with the reincarnation of the materials from which it was manufactured.

This approach is part of the philosophy of green design, or design for the environment, and the products are called *green products*. The kernel of this philosophy is the sustainability of natural resources. Note that green design applies to *all* aspects of the product realization process, as shown in

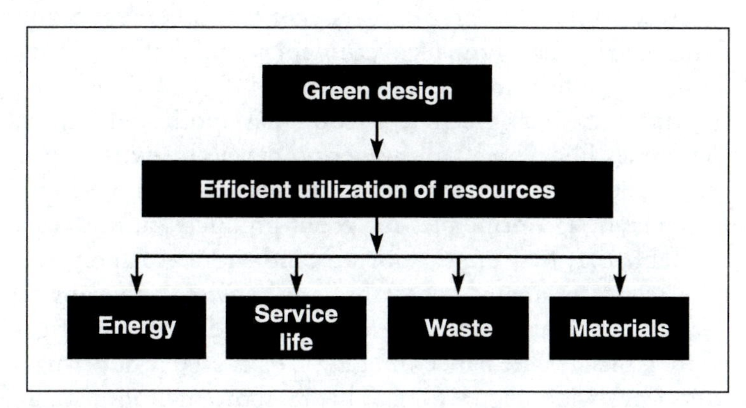

Figure 6.1

The primary attributes of green design.

Figure 6.1. Furthermore, recall from Chapter Five that the conceptual design stage is responsible for over seventy percent of the costs of product development, manufacture, and service. Therefore the design team has a unique role to play in controlling the undesirable effects on the environment caused by a product.

Rules don't exist for determining whether one product is greener than another.

Generic rules exist for environmentally sensitive designs but there is no analytical technique for ascertaining the environmental superiority of one complex product relative to another. The ultimate solution generally involves a series of design compromises.

In order to illustrate these compromises, consider the electric light bulb. The standard incandescent light bulb is a highly inefficient light source that uses only five percent of the energy it consumes to generate light. The remaining energy is wasted as heat. Fluorescent bulbs, however, consume seventy-five percent less energy to generate the same amount of light. Their energy efficiency makes them a much "greener" product. However, fluorescent bulbs contain small amounts of mercury that can damage the human nervous system. Consequently fluorescent bulbs are not so green in the context of heavy metal pollution. This design dilemma is typical in the design of green products.

The expression "Mad as a hatter" (from *Alice in Wonderland*) derives from the dementia sometimes seen among early hat makers who were poisoned by mercury once used to convert fur into felt for hats.

In addition to applying green design rules for manufacturing products, industrial designers must consider legislation controlling their products. For example, legislation proposed by the German government would require all auto makers to collect and recycle their cars at the end of each car's service life. This green legislation has motivated BMW, Mercedes Benz, and Volkswagen to develop new strategies for the design, manufacture, and recycling of their products.

Figure 6.2

Plastic and paper grocery bags. Which is more environmentally friendly?

The design principles for green products, summarized in Table 6.1, may appear obvious and simple to apply, but the reality of many situations is otherwise. Solutions to design problems frequently require considerable investigation. Consider the paper and plastic bags used in supermarkets worldwide (Figure 6.2). Which is more environmentally friendly? The West German environmental protection agency evaluated the environmental friendliness of these two prod-

- Minimize energy consumption
- Minimize use of materials
- Minimize environmental damage
- Maximize product service life

Table 6.1

The basic principles of green design.

ucts and concluded that the plastic product was more friendly than the paper bag. The reasons included the facts that the plastic bag caused less pollution during processing than the paper bag, that it was manufactured using only one-third the energy needed to manufacture the paper bag, and that a plastic bag was more likely to be reused by the consumer. Thus the plastic bag conformed more closely to the principles in Table 6.1 than did the paper bag!

The complexities of environmental design can be further illustrated by considering the reduction of automobile pollution in southern California. A potential solution is an electric automobile that would utilize electrically rechargeable batteries as a power source. Such a solution would reduce the smog in Los Angeles, for example, while also reducing the traffic noise in the city. Although this appears to be a good solution for the local ecology, it has some undesirable global consequences. The current generation of electric cars are less efficient than the gasoline-powered competition, and the additional power required from utility companies in southern California would generate more pollution from additional power plants burning more fossil fuel. This situation would add to the pollution problems on a global scale through the creation of more acid rain and more global warming. To circumvent this problem, electricity might instead be generated from a reusable resource such as the wind, ocean waves, or hydroelectric schemes.

There have been some successes in the evolution of green products. Consider the task of designing green batteries for domestic use and for automobiles. In a five year period during the late 1980s, manufacturers of household

Figure 6.3

Green batteries.

The automobile battery is the largest consumer of lead in the USA.

batteries reduced the level of mercury in these products by almost ninety percent (Figure 6.3). Furthermore, during the past twenty years the General Motors Delco Remy Division has reduced the mass of lead in a typical automobile battery from 14 kilograms to only 9 kilograms while simultaneously improving its performance. This is a significant product enhancement because automobile batteries are the largest single consumer of lead in the United States. When inhaled or absorbed through the skin, this chemical element damages body organs such as the brain, liver, and kidneys. The enhancement is design is partially due to the specification of a different electrode material containing calcium and tin rather than the traditional lead alloy containing antimony and arsenic. Over six million pounds per year of lead waste and over one million pounds of antimony and arsenic waste have been eliminated.

The photocopier industry is actively implementing the principles of green design:

- only 25 different materials
- reuse 60% of parts
- recycle 35% of parts
- discard only 5% of parts

Other successes in the development of green products can be found in the photocopier industry. Canon, Kodak, Océ, Xerox, and other manufacturers now remanufacture copiers at the end of their service life instead of casting them into landfills. This is significant because each year over four million copiers are discarded. These manufacturers collect their products from their customers and dismantle them. Some 60 percent of the parts are reused, 35 percent are recycled, and only 5 percent are discarded. Xerox has enhanced this program by reducing the number of different materials used in their copiers from 100 or so in the previous generations of machines to only 25 or so in the current models. The goal is to reduce this to less than 10.

Reduce, reuse, and...

recycle!

Many progressive companies have initiated broad policies which embrace the slogan *reduce, reuse, and recycle.* For example, the 3M Company has a program entitled *Pollution Prevention Pays* (*3P*) focused on reducing their emissions by ninety percent by the year 2000. The program's structure is focused on ensuring compliance with environmental legislation, enhancing the company's reputation as an environmentally sensitive company, and minimizing potential liabilities. The program is implemented through a recycling initiative, by redesigning products and processes to generate fewer pollutants, and by developing new products which do not pollute.

The cornerstone of the Philips environmental policy is that "prevention is better than cure." This company also offers services to others who wish to develop green products. A similar benevolent philosophy is embraced by IBM, whose environmental policy advocates environmentally-friendly and energy-efficient processes in design and manufacture. The company policy exceeds all relevant government regulations and IBM has established stringent regulations for situations in the company where current legislation does not apply.

Perhaps the leaders of the environmental revolution are in Europe where, for example, eighty percent of Danish companies and over seventy-five percent of companies in the region defined by the former West Germany have a member of the board of directors with the primary responsibility of developing, enforcing, and managing the company's environmental policies.

Table 6.2 summarizes some of the fundamental principles of green design.

- Use all materials efficiently.
- Select materials with minimal pollution qualities during their extraction, processing, deployment, recycling, and disposal.
- Use all energy sources efficiently.
- Ensure that the product has minimal adverse effects on the environment during manufacture, deployment, and disposal.
- Undertake a comprehensive evaluation of product disposal methods.
- Ensure that the service life of the product is environmentally appropriate.

Table 6.2

Fundamental principles of green design.

6.3 Materials Management

The management of materials is important in the design of green products. Central to materials management is the efficient use of materials (Figure 6.4). Three rules are proposed:

- Minimize waste generation during the life cycle of the product.
- Minimize waste during extraction of raw materials.
- Maximize recovery, recycling, and remanufacturing throughout the product service life.

Material selection is a complex environmental problem because the source and processing of materials are just as important as their inherent properties and characteristics.

Material selection is a complex environmental problem because the source and processing of materials is just as important as the inherent properties and characteristics of the materials. The situation is exacerbated by the fact that each year over ten thousand different chemicals are produced in industrial quantities and approximately a thousand new ones are developed, although not all of the latter have potential for industrial production. Thus there are many substances from which a design team can select an appropriate material for a specific application.

Reduce Material Consumption

The first principle of materials management is to reduce the amount of material used in a product.

In materials management, a logical first principle is to reduce the amount of material used in a product. This reduces the consumption of raw materials, it reduces global pollution associated with the extraction and refinement of

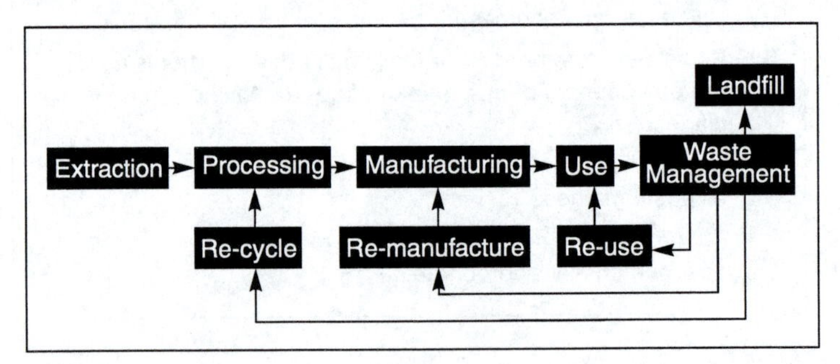

Figure 6.4

Materials management.

material, and it reduces pollution and energy consumption during the subsequent manufacturing phase.

Material consumption can be reduced by, for example, reducing the wall thickness of structural members while their geometry is changed to maintain structural integrity. New can-forming machines generate a superior base profile for aluminum beverage cans (Figure 6.5); this means that the thickness of the sheet metal can be reduced which in turn reduces the mass of aluminum in the can by seven percent. Improvements of this kind have been responsible for the manufacture of aluminum cans which today weigh thirty percent less than they did twenty years ago.

Figure 6.5

Beverage cans; a common product with sophisticated manufacturing.

Similar successes have been achieved with plastic packaging. Advances in plastics technologies have made possible strong plastic grocery bags only seven thousandths of an inch thick. In 1976 plastic grocery bags were 23 thousandths of an inch thick but were neither stronger nor more durable than today's bags (Figure 6.2). That's a saving of 70 percent in material consumption. Similarly, plastic milk jugs weighed 95 grams in the early 1970s, but current jugs weigh only 60 grams—a material saving of 36 percent.

Compared to 20 years ago,

- pop cans weigh 30% less
- grocery bags weigh 70% less
- milk jugs weigh 40% less

Lengthen Service Life

A second principle of materials management is to lengthen the service life of products so that fewer enter landfills each year. If a product is designed to have a life of two years instead of one, then the number of products entering the waste stream would be reduced by fifty percent. This philosophy has design ramifications because it is the designer who determines the factors that limit service life. If the service life is ended by worn bearings, for example, then these parts could be replaced to extend service life. If parts are standardized, then refurbished old parts can be recycled into a new generation of products. Such an approach is environmentally sound because it does not require the reprocessing of materials and therefore uses less energy.

A second principle of materials management is to lengthen the service life of the product. How?

- enhance maintenance protocols
- facilitate upgrades by modular designs

Recycle Materials

A third principle of materials management is to specify recycled materials rather than virgin materials whenever possible.

The third principle of materials management is to specify recycled materials rather than virgin materials whenever possible. The use of recycled rather than new materials reduces the volume of waste and usually is far more energy efficient.

Recycling aluminum waste reduces energy consumption by 95% relative to the processing of bauxite.

For example, the recycling of aluminum reduces energy costs by ninety-five percent relative to the processing of aluminum ore, and the recycling of glass reduces energy costs by seventy-five percent. Recycling materials has some ramifications in design practices because in order to recycle a material, it must first be identified and separated from other materials before being reprocessed. Several philosophies have been proposed to accomplish these objectives including design-for-assembly, design-for-disassembly, and parts consolidation.

Table 6.3 summarizes the application of green design to materials management.

Design rules for recycling materials:

- design for assembly
- design for disassembly
- practice parts consolidation
- use material identification codes

- Minimize the amount of material in each part.
- Lengthen the service life.
- Specify recycled materials where possible.
- Specify energy-efficient materials in manufacture and in service.
- Specify materials which pollute minimally during their extraction, manufacture, use, and disposal.
- Specify readily available materials which do not use declining natural resources.
- Specify materials which are not likely to be affected by new legislation that constrains their deployment, manufacture, or disposal.

Table 6.3

Green design principles for materials management.

6.4 Energy Utilization

Energy conversion is the primary cause of pollution and global warming. It is imperative, therefore, that designers try to create energy-efficient products and seek environmentally-friendly sources of power. Designers of products that require energy for their operation, such as industrial machinery or kitchen equipment, must minimize the power used and the energy lost. This can be achieved by evaluating conceptual designs using an energy-utilization criterion. Alternative materials, parts, insulation, subassemblies, lubrication systems, and heat recovery schemes can be compared. Energy consumption can be reduced by sensors, actuators, and microprocessor controls.

The primary source of pollution and global warming is energy conversion.

The selection of an appropriate energy source or fuel for a product is a challenging task. The decision involves not only the efficiency of the final product but also the efficiency of the method of producing the chosen fuel source. Electricity, for example, is generated by methods that differ in environmental friendliness. Some energy sources, such as geothermal, wind turbines, and solar cells, are environmentally friendly. Energy from ocean waves is still under development. Hydroelectric schemes exploit a renewable resource and nuclear power stations are clean but generate nuclear waste. Coal power stations are inefficient and generate considerable air pollution.

Electricity can be generated by fossil fuels such as:

- coal
- gas
- oil

Electricity can be generated by renewable resources such as:

- geothermal energy
- wind
- waves
- solar energy
- hydroelectric energy

The Environmental Protection Agency has initiated a program to encourage manufacturers of personal computers to create new machines twice as energy efficient as the current ones. Approximately a hundred companies are participating in a program focused on development of computers and monitors which each consume no more than 30 watts. It has been said that the energy saved by these machines annually could power the states of New Hampshire, Maine, and Vermont.

Initiatives of this kind are expected to increase as the United States government introduces labels for products that are more environmentally friendly and as customers

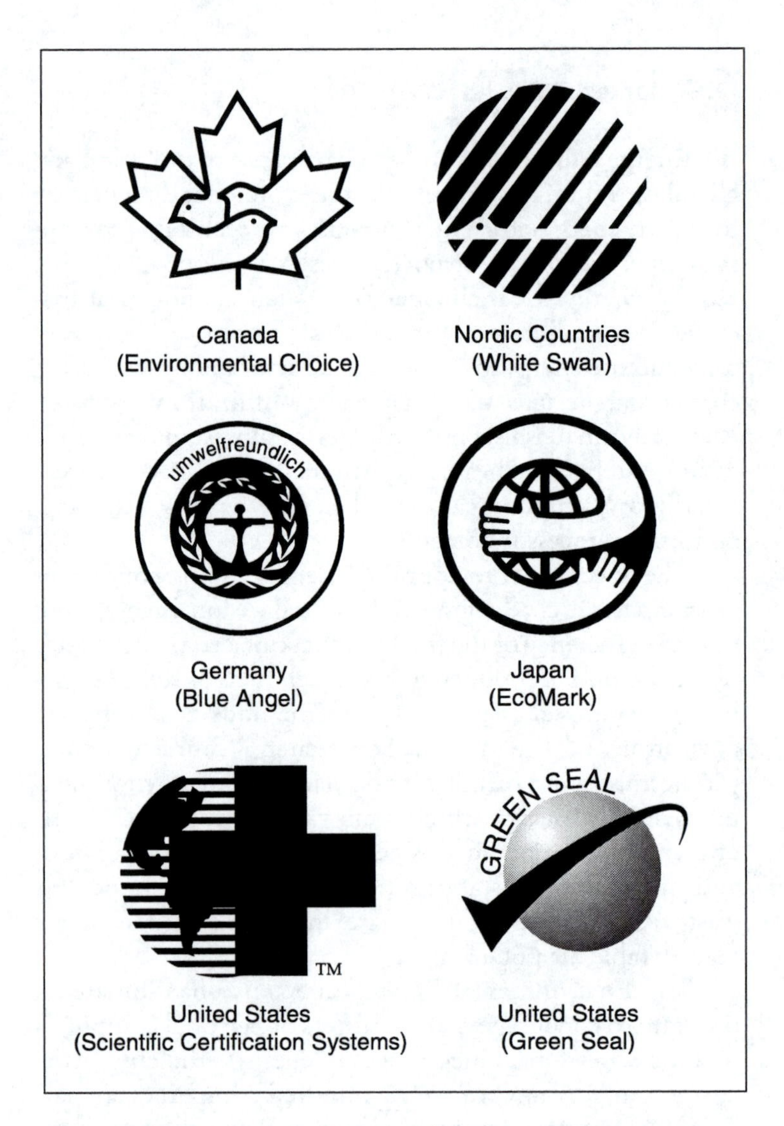

Figure 6.6

Labels for environmentally-friendly products sponsored by several governments.

become more sensitive to environmental issues. Such labels already exist in Canada, Germany, Japan, and the Scandinavian countries. Figure 6.6 shows some of these labels.

Domestic Products

Domestic appliances have been the subject of many energy-efficiency investigations. The domestic refrigerator, for example, consumes one-fifth of the energy supplied to the average U.S. household. The nation's 110 million refrigerators consume 77 billion kilowatt-hours of electricity per year from power stations that will release 50 million metric tons of carbon dioxide. In addition, these refrigerators contain 275 thousand metric tons of ozone-depleting chlorofluorocarbons (CFC's) in their insulation and in the refrigerant. Current design efforts focus on developing better insulation, more efficient motors and compressors, more efficient evaporators and condensers, and superior door seals.

Refrigerators consume one-fifth of the energy of each household in the USA.

These enhancements are being addressed by a number of companies including the Whirlpool Corporation and the Danish company Gram. Technologies developed by Gram have permitted it to market the Gram LER 200 refrigerator which has a power consumption of 0.24 kWh per day and a storage capacity of 196 liters. This performance is an order of magnitude better than that of competitive products.

A review of domestic dishwashers and washing machines reveals that some similar-capacity products consume twice as much energy as others. In energy-efficient machines, the volume of water used by each washing cycle has been reduced, the chemical composition of detergents has been modified, and washing temperatures have been lowered by the cooperative efforts of detergent and washing machine manufacturers.

A new generation of washing machines:

- 50% less detergent
- 50% less water saves over 6000 gallons each year for an average household
- lower water temperature
- less energy used
- less damage to the fabric of the clothes

Progress is also being made in the development of green clothes dryers by the use of energy-efficient control systems. Simple time clocks are being superseded by sensor-based control schemes which measure the dryness of clothes. When a prescribed dryness is achieved, the appliance stops instead of wasting heat on dry clothes.

A new generation of energy-efficient products has been introduced which incorporates sensors and microprocessors to control performance.

As already mentioned, the standard 100-watt incandescent light bulb has a service life of about 250 days, but it uses only five percent of the electricity it consumes to generate

Incandescent light bulbs convert only 5% of the energy used into light.

light. The remaining 95 percent is wasted as heat. A range of new energy-efficient bulbs is now available.

One innovative approach that has emerged from a joint venture between Intersource Technologies and Diablo Research Corporation is the development of an electronic bulb that uses standard sockets but which consumes only 25 percent of the energy of a conventional bulb to generate the same light. The bulb has a service life of approximately fourteen years, during which time it could save the consumer as much as $100 in utility bills. The bulb should retail for approximately $10.

Transportation

Automobile manufacturers must comply with environmental regulations imposed upon the exhaust emissions and fuel consumption of their products.

Cars, aircraft, and ships have always used energy inefficiently. The automobile industry has become the subject of severe environmental legislation focused on exhaust emissions and fuel consumption because of the automobile's contribution to global warming. In response to these legislative initiatives, considerable effort has been made to develop alternative-fueled vehicles and to reduce weight and drag.

These design activities have produced innovative solutions to a number of problems. Consider, for example, the energy expended in accelerating and decelerating a bus in an urban area. Volvo, in Sweden, has developed a hydraulic energy-recovery system for buses which enables a half-full bus to decelerate from 48 km/h to rest and then to be accelerated to 38 km/h before the engine has to be reactivated. The approach uses the kinetic energy of the bus to pump a fluid from a reservoir into an accumulator as the bus stops. The energy stored in the pressurized gas is then used to accelerate the bus before the engine is engaged.

Plastics and composite materials have helped develop energy-efficient products because their strength- and stiffness-to-weight ratios are superior to commercial metals. It has been estimated that an average automobile contains 70 kilograms of plastics. These materials reduce fuel consumption by five percent compared to the fuel consumption of an all-metal automobile.

The packaging industry is the largest single consumer of plastics. Plastic bottles for numerous domestic and industrial fluids and powders afford cost-effective transportation of these commodities and provide recyclable scrap (Figure 6.7). The manufacture of plastics is energy efficient. Indeed, if alternative materials were specified for packaging, then the energy consumed by this segment of the packaging industry would double.

The packaging industry is the largest consumer of plastics.

Figure 6.7

Recyclable plastic bottles.

An internal combustion engine piston with a complex shape can reduce frictional energy loss by about thirty percent and this type of piston can raise the energy efficiency of an engine three percent. Other improvements in the efficiency of vehicles have been accomplished by introducing sensing systems and microprocessors to monitor and control numerous subsystems in the vehicle. Better aerodynamic automobile designs have reduced losses from air turbulence. For example, kits retrofitted to trucks to improve aerodynamic performance can reduce fuel consumption by as much as forty percent for trucks travelling at 105 km/h.

Retrofitting trucks with aerodynamic fairings can reduce their fuel consumption by 40%.

Energy losses in aircraft always have had a major influence on their performance and consequently they have received much attention. Currently Airbus Industrie is contemplating covering all of the surfaces of their A320 commercial aircraft with a textured plastic film designed to reduce drag. Drag reduction is accomplished by microscopic

- Minimize energy consumption, perhaps by employing control systems to ensure efficient energy utilization.
- Minimize energy losses, perhaps by recovering waste heat and other forms of energy.
- Select sustainable fuel sources rather than fossil fuels where appropriate.
- If a fossil fuel must be employed in a product, then determine the potential for a reduction of cost-effectiveness due to new pollution legislation or taxation.

Table 6.4

Design principles for energy utilization.

The idea for the drag-reducing riblet skin for airliners was borrowed from the drag-reducing skin of sharks (see biomimetics in Chapter 9).

ribs and grooves in the film (called riblets) which produce numerous small vortices in the air flow immediately adjacent to the surface of the aircraft. Potential savings in fuel consumption are 1.5 percent for long-distance routes.

Energy losses in the propellers of ships have been investigated extensively by the shipbuilding industry. The Japanese companies Ishikawajima-Harima and Mitsubishi studied losses caused by turbulence and cavitation and proposed the installation of contrarotating propellers in ships. These propeller systems generate less vibration and noise while they improve fuel efficiency about fifteen percent.

The application of green design to energy utilization problems is summarized in Table 6.4.

6.5 Life Cycle Implications

Product Life Cycle:

- extraction of raw materials
- processing of raw materials
- design and manufacture
- service life
- disposal of product

The *life cycle* of a product includes the extraction and processing of its materials, its design and manufacture, its service life, and its disposal. The *service life* of a product is the time during which the product complies with its design specification. The useful life of a product can be affected by such factors as the rate of technological development, new maintenance protocols, and the emergence of new styles and features. At the end of a product's service life, it has traditionally been discarded in a landfill and replaced by a new product.

This philosophy is gradually changing because of public and legislative concern about waste and its disposal. Under these new conditions, designers must reevaluate the concept of the product life cycle and adopt an environmentally-sensitive philosophy. Such action will enable the engineering community to address legislation intended to reduce the waste stream and to increase the recycling of both products and materials.

A compromise typically must be sought between increased life for a product and the environmental effect of the resources necessary to extend its life.

A compromise must be sought between efforts to extend the service life of a product and the effect on the environment of the consumption of additional materials and energy. For example, a compromise may be necessary between a material which ensures that a part has a long energy-efficient service life and the environmental consequences of extraction, processing, and manufacturing of that material. Perhaps a long service life will become more desirable if environmental concerns increase and single-use disposable products become unpopular.

Maintenance

A green product should be easy to maintain in order that the service life is as long as possible. For ease of repair, the product should be made with a design-for-maintenance philosophy and any associated documentation or instruction manuals should be user-friendly.

A green product should be easy to maintain in order to extend its service life.

Repair manuals are important for repair personnel. If user-friendly manuals are not available, then the consumer will often discard a product because repair is too difficult or costly. Investigations of discarded domestic appliances dumped in landfills have revealed that a large percentage of them could have been repaired easily and at little cost to the customer because only minor adjustments or replacement of cheap parts were necessary.

If good repair manuals are available, products are more likely to be repaired rather than thrown away.

Many products such as automobiles and dishwashers now contain microprocessors. These allow appropriate sensors to identify faults and provide advice to the customer or maintenance personnel.

Microprocessors and sensors in products enhance fault identification and the maintenance of the product.

6.6 Recycling

A green product should be designed to facilitate the recycling of parts because at the end of its service life a product generally will contain components which are still state-of-the-art and which are still performing to their design specification. When the service life of a part is longer than the service life of the product, it often can be incorporated into a new product with little reworking. Other parts may require remanufacturing.

Green products should be designed to facilitate the recycling of parts.

Remanufacturing of a product involves the introduction of new parts and the refurbishment of old ones. Products which are most readily remanufactured are those with a small number of design changes each year. Implementation of this green philosophy encourages extension of the service life of a product by recycling materials and components. This philosophy is a form of waste prevention. It has been embraced by many organizations. For example, the United States Department of Defense remanufactures numerous military systems ranging from aircraft to rifles. Arrow Automotive Industries remanufactures automotive parts such as clutches, carburetors, and starter motors. Xerox Corporation remanufactures approximately one million copier parts each year. Indeed Xerox is now standardizing its designs so that many parts can be used in several different products. This *design-for-manufacture principle* is discussed in Chapter 12.

Recycled products satisfy the same design specifications as new products.

Some customers may believe that new products are better than recycled products; some may fail to appreciate that recycled products can satisfy the same design criteria as new ones. Such imponderables of the market are perhaps best addressed by the marketing department of a company as it projects an environmentally-orientated good-citizen image.

The recycling of parts, whether they are to be remanufactured or not, involves disassembly of the product. In the past this has not been a major concern to the design team,

- Minimize the variety of materials in a product.
- Specify compatible materials.
- Consolidate parts.
- Reduce the number of assembly operations.
- Simplify and standardize the fits between parts.
- Identify the separation points between parts.
- Specify water-soluble adhesives whenever possible.
- Use a material identification scheme on parts to simplify separation.

Table 6.5

Design Principles for Design-for-Disassembly.

but it is an important green design consideration. To provide guidance on this activity, General Electric Plastics developed *design-for-disassembly principles* for design teams (Table 6.5).

Recycling of Materials

Many materials can be reused. Table 6.6 presents seven advantages of the use of scrap iron and steel over the manufacture of steel from iron ore, coal, and limestone. Consumption of raw material, energy, and water is reduced. In addition, water, air pollution, and waste production decrease. Today one-third of U.S. steel production is by minimills which primarily use scrap iron.

One-third of current U.S. steel production is by minimills which primarily use scrap iron.

97%	reduction in mining waste
90%	reduction in consumption of virgin materials
86%	reduction in air pollution
76%	reduction in water pollution
74%	reduction in energy consumption
40%	reduction in water consumption

Table 6.6

The environmental benefits of recycled iron and steel scrap. (Source: Institute of Scrap Recycling Industries, Inc.)

The data in Table 6.6 are quite typical of the recycling process for many different materials. Table 6.7 suggests possible energy savings from the use of recycled rather than new materials.

A measure of the maturity of the ferrous and nonferrous recycling activities in the United States is presented in Table 6.8 which contains data on the percentage of United States metal consumption derived from recycled metals.

Material	Savings
Aluminum	95%
Copper	85%
Plastics	>80%
Iron and steel	74%
Lead	65%
Paper	64%
Zinc	60%

Table 6.7

Energy savings from using recycled rather than new materials. (Source: Institute of Scrap Recycling Industries, Inc.)

Metal	Percentage
Steel	50%
Copper	43%
Aluminum	32%
Lead	55%
Zinc	19%

Table 6.8

Percentage of U.S. annual metal consumption derived from recycled metals. (Source: Institute of Scrap Recycling Industries, Inc.)

Recycling is easier when products contain the fewest different materials.

There are thousands of different plastics. Therefore, plastic parts should be identified to facilitate recycling.

Recycling is easier when products contain the smallest number of different materials. Attainment of this goal is naturally dependent upon the type of product and upon the recycling processes available for the materials involved. It helps to ensure that the materials can be easily separated at a recycling facility. This requires the product to be designed using design-for-disassembly methods so that, for example, a steel part can be separated from an aluminum part. However, the identification of the different materials in a product is difficult because thousands of commercially available thermoplastics, thermosets, and monolithic materials are used. In the future, material identification prior to sorting may be simplified by bar codes on each part. This would be particularly useful to identify parts containing expensive materials (platinum in automobile catalytic converters) or dangerous materials (cadmium-coated bolts).

Some monolithic materials can be separated during remelting. For example, the small amount of aluminum in scrap steels can be easily removed by this approach. In other industries this is unnecessary. Consider the apparel industry where shoes are currently manufactured from a mixture of plastic bottles, paper, and trim waste from diaper manufacturers. Handbags, belts, and wallets are being made from industrial scrap rubber and discarded rubber products. Tote bags are manufactured from recycled plastic bottles (Figure 6.8).

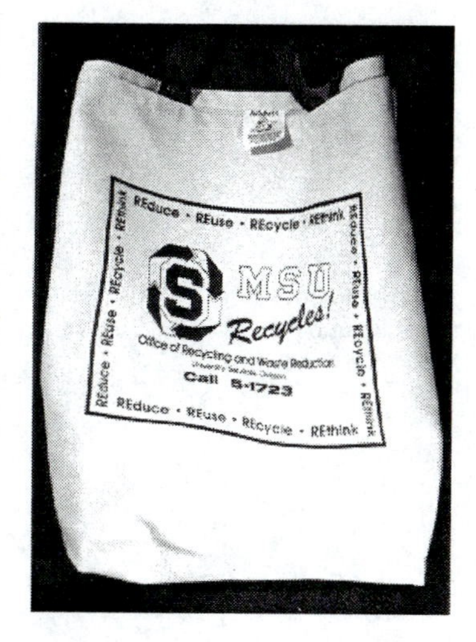

Figure 6.8

A tote bag manufactured from 50% recycled PETE plastic bottles and 50% virgin polyester.

Whenever possible, high quality materials should not be degraded with contaminants during manufacturing processes. Contaminants are often difficult to remove from primary materials, and the desirable characteristics of the primary materials can be progressively degraded during repeated recycling operations. It may be appropriate to repeatedly recycle material into progressively lower quality products before finally incinerating the residual material to recover its intrinsic energy. For example, the plastic polyethylene terephthalate (PET) is readily recycled. It represents 30 percent of the plastic bottle market, but it can be degraded by several polymeric contaminants with approximately the same specific gravity as PET. This makes commercial separation difficult. Similarly, the presence of 0.2 percent copper in steel dramatically reduces the value of scrap steel because it causes cracking.

Economics is naturally an important consideration in recycling processes. Waste disposal costs have continued to increase during the past decades as the number of landfills has decreased. For example, the average cost of waste disposal in 1978 was $15 per ton but in 1994 it was $100 per ton. The economics of recovery and disposal are governed by disposal costs, collection costs, and sorting costs in comparison to the cost of virgin material.

Recycling costs:

- collection costs
- sorting costs
- cleaning or reprocessing costs

Clearly recycling must be economical if it is to be adopted by industry. Packages for consumer products increasingly are manufactured from recycled paper. In 1992 approximately 63 billion aluminum beverage cans made in the United States were recycled compared to 28 billion in 1982 and 1.2 billion in 1972.

Plastics do not easily decompose in landfills and they make up 20% of solid municipal waste.

Plastics are relatively inexpensive to make and are extremely versatile but they pose a waste recovery problem because they do not readily decompose. They amount to 20 percent by volume of solid municipal waste, and they are not easily recycled. Thermosets are not as amenable to recycling as are thermoplastics because they do not soften or melt. However, thermoplastics can be recycled by chemical processes or by thermal decomposition. Thermosetting plastics typically are granulated and mixed as filler in new thermoplastic products.

Code	Plastic	Uses
1 PETE	Polyethylene terephthalate	Used for bottles (30%); food and beverage packages. Recycled into paints, automotive parts, bottles, carpets.
2 HDPE	High density polyethylene (LDPE is low density polyethylene)	Most common plastic in homes. HDPE used for containers for milk, etc. LDPE (new or recycled) used for trash bags and kitchen wraps.
3 PVC	Polyvinyl chloride (PVC) or simply Vinyl (V)	Used for pipes, bottles, windows. Recycled into house siding, tiles, sewer pipes.
4 PP	Polypropylene	Used for auto battery cases, rigid food containers. Recycled into furniture, auto parts, sports equipment.
5 PS	Polystyrene	Used for fast-food packages, cartons, packaging foam. Recycled into office accessories, videotape cassettes.
6 OTHER	Other	Used for high-performance and fiber-reinforced products in many industries. Difficult to recycle.

Table 6.9

Plastics coding system. (Source: The Society of the Plastics Industry.)

To facilitate the recycling of scrap plastics, the Institute of Scrap Recycling Industries and the Society of the Plastics Industry have developed guidelines for recycling plastic bottles.

The plastics industry has also developed an approach to identify the plastics from which plastic bottles are manufactured. A numerical code is either imprinted or molded on the base of the bottle (Table 6.9).

If recycling is not a viable option because of adverse market pressures, then energy recycling should be explored. Incineration of materials in municipal and other waste processing facilities for energy production should be considered. After extraction of heat from waste material, the relatively small volume of ash can be disposed of in landfills. This process has been used by the Dow Chemical Company to process polymeric waste with some success. The heat generated by plastics ranges from 25,000 to 48,000 kJ/kg, which compares favorably with the 50,000 kJ/kg from fuel oil. Design principles for material recycling are summarized in Table 6.10.

Incineration for energy production should be considered if recycling is not an option.

- Minimize the number of different materials in a product.
- Select materials with recycling in mind.
- Ensure that materials can be readily reclaimed by recycling.
- Make it easy to disassemble the product and isolate materials.
- Mark each part with a code that identifies the material from which it was made.

Table 6.10

Green design principles for material recycling.

Product Upgrades

Many products contain a combination of traditional technologies, such as iron castings, and new technologies, such as electronic circuitry. Change in many traditional technologies is slow, but some of the newer ones— micro-

A modular design philosophy can extend the service life of a product. Old subsystems can be replaced with new ones.

processor-based systems, for instance—change rapidly. Under these conditions, a modular design philosophy can extend the service life of a product. Old subsystems can be replaced easily by state-of-the-art subsystems to upgrade the product and prevent it from being discarded.

Modular design has broad applicability, ranging from the installation of more energy-efficient control systems in domestic washing machines to the upgrading of critical elements in commercial photocopying equipment. For example, Agfa-Gevaert worked with the Institut für Produktdauerforschung in Geneva to develop a modular photocopier with a design which has extended the average machine life to eight years. The industry average is four years for a comparable machine. Some design principles for extending product service cycle are summarized in Table 6.11.

- Does the product have an environmentally-optimal life cycle?
- Is repair and maintenance documentation user-friendly?
- Can the life of the product be extended by replacement of old parts with new and possibly improved parts?
- Are parts manufactured from remanufactured or readily recyclable materials?
- Are inherent weaknesses of the product identified and corrected to prevent premature failure?
- Is the product designed for ease of disassembly to facilitate remanufacture or recycling of parts?
- Are all potential hazards associated with the product at the end of its service life identified and addressed?

Table 6.11.

Green design principles for extending the product service cycle.

6.7 The Student Design-Build-Test Project.

Green design is not so important when only one birthday-powered vehicle is being built. However, student design teams did consider several green design issues associated with minimizing energy consumption and minimizing the use of materials. One student design team wrote:

> The amount of materials used in the construction of the vehicle was minimal, so few natural resources were consumed. There was more concern with reducing weight because this affects the performance of the car. This was accomplished by using glue and balsa wood.

Another team of students wrote:

> Efficient energy utilization is important because the energy available to propel the vehicle is limited and frictional losses of various kinds reduce the efficiency of the machine. Energy losses in subassemblies have been identified in order to create more efficient combinations of machine elements. Typical problem areas are bearings and gear sets. The grease was removed from the ball bearings supporting the axles of the car because of the drag it caused, and a silicon spray was used instead for lubrication. That really helped.
>
> This focus upon energy efficiency has also been extended to the combustion process associated with the candles. How can this thermodynamic process be controlled to ensure that the maximum energy is usefully extracted from the candles? Should the flow of air to the candles be carefully adjusted with ducting? How should the candles be arranged relative to each other? Various arrangements were tried using a trial-and-error approach in the Heat Transfer Laboratory.

Summary

Humanity is becoming increasingly concerned about environment pollution and the continual decline of natural resources. These issues can only be addressed by the combined efforts of legislation and the endeavors of the design community. This chapter has addressed the issues of environmentally-friendly design practices. You have learned how "green design" practices can reduce the pollution of land, water, and air by careful management of materials resources and by minimizing energy consumption.

Global pollution is attributed to the misuse of natural resources. The disposal of waste material in landfills, water courses, and the air is of primary concern. However, the task of reducing pollution is complicated by international competition. Products must sell in the marketplace, and factories manufacturing them generate pollution while they also provide employment. Products such as the automobile are highly desirable, but they also pollute the environment. The severity of pollution can be measured by the phenomenon of global warming. The challenge confronting the design community is to create useful products that cause minimal damage to the environment.

You learned that the effect of a product on the environment is directly related to design decisions made during the product realization process. These decisions concern the specification of materials, the duration of service life, and the energy efficiency of the product. Green design has been responsible for the emergence of several slogans such as "reduce, reuse, and recycle."

Materials management is an important part of green design because it tries to minimize waste associated with extraction of the raw materials as well waste generated during the life cycle of a product. Throughout the service life of a product, materials should be recovered, recycled, and remanufactured.

Energy use is the primary culprit for global warming and pollution. To control energy use, new schemes propose better energy control, a reduction in system losses, and the specification of sustainable fuel sources.

Rules have been developed to extend the service life of products. Compromises must be sought between longer product life and the resources needed to accomplish this. An important aspect of this design activity is maintenance.

Products can be more readily upgraded if modular construction facilitates replacement of obsolete parts with new and better modules.

Environmentally friendly design practices permeate many aspects of the design process. In the next chapter you will learn how to develop a design specification for a product and you will see the impact of environmental concerns upon this task.

Key Concepts

Commercial scrap. Recyclable materials that are derived from business activities.

Design-for-the-environment. A design process in which environmental concerns are primary. See also "green" design.

Designers and pollution. The design community has an important role to play in the reduction of pollution.

Energy utilization. An important factor in green design. Minimize energy consumption and use sustainable fuels rather than fossil fuels.

Environmental friendliness. Concerns the effect of a product upon the environment. It advocates protection of ecosystems and natural resources.

Global warming. A phenomenon attributed to the inability of heat to be radiated from the earth into space because of a layer of gas in the atmosphere attributed to pollution—the greenhouse effect. This phenomenon could increase the earth's average temperature and disturb the ecological equilibrium of the planet.

Green design. Design that advocates reducing, reusing, and recycling. It encourages energy conservation, materials management, and waste prevention. Good practice results in environmentally-friendly products with a minimal reduction of functionality, life, and performance. This approach is sometimes termed "design-for-the-environment."

Greenhouse effect. An effect caused by pollution in the earth's atmosphere that traps heat from the sun close to the earth's surface. The name derives from the analogy to the horticultural greenhouse in which a glass structure admits radiant heat and retains it. Heat trapping by the atmosphere is facilitated by ozone, carbon dioxide, and water vapor which severely reduce the radiation of heat from earth into space.

Green products. Products developed by green protocols. These environmentally-sensitive protocols pertain to all aspects of the product realization process.

Industrial scrap. Recyclable materials generated by manufacturing processes or industrial products that are no longer functional, such as broken pallets and carts.

Life cycle. The product life cycle includes the extraction and processing of the materials from which it is manufactured, its design and manufacture, its service life, and its disposal.

Materials management. A primary component of green design based upon minimizing waste associated with a product from the extraction of its raw materials to the remanufacturing or recycling of its parts.

Metrics for green products. There are no metrics for green products—whether one product is greener than another—because of the complex nature of environmental concerns.

Municipal solid waste. Waste generated in a municipality by both residential and commercial activities.

Pollution. The introduction of waste gases, liquids, and solids into the environment. These waste products contaminate the soil, water, and air—and ultimately the ecosystems.

Product upgrades. Modularity in product design can extend the service life because old modules can be easily replaced by newer modules having more up-to-date technologies.

Recycled materials. Material reassimilated in such a manner that it can be used for a purpose identical or similar to that of its first use. A material is only available for recycling after it has been recovered from a waste stream.

Recycled parts. The remanufacturing of parts to a condition in which they perform as well as new parts.

Recycling. The practice of diverting scrap material from the waste stream into the manufacture of new products. This reduces the volume of virgin materials consumed in the manufacture of new products and is characterized by the collection, sorting, and processing of these scrap materials.

Reduce, reuse, and recycle. An environmentally-sensitive approach to the development of products from conception through disposal and beyond.

Sanitary landfill. A depression in the ground into which waste material is dumped prior to being covered beneath a layer of soil. The depression contains an impervious liner to minimize water and soil pollution.

Virgin material. Material which has not been subjected to any industrial processing. Examples are bauxite ore for the manufacture of aluminum, petroleum for the manufacture of plastics, or wood for the manufacture of paper.

Waste. Solid, liquid, and gaseous materials with low economic worth that are disposed of in landfills, discharged into treatment facilities, or vented to the atmosphere. Thus the term "waste" pertains to those products or residuals that cannot be used economically and are not worth saving. They are the cause of pollution.

Waste stream. The confluence, unification, and subsequent flow-through time of the total waste material generated by the numerous sources in a region, a community, a facility, or a residence.

Review Questions

1. Define green design and explain its importance relative to the earth.
2. Why is energy utilization an important aspect of green design?

3. List and explain four important principles of green design.
4. If the service life of a product is extended, what are the green design ramifications?
5. Discuss the use of landfills in the disposal of domestic and industrial waste.
6. What is the role of the design community in minimizing pollution?
7. Which is more environmentally friendly, the paper bag or the plastic bag used by supermarkets? Explain your answer.
8. What are three primary principles of materials management?
9. Write a report, with specific examples, on efforts to reduce the energy used by domestic products.
10. Write a report on efforts to reduce energy consumption in the automotive industry.
11. What factors typically determine when a product should be discarded? How can the designer extend the service life of a product?
12. What are the principles of energy utilization. Consider some illustrative examples.
13. List the principles of design-for-disassembly.
14. What are the principles of material recycling?

Problems

1. How significant do you believe the ten-cent deposit on returnable pop cans in some states in the USA has been in increasing the percentage of aluminum beverage cans recycled? Suggest returnable-deposit schemes for other products to improve recycling and decrease the burden on landfills.

2*. Scrap automotive tires are a significant environmental concern because of their abundance. Few tires are recycled, and this is responsible for the rate of increase of scrap tires being one tire per person annually in the United States. What are the current methods and problems of dealing with scrap tires? Propose new methods to improve the situation. You might consider tire service life, recycling, new materials, new tire designs, new automobile suspension systems, and new public transportation systems.

3*. Air bags are increasingly being used in the automobile industry and statistics indicate the advantages of this safety feature. However there are some environmental concerns with air bags. A typical air bag inflates extremely quickly when a chemical (sodium azide) explodes in a controlled manner. However, this chemical is associated with mutagenic and carcinogenic health problems. Consequently there is an issue of worker health and safety in scrap yard shredders when unspent sodium azide canisters in discarded vehicles are being processed. Sodium azide can contaminate shredder residue and the air bag canister can explode.

   Generate several solutions to the problem of worker safety by proposing a variety of approaches such as by a method to identify which cars have undetonated air bags, by a method to easily remove sodium azide canisters from vehicles, by inflating the air bag with safer chemicals, or by an alternative passenger safety system which satisfies the design specification imposed on the current generation of air bags.

4*. Polystyrene is a recyclable material frequently used in the restaurant industry to package food. However it is often not recycled because of the low density of polystyrene containers. Design a device which can be used by restaurants to compact post-consumer waste polysty-

*Questions two through four are based upon material from the **Design for Recycling—Solving Tomorrow's Problems Today** project in the School of Engineering at Grand Valley State University, Grand Rapids, Michigan.

rene. The device should wash the package to remove food residue. As a competitive benchmark, a device has been designed which generates cylinders of compacted polystyrene 38 cm long and 38 cm in diameter that weigh 18 kg.

5. The gasoline-powered lawn mower is a noisy, air-polluting machine. A 3.5 horsepower mower emits as much hydrocarbon in one hour as a new car driven 340 miles. Develop conceptual designs for a superior class of green mowers. These should include human-powered machines, more efficient gasoline engines, cordless electric mowers, and other less conventional approaches such as the use of solar cells.
6. Current domestic refrigerators consume a large percentage of the power used by most households. Identify the primary sources of energy loss and design a new generation of green refrigerators.
7. The recycling of automotive materials can be facilitated by use of fewer kinds of materials in each product. While it is relatively easy to sort steel and aluminum parts, those made of polymeric materials are not so easily separated. Devise and evaluate ways to identify the chemical composition of plastic parts so that they can be separated.
8. Many buildings contain vending machines that distribute both hot and cold beverages as well as cold foods. Design a green vending machine that could dispense both hot and cold products. This machine might use the heat discharged in the refrigeration cycle of the cold portion to raise the temperature of hot products. In addition, consider other heat transfer attributes such as insulation and door seals. Develop systems for recycling polystyrene cups and aluminum cans.
9. Commuter trains in suburban communities accelerate and decelerate as they move in and out of stations. Instead of dissipating the kinetic energy of motion as heat during braking, devise strategies to conserve this energy. Could you use this energy to generate electricity for illumination or could it be stored and used later to accelerate the train?

10. Design an environmentally-efficient water system for the domestic bathroom by considering the shower stall, toilet, and wash basin. Consider water management and energy utilization.
11. Design a domestic recycling unit that would encourage recycling and reduce the volume of waste transported to a landfill.
12. Design-for-disassembly is important in many industries where green design is practiced. List the primary approaches to this activity and discuss their attributes.
13. How could the "greenness" of a product be measured? The measurement should reflect weight reduction, toxicity, energy consumption, product service life, recycling opportunities, composting, energy recovery, and remanufacturing. Are all of these factors equally important? Should they be weighted to reflect relative environmental and health risks? For example, if a new kind of light bulb is extremely energy efficient but its performance depends upon a small amount of a toxic substance, then is this product more environmentally sound than a less efficient conventional product? Discuss these issues.
14. Eventually products available to the consumer may be labelled with a "green factor" or "green quotient" to inform the consumer of how green the product is—just as food labels now inform the consumer of the percentage of fat, etc. If you were in charge of such a labelling project, what variables would you take into account in the green factor? What weighting and calculations would you use?
15. If you were in charge of the labelling project of the previous question, how would you design a label that would be easily recognizable and easily read? Consider graphic and numeric schemes. Draw a sample of your label.

Further Reading

Green Products by Design: Choices for a Cleaner Environment. Washington, DC: U.S. Congress, Office of Technology Assessment, OTA-E-541, 1992.

Plastics Recycling: A Strategic Vision Beyond Beverage Bottles. Washington, DC: Plastics Recycling Foundation Inc., 1992.

Burall, P. *Green Design.* London: The Design Council, 1991.

Henstock, M. E. *Design for Recyclability.* London: The Institute of Metals, 1988.

Turner, K., ed. *Sustainable Environmental Management.* London: Bellhaven Press, 1988.

Van Weenen, J. C. *Waste Prevention: Theory and Practice.* The Hague: CIP-Gegevens Koninklijke Bibliotheek, 1990.

The Design Specification

7

Mankind always sets itself only such problems it can solve; since, looking at the matter more closely, it will always be found that the task itself arises only when the material conditions for its solution already exist…

Karl Marx
(1818-1883)

Objectives

- To illustrate the role of quality function deployment (QFD) in translating the requirements of the customer into appropriate company requirements.
- To develop an appreciation for design parameters and their relative importance.
- To emphasize that the design specification is a crucial step in the design process.
- To present a list of generic design parameters for developing comprehensive design specifications for products.

Contents

What if...

You are to direct the development of a new hypersonic transport aircraft capable of flying a large number of passengers from New York to Tokyo. The aircraft must be able to take off and land on conventional runways.

However, it is unclear what other characteristics the aircraft must have. For example, how fast should it fly? Mach 5? Mach 20? How many passengers should it carry? How many should be built? What should it look like? Are the necessary technologies developed for such an enterprise? What do the airlines and the customers really need? The problem is not clear.

It appears to you that the first job should be to define the problem and solicit input from potential customers. After all, if you don't define the problem, then you certainly can't solve it! But how will you do this?

7.0 Introduction

In previous chapters we have introduced you to engineering design, its history, and its role in product creation. We have suggested a systematic approach to design problems. Now it is time to examine in detail the first step in the design process: the recognition and subsequent definition of the problem.

We begin with a discussion of quality function deployment (QFD). This process helps the manufacturer identify the needs of the customer and translate these needs into requirements for the product realization process.

You will learn how to develop a *design specification*. This specification formally documents the detailed requirements to be satisfied by a product. Specifications are not only developed for the complete product but, as the product realization process matures, design specifications also are developed for subassemblies and individual parts.

The product *design specification* formally documents the detailed requirements to be satisfied by a product.

A product design specification is developed by systematically considering parameters that constrain the search for solutions to the design problem. Because these design parameters differ in importance, each one can be assigned a numerical weighting factor to quantify its relative significance.

A product design specification is developed by systematically considering the design parameters constraining the search for a solution.

The formulation of a design specification is one of the first steps in product realization. It is sandwiched between the initial efforts of a company to identify a market and the conceptual design stage where a number of different visions of the product are proposed and subsequently evaluated.

7.1 Defining the Problem

Before any problem can be solved, it must be defined. This is true of all problems, whether they are concerned with the purchase of an automobile, the planning of a vacation, or the design of a new product. Each kind of problem has its own set of governing parameters and constraints. Therefore, before the design process is initiated, it is imperative to

The formulation of a problem is far more often essential than its solution, which may be merely a matter of mathematical or experimental skill.

Albert Einstein
(1879-1955)

Design parameters are the essential elements of the problem specification. They impose constraints on the options available and define the boundary of the design task in the solution domain.

clarify the design parameters to be considered in the problem specification. These *design parameters* (Table 7.1)—such as cost, safety, and size—impose constraints on the options available to a design team. They provide a set of criteria that must be met by any conceptual design. The establishment of a design specification, therefore, is an important undertaking because if the problem is defined incorrectly, then the design team will create a solution for the wrong problem!

• Function/performance	• Spatial constraints
• Product cost	• Aesthetics
• Delivery date	• Transportation & packaging
• Quantity	• Personnel
• Environmental issues	• Service life
• Safety	• Noise radiation
• Quality	• Operating instructions
• Energy consumption	• Human factors
• Reliability	• Health issues
• Maintenance procedures	• Government regulations
• Mechanical loading	• Shelf-life storage
• Size	• Operating costs
• Weight	• Environmental conditions

Table 7.1

Some generic design parameters.

The product realization process starts with prospecting in technology, marketing, and business to find an unfulfilled need. A product design specification is then developed to define that need. Technologies required to develop the product are then reviewed and competitive products are scrutinized to establish appropriate benchmarks for evaluating the new design.

When I ask the designer what is a good car, what is a good refrigerator, and what is a good synthetic fiber, most of them cannot answer. It is obvious that they cannot produce good products.

Kaoru Ishikawa
(1915-1989)

In the early 1960s after the Soviet space program had achieved a number of successes which included launching the first satellite and putting a man into earth orbit, President John F. Kennedy saw the need to rejuvenate the morale of the American people. In 1961 he delivered a speech in which he said, "I believe that this nation should commit itself to achieving a goal, before this decade is out, of landing a man on the moon and returning him safely to earth."

The goal was clear, and the time frame was the only constraint imposed upon the engineering community. However, the President did not provide a design specification for a spacecraft that would go to the moon, nor did he stipulate how this problem should be solved. That was the responsibility of engineers.

7.2 Quality Function Deployment

The task of systematically developing a design specification using the parameters in Table 7.1 can be accomplished by employing *quality function deployment.* This process of total company-wide quality control ensures that customer requirements are translated into appropriate technical specifications at each stage of the product realization process. Such systematic planning involves all departments in a company from marketing, design, and development to production, sales, and distribution. It therefore embodies the notion of concurrent engineering described in Chapter 5.

Quality function deployment is a systematic process for translating customer requirements into appropriate technical requirements during all stages of the product realization process.

Before quality function deployment is initiated, the customer must be carefully identified. Who is the customer? In some situations it is clear but in others it is not. Consider this scenario: Flawless Machine Tool, Inc., is contracted to develop a piece of high-speed production machinery for another company, Quality Widgets, Inc. While Quality Widgets is indeed the customer of Flawless Machine Tool, the individuals in Quality Widgets may have quite different opinions about the desirable attributes of the machinery that is to be designed. Thus management personnel in Quality Widgets will probably view the production machinery in terms of productivity, operating costs, and the return on the initial investment, while the machine operators will prob-

The customer must be carefully identified.

ably be more concerned about safety, the location of instrumentation, and the operation of the machine. Responses of these two groups to a quality function deployment initiative would therefore be quite different.

The type of customer can also change. Data compiled by one quality function deployment activity may therefore need to be reevaluated later in response to a changing marketplace. Consider an automobile buyer, for example. In the past this customer group has been dominated by males. However, surveys by Ford Motor Company in the United States in 1992 indicate that women influenced 80 percent of the car-buying decisions in that year. Moreover, in 1991, women purchased 49 percent of new cars—up 36 percent from 1980 figures—and they also bought 23 percent of new trucks—up 15 percent from 1986. This changing customer base was responsible for a new emphasis on color schemes, style, and ergonomic design in vehicles.

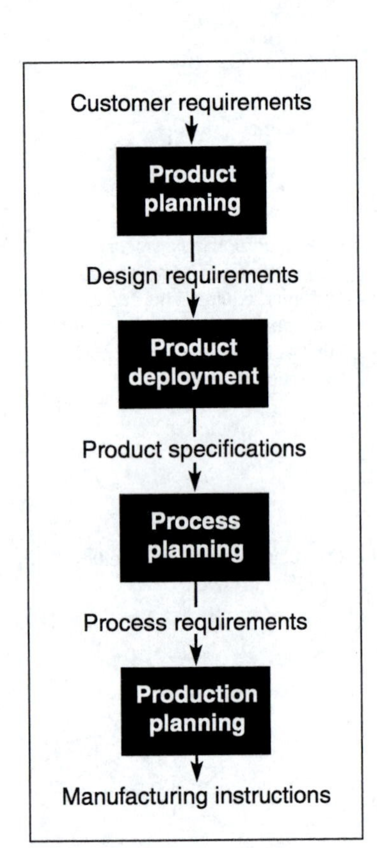

Figure 7.1

The essential elements of quality function deployment.

The principal elements of quality function deployment are illustrated in Figure 7.1. It is evident from this figure that customer requirements initiate a four-phase process involving product planning, where customer requirements are identified and translated into design requirements; product deployment, where design requirements are transformed into product characteristics; process planning, where product characteristics are related to a plethora of manufacturing processes; and production planning, where specific manufacturing sequences and detailed instructions are derived from the characteristics of the manufacturing processes.

Quality function deployment enables the customer's desires, preferences, and requirements to be incorporated into all aspects of the product realization process. During each phase of the process a matrix similar to that in Figure 7.2 is employed.

A quality function deployment matrix facilitates determination of relationships between the customer's requirements and the corresponding activities of the company at all stages of product development. Questions presented on the left side of the matrix represent input to a particular phase of the four-phase QFD program, while responses to those

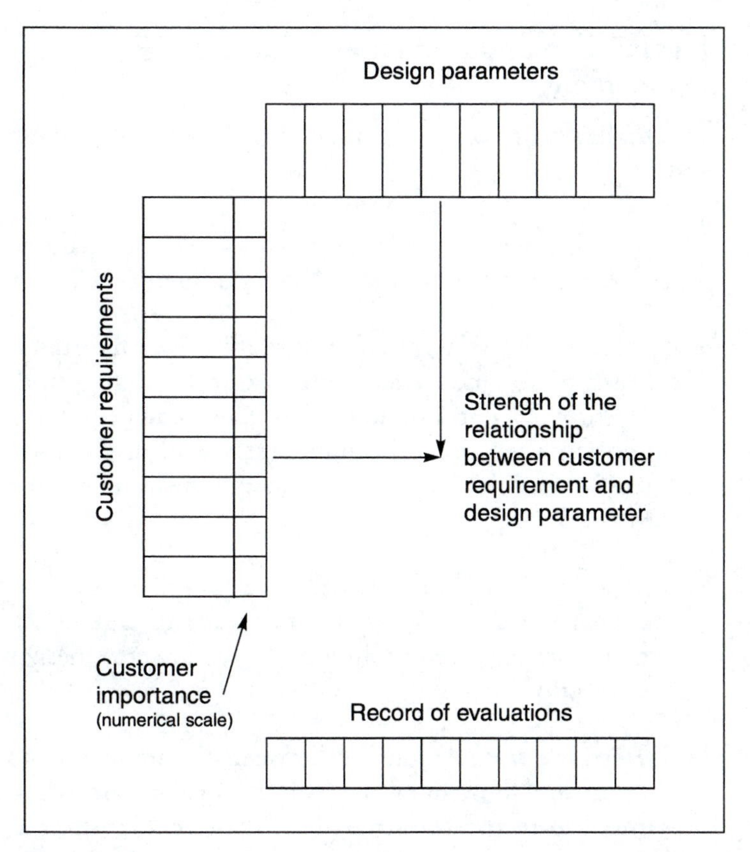

Figure 7.2

A quality function deployment matrix for the product planning phase.

questions (documented along the top of the matrix) represent the output. A QFD matrix for the product planning phase involves transformation of the customer's requests and any other constraints into design parameters. These parameters constitute the design specification.

The QFD matrix helps you determine the relationships between customer requirements and actions of the company to fulfill them.

The Fundamentals of Quality Function Deployment

Quality function deployment involves the following steps:

1. Ascertain the customer requirements and list them down the left side of the matrix, as shown in Figure 7.2.

2. Record in the adjacent column the relative importance of each requirement on a numerical scale ranging from zero to ten. For example, if a requirement is very important then assign the number ten, while if a requirement is of minimal importance, assign the numbers one or two.

3. List across the top of the matrix the company responsibilities necessary to satisfy the customer requirements. These company responsibilities constitute the design specification.

4. Determine the relationship between the items on the vertical and horizontal axes of the matrix and use a symbol to record the strength of these relationships at the appropriate intersection in the matrix. These relationships can be designated as weak, average, or strong.

5. Examine the design requirements along the top of the matrix to find the most desirable way to satisfy the customer requirements, and record these evaluations in the row along the bottom of the matrix.

Customer Requirements and the Design Specification

The implementation of the five steps listed above translates the customer requirements into design requirements. Several activities are involved in this early phase.

- **Identify what the customer wants.**

A primary goal of most manufacturers is to generate profits through product sales. Clearly products must satisfy customers to sustain sales. Therefore a manufacturer must listen to its customers and translate their needs into product attributes. Customer needs can be identified in many ways ranging from industrial sales representatives interacting with their customers, to elaborate marketing surveys.

Customer needs can be identified by marketing surveys and the interaction of sales representatives with their customers.

- **Identify relevant government standards.**

While the customer determines the primary requirements for a product, a second group of requirements is imposed by external standards. These include those imposed by the Environmental Protection Agency (EPA) and the Occupational Safety and Health Administration (OSHA) in the United States, as well as others such as the German DIN (Deutsche Industrie Normen) standards and British Standards (BS). All relevant standards must be added to the requirements in the quality function deployment matrix.

- **Benchmarking.**

Products already on the market should be scrutinized to see how competitors satisfy their customers. This scrutiny should not only include a direct comparison of product performance but also how this performance is achieved and how the product is manufactured. Competitors' products provide target values for the new product's characteristics. This is called *competitive benchmarking.* Data from these investigations can establish benchmark requirements for a new product.

Benchmarking of competitors' products should examine:

- performance
- design
- manufacturing

- **Design requirements.**

Design requirements are established by translating the customers' words into engineering terms useful to a design team. These terms can be organized into categories—energy consumption, ergonomics, and mechanical loading, for example—that can be incorporated into the quality function deployment matrix.

Translate customer needs into design requirements:

- maintenance
- energy consumption
- ergonomics
- cost, etc.

- **Scrutinize relationships between customer requirements and the design requirements.**

Design requirement interrelationships can be:

- complementary
- unrelated
- conflicting

Relationships between customer requirements and the design requirements should be evaluated to ensure that each customer request is associated with *at least* one design parameter. Naturally some design requirements will be related to several customer requirements while others are strongly related to only one.

Design requirements should be reviewed to determine interrelationships between them. Typically these relationships fall into the following categories: complementary, unrelated, and conflicting. All of these activities provide the essential data for establishing a product design specification.

Design parameters are the essence of a design specification.

The kernel of a product design specification is a group of design parameters that are the essential elements of the problem formulation. These parameters identify the primary ingredients of the problem and limit the design task. They interact with each other in both conflicting and complementary ways but, assembled correctly, the completed jigsaw defines the design specification, as presented in Figure 7.3.

7.3 The Structure of a Product Design Specification

A design specification for a product will often have several layers of specifications and while all of these specifications concern the same set of generic design parameters presented in Table 7.1, they occur at different layers and are developed at different times. Parameters associated with a lower level are usually not specified at the higher level. This hierarchy includes:

- Parameters defining the performance of the complete system.
- Parameters associated with major assemblies of the system.
- Parameters associated with subassemblies of the system.

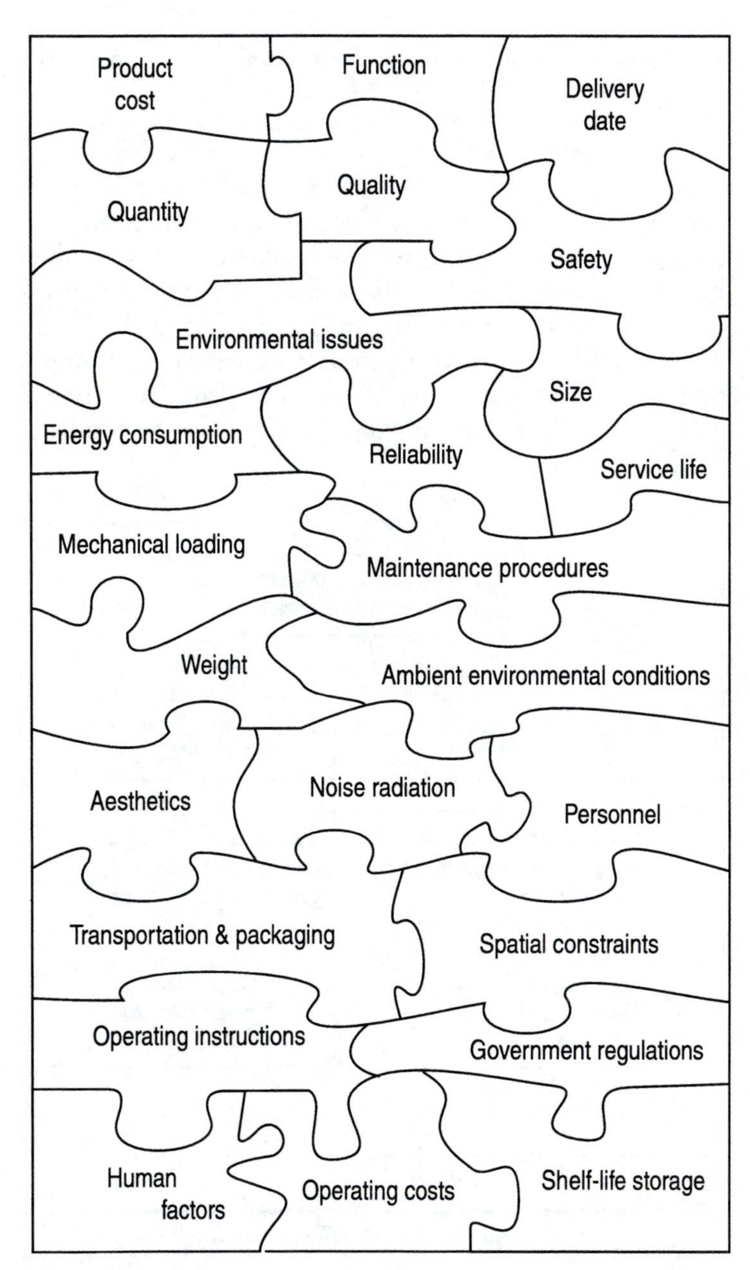

Figure 7.3

The product design specification: interactions between the design parameters.

- Parameters associated with individual components of the system.

First the design specification for the complete product is developed. Subsequently through the process of conceptual design and product design, plans for the product emerge. Often the product will consist of several major assemblies. Each of these must also be designed according to a specification using the same set of parameters presented in Table 7.1. The relative importance of these parameters governing a major assembly may be quite different from the relative importance of the same design parameters for the complete product.

Design specifications must be written for the complete product, the primary assemblies, subassemblies, and individual parts.

Every part, assembly, or product is a response to a design specification.

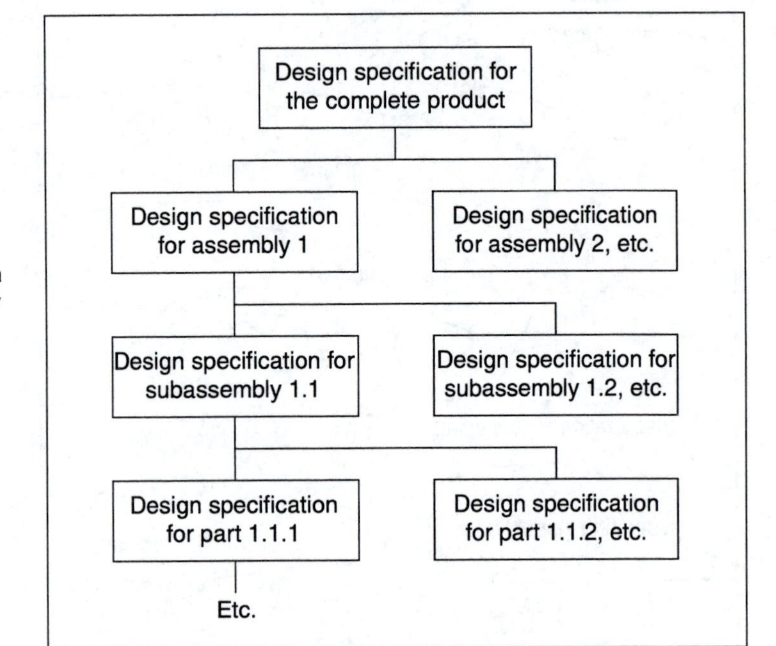

Figure 7.4

A tree of design specifications showing different levels.

Naturally these assemblies consist of subassemblies and individual parts. Every component of a product must satisfy a design specification. Again the relative importance of the design parameters may be quite different from those of the complete product or even for the major assemblies. The different levels of design specifications are illustrated in Figure 7.4.

Consider the design specification for an automobile. At the highest level there is the specification for the complete vehicle. This involves, for example, requirements to accommodate four people with an engine appropriate for a prescribed performance, to warm or cool the interior as needed, to provide a comfortable ride, and to carry reasonable luggage.

Such characteristics are achieved by integrating several major systems at the next level of the design specification. These include the power plant, chassis, drive train, steering, braking, suspension, and heating and cooling systems. Each one must satisfy its own design specification and must work harmoniously with the other major systems so that the vehicle complies with its global design specification.

Finally each of these major elements is made of many subassemblies and parts which must be designed in response to yet more design specifications. An automobile typically contains some 14,000 parts, each of which has its own specification.

Design Parameter Weighting

Consideration of automobile design shows that many design parameters must be considered at various levels. With all design problems, it is first necessary to quantify the importance of each design parameter. Are some more important than others? Can some be neglected for some problems? This evaluation process can be accomplished by assigning each parameter a factor weighting its relative importance. Plotting the characteristics graphically can reinforce the results of the evaluation process and permit the essential requirements to be easily identified as shown in Figure 7.5.

Quantify the importance of each design parameter.

Design parameters constrain creativity.

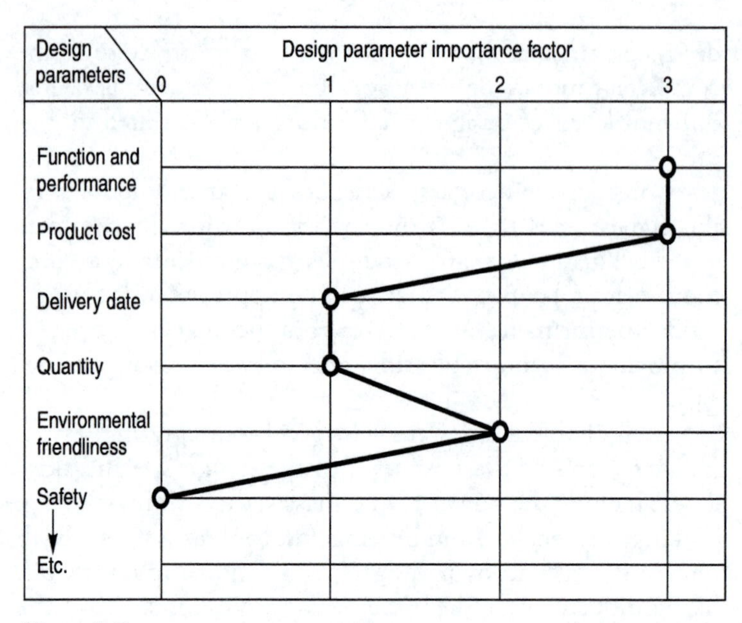

Figure 7.5

Graphical presentation of the design parameter importance factor.

A numerical system can be established for the *design parameter importance factor* using the following criteria:

	Points
Requirements which are essential	3
Requirements which are highly desirable	2
Requirements which are desirable	1
Requirements which are almost irrelevant	0

All requirements assigned two and three points must be satisfied by each conceptual design. Naturally, alternative schemes using more points can be proposed, but the primary objective is to establish and quantify the relative importance of the different design parameters before the conceptualization phase is begun. This completes the definition of the design task.

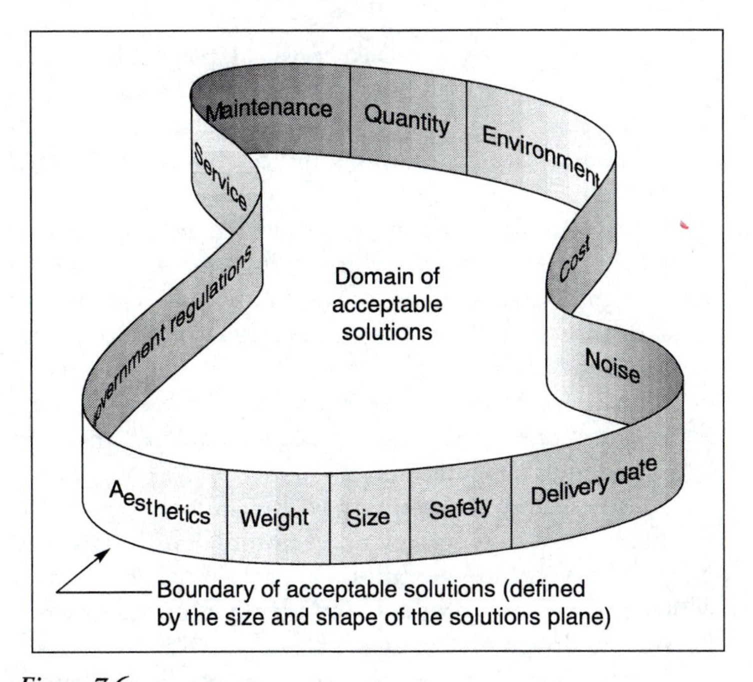

Figure 7.6

Constraints imposed on a conceptual design by the design specification.

7.4 Generic Design Parameters

In the remainder of this chapter we discuss the implications of the design parameters in Table 7.1 as a basis for defining design problems. These parameters collectively define the design specification and they impose limits on the designer's freedom of choice. This philosophy is illustrated in Figure 7.6.

Function and Performance

What must the design do, and what are the tolerances on the performance?

All products must perform a function, and there is invariably a performance specification associated with that function. For example a device may be required to convert a rotary motion into a reciprocating motion while it transmits a prescribed power at a prescribed speed. This design parameter can involve mechanical, pneumatic, hydraulic,

Consider both primary and secondary functions:

- mechanical
- electrical
- electronic
- computational
- chemical
- biological

Quantify and characterize forces, temperature, deflections, voltages, frequencies, pressure gradients, etc.

Does an unrealistic performance specification impose higher costs?

and electrical systems as well as acoustical, magnetic, optical, and thermal properties. Questions may arise about voltages, pressures, field intensities, frequencies, wavelengths, temperatures, heat flux, continuous and discontinuous characteristics, etc.

It may also be necessary to address secondary considerations associated with some primary function. Consider the internal combustion engine of an automobile. A secondary consideration is cooling. The combustion process generates heat, and the engine must be cooled to function effectively. Of course cooling might be unnecessary if the engine were made from one of the new ceramic materials that can withstand high temperatures.

Another kind of secondary consideration involves the mechanical loading imposed upon a product by its environment. These include vibration, impact, and inertial loads which occur when a product is accidently dropped or when a vehicle is driven on a rough road.

Finally, how does the performance specification affect the cost of the product? Is the customer realistic in specifying extremely demanding performance? Would relaxation of some characteristics significantly reduce the cost of the product without affecting performance significantly? A good understanding of the customer's requirements and the motivation for these requirements is essential.

Product Cost

Product cost is a critical design parameter.

Cost reduction strategies should be employed at all stages of the product realization process.

The final cost of a product is a major design consideration because it can be responsible for the commercial success of a product. The marketing department of a company will generally make cost estimates that will involve benchmarks of competitive products. This usually triggers cost reduction activities. However, cost estimates are uncertain when they precede conceptual design. The projected cost of the product should be refined throughout the concept-to-customer phase of the product realization process, and if the product is complex, then target costs should be assigned to each subsystem. Typically, more realistic costs can be estimated in the product design phase when the original concept has been transformed into specific hardware.

Companies should use concurrent engineering strategies to achieve target costs. Design-for-manufacture approaches should be emphasized because they focus on a number of questions that affect product cost. For example, how many products must be manufactured annually? When must the first product be delivered to the customer? What are the financial consequences of specified tolerances, surface finishes, and material selection?

Reduce costs by implementing concurrent engineering practices and design-for-manufacture protocols.

Further cost savings can be attained by selecting a simple working principle at the conceptual design phase, by reducing the number of parts in a product, and by choosing commercially available materials and components as well as conventional manufacturing processes.

When cost is a question, consider the following:

- What parameters govern the product cost? Examine complex systems, subsystems, subassemblies, and individual components.
- The accuracy of the cost estimate depends upon the phase of the design process during which the estimate was developed. An estimate developed at the conceptual design phase is less accurate than one at the product design phase. Accuracy also depends upon the quality of the data used in the estimate.
- Use rule-of-thumb guidelines during the initial selection of materials and manufacturing processes.
- Evaluate the quality of different conceptual designs using the estimated cost as one selection criterion.
- Which components will be manufactured by the company and which will be purchased from suppliers?
- If a component is to be manufactured by the company, then:

  Generate cost information on potential manufacturing processes in the available production facilities. Consider materials, tooling, labor, financial burden, etc.

  Using the cost data, choose the most appropriate manufacturing processes for the required quantity of parts, production rates, etc.
- If a component is to be purchased from a supplier, then:

  Determine costs for out-sourced parts, systems, and modules. Evaluate their quality and cost.

Delivery Date

The delivery date affects design time and the selection of manufacturing processes.

The date when the first product is to be delivered to the customer has a major influence on the time spent by a company developing the product. It also affects the kinds of manufacturing processes used to produce it. If a product must be delivered in a short time, then there will probably be little time for a detailed analysis of the product during the design phase. The product should be produced by manufacturing processes with short lead times such as welding or machining from bar stock rather than casting or forging. A schedule for the product realization process should be agreed upon by all departments in a company in order to avoid penalties for late delivery of products to the customer.

Quantity

Quantity affects process planning and the choice of manufacturing processes.

The number of products to be delivered to the customer affects both their cost and the production processes to be used. Manufacturing considerations in turn influence the product design phase and require use of design-for-manufacture techniques. If the batch size is very small, then general-purpose machine tools and processes are employed to make parts from standard stock materials. However, if the company is to produce a large volume of products, then specialized equipment can be employed. Different production processes require subtly different design characteristics of parts. These are discussed in Chapters 11 through 18.

Use general-purpose machine tools for small batch sizes; use special-purpose machine tools for mass production.

Environmental Friendliness

Environmental issues are becoming increasingly important. They were discussed in Chapter 6.

The impact of a product on society and the environment has recently assumed considerable importance. Debates have focused upon dwindling natural resources and the fragile ecosystems of the earth. Probably environmental friendliness will assume even greater significance in the future as governments impose more stringent laws on products and as the sensitivity of people to environmental issues increases.

Environmental concerns that influence design were examined in Chapter 6. These issues are diverse and they all influence design decisions. Designers have a responsibility to enhance the quality of life without impeding the creation of new products by the next generation of human beings.

Safety

Safety is primarily concerned with the health of human beings in the neighborhood of a product, but it can also pertain to the property and equipment in this neighborhood. All parameters which define this neighborhood are under the control of the concurrent design team which must assume responsibility for recognizing and dealing with potential product hazards throughout its life cycle.

Because safety affects people, equipment, and property, designers must consider hazards associated with their products. Eliminate hazards or provide guards and warning labels.

It is important to note that the majority of accidents occur as a result of an inappropriate use of a product rather than from product defects. Consequently this aspect of the relationship between a product and humans requires considerable attention.

Hazards assume different forms, and in the conceptual design phase they may be poorly defined. However in redesigning or upgrading a product, they will generally be reasonably obvious. There are numerous federal and international regulations that must be complied with by products. Examples are those developed by the Environmental Protection Agency and the Occupational Safety and Health Administration. Some of the more common hazards involve electrical, environmental, ergonomic, stored energy, and relative motion considerations.

EPA: Environmental Protection Agency

OSHA: Occupational Safety and Health Administration

Once hazards are identified, the probability and potential severity of accidents should be assessed. Three approaches exist to eliminate hazards.

- Completely eliminate the hazard from the product. This can be most easily accomplished at the conceptual design phase by selecting an alternate working principle. However this is an expensive undertaking when a product must be redesigned after it is already in production.
- Hazards which cannot be designed out of a product should be insulated from the user by guards, location of machine controls, etc. Sophisticated sensing systems can prevent operation of a machine when an operator is in a dangerous area.

1. Are all moving parts of machinery and power transmission equipment properly guarded?
2. Are edges of components and access openings rounded, or are sharp edges protected?
3. Are audible signals distinctly recognizable and unlikely to be masked by ambient noise?
4. Are fault-warning systems designed to detect weak or failing parts before an emergency occurs?
5. Are all liquid, gas, and steam lines clearly identified and labeled, with warnings of specific hazards to persons or equipment?
6. Are struts and latches provided for hinged and sliding components to keep them from shifting and endangering maintenance personnel?
7. Are drawers and foldout assemblies provided with limit stops to keep them from coming out too far, coming loose, or falling?
8. Are components located and mounted so maintenance personnel can get at them easily without exposure to hazards from electric charges, hot surfaces, sharp points or edges, moving parts, or chemical contamination?
9. Are mechanical components which utilize heavy springs designed so the springs cannot suddenly fly out and inflict injury or damage?
10. Can the equipment controls be locked out to prevent inadvertent start-up during maintenance?
11. Are indicators color-coded to show the normal operating range and danger range?
12. Are warning circuits designed so they actively sense hazardous conditions, rather than passively display control settings?
13. Are the most critical warning indicators grouped within the operator's normal field of view and separated from other, less important indicators?
14. Are go/no-go or fail-safe circuits used wherever possible to ensure that failures do not produce additional hazards and that operators know that a failure has taken place?
15. Are charge bleed-off devices provided for high-voltage capacitors which may be touched by personnel during servicing?
16. Are adjustment screws and commonly replaced parts located away from high voltages or hot surfaces.
17. Are conspicuous warnings placed near high-voltage or high-temperature components?
18. Are safety interlocks used where necessary?

Table 7.2.

Product safety features (From T. A. Hunter, *Engineering Design for Safety*, McGraw-Hill, New York: *1992. Reproduced with permission.*)

- Warning signs should be displayed on dangerous products and instructions for their safe operation should be provided.

A checklist for the design of safe products is presented in Table 7.2.

In 1996 a US District Court awarded $5.93 million in damages against the Digital Equipment Company to three plaintiffs. Digital was faulted for not placing labels on its keyboards warning that excessive and improper use could lead to carpal tunnel syndrome and other injuries.

Quality

Quality is difficult to define; it tends to be a relative term involving a comparison of the characteristics of two alternative products. However, product quality is especially related to cost and reliability. British Standard 4891 defines quality as "the totality of features and characteristics of a product or service that bear on its ability to satisfy stated or implied needs." In other words, quality involves the ability of a product to satisfy the customer's requirements at a price that is acceptable to the customer. There must be effective communication between the customer and the manufacturer for the latter to understand the customer's requirements. This will generally guarantee acceptability of the product by the customer and improve its marketability.

Quality involves the ability of a product to satisfy the customer's requirements at a price that is acceptable to the customer.

ISO stands for the International Standards Organization.

ISO 9000 is an international standard for quality assurance.

The cheapest way to solve quality problems is to eliminate them at the source. (Recall the quality-lever analogy in Figure 5.5.) Quality must be incorporated into both the design and manufacturing phases of the product realization process. If this is not accomplished successfully, there is a high cost to any company whose name becomes associated with low quality products. This cost can be measured by warranty costs from the return of products for correction of failures or by the loss of customer confidence, which can be difficult to rectify.

The U.S. Office of Consumer Affairs has published data showing that 25 percent of the products purchased by U.S. consumers were unsatisfactory because of their poor quality.

In recent times there has been great emphasis on *Taguchi methods,* or quality engineering, to develop products of superior quality. Taguchi methods involve evaluation and improvement of product robustness, evaluation of tolerance specifications, the role of engineering management, and the financial consequences of variation in product characteristics. Taguchi defines quality as the degree of loss to society caused by a product after it has been delivered to the customer.

Taguchi defines quality as the degree of loss to society caused by a product after it has been delivered to the customer.

The Taguchi method tries to reduce the variation in performance of a product operating under variable service conditions.

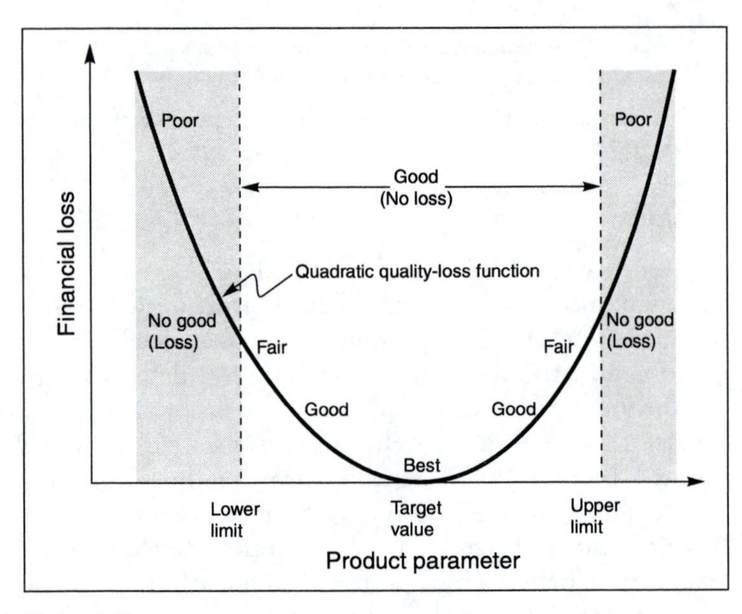

Figure 7.7

A general quality loss function.

Quality loss functions **measure the cost to the manufacturer of variations in product or process characteristics relative to target values.**

The objective of the Taguchi approach is to improve the quality of both product design and process design while cost is simultaneously reduced. The aim is to reduce the variation in performance of a product or manufacturing process so that performance remains as close as possible to a predetermined target value when the product operates under variable service conditions. A *quality loss function* is established which measures the cost of the variation of product or process characteristics relative to some target quantity such as a tolerance or changing environmental conditions. Quality loss functions are quadratic functions in which losses increase by the square of the deviation from the target value (Figure 7.7).

Product standards or specifications are not always thorough; raw materials specifications are not always thorough; and tolerance limits in drawings are not always thorough; do not believe in product standards or quality specifications, raw materials specifications, tolerance limits in drawings, measurement instruments, or chemical analysis. This has been my motto to bespeaking the very fundamentals of quality control.

Kaoru Ishikawa
(1915–1989)

All product and process specifications include a tolerance on each characteristic. There are acceptable upper and lower values, and the characteristic is said to be "in spec" if it lies between these two extremes. Figure 7.8 illustrates three situations in which products meet the specification. In the

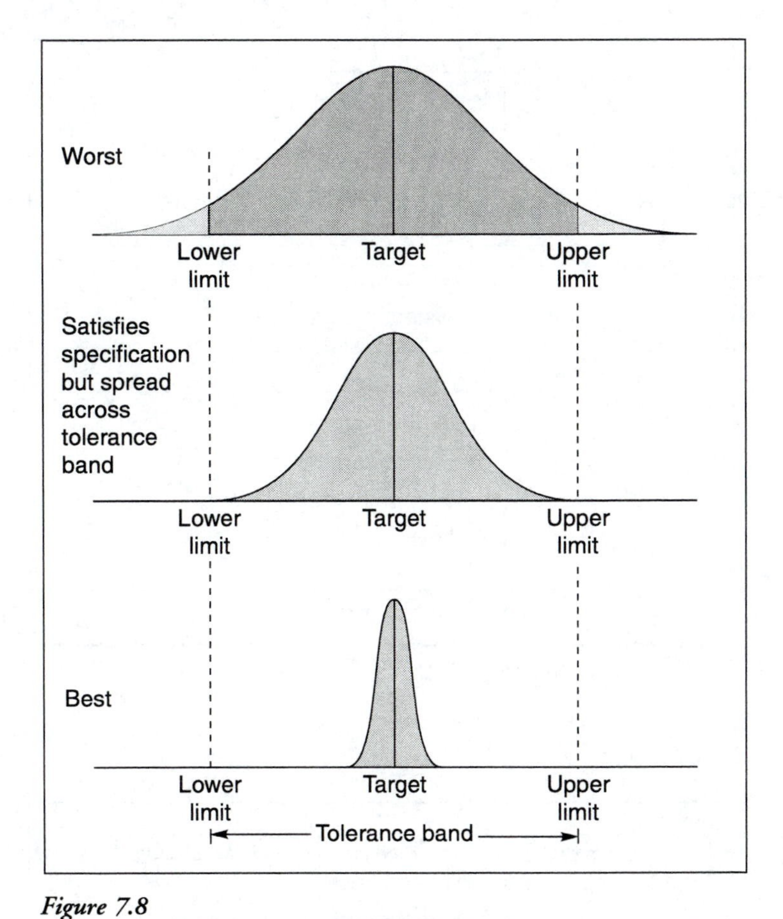

Figure 7.8

Product sample distributions relative to specified tolerances.

first case, some samples are above or below the extremes even though most are within acceptable values. In a second case all samples meet the specification but are spread across the tolerance band. In the third case the products are narrowly scattered about the target value. This third situation represents a higher quality product than the others because there is greater uniformity among the samples.

While the lower two sample sets of Figure 7.8 comply with specifications, they are not of equal utility if products must accurately interface with one another. Under these conditions, all samples must be of high quality, as repre-

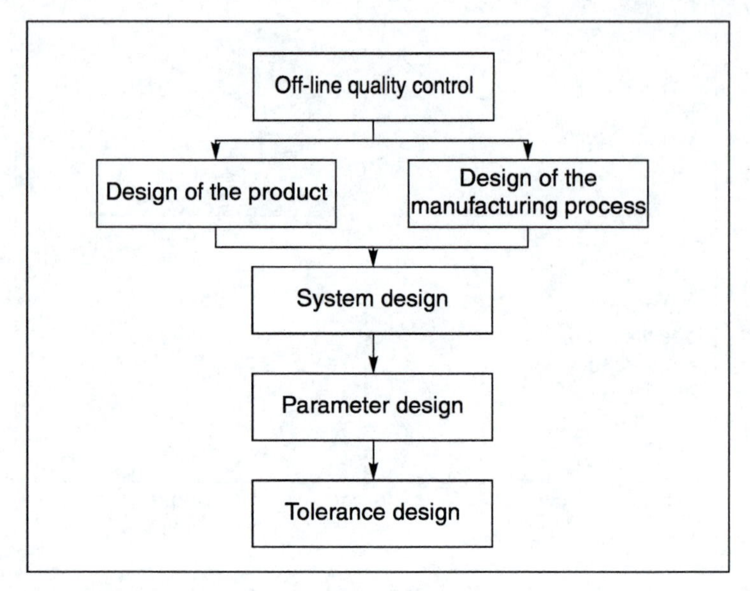

Figure 7.9

The Taguchi design philosophy.

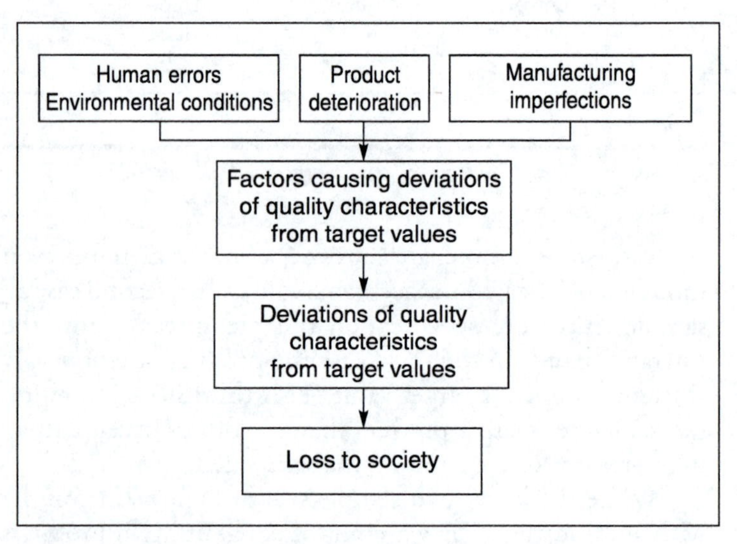

Figure 7.10

The consequences of parameter variations on quality characteristics.

sented by the third graph. The relatively large product variation shown by the center graph of Figure 7.8 makes these products unsuitable for tightly interfaced assemblies.

Taguchi's design philosophy involves three principal stages: system design, parameter design, and tolerance design (Figure 7.9).

Taguchi's method has three stages:

- system design
- parameter design
- tolerance design

The system design phase involves both conceptual design and the preliminary product design phase. Subsequently the parameter design phase is initiated. Here the primary objective is to optimize the system design in order to minimize the variation in performance caused by varying environmental and operating conditions. This is accomplished by defining appropriate quality loss functions as a function of controllable parameters and of external uncontrollable parameters. These parameters include manufacturing errors, material deterioration, and variations in operating conditions as shown in Figure 7.10.

Quality loss functions depend upon both controllable and noncontrollable parameters.

A series of experiments are then established to determine the sensitivity of the quality functions to variations in the parameters. The objective is to identify and control only those parameters that can be cost-effectively controlled, rather than to try to control all parameters associated with the performance of a design. Indeed it may be important to determine from the customer the relative importance of the various quality loss functions in order that the most important function receive the most attention.

Identify and control only those parameters that can be controlled cost-effectively.

Finally the tolerance design phase is initiated in which the objective is to determine the appropriate tolerances for the different parameters that govern the quality characteristics. This involves determining the percentage that a variation of each parameter contributes to the deviation of the quality function from the target value. Subsequently tolerances can be specified for each of the parameters in order that the target value is achieved in the most cost-effective manner.

Determine the sensitivity of the quality loss function to changes in each parameter.

Energy Consumption

Escalating energy costs, environmental issues, and government regulations govern energy consumption.

Increasingly environmentally concerned consumers as well as the escalating cost of energy for industrial and commercial organizations have fueled the demand for energy-efficient products. Government regulations and incentives now affect the design of many products, including automobiles, for which federal fuel economy standards exist. Consequently energy considerations have increasingly influenced the marketability and hence the success of products.

In order to create an energy-efficient product, a design team typically must do the following:

Increase efficiency by isolating consumption losses.

- **Isolate all sources of energy consumption**

  The primary sources of energy consumption typically are electric motors, electronic devices, and combustion engines. Once these sources have been identified, energy consumption should be quantified and methods for reducing it should be considered.

  One approach at the conceptual design level is to devise more energy-efficient alternatives—light bulbs, for example (See Section 6.4.)–or to retrofit a product with more energy-efficient components.

- **Isolate all sources of energy loss**

  The objective is to identify all sources of energy loss and to quantify the losses prior to developing more energy-efficient alternatives. This philosophy has been successfully applied to the design of gas-condensing home furnaces in which heat losses are reduced by waste heat recovery. A second heat exchanger is used to absorb heat that would otherwise escape with the flue gases. These furnaces achieve an energy efficiency in excess of 90 percent compared with 74 percent for older furnaces.

Energy-efficient designs reflect:

- energy consumption
- energy losses
- efficiency

- **Increase the efficiency of the product**

  The energy efficiency of a product depends upon both its energy consumption and the attendant losses. Energy consumption might be reduced with better energy sensors and controllers.

  Energy losses might be reduced with insulating materials or alternative components. Compare domestic refrigerators and freezers in Great Britain and elsewhere. British refrigerators

> are thermally inefficient compared to competitors' products. They consume two to three kilowatt hours of electricity per liter of storage space. That represents annually eight percent of the total electricity generated in Britain. Increasing the thickness and quality of insulation as well as increasing the efficiency of the motor and heat exchanger would reduce energy consumption to about that of the competition, which is approximately 0.3 kilowatt hour per liter of storage space.

Reliability

The *reliability* of a product can be defined as the probability that it will perform its intended functions under prescribed operating conditions. Thus the reliability of a product is closely related to its quality because a highly reliable product must be a high quality product. The intended performance and prescribed operating conditions must be comprehensively described in the design specification. These statements impose tolerances on the behavior of the product, such as whether it must operate continuously for a prescribed time or whether it must operate discontinuously at a prescribed frequency under given conditions.

The *reliability* of a product can be defined as the probability that it will perform its intended functions under prescribed operating conditions.

Reliability requirements generally are dictated by the customer. Achievement of the intended reliability depends upon the product design and on the manufacturing processes that are used. For a product to operate reliably, every part must function in compliance with its design specification. No part is unimportant. All details associated with both design and the manufacturing processes must be examined meticulously.

No part is unimportant; every part must comply with its specification if the product is to function reliably.

The importance of attention to detail is illustrated by the explosion of the space shuttle *Challenger* in 1986. This disaster was caused by the failure of an O-ring seal on a booster rocket. The failure of this simple item was responsible for the loss of seven lives, the destruction of the space shuttle, a two-year delay in shuttle flights and the loss of significant national pride and prestige.

Reliability feedback from the customer to the manufacturer is important in the evolution of products. All manufac-

Federal standards
International standards
Failure analysis
Bought-out items
Manufacturing protocols
Prototype testing
Transportation
Installation
Operating instructions
Maintenance
Servicing

Table 7.3

Some parameters affecting product reliability.

turers should establish mechanisms to ensure that the same mistakes are not be repeated.

Throughout the product realization process numerous parameters can influence product reliability. However, the primary parameters affecting reliability are embedded in the design phase. Table 7.3 lists some of them. They include:

- **Legislation**

  Federal and possibly international standards must be considered during the specification phase of product realization. Many of these standards influence the reliability of a product and safety legislation frequently has a major influence on product reliability.

- **Failure analysis**

  During the product design phase, the design team should undertake failure analyses using simulation software tools rather than wait until the design is more mature and hardware prototypes have been built. The latter approach is much more expensive.

  While one or more prototypes must be built to test the product under service conditions, the probability of success of this phase can be increased by the use of computers and computer-aided design techniques. Prototype test results must be evaluated cautiously if the prototype has been constructed by highly-skilled individuals because it may not represent products manufactured under normal production conditions.

- **Manufacturing processes**

  The reliability of a product depends upon the processes used to manufacture it. Typically each process has its own advantages and disadvantages, so processes frequently generate distinct attributes in the product. Some of these attributes are discussed in Chapters 11 through 18.

  If the appropriate manufacturing skills or plant facilities are unavailable in a company, then the design team must decide either to change the specified manufacturing processes or else to subcontract the work to a vendor. Bought-out parts must be tested because their quality affects product reliability.

- **Bought-out parts**

  Most new products contain a variety of bought-out items ranging from standard nuts and screws to special-purpose castings. All of these can affect the reliability of the product if they do not comply with design specifications. Therefore the design team should ensure that all bought-out items are tested.

- **Prototype testing**

  Prototypes are made to learn the characteristics of the product. Fabrication and assembly processes can be evaluated, but the primary motivation is to evaluate performance relative to the design specification and to discover the reliability and failure modes. Various testing procedures exist, including accelerated test programs in which loads or speeds are increased beyond the normal operating range. Naturally prototypes should also be tested in their normal operating range to replicate the behavior of the product in practice. Data from these tests should guide the design team in making improvements.

A *prototype* is an original or experimental model on which later copies are based.

- **Maintenance and operating instructions**

  The design and manufacture of a reliable product does not mean the end of the concurrent-design team's reliability responsibilities. The team is responsible for ensuring that reliability is not adversely affected during storage prior to delivery to the customer, that the product is packaged and transported without damage, and that the product is installed properly.

  The reliability of a satisfactory product can be compromised by a variety of actions. To ensure that a product operates reliably for a prescribed time, an instruction manual is frequently required. The user needs to know the product limitations as well as procedures for its operation, maintenance, and troubleshooting. The design team should supervise the creation of this document.

Maintenance Protocols

Maintenance protocols describe what must be done to ensure that a product continues to function within its design specification. They are especially important for certain kinds of products. For these products it is important for the design team to include individuals with maintenance experience.

A *protocol* is a step-by-step plan that specifies how something is to be done.

Maintenance requirements should be identified and addressed early in the evolution of a product.

Typical tasks include bearing lubrication, parts cleaning, checking bolt preloads, and replacement of worn parts. Depending upon the product and the nature of the maintenance procedure, these tasks may be undertaken three ways: by the customer, by the manufacturer, or by an independent service company. Each way imposes different constraints on the development of the product.

Maintenance can be undertaken by:

- the customer
- the manufacturer
- an independent organization

If it is uneconomical to repair a product, the customer can simply replace it. If the malfunctioning product is an electric coffee maker, for example, then replacement is easy. In other situations some simple dismantling may be necessary. Replacement of automobile windshield wipers, for example, is easy. More complex tasks can be done by trained personnel in a service organization such as in automobile repair. When specialized facilities are required, the product usually must be returned to the manufacturer.

Manuals should use numerous schematic diagrams and sketches instead of pages of text to explain procedures.

For low cost products, maintenance is not an important design issue. However, for products that must have a long service life, a design-for-maintenance philosophy should be employed because lifetime maintenance may cost more than the initial cost of the product. Here the manufacturer should establish maintenance procedures and training schemes for maintenance personnel. Maintenance manuals should be written to provide advice on fault recognition and rectification along with routine maintenance procedures. Video tapes are also useful.

Maintenance manuals provide advice on:

- fault recognition
- fault rectification
- routine maintenance

The design team also is responsible for producing user guides and safety instructions. They can range from labels on a product to comprehensive instruction manuals. These provide protection in legal disputes. Some design rules for product maintenance are listed in Table 7.4.

- Make maintenance as quick and easy as possible.
- Avoid parts that require maintenance.
- Provide easy access to regions that require attention.
- There should be only one way to disassemble and reassemble a product to perform maintenance.
- Wherever possible, use standard fasteners that require only common tools.
- Require the fewest possible skills of the person doing the maintenance.
- Parts replacement should be easy.
- Use self-adjusting parts where possible. Adjustments should be easy to do.
- Where appropriate, use sensing systems or indicators.
- In complex products use artificial intelligence procedures to identify faults.
- Minimize the study required to perform maintenance.

Table 7.4

Design for product maintenance.

Size

Economic considerations dictate that products should be as small as possible within the constraints of the particular class of products. Orders-of-magnitude size constraints should be established to provide targets for the design process even when this is not of primary concern. However, size should be studied carefully when there are specific spatial constraints. Must the product fit inside a space of known size or fit into space having a complex shape?

Size and weight are important parameters in designing parts for the transportation industries.

For example, if a mechanism must move an automobile window up and down, then that mechanism must fit within the door. An artificial organ that must be implanted in the human body—an insulin infusion pump, for instance—clearly has a size constraint.

The downsizing of modern automobiles has been responsible for a number of challenging size-dominated design problems, especially under the hood. The shipbuilding and aerospace industries have similar size constraints.

Weight

Product weight affects:

- Manufacturing
- Handling
- Transportation
- Operating costs
- Performance

The weight of a product should be limited for economic reasons because cost generally is proportional to the weight of a product or to the volume of material it contains. The weight of the product also influences handling during its manufacture, installation, and use. A heavy product may require attachment points for slings to a hoist. Weight also can affect transportation costs which frequently are directly proportional to the product weight. Weight can affect operating costs of a moving product because energy consumption is directly related to the weight of moving parts.

Weight assumes considerable significance in the transportation industry because it affects operating costs. This has led to the development of lightweight construction techniques in the automobile, aerospace, and shipbuilding industries. The primary focus usually is the design of high-strength, high-stiffness parts of low weight using high quality steels, aluminum alloys, polymeric composites, and honeycomb sandwich structures.

Lightweight construction may incorporate:

- aluminum alloys
- polymeric composites
- high quality steels
- honeycomb structures

Weight is of paramount importance in the airline industry. This is seen in the color scheme of American Airlines planes which have a polished aluminum skin with minimal paint work. The motivation for this is that each pound of paint adds $30 per year to the fuel costs in 1993 dollars. If the airline elected to paint its aircraft, then a McDonnell–Douglas DC-10 would weigh an additional 390 pounds, a Boeing 727-200 would weigh an additional 200 pounds, and a Boeing 727-100 would weigh an additional 135 pounds.

In the design of warships, the U.S. navy is aggressively trying to eliminate thousands of tons of weight from each vessel by fabricating parts from composite materials. These materials can reduce operating and maintenance costs while at the same time improving speed, payload, and the stability of the vessel. Stability can be improved by reducing the weight of topside structures.

Aesthetics

Aesthetics is concerned with those characteristics of a product that cause it to appeal to the customer—characteristics perceptible through the human senses of sight, touch, smell, hearing, and taste. These are important marketing characteristics of many products, but they are especially significant in the consumer goods industries where color, surface texture, and visual appeal is important for success. Aesthetically appealing products have a strong psychological attractiveness which is subtly blended with ergonomic attributes and appropriate manufacturing techniques.

Aesthetics involves:

- Visual appeal
- Texture
- Color
- Smell

Aesthetics is a qualitative design parameter that must be established by market research focused on customers, design trends, and competitors' products. The aesthetic qualities of a product designed for different age groups may be quite different. A senior citizen has different product evaluation criteria than an adolescent. Similarly there is a segment of the population that purchases products based upon cultural and social criteria. Examples include the Mercedes Benz automobile and the Rolex watch which not only provide transportation and the time but also convey a notion of status in the community.

Aesthetic opinions of a customer depend upon:

- age group
- culture
- social class

Ambient Environmental Conditions

Temperature, humidity, chemicals, air quality, and many other environmental conditions affect products and consequently influence their design. Materials vary in their absorption of moisture and in their coefficients of thermal expansion. Corrosion may be a problem in a moist environment. Many products—from a space craft to a fire hose—must withstand both high temperatures and extreme cold as well as wet or dry climates. Air quality enters into the design of products in several ways. Products such as the automobile contribute to air pollution while others suffer from airborne pollutants such as sulfuric acid or airborne dirt and grit.

Environmental conditions that affect product design:

- temperature
- humidity
- chemicals
- particulates
- bacteria

Packaging and Transportation

Ensure that the product reaches the customer in a pristine state. Consider:

- warehousing
- packaging
- handling
- transportation

Subsequent to manufacture, a product usually is shipped to a warehouse where the environment can influence product design in a number of ways. In order to deliver the product to the customer in pristine condition, packaging must protect the product during storage as well as during transportation. Markings should identify hazards to personnel handling the package and to the consumer.

Many products are packaged in protective polystyrene foam within a paper board box. Delicate parts may need to be locked down. Large objects may require a wooden crate, and extremely large products usually are dismantled before shipment.

Some products—bookshelves and office furniture, for instance—can be shipped knocked down for on-site assembly by the customer. The design of a product must make any on-site assembly easy, and consumer-oriented instructions must be developed.

Human Factors

Human factors are design parameters concerned with the interface between a product and its human user.

Human factors engineering is an activity concerned with the interactions between a human and a machine or product.

Many products must function in harmony with a human being. Human factors are design parameters concerned with the interface between a product and its human operator. These factors relate to all stages of the product realization process from the early design stage through manufacture to servicing and product disposal or recycling. Human factors are either processes or knowledge associated with a field of engineering called *human factors engineering* or *human engineering* in the United States but called *ergonomics* elsewhere in the world.

Human factors engineering is concerned with the interactions between a human operator or user and a single machine or product. The human and the product are the primary ingredients of a system which will have a global design specification and which is typically subjected to a variety of environmental conditions involving temperature, vibration, noise, and gaseous chemicals, for example. Since the global system must attain prescribed operating characteristics, the impact of these environmental conditions upon the performance of the human operator must be carefully assessed. Operator performance can be measured by work performed, the monitoring of events, and the processing of data prior to taking appropriate action.

Determine the influence of the environment upon the performance of the human operator.

Select sensors and appropriate means of communication with the operator.

This aspect of design has in recent years assumed considerable significance as machines have become increasingly complex and greater demands have been imposed upon people. Typical examples of these machines include high-speed aircraft, military systems, computer terminals, trains, automobiles, robots, automated factories, and telephones.

To illustrate the tasks associated with the interaction between a human operator and a machine, consider the situation in which the state of the product is discerned by the operator from a sensing system—an audible alarm, a flashing light, a readout pointer, or a paper printout, for example. The design team is responsible for selecting an appropriate sensor and means of communicating with the operator.

The information provided by the sensing system must be processed by the operator prior to making a decision. This decision involves the operator's skills and experience. Usually the operator initiates some action to control the product by means of a pedal, lever, push button, knob, hand wheel, or some similar device. The resulting change in the product may be registered by the product sensing systems and detected by the operator. This action closes the control loop governing the interaction between the operator and the product as these two subsystems function cooperatively. Figure 7.11 illustrates this control system.

As the design team reviews this situation, it must possess several sets of data in order to create a good system. First, the team must understand the mental and physical behavior of a typical human operator under the operating conditions. Second, the design team must possess data describing the characteristics of typical sensors and their interactions with people. Finally, the design team must have similar data for the controls, which typically involve using the hands or feet.

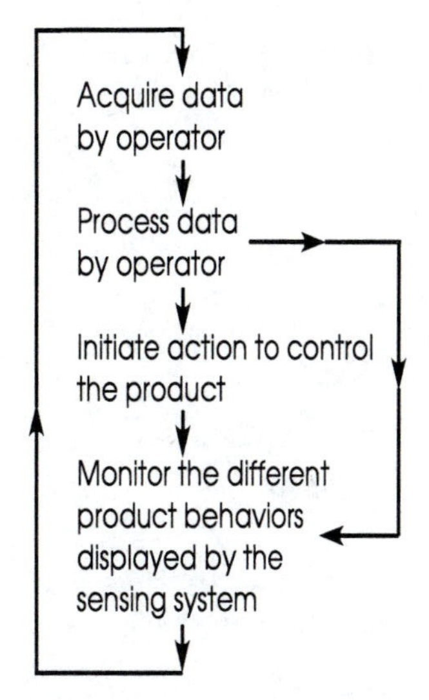

Figure 7.11

A basic human factors engineering control system.

Information about the mental and physical attributes of human beings has been assembled by at least two groups: the Society of Automobile Engineers (SAE) and the space and military services of the United States government. This research has resulted in the publication of numerous documents including *Human Physical Dimensions*, SAE J833; *Man-System Integration Standards*, NASA STD 3000; and *Human Factors Engineering Design for Army Materiel*, MIL-HDBK-759A.

These documents are comprehensive resources for information on human factors data. For example these documents provide tables, charts, and diagrams of anthropometric data on the principal dimensions of the human body obtained from measurements of large groups of people. Data provide detailed information on the face, hands, and feet, for example. Furthermore, data exist for the forces that can be exerted by the articulation of hands, fingers, and arms, and there are recommended levels of illumination for diverse tasks.

Anthropometry: The measurement of the size and proportions of the human body.

Approximately eighty percent of information is acquired visually. Consequently, the visual display of data is important. The design team must decide on what quantitative and qualitative information the operator needs and then must select appropriate data displays.

80% of information acquired by humans is obtained visually.

After receiving and processing data, the operator must decide upon appropriate action and activate one or more controls to change the performance of the product. The motion of the control should be in harmony with the anticipated response of the product in order to fully integrate the product and the operator into a unified human-product system. Thus aircraft pilots pull the control column toward themselves to climb but push forward to loose altitude. The controls and the human-product system are in harmony.

Service Life

The life of a product under normal service conditions must be determined before design begins. How long must the product function within the design specification? Is this time measured in minutes, days, or years? The answer will affect design strategies, the working principle, material selection, and manufacturing considerations such as tolerances and surface finish. All parts of a product need not have the same service life; some parts may need replacement earlier than others. These parts should be designed for easy replacement using ideas presented earlier on maintenance protocols as well as the green design principles presented in Chapter 6.

Service life:

The life of the product operating under normal conditions while complying with the design specification.

Noise

The noise generated by a product is highly undesirable if it must operate close to people. Consequently, noise adversely affects the marketability of many products ranging from dishwashers to automobiles. Increasingly severe regulation has been imposed by the Occupational Safety and Health Administration upon products that emit loud noise. Consequently the design specification must incorporate these regulations and noise control treatments may be necessary in the product.

Noise is a form of pollution that affects adversely the marketability of a product.

Noise is generated by the vibration of a solid or a fluid and propagates through any medium. The human ear can detect acoustic vibrations from about 20 Hz to around 20,000 Hz. The associated sound pressure level is measured in decibels, and international standards have been established to protect employees from sustaining hearing damage at high intensities. An energy level of 140 decibels will cause pain while a person exposed to a continuous level of 85 decibels will suffer permanent hearing loss.

The human ear can detect acoustical vibrations in the frequency range 20–20 kHz.

140 dB pain
110 dB jet aircraft
100 dB lawn mower
10 dB rustling leaves

For a frame of reference, the sound pressure level of a jet aircraft is approximately 110 decibels, a gasoline lawn mower produces about 100 decibels, and wind rustling leaves on a tree is around 10 decibels.

A continuous level of 85 dB results in permanent hearing loss.

The physiological effects of sound are well known. Sound from a product may cause employee fatigue, boredom, or relaxation depending upon its characteristics and the tasks the employee must accomplish, but sudden sounds usually reduce performance. High noise levels increase muscular tension and impede tasks that require mental alertness or manual dexterity.

The designer can use well established methods to reduce noise. They include use of vibration-suppression supports for machines, the tailoring of fluid flow to reduce turbulence, use of acoustic barriers, and specification of polymeric composite materials with high damping capacities.

Governmental Regulations

Products must comply with the governmental regulations of the countries in which they are sold.

Federal and international regulations impose constraints upon product design. These regulations often motivate the creation of new products because older ones do not comply with requirements. This particularly applies to U.S. automobiles which must meet increasingly stringent fuel economy and exhaust emission standards. With the rapid growth of international marketplaces it is imperative to ensure that the product complies with all relevant regulations of the country in which the product is to be sold. Thus the design team must first determine which countries the product will be sold in and then must integrate the relevant governmental standards into the design specification. Failure to do this may necessitate the redesign of a product. This is an expensive, time-consuming activity.

ISO 9000 standards are important in Europe and are being increasingly accepted throughout the world.

Governmental regulations pertain to a plethora of design parameters for seemingly mundane things such as the forms of wall electrical power outlets, the voltage and frequency of electric power, and the noise level that the operator of a product can be exposed to in a day.

Operating Costs

Operating costs:

- energy consumption
- reliability
- maintenance
- service life
- personnel

Operating costs are an important design parameter. Because energy consumption is the primary ingredient of operating costs for many products, energy efficiency should be improved by minimizing both energy consumption and energy losses. This was discussed in the green design chapter. In the aircraft industry, for example, there is a never-ending effort to increase the fuel efficiency of engines, to lighten the aircraft by use of new materials, and to reduce drag by better aircraft geometry or by covering the aircraft with a drag-reducing skin.

Operating costs also depend upon the state of the product in service. Has it been damaged in transit or by inappropriate operator actions and does it still comply with the performance specification? If the product is not functioning properly, then the reason must be identified and repairs made. These actions contribute to operating costs, although they may be embedded in other design parameters such as transportation, maintenance, service life, reliability, and personnel.

7.5 The Student Design-Build-Test Project

As student design teams began to develop a design specification for their complete vehicles, they started to consider each design parameter systematically and examine how it influenced the creation of the vehicle. They began to think about the importance of each design parameter and quantify it. Later, when they had decided upon the best concept, a second series of design specifications were developed for the individual assemblies and then the individual parts of the car.

A design specification extracted from one student group is presented below:

> Three challenges face the designers of a candle-powered vehicle. First, the vehicle must have the ability to translate a finite amount of heat energy into a maximum horizontal translation. Second, the vehicle must begin horizontal translation without any external input. Third, the vehicle must be able to travel in a linear path or else be self-correcting.

Other parameters, discussed in class, were added to create a design specification for the project. Some notes by students are listed below.

Function and Performance

The function of the birthday candle powered vehicle (BCPV) is to travel farther on a horizontal surface than any other team's car using only the energy provided by ten burning birthday candles. The vehicle must be self starting and the change in potential energy shall be zero between the initiation of motion and the final position at the completion of the test run. Other issues on performance are given in the previous paragraphs. When in contact with the ground, the chassis of the vehicle and the wheels must support its weight. In addition, the chassis and other systems must withstand any vibrational loads imposed by irregularities in the floor of the test track. The part of the vehicle near the burning candles must not be flammable; otherwise the car could be destroyed.

Product Cost

The total cost of materials for the BCPV must not exceed $50 although more than this will be spent during development and testing.

Delivery Date

The delivery date for the working BCPV is the end of the semester at the Mechanical Engineering Department's Student Design Conference. A Gantt chart will be developed to ensure that the goals of the program will be achieved throughout the semester.

Quantity

There will only be one vehicle built, so we will be able to custom-build any parts needed if they are not otherwise available. The production processes available to us in the College of Engineering Student Shop are all focused on one-off production; there are no mass production facilities.

Environmental Friendliness

The main concern of our design is the burning candle. However the products of combustion are small and of no consequence. We intend to use components that are recyclable for manufacturing the BCPV.

Safety

The BCPV must be safe to operate. We need to be concerned with such things as the bursting of pressure vessels, combustible materials, and trapped and burned fingers

Quality

The quality of the BCPV must be high enough to ensure that the vehicle can be thoroughly tested and also compete in the design competition at the end of the semester. It will be manufactured and assembled by our team so quality control should not be a problem.

Energy Consumption

This is an important design parameter. The design must utilize the energy of the candles efficiently. The losses in the vehicle must be low enough to ensure that the vehicle will win the design competition in front of the cheering crowds on design day.

Reliability

In order for the vehicle to function properly, all of the components must be reliable. Since the vehicle has only one chance to compete in the race, the car must function properly when the time comes to go to the starting line. There is no second chance.

Maintenance Protocols

The car will be designed for ease-of-maintenance. Shafts will require lubrication, and the candles must be replaced after each test. Since our team will have built the car, the replacement or modification of parts should be trouble-free.

Size

The car should be small for reasons of energy efficiency. In order to travel in a straight line, it might need to be longer in the direction of motion. The relative location of the candles for the best burning characteristics might impose constraints on the size and shape of the car. This will be determined experimentally.

Weight

Mass should be minimized so that the car will travel the furthest distance.

Aesthetics

Since the vehicle is not to be marketed, this parameter is not of any concern to us.

Ambient Environmental Conditions

The vehicle will operate at room temperature in air. The influence of chemicals and humidity is of no consequence, except for any candle wax on the floor from another car. This may cause the car's drive wheels to slip or the steering system to be upset.

Packaging and transportation

The car is going to be fairly delicate, so a transportation fixture is probably desirable to prevent damage.

Service Life

The vehicle must be able to withstand numerous test runs during the refinement of the concept and still be in good shape to compete in the contest at the Student Design Conference.

Noise

This design parameter has negligible importance. However, if the vehicle is noisy then this implies that energy it is being lost and is not being converted into vehicle motion. Our final design should be quiet.

Summary

In this chapter you learned how to develop a comprehensive product design specification. This important phase of the product realization process is responsible for defining the problem that must be solved. Clarity can be improved by quality function deployment procedures that translate customer requirements into appropriate company requirements at all stages of product realization.

The design specification is formulated as a list of design parameters that collectively define the product. This specification constrains the activities of the design team. Design specifications must be developed for the complete product as well as for individual parts and subassemblies. These design parameters may interact in conflicting ways that require trade-offs and compromises. The imposition of weighting factors on design parameters can be useful to represent the relative merits of different options.

Design specifications can be developed by systematic consideration of generic design parameters. The product must perform a specific function at a targeted cost. This is influenced by many design parameters including the date for delivering the first product, the quality of the product, the number of products required, and the manufacturing processes used.

Production quality can be addressed by Taguchi methods which increase the competitiveness of products by reducing their cost and increasing their quality. These methods also improve the reliability of the product, which is another design parameter. Product reliability can be influenced by many other things such as legislation, manufacturing, maintenance procedures, and the sophistication of analysis.

Environmental friendliness is a design parameter that affects other parameters. It requires careful use of natural resources and judicious selection of materials. Environmental friendliness calls for reduction of the weight of parts, the noise generated by the product, and the amount of energy consumed. It requires a long service life. Some environmental factors are affected by governmental regulations.

The weight of a product is an important design parameter in many fields because it affects performance, cost, and marketability. The size and weight of a product can influence its packaging and mode of delivery to the customer.

Operating costs depend upon several design parameters including maintenance requirements, operating conditions, and energy consumption.

Human factors and safety are related design parameters. Both are concerned with interaction between the product and people. Aesthetics is related to this interaction too because it is concerned with product characteristics that are perceived by people.

We have learned that the design specification involves diverse but sometimes related parameters which govern the solution of design problems. The next step in the product realization process is to create a number of different solutions to the problem posed by the design specification. This is accomplished in the conceptual design stage discussed in Chapter Nine. Before this stage is discussed, we shall examine in Chapter Eight the thought processes associated with creativity and the impediments to creative endeavors.

Key Concepts

Aesthetics. The branch of philosophy that provides a theory of the beautiful. In design, it is concerned with the attractiveness of a product.

Ambient environmental conditions. The environment in which a product must operate. This is typically characterized by temperature, humidity, pressure, chemicals, and air quality.

Anthropometry. The measurement of the size and proportions of the human body.

Benchmarking. The evaluation of a company's products by comparing their value, quality, etc., with competitors' products.

Delivery date. The date by which a product is to be delivered to the customer. This date influences the time and resources to be expended, the design process, and the manufacturing processes to be used in production.

Define the problem. The problem must be defined before a solution is sought. The definition of the problem is called the design specification.

Design parameter. A design requirement that must ultimately be satisfied by the product.

Design parameter importance factor. A numerical weighting assigned to a design parameter in order to quantify its relative importance.

Design specification. A comprehensive statement of the requirements that must be satisfied by the product. This statement formally documents the design requirements but does not suggest methods of satisfying these requirements. Its generation requires a systematic analysis of the situation.

Energy consumption. The demand for energy-efficient products is motivated by the cost of energy and from environmental concerns. This design parameter requires the isolation of all sources of energy consumption, the isolation of all sources of energy losses, and the increase of product efficiency.

Environmental friendliness. Today's values call for protection of ecosystems and preservation of natural resources. The result is the green design philosophy of "reduce, reuse, and recycle."

Ergonomics. Those aspects of design focused upon the interface between a product and the operator. They involve anthropometric, psychological, and physiological considerations.

Function. What a product must do. There is typically a performance specification associated with this attribute.

Governmental regulations. Products must comply with laws.

Hierarchy of design specifications. Complex products are developed using design specifications for the complete product, the major assemblies, the subassemblies, and the individual parts

Human factors. This design parameter is concerned with the an interface between a product and the human operator through consideration of the inherent physical and mental limitations, capabilities, and characteristics of people.

Maintenance protocols. These are the tasks which must be performed to ensure that a product continues to comply with the design specification.

Noise. Loud, discordant, or disagreeable sounds are increasingly regarded as undesirable characteristics. If the product must operate in the neighborhood of people, then noise is subject to federal regulations.

Operating costs. The cost of operating a product within the performance specification. This cost is influenced by energy consumption, maintenance, reliability, and service life.

Operating instructions. Documentation provided to the customer on operating procedures, fault recognition, and fault rectification.

Packaging and transportation. These design parameters are concerned with ensuring that the customer receives the product in the same condition in which it passed the final inspection of the quality control process.

Product cost. The total cost of the product. Typical components include the cost of materials, manufacture, transportation, packaging, and financial overhead.

Quality. The characteristics of a product or service that bear on its ability to satisfy stated or implied needs.

Quality function deployment. A systematic process for translating customer requirements into appropriate technical requirements during all stages of the product realization process. Product attributes desired by the customer are identified before benchmarking them relative to the competition and before developing attributes that offer a competitive advantage.

Quality loss function. A measure of the cost of any variation from a target value for any product characteristic.

Quantity. The number of products requested by the customers dictates the type and cost of the manufacturing process.

Reliability. The probability that a product will perform its function while being subjected to prescribed operating conditions. Product reliability is affected by legislation, the analytical tools employed in the design process, manufacturing processes, prototype testing, and maintenance protocols.

Safety. This design parameter is primarily concerned with the health of human beings in the neighborhood of products, but it can also pertain to equipment and property.

Service life. The life of the product when operating under normal service conditions.

Size. The spatial dimensions of a product. The cost of materials mandates that parts and products should be as small as possible.

Taguchi methods. A variety of methods that determine the required features of a design to render it robust against uncertainties, variations, and disturbances in the governing parameters. The objective is to deliver high quality products at low cost.

Weight. The weight of parts and products should generally be minimized since additional weight can increase costs. These costs include manufacturing, transportation, and energy consumption.

Review Questions

1. Define quality function deployment. What are the advantages of quality function deployment?
2. What is a design specification? Why is it important in the product realization process?
3. What are the primary ingredients of a design specification, and how do they influence the creation of a product?
4. List four major ingredients of quality function deployment in the product realization process. Characterize each of them.
5. Describe the stages in the generic quality function deployment philosophy.
6. Describe several different approaches for translating customer requests into a specification for a product.
7. Discuss the weighting of design parameters.

8. List some generic design parameters that can be used to develop a product design specification.
9. Discuss the role of function and performance on the conceptualization of a product.
10. What parameters dictate the cost of a product?
11. Discuss the development of cost estimation models.
12. How does the product delivery date affect the product realization process?
13. How does the size of the customer order, or the magnitude of monthly sales, influence the product realization process?
14. What is environmental friendliness and how does it influence the design process?
15. Discuss product hazards and three approaches for their elimination.
16. Define quality and describe how this design parameter affects design and manufacturing activities. What is the quality lever analogy?
17. What is the primary objective of Taguchi methods? How are quality loss functions involved in these methods?
18. Why are products with parameter characteristics in the neighborhood of target values of greater utility in practice than those with parameter characteristics spread across the complete tolerance band?
19. Describe the three primary stages of the Taguchi design philosophy. What is the motivation for conducting parameter and quality function experiments?
20. Discuss methods that reduce energy consumption.
21. What is a reliable product? Identify and discuss parameters that influence reliability.
22. How does maintenance affect the design of a product? Discuss three methods of accomplishing product maintenance. List ten rules for completing maintenance tasks.
23. Discuss how size affects the creation of a product. Why is size important in the automobile industry?
24. Why is weight an important design element in the transportation industry?

25. Define the aesthetic attributes of a product. How do these attributes influence the success of a product?
26. Discuss the influence of ambient environmental conditions on product design.
27. How is product design influenced by the need to deliver a product to the customer in an undamaged state?
28. Define human factors engineering. What role does anthropometry play in the field of ergonomics?
29. Define the service life of a product and explain how it is influenced by the philosophy of green design.
30. Discuss noise in the creation of products. List some ways to control noise.
31. What role do government regulations play in the design of products?
32. Define product operating costs and list some factors that contribute to them.

Problems

1. Students frequently look forward to graduation and the thought of buying a new car. The market place is populated by a diverse range of products from sports cars to four family station wagons. These products are all based upon a set of design specifications and are potential solutions to your search for a new vehicle. Develop a specification for the type of vehicle to wish to purchase upon graduation and compare the attributes of four potential candidates.
2. The toaster is a common domestic appliance. Develop a comprehensive design specification for a toaster that can simultaneously heat four thick slices of bread, bagels, or waffles. Assign weighting factors to the design parameters.
3. Develop a specification for a personal computer that you would like to purchase. Compare the attributes of four possible systems that satisfy your specification and select the best.

4. Develop a design specification for a new wheelbarrow or garden cart. The product must be able to carry 50 kg up a 15 percent incline. The fully loaded wheelbarrow must not tip over on a similar slope. It must be able to easily negotiate a 60 mm step. The wheelbarrow must not degrade seriously when stored outside where it is exposed to the elements, and it must operate over soft ground without damaging a lawn. Since a wheelbarrow or garden cart is a common seasonal product, assume that it will be manufactured annually in large batches.
5. Assume that a customer asks you to supply a pump. It is to be located in a vertical shaft and must raise 100 liters per second of liquid through a height of 100 meters. This is all the information that you are given! Clearly you will need answers to many other questions before you will be able to design a pump or to supply one off-the-shelf. Develop a list of questions that you will need to have answered before you can proceed.
6. Develop design specifications for items found in a home kitchen such as bottle openers or vegetable peelers.
7. Write a comprehensive design specification for a collapsible umbrella that will fit inside a coat pocket.

Further Reading

Akao, Y., ed. *Quality Function Deployment.* Cambridge, MA: Productivity Press, 1990.

Burgess, J. H. *Designing for Humans: The Human Factor in Engineering.* Princeton, NJ: Petrocelli Books, 1986.

Chow, W., *Cost Reduction in Product Design.* New York: Van Nostrand Reinhold, 1978.

Cohen, L., *Quality Function Deployment.* New York: Addison-Wesley, 1995.

Ealey, L. A. *Quality by Design: Taguchi Methods and U.S. Industry*. Dearborn, MI: ASI Press, 1988.

Hollins, W. and S. Pugh. *Successful Product Design*. London: Butterworths, 1990.

Hunter, T. A. *Engineering Design for Safety*. New York: McGraw-Hill, 1992.

McCormick, E. J., and M. S. Sanders. *Human Factors in Engineering and Design*. New York: McGraw-Hill, 1982.

Michaels, J. V. and W. P. Wood. *Design to Cost*. New York: Wiley, 1989.

Pahl, G. and W. Beitz. (English edition edited by K. Wallace). *Engineering Design*, London: The Design Council, 1984

Ray, M. S. *Elements of Engineering Design*. Englewood Cliffs, NJ: Prentice-Hall, 1985.

Sheppard, A. *Aesthetics: An Introduction to the Philosophy of Art*. Oxford: Oxford University Press, 1987

Design Creativity

8

Imagination is more important than knowledge, for knowledge is limited.

Albert Einstein
(1879-1955)

Objectives

- To discuss the role of creativity in the design process.
- To introduce convergent and divergent thinking.
- To discuss some impediments to design creativity.

Contents

What If...

It is the 1930s and you have an idea for a new aircraft engine. Indeed this is the topic of your final year college thesis. Your engine has the potential for a higher thrust-to-weight ratio than state-of-the-art piston engines and this will permit aircraft to fly faster and higher than current fighter aircraft. You are excited about your idea but few people share your optimism. Experts in the field are skeptical. The Air Ministry thinks that the concept is impractical. How could it work? It does not have a propeller! Rather, it would be a turbojet with a turbine and a compressor.

Despite the skepticism of the aircraft engine community, you persist. You are highly motivated, you shrug off the criticisms, you overcome the disappointments, you suppress the emotional impediments, and you surmount the technical hurdles. You persevere using the classical design cycle involving convergent thinking and divergent thinking—and eventually you succeed in building the first jet engine. Tenacity prevails! This happened to Frank Whittle in England.

8.0 Introduction

In the previous chapter we examined the first task of creating a product—namely the specification of the design problem. Next comes the conceptual design stage that is dominated by the creative talents of the designer. But before going on to conceptual design itself, we pause in this chapter to consider the basic thought processes associated with creative thinking. To design something new and useful to society requires imagination, skill, and insight. It is an art that is partly a gift but about which much can be taught and much can be learned.

This chapter focuses on the creative thinking that is relevant to all aspects of the product realization process but which is absolutely crucial to the formation of a useful set of conceptual designs. The quality of these designs depends upon the quality of the creative thinking that goes into them. This observation motivates the discussion on creativity in this chapter and the methods proposed in the next chapter for improving conceptual design.

Creativity is inventing, experimenting, growing, taking risks, breaking rules, making mistakes, and having fun.

Mary Lou Cook (1918–)

We begin with a summary of the primary psychological ingredients of design creativity. You will learn about two quite different thought processes: *creative thinking* and *deductive thinking*. Next you will see some of the psychological impediments to creativity; we will suggest how to avoid them. You will realize perhaps that many creative strategies are quite general and can be useful to you in fields far from engineering design.

Creative thinking has great utility in both the arts and the sciences... indeed in all aspects of life.

8.1 Psychological Factors in Design Creativity

In both the arts and the sciences, it can be useful to think of design as a cycle of thought processes involving the creation, analysis, and evaluation of ideas (Figure 8.1). These processes evolve in an iterative manner, as was described in Chapter Four.

Design thinking requires the iteration of creation, analysis, and evaluation.

The design engineer must be able to think logically and creatively.

The design process could be described as controlled schizophrenia because of the repeated switching between creative thinking on the right side of the brain and deductive thinking on the left.

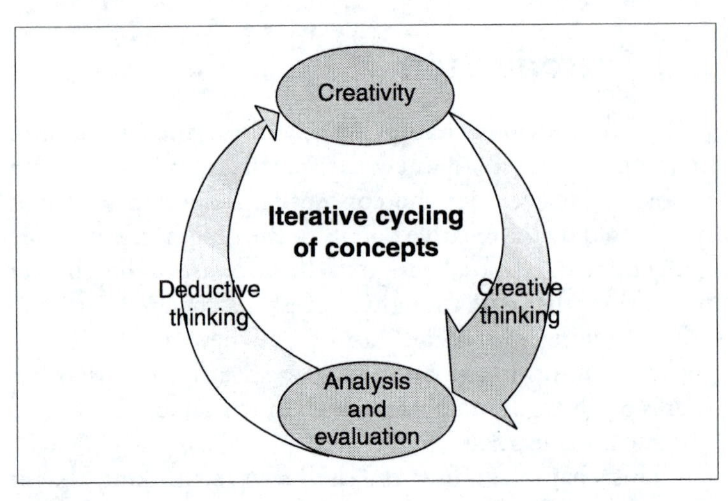

Figure 8.1

The design cycle: a method for solving all classes of open-ended problems.

Some synonyms:

Creative thinking:
- Divergent thinking
- Associative thinking
- Lateral thinking
- Artistic thinking

Deductive thinking:
- Convergent thinking
- Vertical thinking
- Realistic thinking

It is apparent from Figure 8.1 that a designer must be able to employ two quite different kinds of thought processes in order to function effectively. These processes are *creative thinking* and *deductive thinking*. If only creative thought processes are applied to a design problem, then an individual will continue to generate ideas and alternative schemes without ever evaluating them, and a good final solution may never be developed. Deductive thinking is required to analyze and evaluate the numerous concepts generated by creative thought. Both kinds of thinking are necessary to reach the best—or at least a good—final solution to a problem.

8.2 Deductive Thinking

Deductive thinking involves logic, analysis, judgment, and evaluation.

Deductive thinking occurs in the left side of the brain. Because it involves selecting concepts and making decisions, it is associated with logic, analysis, judgment, and evaluation. Thus deductive thinking is generally associated with the sciences and mathematics where there is usually only one correct solution to a problem.

In deductive thinking, the brain assembles and organizes information pertaining to a problem. An abstract model of the problem is proposed and an appropriate solution strategy is selected. The solution may be based upon a scientific theory that incorporates a number of approximating assumptions. This strategy generally produces a single solution that is evaluated and scrutinized and—depending upon the outcome—a more sophisticated model is then proposed. Important characteristics of deductive thinking are listed in Table 8.1.

Analytical, logical
Leads to only one solution
Considers only relevant information
Thinking processes employ rules
Embodies scientific principles
Only pursues the most plausible solution strategy
Thinking halts if a logical step is unattainable
Groups are sacrosanct

Table 8.1

Some characteristics of deductive thinking.

8.3 Creative Thinking

Creative thinking occurs in the right brain and is associated with randomness, synthesis, nonjudgmental scrutiny, irregularity, sensuality, imagination, and visualization. It is the antithesis of deductive thinking.

Creative thinking involves imagination, visualization, irregularity, and chaos.

These characteristics are frequently associated with individuals who study the arts rather than the sciences because they are proficient at addressing open-ended problems for which there is no single "correct" solution. They are comfortable generating numerous ideas that are potential solutions to problems presented to them. Creative thinking, in

Creative thinking involves generating different combinations of artifacts, information, and experiences.

the context of engineering design, involves the manipulation of experiences, artifacts, and information into new patterns. Such patterns are, of course, subject to numerous pragmatic constraints such as functionality and economics. This can be a significant challenge to the engineering designer. Constraints of this magnitude and severity are not imposed upon the artist. Important characteristics of creative thinking are presented in Table 8.2.

A #2 pencil and a dream can take you anywhere.

Joyce A. Myers (1948–)

Some characteristics of creative thinking
Nonjudgmental
Develops numerous ideas
Considers all information
Stochastic and chaotic
Can involve bizarre mental imagery
Explores all possible solution strategies
The idea-generation processes never cease
Groups are continually redefined

Table 8.2

Some characteristics of creative thinking.

Engineering undergraduates often have minimal experience with situations requiring creative thinking although they are extensively exposed to deductive thinking in the engineering curriculum.

Thank goodness I was never sent to school; it would have rubbed off some of the originality.

Beatrix Potter (1866–1943)

The traditional K-12 education in the United States is dominated by training in logical thinking. There is generally only one solution to a problem posed in a science class, for example. Classes which do permit the student to develop creative skills, such as painting or creative writing, are generally not emphasized, perhaps because they are difficult to grade. This situation prevails through many engineering colleges and is further exacerbated by the high societal values associated with activities requiring logic, sound reasoning, and good judgment. Thus there are several reasons why engineers are often ill-equipped to undertake the creative thinking required in effective design work in professional practice.

Creative thinking, whether it pertains to the arts or the sciences, comprises four distinct phases: *preparation, incubation, illumination,* and *verification.* The thinking process is initiated by the careful evaluation of the problem and the development of a clear specification of the task. Resources are collected and subsequently scrutinized. Challenging design problems require a series of sessions focused on the problem because a single session usually will not yield a good solution. Between these sessions, the problem is submerged in the human subconscious, while the designer undertakes unrelated activities. A period of incubation then follows while the mind explores alternative combinations of ideas. It is important that this activity be undertaken without any extraneous constraints or preconceived notions because these conditions impede the creation of workable solutions. Subsequently, the different combinations of resources arrive at a solution to the original task. This illumination phase often occurs at an unprovoked moment when the designer is involved in an unrelated activity. It is then followed by the verification phase in which functionality is proven and modifications are made to the embryonic concept.

Usual stages in creative thinking:
- preparation
- incubation
- illumination
- verification

Design problems often require a series of sessions before a potential solution is found.

A man would do well to carry a pencil in his pocket and write down the thought of the moment. Those that come unsought for are commonly the most valuable and should be secured because they seldom return.

Francis Bacon (1561–1626)

The illumination phase of creative thinking can be illustrated by the legend associated with the Greek mathematician Archimedes who lived in Syracuse on the Mediterranean island of Sicily about 200 B.C. He was assigned the task of determining whether King Hiero's crown was pure gold or, as the king suspected, whether the goldsmith had cheated by adding less expensive silver. Archimedes' solution included the assumption that a crown of pure gold would have the same volume as a piece of gold of the same weight. Furthermore, because gold has a greater density than silver, an ounce of gold would occupy a smaller volume than an ounce of silver. Consequently a crown made of pure gold would occupy less volume than one made from an alloy of gold and silver.

Thus, the task is, not so much to see what no one has yet seen; but to think what nobody has yet thought, about that which everyone sees.

Erwin Schrodinger (1887–1961)

A crown has a complex geometry and the task of measuring its volume was a challenge with no obvious solution to Archimedes. His subconscious mind wrestled with the task until a solution suddenly became apparent when Archimedes immersed himself in a tub filled to the brim with water at the public baths. The water overflowed. He deduced that he might collect and measure the water spilling over the sides of a container full of water when the crown was added. The volume of displaced water should equal the volume of the king's crown. Archimedes' excitement generated by this realization is said to have caused him to run naked from the baths into the streets of Syracuse shouting "Eureka!" The subsequent application of this approach proved that the goldsmith had indeed cheated.

Eureka means "I have found it!"

It is evident that creative thinking is quite different from deductive thinking. Furthermore this unstructured mental activity, which is a primary constituent of engineering design, is subject to a number of psychological impediments. Once these impediments are recognized, you can take steps to address them.

8.4 Impediments to Creativity

Creativity is an essential component of the engineering design process. Indeed, a thread of creativity is woven throughout the design fabric. It is important in product design where the concept is transformed into something that can be manufactured but its greatest importance is in conceptual design. Creativity is the crucial first step in design and consequently is the basis for the success of industrial organizations and ultimately the prosperity of nations.

Creative thinking is not a type of thinking that is only undertaken by a few special individuals. It can be undertaken by anyone. However, individuals with the personality characteristics presented in Table 8.3 have a higher aptitude for undertaking original creative work than other members of society.

Clearly all individuals do not possess the same ability to generate creative solutions to engineering design problems. Nevertheless, the ability of an individual to be creative can be stimulated or enhanced by some of the strategies pre-

Imaginative
Persistent, determined, tenacious
Rebellious, adventuresome, risk-seeking, aggressive
Open-minded, receptive to change
Concerned about implications, significance, and meanings but without initial concern for practicalities
Intuitively perceptive
Able to recognize similarities and associations among concepts and objects
Irrational, pursuers of an ideal
Uninhibited, self-confident

Table 8.3

Some characteristics of individuals with higher aptitude for creative thinking.

sented in the next chapter. Even with this toolkit of techniques, there are a number of impediments that may still hinder the creative efforts of a designer.

These impediments prevent individuals from correctly understanding a design problem or from creating a solution to that problem. They include cultural, environmental, emotional, intellectual, and perceptual impediments. In engineering, creativity also can be hindered by a number of technical obstacles. These six impediments are summarized in Table 8.4.

These impediments to creativity are represented by the shaded areas on the solutions plane in Figure 8.2. Each point on the plane represents a potential solution. Psychological impediments prevent the designer from recognizing solutions in the dark areas, and consequently the number and quality of solutions generated by an individual with these impediments is reduced.

The most pathetic person in the world is someone who has sight, but has no vision.

Helen Keller
(1880–1968)

An appreciation of psychological impediments provides the basis for overcoming them.

Impediments to engineering creativity include:

- Cultural
- Environmental
- Emotional
- Intellectual
- Perceptual
- Technical

Cultural
Suppression of an enquiring mind
Adherence to traditions

Environmental
Inappropriate physical surroundings
Associates who are critical of ideas

Emotional
Reluctant to take calculated risks
Reluctant to let ideas incubate
Quick to criticize ideas
Unmotivated

Intellectual
Inappropriate solution strategy
Limited educational background
Poor communication skills
Incorrect information

Perceptual
Inability to use all human senses
Wrong problem definition
Stereotypical thinking
Failure to study a problem from several perspectives

Technical
Inaccessible computational tools
Immature or outmoded technologies
Immature or outmoded manufacturing processes

Table 8.4

Impediments to engineering creativity.

Cultural Impediments

Cultural impediments to creative problem solving are imposed upon an individual by societal values. Indeed, they vary from one country to another. Some examples follow.

The Suppression of an Enquiring Attitude

Cultural inpediments:

- Suppression of an enquiring mind
- Adherence to traditions

The hallmark of most creative individuals is an inclination to ask questions such as "how?" and "why?" However, a question is often assumed to be an admission of ignorance in western societies where it is desirable to be perceived as knowledgeable and well educated. Therefore the enquiring mind—possessed by most small children—is gradually suppressed as they mature. Individuals who manage to retain an

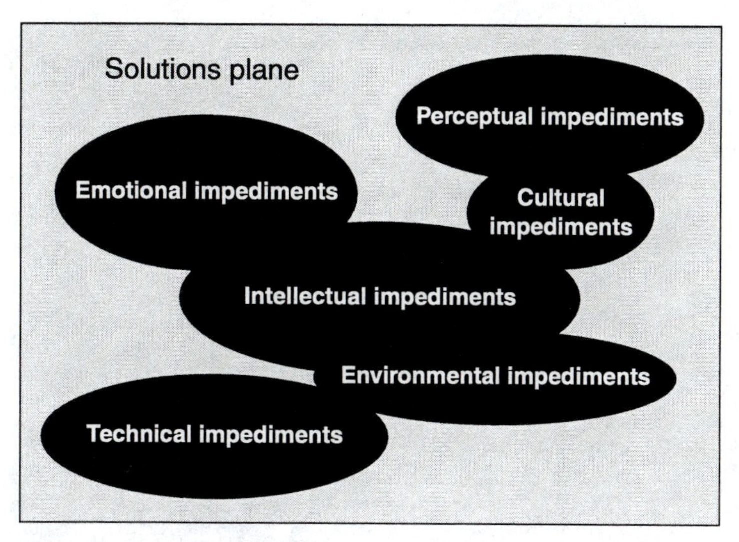

Figure 8.2

The solutions plane and domains of psychological impediments.

enquiring mind are nevertheless still subject to cynical and critical comments from coworkers. Some comments are recorded in Table 8.5

Traditions

Frequently, individuals and organizations are reluctant to change or to take risks. Many industries consequently treat new ideas with skepticism, and traditional approaches are the norm. Novices or individuals from unrelated disciplines sometimes provide new insight and creative solutions to problems. These individuals are not constrained to use the traditional approaches of recognized experts in the field.

The development of the turbojet aircraft engine by Frank Whittle (1907-1996) shortly before World War II is an example. Whittle's senior thesis when he was a twenty-one year old Flight Cadet at Britain's Royal Air Force College Cranwell was entitled "Future Developments in Aircraft Design." Figure 8.3 reproduces a page of his thesis. He wrote:

Accurately predicting the future is not easy. History is full of opinions and forecasts that turned out to be quite wrong. Here are some:

Everything that can be invented has been invented.
Charles H. Duell
Commissioner, US Office of Patents
(1899)

On computers:

I think there is a world market for maybe five computers.
Thomas Watson
Chairman IBM
(1943)

There is no reason for any individuals to have a computer in their home.
Ken Olsen
Founder, Digital Equipment Corp.
(1977)

On airplanes:

Heavier-than-air flying machines are impossible.
Lord Kelvin, President, Royal Society (1895)

Airplanes are interesting toys but of no military value.
Marshal Foch, War Academy, (World War I)

On communication:

Who the hell wants to hear actors talk?
Harry M. Warner, Warner Bros., (1927)

The wireless music box has no imaginable commercial value. Who would pay for a message sent to nobody in particular?
David Sarnoff
(1920s)

Table 8.5

Predictions of the future.

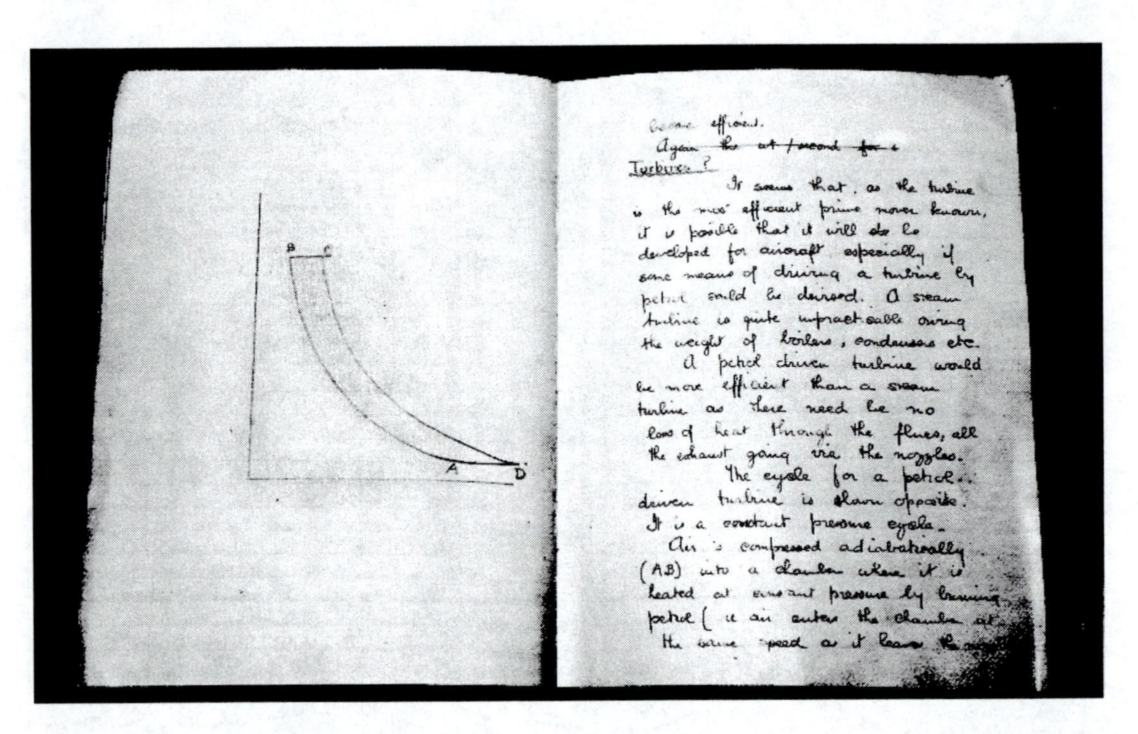

become efficient.
~~Again the air / second for a~~
Turbines?
It seems that, as the turbine
is the most efficient prime mover known,
it is possible that it will be
developed for aircraft, especially if
some means of driving a turbine by
petrol could be devised. A steam
turbine is quite impracticable owing
the weight of boilers, condensers etc.
A petrol driven turbine would
be more efficient than a steam
turbine as there need be no
loss of heat through the flues, all
the exhaust going via the nozzles.
The cycle for a petrol
driven turbine is shown opposite.
It is a constant pressure cycle.
Air is compressed adiabatically
(AB) into a chamber where it is
heated at constant pressure by burning
petrol (i.e. air enters the chamber at
the same speed as it leaves the

Figure 8.3

A page from Cadet Whittle's original essay showing a pressure-volume diagram and the beginning of a section on turbines. It reads, "…It seems that, as the turbine is the most efficient prime mover known, it is possible that it will be developed for aircraft, especially if some means of driving a turbine by petrol could be devised. A steam turbine is quite impracticable owing [to] the weight of boilers, condensers, etc.

"A petrol driven turbine would be more efficient than a steam turbine as there need be no loss of heat through the flues, all the exhaust going via the nozzles.

"The cycle for a petrol driven turbine is shown opposite. It is a constant pressure cycle.

"Air is compressed adiabatically (AB) into a chamber where it is heated at constant pressure by burning petrol (i.e., air enters the chamber at the same speed as it leaves the nozzles)" (Courtesy of Airlife Publishing Ltd., Shrewsbury, England)

I came to the general conclusion that if very high speeds were to be combined with very long range, it would be necessary to fly at very great height where the low [air] density would greatly reduce resistance in proportion to speed.… It seemed to me unlikely that the conventional piston engine and propeller combination would meet the power plant needs of the kind of high speed/high altitude aircraft I had in mind.…

Having recognized the need for a new generation of engines, Whittle employed associative thinking by reviewing all known power plants (Figure 8.3) before he developed the turbojet. He was at the conceptual design phase of the product realization process.

The jet engine revolutionized the aircraft business worldwide. The dawn of a new era of transportation had arrived. As Victor Hugo (1802-1885) stated, "No army can withstand the strength of an idea whose time has come." Clearly, the time of the jet engine had come, as evidenced by the rapid development of jet planes in the postwar years. An appreciation of this revolution can be gleaned from Figures 8.4, 8.5, and 8.6. Figures 8.4 and 8.5 show a state-of-the-art 1940s piston engine and an early 1940s derivative of the original Whittle turbojet engine. Figure 8.5 illustrates the

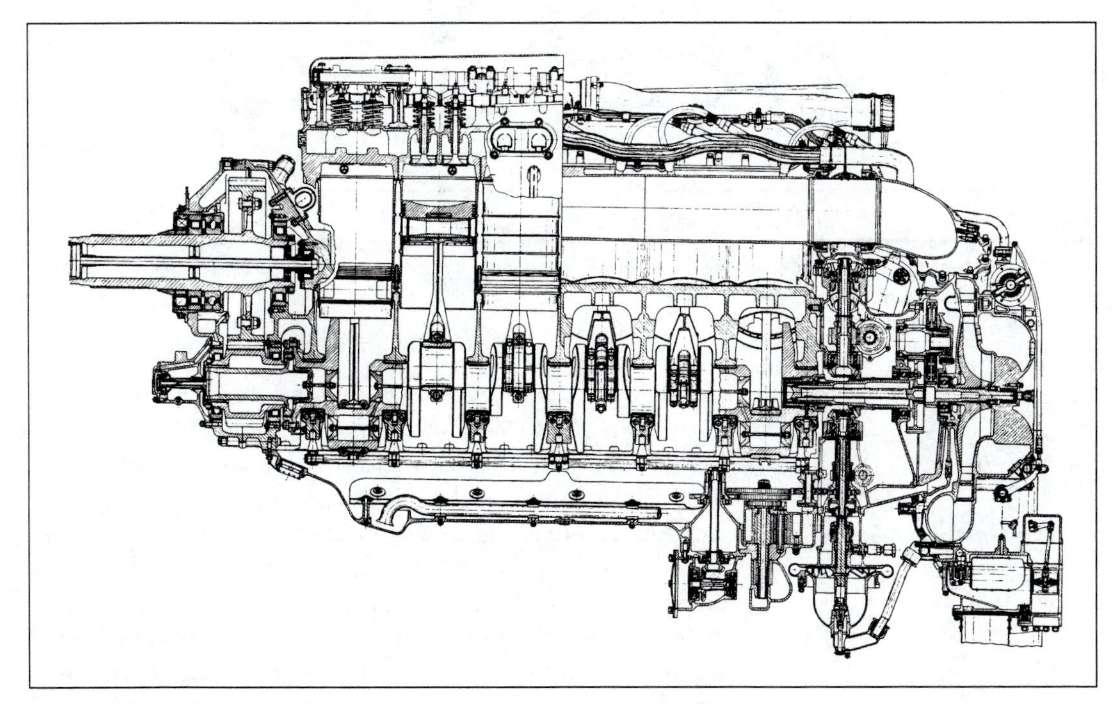

Figure 8.4

A 1940s state-of-the-art piston engine: a Rolls Royce Merlin II series. (Courtesy of Rolls Royce plc, Derby, England)

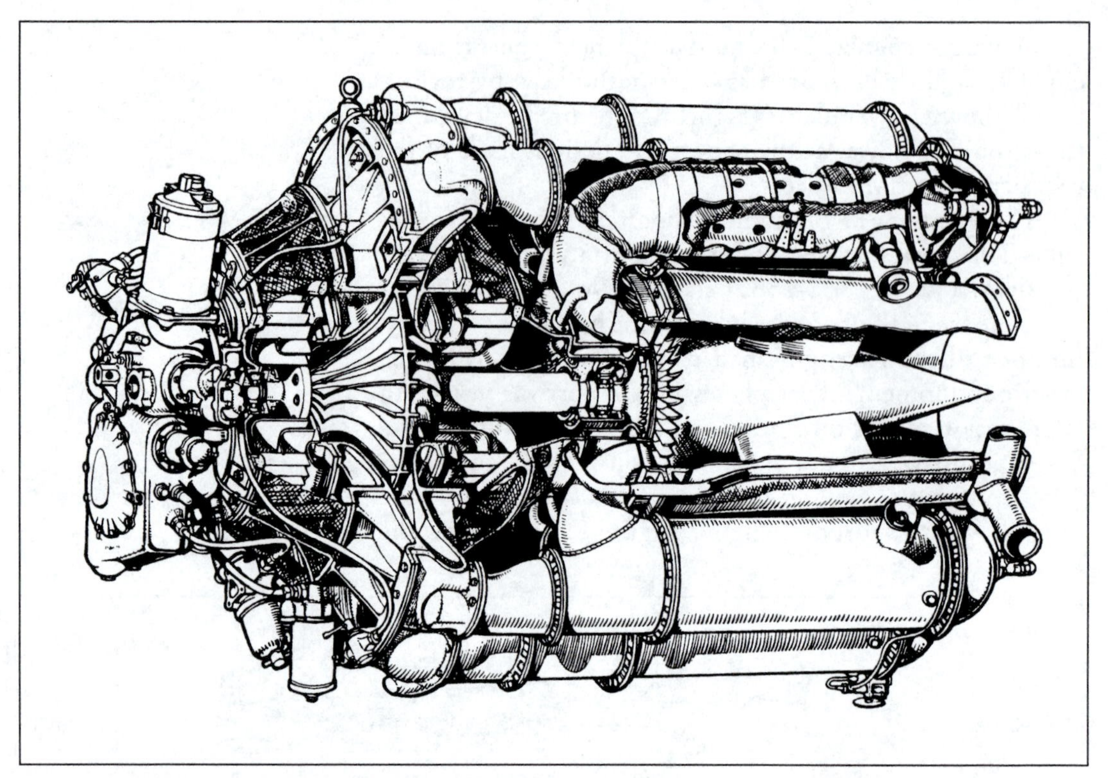

Figure 8.5

A 1940s state-of-the-art turbojet engine based on Whittle's original design: a Rolls Royce W2/B23 Welland. (Courtesy of Rolls Royce plc, Derby, England)

simplicity of the new power plants and their dramatic difference from piston engines. Contrast the radial compressor design of Figure 8.5 with the state-of-the-art Rolls Royce Trent 800 engine with an axial compressor shown in Figure 8.6.

The turbojet engine was developed by a group of engineers whose experience was either not connected with the aircraft industry or whose background was in the design of airframes rather than aircraft engines. Furthermore, their visionary thinking received a lukewarm reception and disdain from the acknowledged experts in the design of aircraft engines, who at that time were focused on the development of turbochargers for propeller-driven aircraft (Figure 8.7).

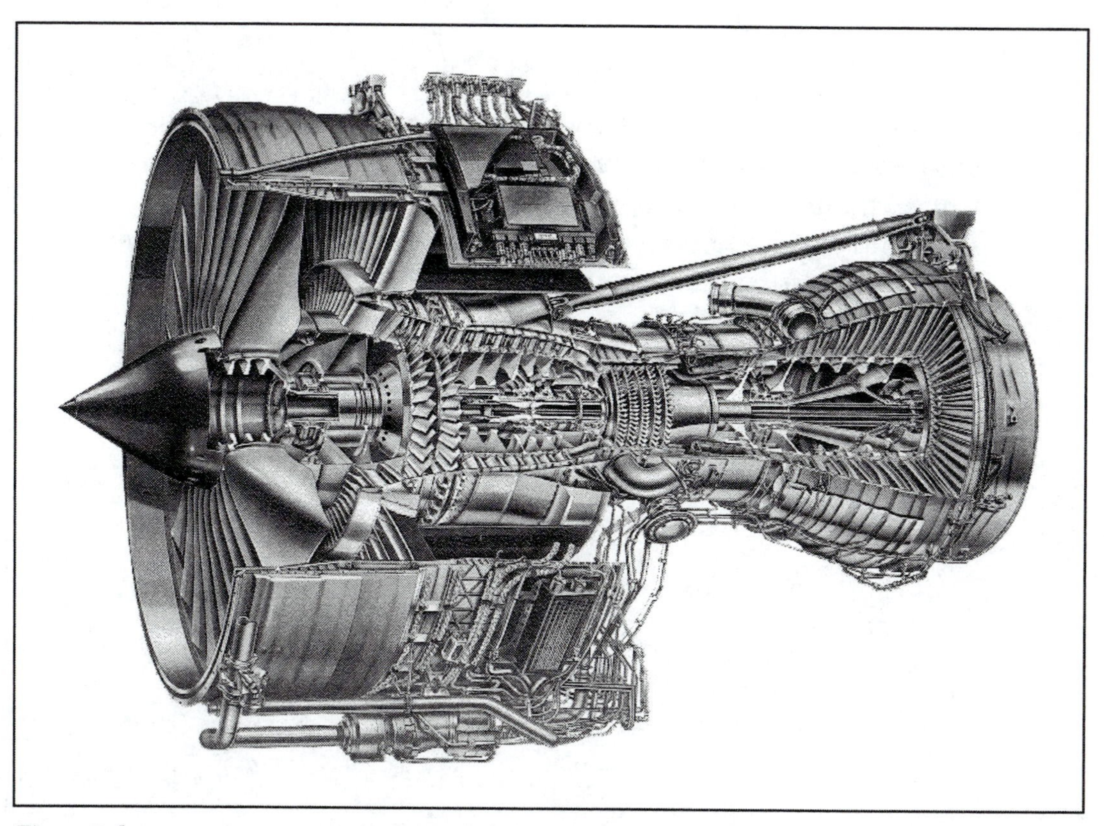

Figure 8.6

A 1990s state-of-the-art turbofan engine: a Rolls Royce Trent 800 for the Boeing 777. (Courtesy of Rolls Royce plc, Derby, England)

However, these new aircraft engines ultimately possessed much higher power-to-weight ratios than traditional aircraft piston engines and, because of this, the performance of jet-propelled aircraft was superior to propeller-driven designs.

Environmental Impediments

The environment in which creative solutions to design problems are sought can greatly impede the problem-solving process. These impediments include the physical working conditions as well as the working relationships with colleagues.

Environmental impediments:
- working conditions
- criticism

"IN ITS PRESENT STATE, AND EVEN CONSIDERING THE IMPROVEMENTS POSSIBLE WHEN ADOPTING THE HIGHER TEMPERATURES PROPOSED FOR THE IMMEDIATE FUTURE, THE GAS TURBINE ENGINE COULD HARDLY BE CONSIDERED A FEASIBLE APPLICATION TO AIRPLANES MAINLY BECAUSE OF THE DIFFICULTY IN COMPLYING WITH THE STRINGENT WEIGHT REQUIREMENTS IMPOSED BY AERONAUTICS.

"THE PRESENT INTERNAL COMBUSTION ENGINE EQUIPMENT USED IN AIRPLANES WEIGHS ABOUT 1.1 POUNDS PER HORSEPOWER, AND TO APPROACH SUCH A FIGURE WITH A GAS TURBINE SEEMS BEYOND THE REALM OF POSSIBILITY WITH EXISTING MATERIALS."

THE COMMITTEE ON GAS TURBINES
appointed by
THE NATIONAL ACADEMY OF SCIENCES
June 10, 1940

Good thing I was too stupid to know this
Frank Whittle

Figure 8.7

Tradition-bound authorities believed Whittle's turbojet aircraft engine was "beyond the realm of possibility." Having successfully started the first turbojet engine three years earlier, Whittle wrote, "Good thing I was too stupid to know this." (Courtesy of Airlife Publishing Ltd., Shrewsbury, England)

Physical Surroundings

Where and when one works does matter. Does the telephone ring too often? Is it too hot or too cold? Does one person work best in quiet while another prefers background music? Each individual should determine his or her most creative conditions and most productive time of day. The ideal is not always possible but it is sometimes surprising how much of the environment can be altered by self-knowledge and effort.

Criticism

During the conceptualization phase of the design process, a nurturing environment is essential for the maturing of embryonic concepts. The environment should be characterized by support and trust. Unfortunately this sometimes is not the case in engineering practice because of the presence of left-brain-dominant individuals such as autocratic managers and critical colleagues. Such an atmosphere is unhealthy for the conceptualization process because most people do not appreciate unsympathetic criticism—especially of tentative new ideas. Premature criticism will often curtail any notion of creativity.

A new idea is delicate. It can be killed by a sneer or a yawn; it can be stabbed to death by a quip and worried to death by a frown.

Brower

Emotional Impediments

Emotional impediments are concerned with the psychological safety of the individual. They involve the desire for security, the fear of making a mistake, and the patience to permit embryonic ideas to mature.

Emotional impediments:

- security
- risk-taking
- incubation
- motivation

Much experimental data exhibits a Gaussian distribution that can be represented by a bell-shaped curve. The quality of conceptual designs fits this pattern, as shown in Figure 8.8. Many conventional ideas of average quality straddle the apex of the curve. A few excellent ones appear at the extreme right-hand portion of the distribution while a few inferior ideas appear at the extreme left. Remember, however, that the merits of a design solution may only be clear much later in the design process.

When is an apparently bad idea for a product really a very good idea? Conceptual designs are embryonic ideas whose full potential is not immediately obvious to the designer. Therefore it is difficult to judge what is good and what is bad. What *can* be asserted is that innovative designs are not found among the conventional designs near the apex of the curve in Figure 8.8. They occur at the extremes of the Gaussian distribution where ideas are either extraordinarily good or very poor. The difficulty in this situation is the risk of mediocre success with conventional ideas in the middle of the bell curve compared with the risk of catastrophic failure of unconventional ideas.

Genius, in truth, means little more than the faculty of perceiving in an unhabitual way.

William James
(1842–1910)

Unconventional ideas lie at the extremes of the bell-shaped curve.

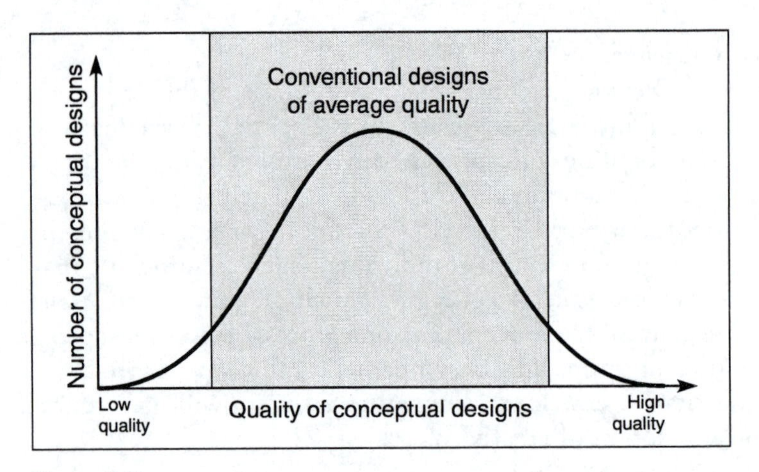

Figure 8.8

The quality distribution of conceptual designs.

A vivid representation of the potential for creating innovative products can be found in an analogy to a fried egg. In the yolk are many ordinary ideas, but the very good ones and very poor ones both lie out near the edge of the egg white (Figure 8.9).

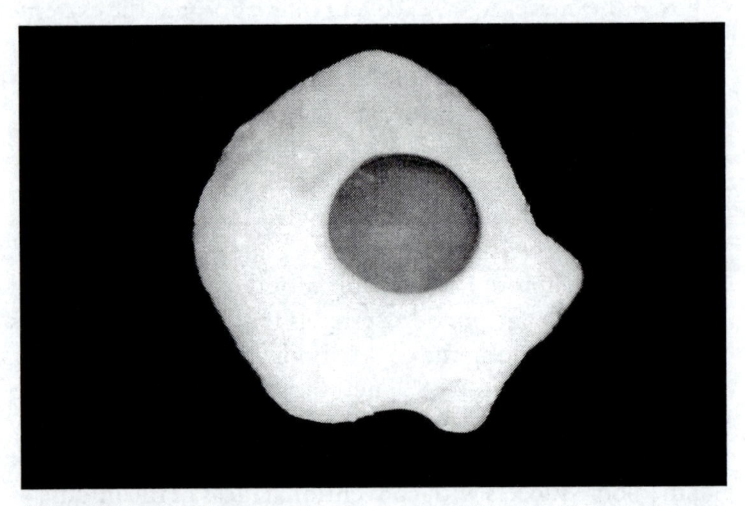

Figure 8.9

The fried egg analogy: prospecting for solutions to complex design problems.

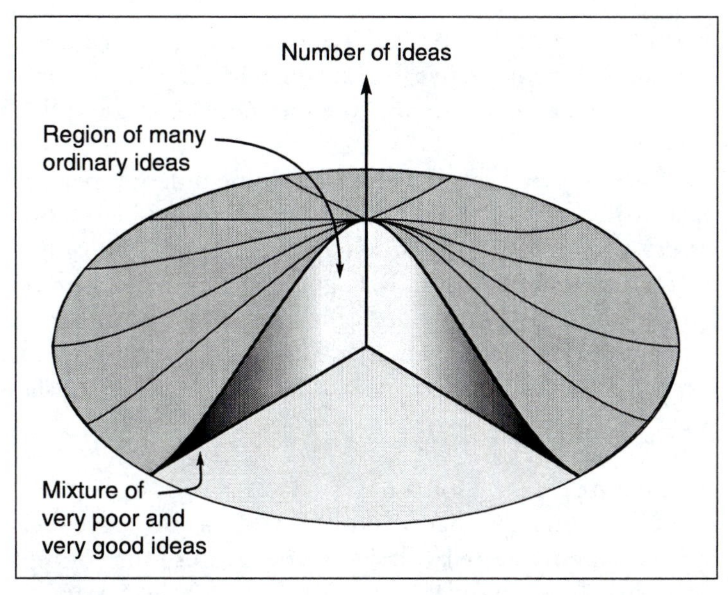

Prospect for innovative solutions in the white of the egg.

Figure 8.10

The relationship between quantity and potential quality of conceptual designs. Compare with the fried egg analogy: many ordinary ideas are in the yolk but the very good and very poor ones lie out near the edge in the albumen.

In a conventional three-dimensional graph, the origin of the coordinate system should be established at the maximum of the bell shaped curve—the center of the yolk—as shown in Figure 8.10. The designer should strive to develop conceptual designs at the extremes of the radius vector where the potential for creating quality designs is highest. After a period of incubation, the excellent ideas must be sorted from the poor ones since this annular region contains both extremes.

Risk-taking

The fear of taking a risk, of making a mistake by proposing a new idea which is ultimately proven to be inappropriate, is a major impediment to individuals contemplating a creative endeavor. This fear compels many to adhere to safe strategies because society rewards a conven-

And the trouble is, if you don't risk anything, you risk even more.

Erica Jong
(1942–)

tional answer while it punishes the individual who generates a potentially innovative but ultimately unworkable solution. In these circumstances the individual appears to be a failure.

The kernel of this psychological impediment is personal insecurity. Under these conditions an individual should assess realistically the negative consequences of proposing an unconventional solution. Risk goes hand-in-hand with scientific and engineering innovation. Even if the unconventional solution is accepted, its creator may face considerable pressure because anything new usually threatens established ways.

Our doubts are traitors, and make us lose the good we oft might win, by fearing to attempt.
William Shakespeare (1564–1616)

Between two evils, I always pick the one I haven't tried before.
Mae West

Incubation Takes Time

Unfortunately many innovative concepts have been rejected prematurely by design teams because the concepts have not been permitted to mature. Whenever possible, allow an incubation period to occur before evaluation. The formulation of a creative solution to a challenging design problem requires time for the mind to incubate, both consciously and subconsciously, the embryonic ideas.

Genius is one percent inspiration and ninety-nine per cent perspiration.
Thomas Edison (1847–1931)

Complex design problems usually have complex solutions that do not materialize suddenly. Instead, a complex idea evolves through a series of elementary ideas or more basic components of the solution. This takes time. Studies of creative problem solving strategies suggest that innovative solutions usually emerge through a series of small ideas rather than as a single stroke of inspired genius. Creative solutions are the result of tenacious, methodical work rather than sudden inspiration.

The one quality which sets one man apart from another—the key which lifts one to every aspiration while others are caught up in the mire of mediocrity—is not talent, formal education nor brightness—it is self-discipline. With self-discipline, all things are possible. Without it, even the simplest goal can seem like the impossible dream.
President Theodore Roosevelt (1858–1919)

However, in most industrial settings, the solution to a design problem is required today or even yesterday. Never tomorrow! Under such pressure, the first solution is frequently accepted without waiting for a better one. This behavior should be avoided.

Design involves a repeating cycle of creativity, evaluation, and analysis (Figure 8.1). With the predominance of left-brained individuals in engineering organizations and their critical view of new concepts, there is a tendency to evaluate ideas too early in the design process. Many ideas that could evolve into innovative products are prematurely rejected. Evaluation should be delayed because immature ideas often lead to other ideas, better ideas, and ultimately better products.

The creative man must have in himself the incentive and the self-discipline to do the thing that he sees needs to be done.

Arthur H. Compton (1892–1962)

Motivation

Motivation is an important component of the search for creative solutions. Some individuals are intrinsically motivated to solve challenging problems and will relentlessly seek a solution in order to gain personal satisfaction. Such persons often develop the most creative design solutions. However, in the real world, engineering organizations often assign work to individuals without considering whether the work is interesting to the worker. An engineer confronted with a design task may therefore need some external motivation or reward to develop creative solutions. The design team also needs to understand the criteria to be used to evaluate their work.

What moves men, or rather, what inspires their work, is not new ideas, but their obsession with the idea that what has already been said is still not enough.

Eugene Delacroix (1798–1863)

Many people with great accomplishments have demonstrated their ability to overcome failures. Consider President Lincoln:

1832 Lost his job as a clerk
1832 Defeated when he ran for the state legislature
1833 His business failed
1834 Succeeded in being elected to the state legislature
1838 Defeated in his attempt to become the speaker
1843 Defeated in his attempt to be nominated for Congress
1844 Defeated again in a run for Congress
1846 Finally elected to the United States Congress
1849 Failed to become a land officer
1854 Defeated in a race for the United States Senate
1856 Failed to be nominated for Vice-president
1858 Again defeated in his attempt to join the Senate
1860 Elected President

How often, even before we began, have we declared a task impossible? And how often have we construed a picture of ourselves as being inadequate? A great deal depends upon the thought patterns we choose and on the persistence with which we affirm them.

Piero Ferucci

It is evident that while Lincoln had successes, he also had his share of failures. Motivation to succeed is a desirable personal asset in engineering design too.

Only the wise possess ideas; the greater part of mankind are possessed by them.

Samuel Taylor Coleridge (1772–1834)

Intellectual Impediments

Intellectual impediments:

- limited education
- wrong strategy
- poor communication
- incorrect information

Intellectual impediments to creative problem-solving are numerous. They include the choice of an inappropriate strategy, limited skills of the individual, an inability to communicate ideas effectively, and the use of incorrect information in the solution process.

Inappropriate Solution Strategy

A design problem can be solved using several different solution strategies; it is important to select the right one. For example, the selection of a mathematical strategy or a picto-

Relationship:
Compare, associate, organize, classify, combine, separate, systemize, generalize, randomize, adapt, select, substitute, cycle, plan

Communication:
Relate, verbalize, exaggerate, understate, list

Process:
Change, vary, repeat, copy, define, force, release, purge

Conjecture:
Hypothesize, guess, predict, imagine

Transformation:
Reduce, expand, build up, translate, interpret, simulate

Illustration:
Symbolize, visualize, exemplify

Evaluation:
Check, test, question, assume, eliminate

Display:
Chart, diagram, display, manipulate

Storage:
Memorize, commit, record, search, recall, retrieve

Focus:
Leap in, hold back, work forward, work backward, defer

Psychological:
Relax, dream, play, incubate, focus

Table 8.6

Strategies for solving problems creatively.

rial strategy depends on the individual's knowledge and experience. Interaction Associates, of San Francisco, market a *Strategy Notebook* that lists words which suggest problem-solving strategies. Table 8.6 lists some of those words; they may trigger useful problem-solving ideas. The designer may be helped by developing experience using these strategies in conceptual design activities.

Limited Educational Background

Designers who are prolific generators of creative designs are likely to be generalists with broad educational backgrounds because design problems are not constrained by artificial boundaries. A cost-effective solution to a design problem may require the combined talents of both a mechanical engineer and an electronics engineer. A solution by either individual working alone may be totally inappropriate for the marketplace. Thus a mechanical engineer who is uncomfortable with electronics will typically generate a mechanical solution to the problem when perhaps an electronic solution is more elegant. The advantages of being able to think and reason in many different disciplines is obvious.

Education is what remains after the information that has been taught has been forgotten.

Benjamin Franklin (1706–1790)

Creative minds have always been known to survive any kind of bad training.

Anna Freud (1895–1983)

Creativity involves the association of objects and knowledge that initially appear unrelated. The ability to juxtapose that information is facilitated by access to many kinds of literature through personal interactions, public libraries, and through the Internet. Creative thinking requires high quality information. Creative individuals generally expose themselves to a rich environment, and engineering designers should, therefore, read widely in academic journals and other literature and should cultivate diverse professional contacts.

Assemble information from

- literature
- personal interaction
- libraries
- computer networks

A talented, creative individual often can independently produce an appropriate solution to a small design problem. However, when an individual is confronted with a large, complex problem, then the desired design quality may be unattainable because it is beyond one individual's intellectual and creative abilities. Under these conditions, a design team is called for. Colleagues can provide emotional support

Colleagues can enhance productivity by providing

- knowledge
- skills
- emotional support
- motivation

and motivation as well as diverse skills and knowledge; they thereby enhance creativity and productivity.

Communication

Language is not only the vehicle of thought, it is a great and efficient instrument in thinking.

Sir Humphry Davy (1778–1829)

Communication involves the transfer of information. This transfer might be done by pencil sketches on a sheet of paper, by telephone, or by more abstract approaches such as mathematics. Communication is valuable partly because the process of describing a design problem to another individual requires the designer to first clarify his or her idea. A presentation can be oral or written. In either case the inability to write or speak effectively will hinder interaction between an audience or readership and the designer.

Communication often is a first step toward a solution. If a solution is the focus of attention, then communication can result in suggestions for improvements. Through putting ideas on paper the inherent weaknesses of the concept become evident.

Some of the earliest images are cave paintings from the Mesolithic Period (Figure 2.3).

Imagery is one of the oldest forms of knowledge dissemination. An image of a concept—the basis of a design—can trigger a myriad of words. Hence the adage that "a picture is worth a thousand words." Images are important for creative engineering design because all products have three-dimensional form, unlike the solutions to some problems in the pure sciences.

During the creative process, the right brain iteratively reconfigures images.

Designers can improve their ability to think visually by learning to use computer drawing programs and by learning some principles of freehand sketching. There are a number of advantages to sketching. It requires the individual to focus on the important aspects of the problem. It leads to appropriate names for parts, and results in appropriate views that emphasize the working principle of the product. Furthermore, pencil and paper provide an extension of short-term memory and permit more information to be processed effectively, especially the three-dimensional aspects of a problem.

Incorrect Information

Correct information is essential to the formulation of a design specification. Solutions to design problems involve associative thinking in which images in the brain are combined and reconfigured iteratively during the process of creativity. If these images are derived from incorrect information, then the resulting "solutions" will be incorrect and the time and finances invested in conceptualization will be wasted. The advantages of interacting with the customer through the application of quality function deployment techniques are clear.

Perceptual Impediments

Perceptual impediments to problem solving prevent individuals from correctly defining the problem or from recognizing the information needed to solve it. We shall examine several.

Perceptual impediments:

- all senses not used
- incorrect interpretation of problem
- stereotypical thinking
- single perspective

Activation of All Human Senses

While we obtain most information visually, all human senses can be used to collect information pertaining to a problem. Whenever appropriate, sound, smell, taste, and touch can help the designer generate more diverse concepts. This is especially important when the product impinges directly on human physiology. For instance, the designer of a noise-control treatment for an automobile should ride a car around a test track to appreciate the noise level.

Problem Definition

The definition of a problem—any problem—defines the purpose of the creative process. If the problem is incorrectly defined, then its subsequent solution will also be incorrect. That is expensive. In an industrial environment design problems frequently emerge from an apparently chaotic environment complicated by pressure to solve the problem quickly. Consequently minimal effort may be spent formulating the problem correctly. The preparatory phase of a challenging design task mandates careful evaluation of the problem and development of clear goals.

Avoid subconsciously imposing unreal constraints on the solution to the problem. Meticulously and systematically scrutinize the problem.

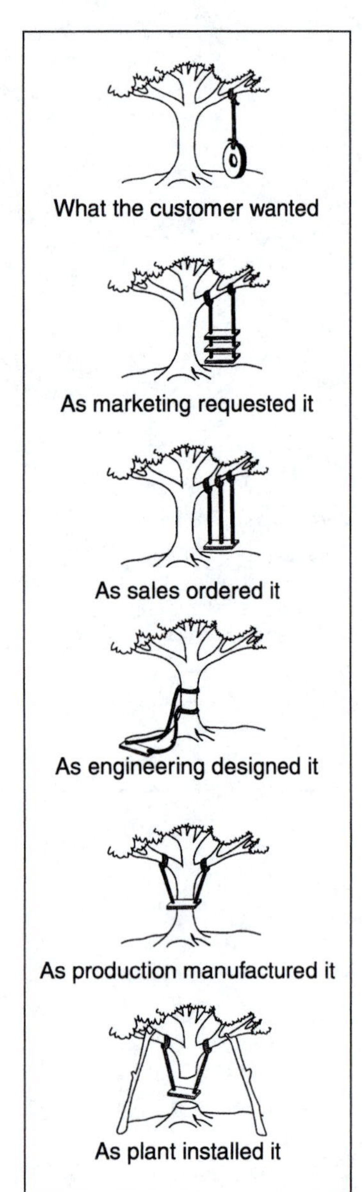

Figure 8.11

Multiple perspectives of a single problem definition. (Reprinted with permission of the Society of Manufacturing Engineers)

Individuals frequently subconsciously impose artificial limits on problems. Generally the less-constrained the problem, the greater the freedom for creativity and the better the ideas generated.

This situation is illustrated by the design of early locomotives. In 1812 John Blenkinsop invented the first practical steam-driven locomotive having two cylinders and a rack and pinion propulsion mechanism. A rack and pinion was used because Blenkinsop erroneously assumed that the locomotive would not have enough traction when its smooth wheels interacted with smooth wooden rails. In 1813 George Stephenson visited the coal mine where Blankinsop's engine was operating and recognized that a rack and pinion was unnecessary. His subsequent activities were responsible for the birth of the railroad.

Stereotypical Ideas

According to psychological theories, the brain classifies and stores information in labelled groups. When new information is assimilated, it is compared with established categories and is assigned to the appropriate group. This process of *stereotyping ideas* clearly hinders creative thinking because it imposes preconceptions on mental images and does not automatically allow for critical judgement or individuality. This psychological impediment hinders the combination of apparently unrelated images into a new entity that constitutes a creative solution to a problem.

Multiple Perspectives

A problem definition is converted in the mind to images and these images are then employed to develop potential solutions. The same problem definition can trigger different perspectives by individuals with different responsibilities in an organization. For example, an accountant, a manufacturing engineer, and a materials scientist will generate different images when confronted by the same problem definition. Therefore it is advantageous in the conceptualization phase to study a problem from several different perspectives in order to develop a larger number of images. These multiple images provide the basis for associative thinking involving the adaptation, modification, or substitution of ideas.

Representing a design problem in alternative modes of expression such as physical and abstract forms is often a useful first step. Thus, mathematical and verbal representations together constitute a more powerful problem-solving tool than implementing either representation alone.

The cartoon in Figure 8.11 is a commentary on a broad range of situations in engineering practice. It illustrates the response of various groups in an organization who each develope a different perspective of the same problem.

Challenging design problems are often interdisciplinary and sources of inspiration are not immediately clear. Under these conditions it is advisable to skirt the perceived boundaries of the main problem and prospect for ideas in related fields. This philosophy is illustrated in Figure 8.12

Creativity involves not only the identification of diverse factors pertaining to a problem but also their fusion into a solution through a set of relationships. Thus extensive knowledge is a necessary—but not a sufficient—condition for a productive mind.

One does not discover new lands without consenting to lose sight of the shore for a very long time.

Andre Gide
(1869–1951)

Many design problems are interdisciplinary problems. Extend the search area to trigger new solutions.

Look and you will find it. What is unsought will go undetected.

Sophocles
(496–406 BC)

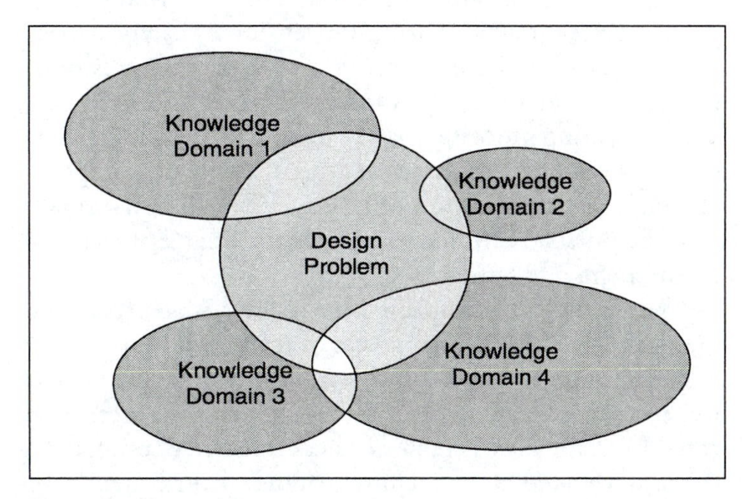

Figure 8.12

Prospecting for solutions to complex design problems.

Analogies provide different perspectives on the problem.

To establish relationships among the facets of a problem, it can be explored with an analogy. The mechanism of analogies establishes a similarity between two relationships, one being directly associated with the problem and the other possibly somewhat distant. Different classes of analogies are useful tools to stimulate the process of creation.

This philosophy was successfully employed by Louis Pasteur (1822-1895), the eminent chemist and bacteriologist, who observed that grapes ferment only when their skins have been torn to expose the flesh of the fruit. This observation was responsible for his postulating that infections in the human body are caused by external rather than internal factors.

Technical Impediments

Technical impediments include inadequate:

- computer software
- computer hardware
- technology
- materials
- manufacturing

The generation of creative solutions to engineering design problems can be impeded by inappropriate computational tools, technological limitations, materials limitations and immature manufacturing processes.

A design team can be seriously handicapped if it does not have access to computer simulation facilities. Many design problems are intractable using conventional analytical methods and the designer must resort more and more to simulations. Furthermore, state-of-the-art technologies and manufacturing processes may be unable to furnish the materials or the subsystems necessary to satisfy the design specification. Under these conditions, alternative solution strategies must be pursued until new technologies are mature enough to enter the marketplace.

The vehicles developed by the former Soviet space program illustrate the consequences of technical impediments. This program was seriously handicapped by inadequate materials and electronics. Soviet rocket engines were fabricated from metals with inferior heat resistance compared to those developed in the United States. Consequently the Soviets were compelled to use twenty low-temperature engines in the first stage of their Vostok missiles (Figure 8.13) whereas NASA used fewer but more powerful high-temperature engines in the Saturn V rocket (Figure 8.14). The Russian rockets were therefore compelled to carry more fuel

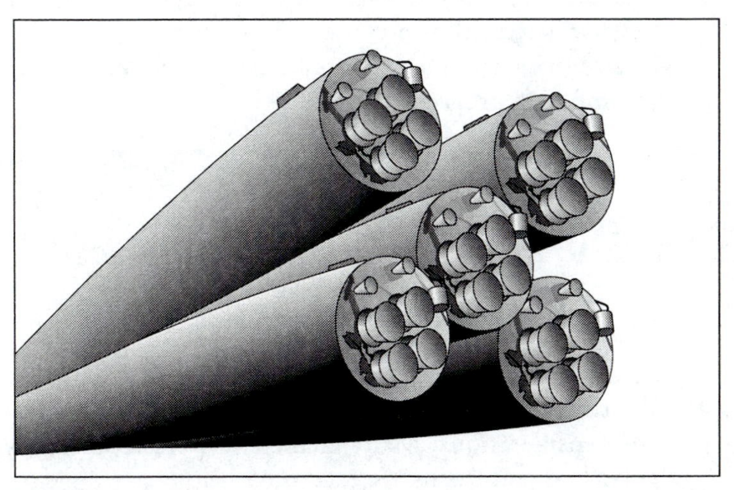

Figure 8.13

The configuration of the twenty engines on the Soviet Vostok launch vehicle.

Figure 8.14

The U. S. Saturn V launch vehicle containing five engines. (Photo courtesy of NASA.)

and were heavier than their American counterparts. The additional weight impeded the evolution of the Soviet space program. In addition to a lack of advanced materials, the Soviet space program, like its aircraft industry, suffered from a somewhat dated electronics industry. Constraints imposed by the size, weight, and performance of Soviet electronics further impeded progress.

8.5 The Student Design-Build-Test Project

Creativity is an essential ingredient of the design process and the psychological impediments to this activity are important considerations. Many challenges were reported by students. Two comments are excerpted below:

> From our historical research it appeared that the obvious way to power the car was to use the heat to expand a gas in a piston and cylinder. This is the conventional way to propel a vehicle. It was difficult to think of some other way of doing this. However, from class notes, the fried egg analogy required us to consider other unconventional things instead. This was very difficult for the group despite our trying the brainstorming approach. Our crazy ideas in the white of the yolk could not be transformed into useful solutions. It was frustrating.
>
> Taking risks with design ideas is supposed to be a part of the design process for creating revolutionary ideas for products. That's OK but we need to make our vehicle work and have it compete successfully on the day of the competition. If we can't perform successfully, then our grade will suffer. While we think that we had many good ideas, we could not develop some of them because it would clearly take too long to complete the development phase and we only have a semester to complete this project. The instructor said that the subconscious mind takes time to incubate ideas. Clearly we are not very good at this!

Summary

Creativity is one of the three primary activities associated with design. This triumvirate comprises creativity, analysis, and evaluation. The thought processes embedded in these three activities are quite different. Creativity employs divergent thinking typically characterized by the generation of numerous ideas, a nonjudgmental approach, chaos, and stochastic processes. Analysis and evaluation employ deductive thinking characterized by logical processes, rule-based approaches, and the development of only one solution to a problem.

In spite of an apparently clear process for solving complex problems there are a number of psychological impediments to the creation of innovative solutions. These include issues associated with culture, environment, emotion, intellect, and perception.

Cultural impediments are imposed upon an individual by the country in which he or she lives. These impediments include the suppression of the enquiring mind in western cultures, a resistance to change, and a reluctance to take calculated risks.

Environmental impediments include the physical surroundings in which individuals must work as well as excessive criticism of embryonic ideas before they mature.

Designers can be subjected to emotional impediments. These include a fear of taking risks, a need for safety and security, a lack of tenacity, and an inability to let an idea mature.

Intellectual impediments to creativity are associated with an individual's inability to perceive relationships, to think, and to reason. Such shortcomings manifest themselves in the selection of inappropriate solution strategies, an inability to communicate effectively, a limited educational background, and the use of incorrect information.

Perceptual impediments are primarily concerned with an incorrectly defined design problem or a failure to recognize the information necessary to solve a problem. The first step is to correctly formulate the problem. Subsequently, all possible human senses should be used to enhance creativity, multiple perspectives should be encouraged, and the individual should avoid stereotypical thinking.

Finally, technological limitations can impede the creation of innovative solutions. These limitations can be caused by immature manufacturing processes, inadequate computational tools and hardware, and unavailable commercial materials.

Chapter Nine builds upon the basic exposition of creativity presented in this chapter by examining how you can become more creative when practicing engineering design. Principles are presented for systematically generating ideas for the conceptual design phase of the product realization process.

Key Concepts

Associative thinking. See creative thinking.

Creative thinking. Creative thinking involves randomness, imagination, visualization, synthesis, nonjudgmental scrutiny, irregularity, and sensuality. It is the antithesis of deductive thinking and occurs in the right brain. It is also called associative thinking, divergent thinking, or lateral thinking.

Convergent thinking. See deductive thinking.

Cultural impediments. Cultural impediments to creative thinking are imposed upon an individual by the societal values of the country in which the designer lives. Traditions are one type of cultural impediment.

Design cycle. Open-ended design problems should be solved using creativity, analysis, and evaluation in an iterative approach.

Deductive thinking. Deductive thinking is associated with logic, analysis, judgement, and evaluation. It is also called convergent thinking or vertical thinking and occurs in the left brain. It is the antithesis of creative thinking.

Divergent thinking. See creative thinking.

Emotional impediments. These impediments to creative thinking are associated with the psychological safety of the individual. They include the desire of the individual for security, fear of making a mistake, and the patience to let embryonic ideas mature.

Environmental impediments. These impediments to creative thinking are imposed upon individuals by the environment in which they work. Environmental impediments include one's physical surroundings and one's relationships with colleagues.

Innovation. The introduction of novel ideas into products, organizations, or systems.

Intellectual impediments. These impediments to creative thinking are associated with the inability of an individual to reason, understand, or to perceive relationships and differences. They include inappropriate solution strategies, limited educational background, and the inability to communicate effectively.

Invention. Invention creates something that has never existed before, such as a new device, process, or product. It is a combination of knowledge and skill applied to various discoveries and observations. Invention begins with a concept and, upon refinement, this concept matures into a device, process, or product.

Perceptual impediments. Perceptual impediments are associated with an inability to correctly define a design problem or an inability to recognize information needed to define it. They are connected with the activation of all human senses, the stereotyping of ideas, and the inability to consider the problem from several different perspectives.

Technological impediments. Technological impediments to engineering problems can be inappropriate computational tools, technological limitations, materials limitations, or immature manufacturing processes.

Review Questions

1. What are two types of thought processes associated with design creativity?

2. List the essential stages of the generic algorithm for solving open-ended design problems.

3. List some alternative terms for deductive thinking and describe the primary characteristics of this type of thought process.

4. List some alternative terms for creative thinking and describe the primary characteristics of this type of thought process.

5. What are some personality traits of individuals with a high aptitude for creative thinking?

6. Identify six impediments to creative thinking for design engineers.

7. Define each of the following kinds of impediments to creative thinking and list the primary attributes.
   a. cultural
   b. environmental
   c. intellectual
   d. emotional
   e. perceptual
   f. technical

Problems

1. Without lifting the pencil from the paper, draw four straight lines through the centers of all nine dots in the diagram below.

2. Ten balls are arranged on a pool table in the pattern shown below. Convert the pattern on the left to the one on the right by moving just three balls:

3. Imagine that you have a cylindrical cake. Using just three straight cuts with a knife, how would you divide this cake into eight pieces of identical shape and size?

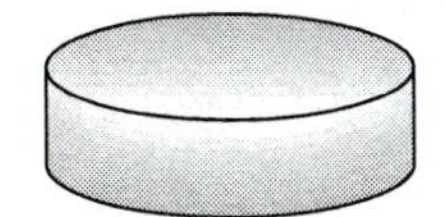

4. Combine four lengths of metal chain, each consisting of three links, into a single continuous loop. The cost of opening a link is 3 cents and the cost of closing a link is 2 cents. Complete the task for 15 cents.

5. At how many places on Earth, other than the North Pole, is it possible to travel one mile south, one mile west, and then one mile north, and end up at the point from which you started?

6. Six matches are arranged in a regular hexagon. Rearrange the matches to create exactly four equal-size equilateral triangles.

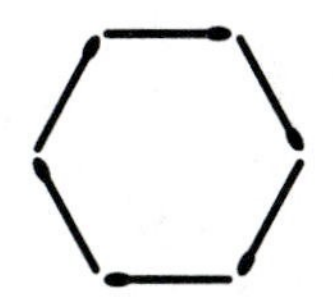

7. Two colleagues of different height are taking a lunch-time stroll together. They start walking at the same instant side by side; their first step is with their left foot. The taller person has a stride of 90 cm and the smaller person has a stride of 60 cm. When will they both use the right foot to take a step simultaneously?

8. The following nine coins are arranged in eight rows of three. By moving just two coins, rearrange them into ten rows of three.

• • •

• • •

• • •

9. A 5 meter metal chain is fastened securely around the neck of a hungry polar bear. On the ice-pack 10 meters away from the bear lies a dead seal pup. How does the bear satisfy his hunger?

10. In a rectangular room, 30 x 12 x 12 feet, a spider is 1 foot from the ceiling at the center of an end wall. A fly is at the center of the opposite end wall 1 foot from the floor. The fly is so frightened it cannot move. What is the shortest distance the spider must crawl in order to reach the fly?

11. A 3.5 inch 16d bright common nail protrudes 3 inches vertically from a piece of wood. Devise a method for supporting six other 3.5 inch 16d nails on the top of the protruding nail. No tape, adhesive, string, or other materials can be used. Test your method experimentally.

Further Reading

Adams, J. L. *Conceptual Blockbusting.* Reading, MA: Addison-Wesley, 1974.

Alger, J. R. M. and C. V. Hayes. *Creative Synthesis in Design.* Englewood Cliffs, NJ: Prentice-Hall, 1964.

Buhl, H. R. *Creative Engineering Design.* Ames, Iowa: The Iowa State University Press, 1968.

Cox, C. M. *The Early Mental Traits of Three Hundred Geniuses.* Stanford, CA: Stanford University Press, 1926.

DeBono, E. *Lateral Thinking.* New York, Harper & Row, 1970.

Fritz, R. *Creativity.* New York: Fawcett, 1991.

Gordon, W. J. J. *Synectics, the Development of Creative Capacity.* New York: Harper Brothers, 1961.

Kim, S. H. *Essence of Creativity.* New York: Oxford University Press, 1990.

Lumsdaine, E. and M. Lumsdaine. *Creative Problem Solving: Thinking Skills for a Changing World.* New York: McGraw-Hill, 1995.

Mayer, R. E. *Thinking, Problem Solving, Cognition.* New York: W. H. Freeman Company, 1983.

Orstein, R. and R. F. Thompson. *The Amazing Brain.* London: The Hogarth Press, 1985.

Osborn, A. F. *Applied Imagination. Principles and Procedures of Creative Problem Solving.* New York: Scribner's and Sons, 1953.

Parnes, S. J. and H. F. Harding. *A Source Book for Creative Thinking.* New York: Scribner's and Sons, 1962.

Poincaré, H. *Science and Method.* London: Nelson, 1914.

Weisberg, R. W. *Creativity: Beyond the Myth of Genius.* New York: W. H. Freeman Company, 1993.5.

Conceptual Design

9

'Tis a thing impossible, to frame
Conceptions equal to the soul's desire

William Wordsworth
(1770–1850)

Objectives

- **To define conceptual design.**
- **To present generic strategies for enhancing design creativity.**

Contents

What If...

You are a student in a design projects class and your group is required to design, build, and test a vehicle that will travel the greatest distance when powered by ten small red, white, and blue birthday candles. You define the problem meticulously, but there is no obvious solution. How can you convert heat and light into motion? How do you handle those laws of thermodynamics? What about entropy? And what did Newton say about motion? You wish you'd not slept during those lectures!

But how can these laws from your various engineering science classes help you when you've nothing to analyze! You need to create something. Then you vaguely recall the design prof saying something about creative thinking and that there were several different methods to help you think up new solutions to design problems. Methods like brainstorming, checklists, analogy, morphology, and biomimetics. You review your lecture notes and soon you start making progress. Designs are conceived!

9.0 Introduction

This chapter focuses on the most creative step in the design process: conceptual design. In this phase of the design process various solutions to the problem are developed by a combination of creative and deductive thinking. Evaluation of these solutions is the subject of Chapter 10.

The conceptual design phase is crucial to the success of a product because, as noted in Chapter 5, approximately 75 percent of manufacturing costs are committed by the end of this phase.

9.1 Conceptual Design

Conceptual design begins with the design specification. Various ideas are then conceived and recorded in notes or sketches showing how the product will function. These ideas are then analyzed and compared with the design specification to select the most promising concepts for further development. This conceptualization process can be applied to the complete product, to major subassemblies, or to individual parts of the product.

Conceptual design activities yield numerous different visions of the product.

The result of the conceptual design phase is largely responsible for the product characteristics and its ultimate quality. A poor conceptual design can rarely be transformed into a successful product regardless of ingenuity at the product design stage.

A child from the time he goes to school is taught it is very dangerous to fail. The inventor fails 9,999 times, and if he succeeds once, he is in.

C. F. Kettering
(1876–1958)

Although a combination of both creative and deductive thinking is required throughout the design process, conceptual design requires creative thinking. Without it there will be no concept to analyze. It is creative thinking that generates the initial concept.

Conceptual design activities can be classified by the type of product being created. For example, Pahl and Beitz (See *Further Reading*) propose three kinds of conceptual design. The first is called *original design*, in which a radically new product or concept is created. This is the most challenging

Classification of conceptual designs:

- original design
- variant design
- adaptive design

Original design: a conceptual design for a new generation of products

Variant design: a conceptual design where the size or configuration is changed but the operating principle is not.

Adaptive design: a conceptual design where the product is changed to satisfy a different specification.

form of conceptual design because solutions are not obvious and the product is the first of a new generation. Examples include the first telephone system, the first commercial semiconductor device, and the first NASA space shuttle.

The second conceptual design activity is *variant design*, in which the size or configuration of a product is changed but the operating principle and the function remain unchanged. An example is the replacement of an off-the-shelf gearbox with another of different power rating. This kind of conceptual design activity is the least demanding because it involves only modification of a product—perhaps in response to parts failures during service.

The third class of conceptual design activities is *adaptive design*, in which a product is changed to solve a different problem but the original solution principle remains unchanged. This usually requires the creation of new parts. Examples include automobiles which are marketed with different options, such as different engine sizes or transmissions, yet they are all built on the same vehicle platform.

A survey of conceptual design activities in German industry undertaken by the German Association of Mechanical Engineering Companies concluded that 20 percent of conceptual design effort was variant design, 55 percent was adaptive design, and 25 percent was original design. While the boundaries between these domains are fuzzy, evidently only a fourth of conceptual design effort is spent on fundamentally new designs. Of course the remaining three-quarters of the conceptual design effort still requires considerable creativity and ingenuity.

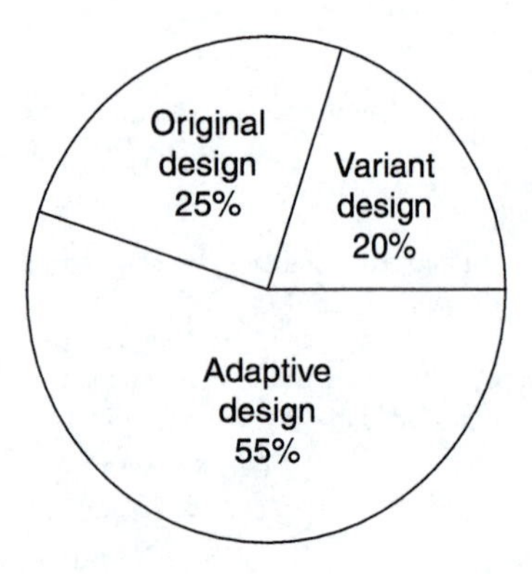

Conceptual design activities

This chapter focuses on strategies to enhance the creative process. Some of these can be used by individual designers while others must involve a group. We suggest eleven strategies in Table 9.1.

The First Strategy: Duplication

An obvious strategy in solving a design problem—useful for either individuals or teams—is to determine whether someone else has solved a similar problem. An existing

Strategies for Individuals or Groups	Strategies Only for Groups
Duplication	Brainstorming
Combinations of elements	Synectics
Historical evolution	
Hypothesize and test	
Analogy	
Morphology	
Empathy	
Checklists	
Biomimetics	

Table 9.1

Strategies for creative thinking.

Think wrongly, if you please, but in all cases think for yourself.

G. E. Lessing
(1729–1781)

solution can perhaps be used with little or no modification for the problem at hand. It's an easy way out to copy someone else's design. Of course time can be saved and "reinventing the wheel" avoided. Duplication will also satisfy the innate desire of people to avoid the *thinking* needed to solve challenging problems!

Most people would sooner die than think; in fact, they do so.

Bertrand Russell
(1872–1970)

Duplication requires a designer to recognize similarities between the problem and similar ones with proven solutions. Work often begins with a search of both related and unrelated scientific and technological disciplines. The mind must be unbiased and receptive to associations between seemingly unrelated ideas by focusing on the underlying principle of a product rather than on its intended use. For example, a ball point pen might provide the inspiration for a roll-on deodorant applicator. In both devices the process is the transfer of a fluid to a surface—paper or skin—by a partially encapsulated sphere. In practice, the designer must generate a modified version that avoids patent infringements.

Recognize similarities between the design problem and problems with proven solutions

Another example of the strategy of duplication was Herman Hollerith's use of Jacquard's loom pattern cards to record information gathered in the 1890 Census. Hollerith cards were used later to input data into computers.

A standard practice in many industries is to analyze competitor's products. These must be obtained and disassembled to study working principles, design attributes, subassemblies, materials, and shapes as well as to infer what manufacturing processes were used. The process is often called *reverse engineering* and has the four stages: design evaluation, technical data collection, design modification and verification, and implementation. While the exploitation of another company's products can be informative, it can so polarize the designer's thought processes that alternative and possibly superior solutions may be hard to see.

Reverse engineering is the process of studying or analyzing a device in order to learn details of its design, construction, and operation—perhaps to produce a copy or an improved version.

Ideas for solutions that might be duplicated come from many sources. Compared to someone who reads little, a designer who reads widely has distinct advantages in the creation of quality designs. This is because creativity involves the association of seemingly unrelated ideas. At a basic level, designers should continually read current technical journals to appreciate state-of-the-art developments in their field, to be aware of new analytical procedures, and to recognize new machine elements.

Prospect for new ideas in professional publications; visit trade shows and exhibitions; consult the patent literature.

Exhibitions and trade shows are an additional source of valuable information because they present the latest developments and because the designer can interact directly with company representatives. Furthermore, these shows are ideal for comparing the products of a variety of vendors.

By studying biology—a field seemingly unrelated to engineering—engineers have discovered a rich source of ideas that can be duplicated in solutions to engineering design problems. This topic is discussed further in the Eleventh Strategy.

The Second Strategy: Combinations of Conventional Elements

A creative solution to a design problem can arise when a designer recognizes a new relationship among familiar objects. This strategy can be used by design teams or individuals—especially for variant or adaptive design. The prerequisite to use of this strategy is a large knowledge-base of ideas that can be regrouped or reconfigured by the subconscious mind. This knowledge-base can be established by individuals who read widely (Figure 9.1) and collect information from many sources. Some of these include:

- standard engineering periodicals and textbooks
- product data sheets
- brochures
- patent records
- scientific journals
- computer databases
- exhibitions at professional meetings and trade shows

When I get a little money, I buy books; and if any is left, I buy food and clothes.
Desiderius Erasmus
(1468–1536)

Reading is to the mind what exercise is to the body.
Sir Richard Steele
(1672–1729)

There will inevitably be many problems of which the designer has minimal knowledge, and a literature search must then be undertaken. When the information to be sought is clearly defined, the search can begin in any of several ways including seeking help from librarians, consulting *Engineering Index*, *Science Citation Index,* the patent literature, encyclopedias, etc.

As an example, consider the task of converting a straight-line motion in the vertical direction into a straight-line motion in the horizontal direction. This task can be initiated by reviewing off-the-shelf machine elements such as those presented in Figure 9.2 and those listed in Tables 9.2. and 9.3.

Figure 9.1

Design Literature.

Consider the Internet as an aid to literature searches.

Figure 9.2

Some off-the-shelf conventional elements. (Courtesy of W. M. Berg, Inc., East Rockaway, NY)

Connections/Interfaces
Hydraulic elements
Pneumatic elements
Electrical elements
Optical elements
Acoustical elements
Mechanical elements

Table 9.2

Some conventional machine elements

Bearings	Flywheels
Cams	Clutches
Screws	Brakes
Gears	Valves
Levers	Fasteners
Ratchets	Springs
Couplings	

Table 9.3

Some conventional mechanical elements

Can the infusion of new technologies into a dated product yield a new generation of products?

Different combinations of these machine elements result in many potential solutions. Some of these are sketched in Figure 9.3.

Numerous inventions have used the strategy of combining conventional elements. For example, in 1885 Gottlieb Daimler, a German engineer, combined two existing products—a four-stroke piston engine and a wooden-frame bicycle—to create the motorcycle.

In 1450, Johann Gutenberg invented printing with movable type by combining ideas from several fields: a press similar to those used for wine making, an oil-based ink, a method of making reusable metal slugs with a single letter on one face, an alloy for casting the slugs, paper making technologies, and bookbinding techniques. None of these attributes existed before in a single process, even though the Chinese and Koreans had been printing text and pictures using wooden blocks since the eighth century. The complexity of their ideographic language prevented them from developing this idea further and it lay dormant until Gutenberg's time.

The Third Strategy: Historical Evolution

Solutions to design problems are influenced by the technological and scientific limitations of their period. A state-of-the-art solution today will invariably lose its luster tomorrow because of new technical advances. Therefore new solutions can be created by the infusion of new technologies into old solutions. The historical evolution of many products is characterized by the return to the marketplace of old products in significantly improved forms with new qualities and features. Therefore the design engineer should review the historical development of the field in order to identify characteristics which could be used in a new generation of products.

An example is the evolution of the large converters used to change pig iron into steel. The original converters, developed in 1850 by Sir Henry Bessemer, used air to oxidize impurities. While oxidation with air was much less efficient

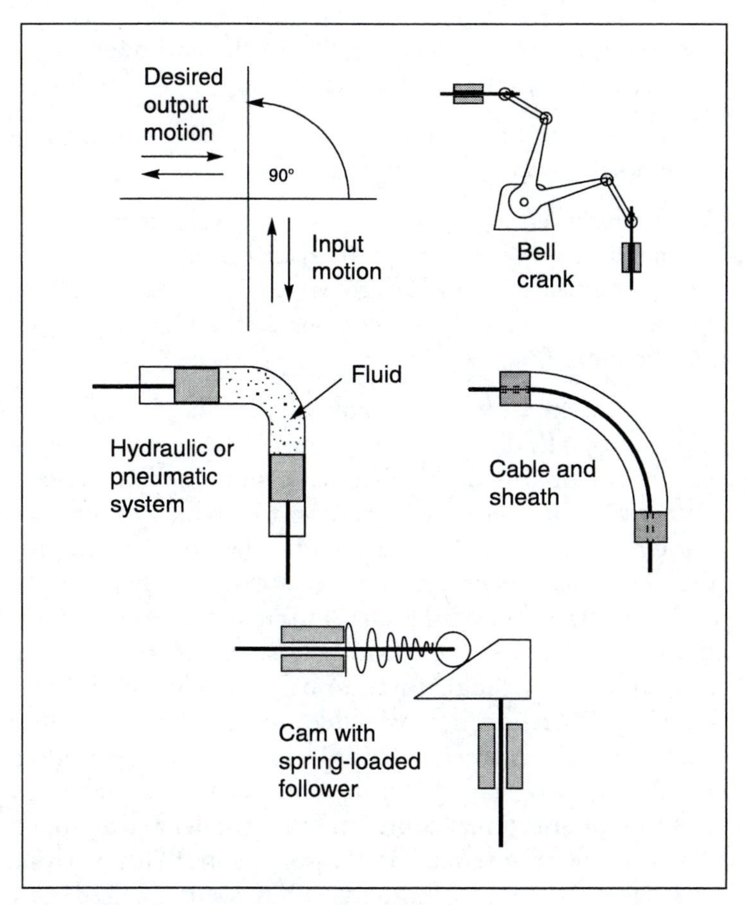

Figure 9.3

Some conceptual designs to convert a straight line vertical motion into an equivalent horizontal motion.

than with pure oxygen, the latter was prohibitively expensive. Subsequent developments in refrigeration techniques so reduced the price of oxygen that now an oxygen-based steelmaking process is standard practice.

Some important approaches in the implementation of this historical evolution strategy include:

Recognize trends, identify weaknesses, and consider emerging technologies.

- Recognize trends in the characteristics of products in order to extrapolate into the future.
- Identify features in a product that have not been improved upon for some time.
- Examine emerging technological developments for possible incorporation into existing product designs. Fertile fields from which to harvest new ideas include advanced materials, information technology, and biotechnology.

Consult the patent literature for the history of a particular field.

Since there are very few books on the evolution of classes of products, the designer might consult the patent literature for information about the historical evolution of the subject. Patent files are arranged chronologically which facilitates a rapid review. Furthermore, a patent review helps protect the designer from reinventing a product that already exists.

This strategy of synthesis is illustrated in the evolution of the airship. In 1783 the Montgolfier brothers built and successfully flew the first hot-air balloon. Their design was based upon Archimedes' principle and used heat to create a machine with an average density less than the surrounding air. Almost a century later, the concept was improved by addition of directional control using a rudder and a propeller driven by an internal combustion engine. This machine was called an *airship*. Nonrigid airships were superseded by rigid airships with aluminum frames. The frame contained gas bags filled with hydrogen rather than hot air to generate buoyancy. Nonflammable helium has now replaced hydrogen, and new composite materials can replace aluminum in the frame.

Addition of directional control to a balloon led to what is now called a dirigible.

The Fourth Strategy: Hypothesize-and-Test

The hypothesize-and-test strategy appears in several forms in scientific and technical investigations. Products are based upon a conceptual design hypothesis that is transformed by product design into plans for a prototype. The prototype is manufactured and tested to determine whether it satisfies the design specification.

This same strategy applies to the conceptual design phase (Figure 9.4). An idea is proposed as the concept upon which the product should be based. It is then tested for credibility. The critical step is the generation of the original hypothesis. Naturally hypotheses can be based upon absolutely no scientific thought whatsoever, but these are generally quite useless. Better hypotheses arise after meticulous thought. The first trial-and-error category admits all ideas no matter how crazy, while the second more scientific group will have greater potential for success.

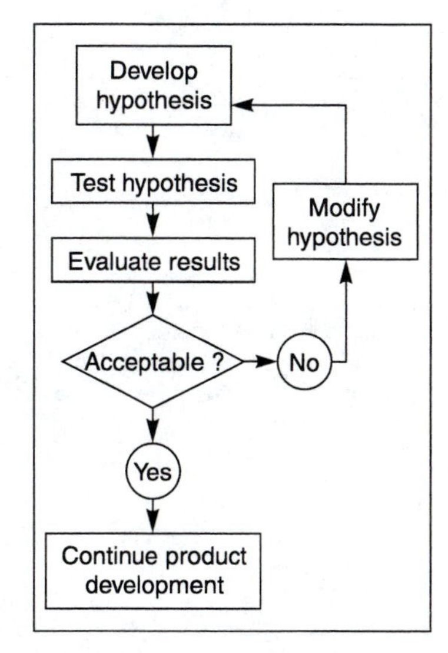

Figure 9.4

The hypothesize-and-test methodology.

Experimental investigations have been a multifaceted source of inventions, discoveries and creative endeavors in science and technology. If theories are not well developed or there is minimal experimental data pertaining to a phenomenon then the route to a solution is often impeded. This impediment can be removed if the designer develops a relevant experimental program. In this scenario a really detailed plan of work must be developed with access to appropriate facilities and instrumentation. Detailed records must be kept during the experimental program and the resulting data carefully scrutinized.

Experimental tests must be designed carefully to produce good data.

Significant knowledge and skill can be acquired by the most rudimentary hypothesize-and-test strategy in which the designer poses a series of "what if" questions. While this is the simplest strategy, it is also the least effective because many trials may be required. Furthermore, the ability to ask the most appropriate set of questions and to evaluate the results effectively before proposing a new set of questions is important in this approach. Thomas Edison (1856–1931), who was one of the greatest inventors of all time, employed this strategy with much success.

Employ experimental investigations when mathematical models are inadequate.

For a successful technology, reality must take precedence over public relations, for Nature cannot be fooled.*

Richard Feynman
Nobel Laureate in Physics
(1918–1988)

*Report of the Presidential Commission on the Space Shuttle Challenger Accident, Washington, D.C., 1986

Trial-and-error is an inefficient approach to solving problems. This strategy is sometimes referred to as "no stone unturned."

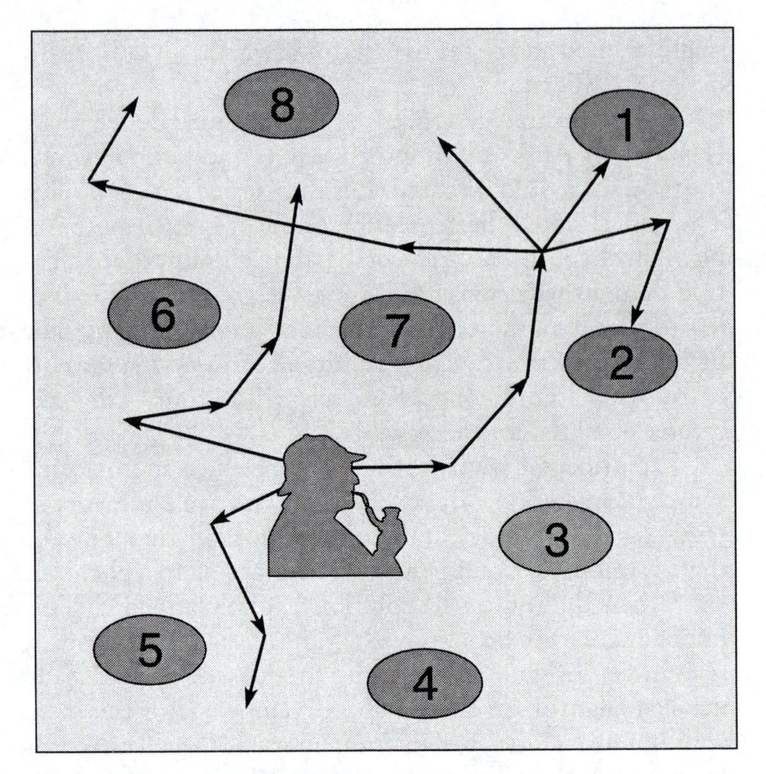

Figure 9.5

The characteristics of the trial-and-error strategy on the solutions plane as undertaken by one person with solutions 1 through 8. Only solutions 1 and 2 are discovered.

Trial-and-error approaches do not permit the solutions plane to be thoroughly investigated; the prospecting process is inefficient.

Figure 9.5 depicts this trial-and-error strategy on the solutions plane. A series of arrows depict the progression of the designer's thoughts through the solutions plane. Of the eight possible solutions represented in the diagram, only two are discovered. Trial-and-error is characterized by the proposal of quite different approaches to a potential solution with distinct discontinuities in the evolution of ideas. Since this approach is generally undertaken by only one person, the solutions plane is not exhaustively searched.

A number of well known products have been developed by trial-and-error approaches. For example, in 1910 a German metallurgist, Alfred Wilm, patented the aluminum alloy Duraluminum—a light but strong metal now widely used in aircraft. It has a relatively high strength-to-weight ratio compared to steel. Wilm created Duraluminum after approximately 10,000 experimental trials!

A more recent discovery attributed to the hypothesize-and-test strategy is WD-40, a familiar product in many factories, shops, and households. WD-40 was launched in 1953 by the Rocket Chemical Company in response to a request from Convair, an aerospace and defense contractor, for a water displacement (WD) formulation with the ability to inhibit metal corrosion and thereby to protect the skins of rockets. The product was successfully created on the fortieth attempt; hence the name WD-40 (Figure 9.6).

Another product of the hypothesize-and-test strategy is the parachute. Its performance characteristics still are not well understood despite recent aerodynamic studies at the Sandia National Laboratories using supercomputers in tandem. The original device appeared in the notebooks of Leonardo da Vinci in 1495 but the idea lay dormant until the late 1700s when several individuals attempted parachute descents—with disastrous results! It was the Frenchman Andre-Jacques Garnerin who, by making repeated descents in 1797, began to develop the product successfully. Today's powerful computers and sophisticated analytical techniques are beginning to shed light on how a parachute unfurls to become a canopy. During this change in geometry, the air flow near the fabric changes rapidly and it is this interaction that defines the fluid mechanics problem.

The hypothesize-and-test strategy has proved to be of great utility to both scientists and engineers in making serendipitous discoveries. However, such discoveries generally are preceded by an intense period of investigation during which the designer gains considerable exposure to the problem. That experience prepares the investigator to recognize the significance of unexpected phenomena.

Serendipitous discoveries certainly may help solve a designer's original problem but they often uncover new problems whose solutions prove to be more valuable than the original one. Unexpected phenomena should not be

What is defeat? Nothing but education; nothing but the first step to something better.

Wendell Phillips (1811–1884)

Figure 9.6

WD-40: a product synthesized by the hypothesize-and-test strategy. (Courtesy of WD-40 Company)

In the fields of observation, chance favors only the prepared mind.

Louis Pasteur (1822–1895)

The word *serendipity* was coined by Horace Walpole in a letter written in 1754. Walpole was intrigued by a fairy tale about "The Three Princes of Serendip" who often made accidental discoveries of things they were not seeking.

Serendip is an ancient name for Sri Lanka.

dismissed without careful evaluation. In 1844, Charles Goodyear (1800–1860) was granted a patent for the vulcanization process which imparts strength, greater elasticity, and durability to rubber when it is heated with sulfur. Goodyear discovered vulcanization serendipitously after he had tried for five years to improve rubber. He accidentally dropped a mixture of crude rubber and sulphur onto the top of a hot kitchen stove and observed that instead of burning, the rubber flowed into a disc shape and was quite flexible when he removed it from the hot surface.

The Fifth Strategy: Brainstorming

Brainstorming is a group approach involving individuals with diverse backgrounds who briefly suggest potential solutions. Ideas are recorded and later evaluated.

Brainstorming, which was proposed in 1953 by the United States psychologist A. F. Osborn, is a group approach to creative thinking. It is based upon a trial-and-error philosophy but embodies some enhancements. Brainstorming has the advantage of covering the solutions plane more uniformly than a single individual does because members of a group often view problems quite differently, as was illustrated in Figure 8.11. Furthermore, the number of ideas generated in unit time is larger because more people participate.

Many ideas grow better when transplanted into another mind than in the one where they sprung up.

Oliver W. Holmes (1809–1894)

This strategy is based upon the notion that new ideas will be generated when individuals with varied backgrounds simultaneously participate in the conceptualization process. They generate a pool of many solutions that is more likely to contain at least one high quality solution than if an individual worked alone. It is important to note the utility of the ideas generated, however—not just their number.

Brainstorming can be employed to solve both large conceptualization problems and the somewhat more focused problems of detail design. The primary components of brainstorming are the following.

1. In the idea-generating phase, a group of four to eight people with expertise in several fields simultaneously think about a problem for 30 to 45 minutes. Each participant expresses his or her ideas briefly—perhaps

for one minute—no matter how ridiculous the ideas may seem. Individuals are encouraged to build upon the ideas presented by others. Unconventional, outrageous ideas as well as ones that integrate schemes suggested by others should be expressed.

2. To support free expression, criticisms or remarks discrediting a proposal should not permitted during brainstorming.

3. An independent person should record all ideas and distribute them to participants for later development and evaluation. People who generate ideas must not evaluate them during the brainstorming session. This avoids criticism of individuals, which can stifle their activities, and it avoids premature rejection of embryonic ideas as they are first enunciated. Credit for successful brainstorming is shared equally among the group participants.

It hinders the creative work of the mind if the intellect examines too closely the ideas already pouring in.
Morris Stein

Figure 9.7 illustrates how a brainstorming group of five participants function in a solutions plane containing four solution domains. Brainstorming involves the prompting of thought by one individual in another—that is, it involves a *synergistic* mechanism, not just five individuals increasing the probability of solution by a factor of five. Here the whole is greater than the sum of its parts. Figure 9.7, in contrast to Figure 9.5, shows more arrows coming together and progressing to a solution. The tip of each arrow represents a potential solution and the length and number of the arrows represent the difference between successive solutions.

Synergy is the mutual reinforcement of elements that produces a total effect greater than the sum of the individual elements.

The Sixth Strategy: Analogy

The First Strategy (duplication) can be extended to stimulate thinking by use of analogy. This involves recognition of similarities between two situations. By adopting the strategy of analogy, ideas can be generated to solve a specific problem by prospecting for similar problems in other fields of engineering or in fields completely unrelated to engineering. A relationship coupling a design specification and its

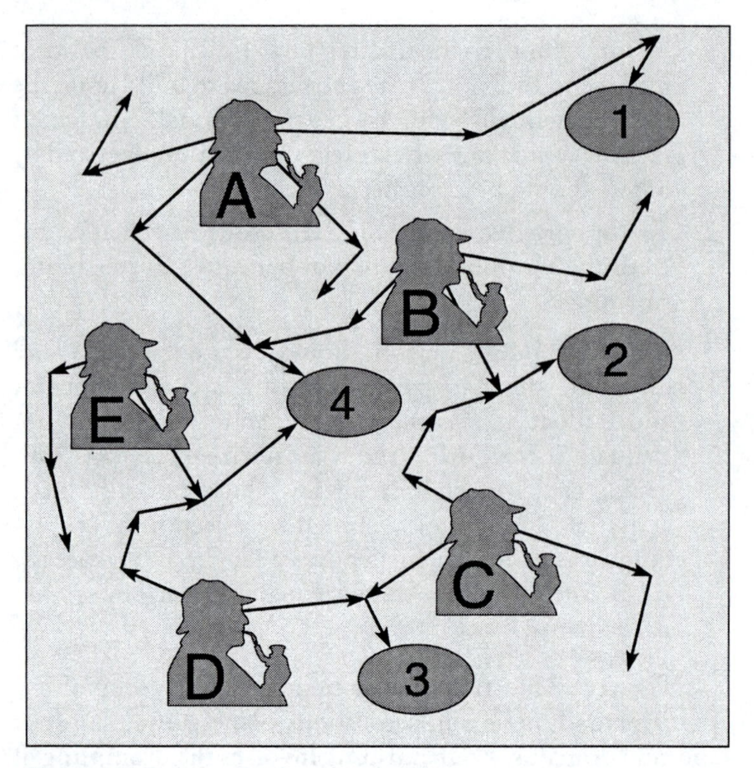

Figure 9.7

Characteristics of a brainstorming session involving five participants with solutions 1, 2, 3, and 4.

solution in one domain may suggest a solution in a second domain based upon the same relationship. The concept is presented in the schematic diagram of Figure 9.8.

Recognize similarities between two apparently dissimilar things.

One kind of analogical thinking involves recognizing the similarities between an existing product and its design specification and the design specification of the desired product. The similarity between the two design specifications permits the designer to modify the existing product to solve the design problem that is without a solution.

An electromechanical graph plotter is analogous (on a far different scale) to an overhead crane in a scrap yard.

Consider a mechanism for an electromechanical plotting device which uses a felt-tip pen or an ink jet to draw graphs on sheets of paper. The pen must be able to move unhindered over the entire surface area of the paper. This task is analogous to the overhead cranes used in machine

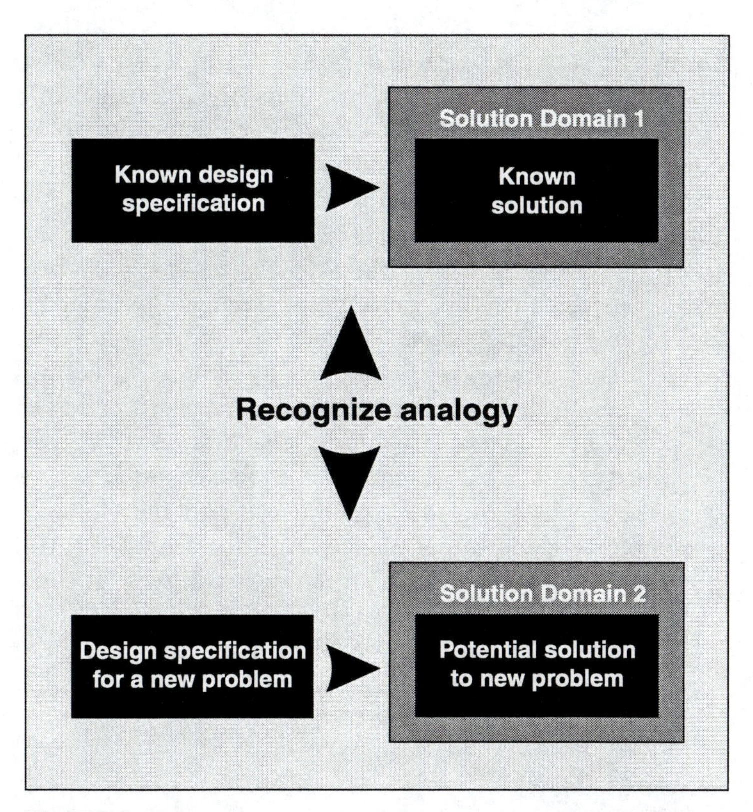

Figure 9.8

The basis for the solution strategy involving a direct analogy.

shops and scrap yards to carry a load from one point to another.

A second kind of analogical thinking involves recognizing the similarities between naturally-occurring events and the design specification for a product. This analogy between biological systems and engineering systems is called *biomimetics,* where biological systems are mimicked by engineering systems. This powerful approach is presented as the Eleventh Strategy.

Biomimetics is the mimicking of a biological system by an engineering system.

The first recorded proposition that all matter consists of atoms was based on analogical thinking. In approximately 400 BC, Leucippus and his pupil Democritus were walking along the sandy beach of the Aegean Sea discussing the composition of the beach and the sea. Their debate focused upon whether seawater is continuous or is divisible into small drops. After all, the sandy beach appeared to be continuous when viewed from a distance but in reality it is composed of many small grains of sand. Could this also be the case with sea water? The outcome of their discussion was the proposition by Democritus that water is made up of small particles that, like sand grains, seem indivisible. He called these particles *atoms.*

Nothing exists except atoms and empty space; everything else is opinion.

Democritus (c. 460–370 BC)

Atoms were first proposed in 400 BC by Democritus whose hypothesis was based upon an analogy.

Similar thinking was employed by Ernest Rutherford (1871–1937), the New Zealand nuclear physicist, when he proposed his concept of atomic structure in 1911. He described the structure of the atom as like the solar system with the nucleus analogous to the sun and the electrons analogous to the orbiting planets. Although incorrect, this planetary analogy of atomic structure is still seen in some elementary science textbooks today.

In 1911 Ernest Rutherford proposed that structure of an atom was analogous to the solar system.

The consideration of a problem from many different viewpoints can be fostered by using a variety of analogies: direct, symbolic, and fantasy.

There are three different analogies:

- direct
- symbolic
- fantasy

Direct Analogy

A direct analogy involves a comparison of similarities in different fields.

Direct analogy involves a comparison of similar techniques, technologies, or knowledge domains in different fields. It is perhaps the most common type of analogy. There are, for example, numerous direct analogies in engineering and nature. Sir Marc Isambard Brunel (1769–1849), the great engineer and inventor, was confronted by the problem of building civil engineering structures under water. He solved the problem serendipitously when he observed a shipworm tunnelling in timber. This long creeping animal constructed a tube for itself as it moved forward. This observation was responsible for Brunel developing caissons, which are watertight enclosures inside which individuals can undertake engineering construction activities under water.

Direct analogy was employed by Alexander Graham Bell (1847–1922) in the conceptual design for the telephone which he patented in 1876. Bell recalled, "It struck me that the bones of the human ear were very massive, indeed, as compared with the delicate thin membrane that operated them, and the thought occurred that if a membrane so delicate could move bones relatively so massive, why should not a thicker and stouter piece of membrane move my piece of steel. And the telephone was conceived."

Symbolic Analogy

Symbolic analogy involves transforming the problem from a group of sentences into an impersonal symbolic format involving mathematics or some type of pictorial representation. Cartographers employ this approach on topographical maps by representing features with symbols. Similarly, many automatic control schemes are represented by block diagrams.

A symbolic analogy involves transferring a problem from language to a mathematical or symbolic representation.

Fantasy Analogy

Fantasy analogy involves identifying the characteristics of the ideal solution and then endeavoring to achieve that solution. In the first phase of this two part process, the designer suppresses all knowledge of the laws of science and then fantasizes. Science fiction writers continually employ this approach. Some schemes proposed by these individuals have become reality.

A fantasy analogy involves considering the ideal solution and its realization in practice.

The Seventh Strategy: Morphology

The morphological strategy requires the designer to identify the primary features or subsystems required to solve the problem. A number of different solutions for these subsystems are proposed and then combinations of these solutions are systematically generated and explored as solutions to the overall problem. The objective is not to miss any opportunity in the field of solutions.

Recognize the primary subproblems and develop a list of potential solutions for each. Then develop a morphological box with these solutions as the primary axes.

The method involves the following steps:

1. Define and list all of the primary attributes or subproblems of the primary problem.
2. Define the primary problem in a minimum number of these attributes—typically two or three. If the final

number is greater, then several morphological iterations should be employed in the search for solutions.

3. Now develop a list of possible approaches to achieve these attributes.

Develop a morphological box for the global design problem. Then develop more boxes for the individual cells that appear most promising.

4. Generate a matrix, or morphological box, in order to identify all possible combinations of these potential approaches. Each side of the box is associated with an attribute and the side is divided into the different approaches. This box provides a systematic mechanism to identify innovative or unusual solutions that might be overlooked.
5. The morphological approach can also be employed to prospect for solutions in individual cells in the box that appear to offer promising opportunities.

Consider devising a transportation system for only one person. This task was addressed by John Arnold (see Parnes and Harding's *A Source Book for Creative Thinking*.) The

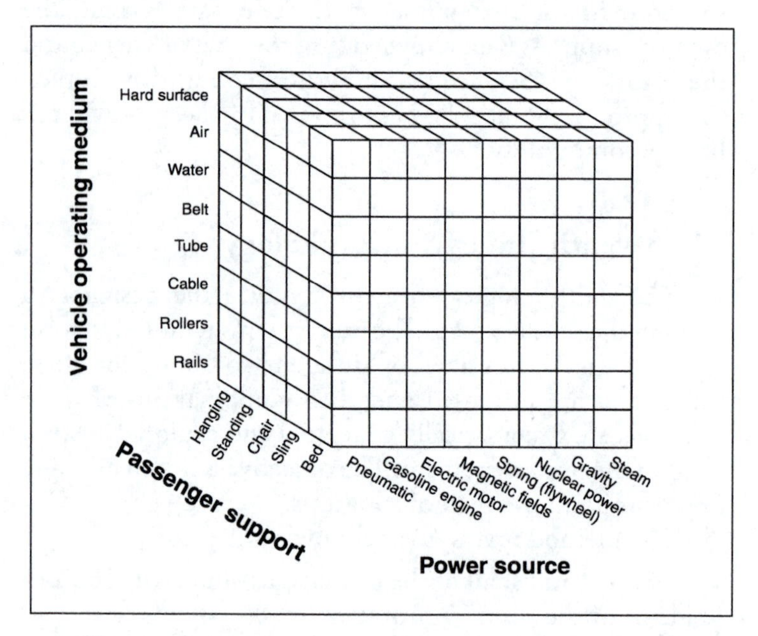

Figure 9.9

A matrix for the morphological analysis of a personal transportation system.

primary attributes of the problem were identified as the vehicle operating environment, the power source, and the passenger-vehicle interface (Figure 9.9).

A number of potential solutions to each of these three subproblems were generated and listed on three orthogonal axes to define a three-dimensional matrix or morphological box. A portion of the matrix is shown in Table 9.4. Each cell in this matrix defines a potential solution to the problem because it is associated with a combination of the three primary parameters of the problem. For the problem posed, there are 5 × 8 × 8 combinations or 320 potential solutions presented in Figure 9.9.

Vehicle operating medium	Power source	Passenger support	Result
Rails	Diesel engine	Seat	Train
Road	Gasoline engine	Seat	Automobile
Cable	Electric motor	Sling	Ski lift

Table 9.4

Three examples from the 320 potential solutions.

A review of the 320 different combinations of the matrix elements reveals some standard solutions (Table 9.4) such as the train (rails, diesel engine, seat), the automobile (road, gasoline engine, seat), and the ski lift (cable, electric motor, sling).

The morphological analysis can be applied to individual elements in the matrix. For example, if the matrix element characterized by the features "air, gasoline engine, and seat" appears promising, then each of these attributes can be decomposed into other elements. A new matrix similar to Figure 9.9 is then created. The gasoline engine can be described by the number of cylinders, the configuration of the cylinders, the fuel supply system, the engine type, etc.; the transportation medium can be a dirigible, a fixed wing aircraft, a helicopter, or an aircraft with moving wings; and the seat can be defined by the number of support points, the number of degrees of freedom of seat adjustment, the back support, etc.

The Eighth Strategy: Empathy

Imagine yourself in the situation for which a solution is sought.

This approach requires designers to imagine themselves subjected to the problem environment before problem-solving paths of action are sought. This compels the designer to become personally involved with the problem, and this triggers creative thinking. For example, positions, sizes, etc., can be changed and the consequences observed.

This creative strategy has been employed by a number of individuals including Albert Einstein who, during his development of the theory of relativity, is said to have imagined what it would feel like to ride a photon through time and space.

The Ninth Strategy: Checklist of Questions

Prospect for solutions by using lists of questions.

Checklists are particularly useful when confronted by variant or adaptive design.

A. F. Osborn presents a list of idea-generating questions in his book, *Applied Imagination: Principles and Procedures of Creative Problem Solving.* Osborn's list is reproduced in Table 9.5. When a designer is confronted by an open-ended problem that demands creative thinking, a review of this list of questions can stimulate a number of potential solutions. This strategy has its greatest utility when it is applied to an existing design, as in adaptive or variant design.

Osborn's list of questions has broad utility in all creative phases of the design process and it provides the stimulus for some of the illustrative examples presented later in this chapter. It would be useful in the detail design phase of a product in which hardware considerations are the primary concern. An example is the creation of a gearbox of a certain size from an assortment of shafts, gears, seals, bearings, and housings. Osborn's questions would be useful in the conceptual design of a product that involves adapting a previous product to perform a new task. The design is largely unchanged; parts are simply rearranged and perhaps some new ones added. We shall now consider several examples where Osborne's list has provided the basis for new generations of products.

Osborn's question "Reverse?" involves the exchange of positions, functions, or movements of the elements of the system under development. Reverse thinking is relevant to the evolution of typewriter designs. The mechanical type-

Put to other uses?
New ways to use as is?
Other uses if modified?
Adapt?
What else is like this?
What other idea does this suggest?
Does the past offer a parallel?
What could be copied?
Who could be emulated?
Modify?
New twist?
Change meaning, color, motion, sound, odor, form, shape?
Other changes?
Magnify?
What to add?
More time?
Greater frequency?
Stronger?
Higher?
Longer?
Thicker?
Extra value?
Plus ingredient?
Duplicate?
Multiply?
Exaggerate?
Minify?
What to subtract?
Smaller?
Condensed?
Miniature?
Lower?
Shorter?
Lighter?
Omit?
Streamline?
Split up?
Understate?
Substitute?
Who else instead?
What else instead?
Other ingredients?
Other materials?
Other processes?
Other power?
Other place?
Other approach?
Other tone of voice?
Rearrange?
Interchange components?
Other pattern?
Other layout?
Other sequence?
Transpose cause and effect?
Change pace?
Change schedule?
Reverse?
Transpose positive and negative?
How about opposites?
Turn it backward?
Turn it upside down?
Reverse roles?
Change shoes?
Turn tables?
Turn the other cheek?
Combine?
How about a blend, an alloy, an assortment?
Combine units?
Combine purposes?
Combine appeals?
Combine ideas?

Table 9.5

Osborn's idea-generating questions.

writer, designed about 1860, contains a linkage system that drives a horizontally fixed type bar against the paper. Then a heavy carriage moves the paper to the next character position (Figure 9.10). The performance of the machine is limited by the time taken for the linkage to articulate and the time taken to move the heavy carriage. This long-used design was superseded in the 1950s by a new concept, the IBM *Selectric* typewriter, which was created by *reversing* the roles of the type element and the paper carriage (Figure 9.11). In the *Selectric* typewriter the paper only rolls up and down and a lightweight spherical type element moves across it at high speed.

An idea, like a ghost, according to the common notion of ghosts, must be spoken to a little before it will explain itself.

Charles Dickens
(1812–1870)

Osborn's question "Duplicate?" can suggest that several products be replicated, combined, and operated in unison. In 1884, Charles Parsons, an English engineer, applied the duplication approach to the development of the steam turbine by creating a multistage turbine that extracted more energy from the inlet steam than a single stage turbine.

Multihead drilling machines for mass production manufacturing operations use this strategy to simultaneously drill multiple holes in parts. Duplication also has been used in the

Figure 9.10

The type bars of an Underwood *typewriter from the 1930s.*

Figure 9.11

The spherical type element of an IBM Selectric *typewriter.*

creation of weapons with multiple gun tubes that fire in rapid sequence. Each gun barrel operates through a sequence of actions (charging, closing, shooting, etc.) at the pace of a single weapon, but the firing sequence is displaced among the barrels so that the combined effect is a very rapid fire.

Minification (Table 9.5) is illustrated in two generations of instruments for measuring the blood glucose levels of diabetics (Figures 9.12 and 9.13). People who suffer from

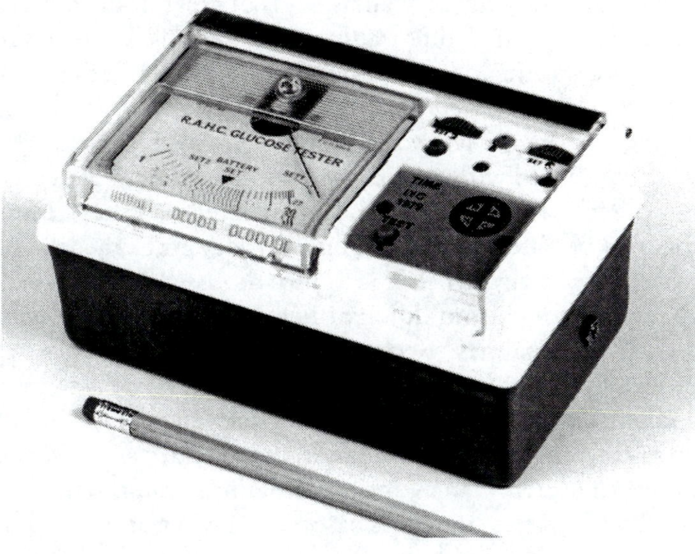

Figure 9.12

A 1980 portable blood glucose monitor (Courtesy BIO Transducers, Ltd.).

this disease need to adjust their food intake, exercise level, or insulin dosage several times a day to keep their blood glucose between prescribed limits. Before 1980 it was not easy to monitor blood glucose levels because tests could only be performed in doctors' offices or laboratories. There were no portable devices on the market. The device shown in Figure 9.12 was one of the first portable blood glucose monitors. It was manufactured by the Australian firm *BIO Transducers, Ltd.* This pioneering device had an analogue display, weighed about 700 grams, and measured 17 cm by 10 cm by 7.5 cm.

Naturally a glucose meter of this size was not convenient to carry and use several times a day. Eventual minification led to the small penlike device with a digital display shown in Figure 9.13. This state-of-the-art instrument weighs only 30 grams, is 14 cm long, just 1 cm in diameter, and can be carried like a pen.

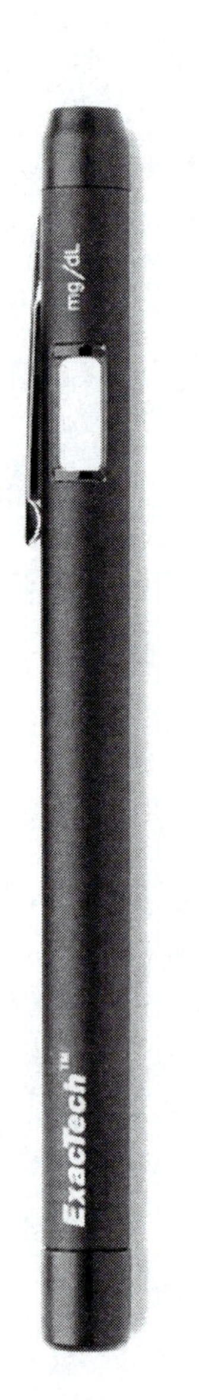

Figure 9.13

A 1995 pen-size blood glucose monitor (Courtesy Medisense, Inc.).

The Tenth Strategy: Synectics

Synectics is a group strategy for creative thinking that combines Strategies Six and Eight. It involves analogy, metaphor, empathy, and fantasy. It is of great utility when a radical departure from traditional solutions is required. Synectics was first described by W. J. J. Gordon in the book *Synectics: the Development of Creative Capacity* and was developed originally by a group at Arthur D. Little who were concerned with how people think creatively. Synectics is a Greek word that can be interpreted as "the joining together of different and seemingly irrelevant elements." As a creative strategy, synectics requires the assembly of a small group of people from different fields of expertise and with diverse personalities. Recent versions of the synectics approach have involved a facilitator who directs the activities, a client or customer who provides relevant details of the problem, and sometimes individuals with special skills. It is more structured than a group at a brainstorming session.

The synectics strategy is based upon two principles. The first is to understand the problem—to make the strange familiar. The second is to make the familiar strange. Once a design group understands the problem, the group tries to consider it in strange or unusual ways, particularly by employing analogies. This approach is intended to produce innovative ideas and new concepts.

Synectics: The study of creative processes, especially as applied to the solution of problems, by a group of diverse individuals.

Synectics is the joining together of different and seemingly irrelevant elements.

A synectics problem-solving session has the following characteristics:

Problem Definition

The task for the group is briefly presented. The customer may provide supporting information.

Understanding the Problem.

The problem is considered from several different perspectives to avoid the perceptual impediments described in Chapter 8. Perspectives can be remote or detached or intimate where a deeper understanding is sought. The personal involvement of an intimate perspective triggers creative thought patterns.

Man can learn nothing except by going from the known to the unknown.

Claude Benard
(1813–1878)

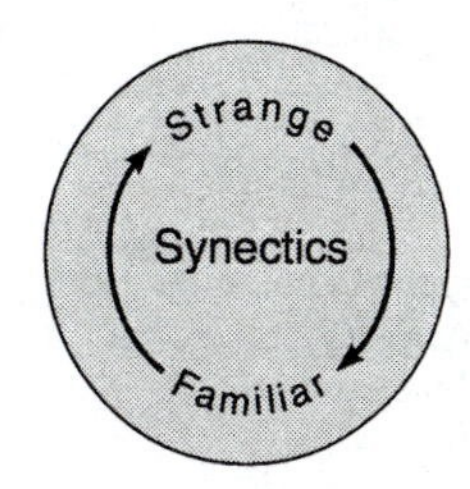

Steps in the synectics strategy:

- Problem definition
- Understand problem
- Seek solutions using analogies
- Evaluate

Prospecting for Solutions

Having ensured that the group members have developed an appreciation for the problem, the facilitator can then initiate a diversionary task of several minutes duration in which all of the group members participate. While the members focus upon this task, their subconscious minds continue to evaluate the original problem to avoid the emotional impediments described in Chapter 8. Subsequently, the group members are required to combine the attributes of the excursion with those of the original problem in order to trigger new ideas or directions for further thought.

The same comments regarding criticism too early in the procedure apply here just as they did in the discussion on brainstorming.

Evaluation

The group facilitator orchestrates the interactions among the group as they generate ideas through the discussion of analogies. These provide a series of new perspectives on the problem. The potential solutions must be recorded and evaluated in consultation with the customer before being refined by further synectics sessions.

Although human genius through various inventions makes instruments corresponding to the same ends, it will never discover an invention more beautiful nor more ready nor more economical than does nature because in her inventions nothing is lacking and nothing is superfluous.

Leonardo da Vinci
(1452–1519)

The Eleventh Strategy: Biomimetics

Study botany, zoology, biology, horticulture, ornithology, entomology, anatomy... for therein lies economy and efficiency of design.

Creative designers often can develop solutions to engineering problems by mimicking biological systems. This field is called *biomimetics.* It uses a direct analogy between a naturally occurring system and an engineering system. Most living systems are highly developed because they result from a long process of evolution and optimization. Therefore living organisms should be studied carefully to understand the subtleties of their designs and processes. Engineering students are encouraged to study botany, zoology, horticulture, ornithology, entomology, and the anatomy and physiology of plants and animals to discover potential biomimetic strategies for engineering design.

Neurological studies have provided a basis for new generations of computers.

Biological systems are a rich source of ideas for creativity in design. Consider the evolution of neural networks in which a large number of imperfect and functionally slow nerve cells together perform complex tasks. A study of neuron activity in the human brain helped inspire a new generation of computers that perform tasks in a parallel fashion rather than in a simple sequence characteristic of von Neumann computers.

We shall now examine how the study of biological systems has influenced the design of several other products:

- Airfoil sections mimic the geometry of the brown trout
- Designs for aircraft wings mimic features of bird wings
- The umbrella mimics the kinematics of flower petals
- Polymeric composite materials mimic biological composites
- Lightweight hexagonal cell structures mimic the honeycomb
- Hook-and-loop fasteners mimic the cocklebur

Airfoil Sections Mimic the Geometry of the Brown Trout (*Salmo trutta*)

The challenge that confronted Sir George Cayley around 1810 was the design of a low-drag shape for his fixed-wing flying machine. This brilliant designer exploited his knowledge of fish to propose that the geometry of the wing cross-section should mimic the sectional shape of the brown trout *Salmo trutta* (Figure 9.14). This sectional shape is obtained by making an incision at right angles to the plane of Figure 9.14 along the length of the fish. For reasons of efficiency

Ichthyology: The study of fish.

Figure 9.14

A sketch of the brown trout Salmo trutta.

and therefore survival, these fish have evolved to an ideal streamlined low-drag state. The upper drawing in Figure 9.15 reproduces Cayley's original sketch of the profile of a trout. The lower drawing contains points (denoted by circles) transferred from the original sketch and superimposed upon the outline of a modern low-drag airfoil section that was proposed some 150 years later. Note the remarkable correlation between the low-drag airfoil section and the low-drag section of the trout!

Characteristics of Cayley's work:

- he worked in isolation
- he rejected the in vogue notion of flapping wings and concentrated on fixed wings
- he first identified the aeronautical principles of lift, propulsion, and control
- he tested his hypotheses experimentally

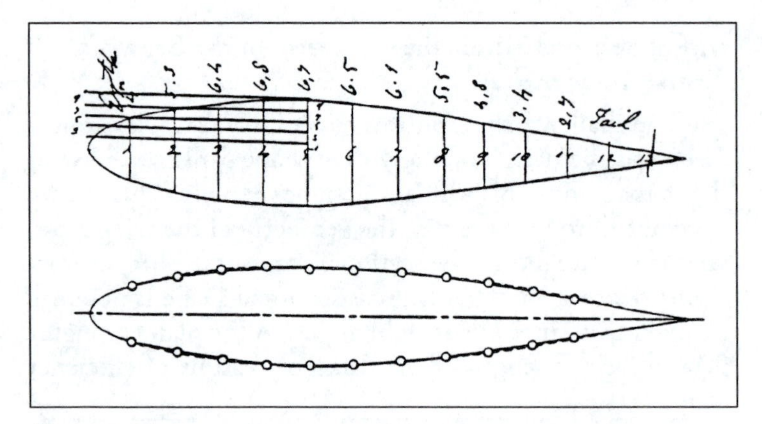

Figure 9.15

The upper drawing reproduces Sir George Cayley's original sketch of the cross-section of a brown trout. The lower drawing overlays the trout profile (represented by circles) on a modern low-drag airfoil section, NACA 63A016.

Design Methodologies for Aircraft Wings Mimic Features of Bird Wings

"My observation of the flight of buzzards leads me to believe that they regain their lateral balance, when partly overturned by a gust of wind, by a torsion of the tips of the wings..." These words were penned by Wilbur Wright to Octave Chanute on May 13, 1900. This observation was responsible for the Wright brothers' inventing the aileron, which is a hinged section of a wing along the trailing edge that permits aircraft to achieve lateral control. This innovation contributed significantly to the first powered flight in 1903.

Birds have profoundly influenced aircraft design in other areas too. In particular, the configurations of groups of feathers in the wings of various birds have been mimicked in the design of modern aircraft wings. For instance, the leading edge slats on aircraft such as the General Dynamics F111 swing-wing bomber are analogous to the wing configuration of woodland birds such as the ring-necked pheasant, *Phasianus colchicus*, which contain a thumb pinion or alula, as shown in Figure 9.16. The trailing edge flaps on the

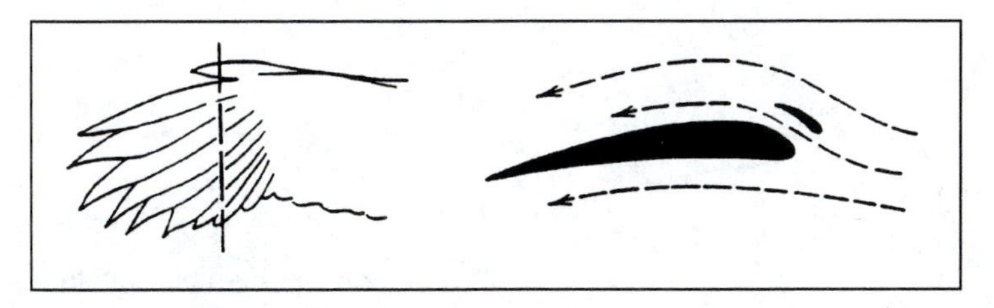

Figure 9.16

Schematic of the alula or thumb pinion of a pheasant which modifies the aerodynamic performance like a leading-edge slot.

multinational Panavian Tornado fighter bomber mimic the wing configuration of the split-tailed falcon. The multisurface airfoils of aircraft such as Boeing's 747 configured for takeoff or landing mimic the layered wing feathers of the gull family, *Laridae*, as shown in Figures 9.17 and 9.18. The winglets on the wing tips of Boeing 747–400s, Airbus 320s, and Lear jets mimic the drag-reducing fingered tip feathers

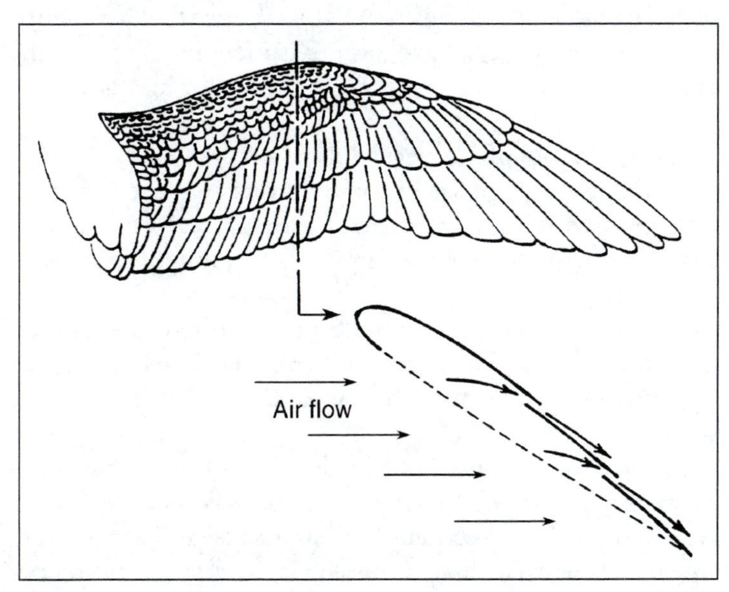

Figure 9.17

The layered wing feathers of a gull are replicated in multisurface aircraft wings.

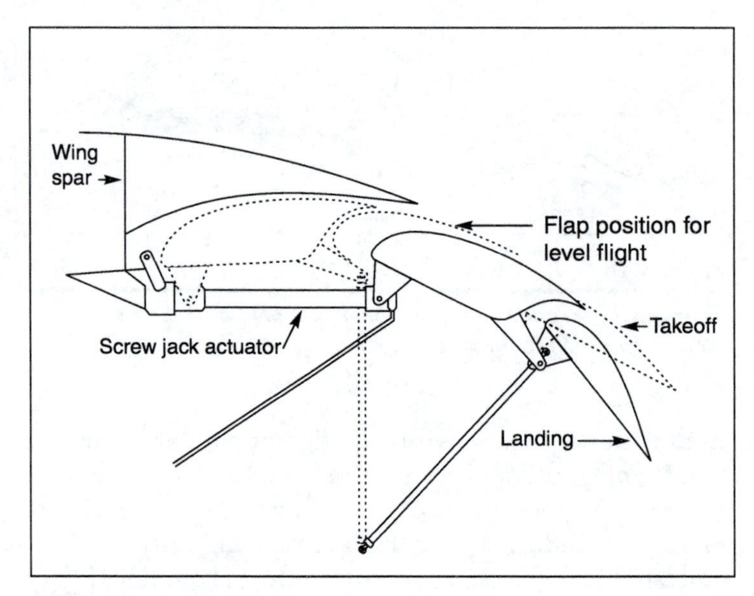

Figure 9.18

An aircraft flap system.

of the hawk family *Accipitridae*. It is evident that aeronautical design engineers have been inspired by birds in the creation of complex wing configurations for aircraft.

The Kinematics of Flower Petals

At sunset, the petals of some flowers coalesce to form a tight bud with a relatively small surface area. They unfold again at sunrise to expose a large area to sunlight for photosynthesis.

This biological process has been replicated by engineers in several fields. Consider the umbrella which is carried tightly packed but which unfolds for protection from rain or sun.

A second example is from the field of aerospace engineering. The payload envelope of most launch vehicles is relatively small. Consequently, designers of large space structures such as solar arrays, antennas, and space telescopes must package their structures densely for the launch. Once in orbit, the packed structures unfold like flowers exposed to sunlight.

Composite Materials

Biological systems have been emulated by materials scientists as well as by engineers. Two important fields have emerged from these studies. The first is the development of artificial body organs to enhance the quality of human life.

Polymeric composite materials mimic the anatomy of natural materials.

The second is the field of composite materials. Engineered synthetic materials are formed of a load-bearing material housed in a relatively weak protective matrix that mimics the microstructure of biological systems. Figure 9.19 shows a sectional view of a generic fibrous composite material with load-bearing fibers embedded in a supportive matrix. All naturally-occurring materials, from the human thigh bone to the husk of a coconut (Figure 9.20) are composite materials. None of these naturally-occurring objects are composed of a monolithic material.

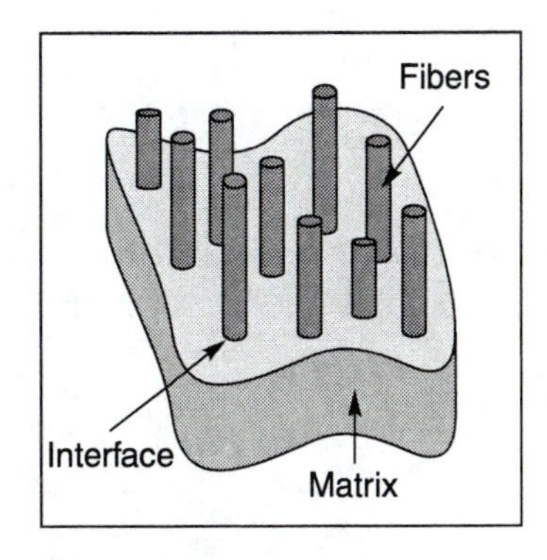

Figure 9.19

The principal features of a fibrous composite material.

There is a sharp contrast between the types of materials commonly specified by engineers and those found in plants and animals. However, by emulating biological materials, engineers have synthesized composite materials such as graphite-epoxy laminates and glass-polyester systems that have revolutionized the automotive, aerospace, and sporting goods industries.

Natural biological solids are composite (not monolithic) materials.

Typical engineering composites are graphite-epoxy and glass-polyester.

The importance of searching for new materials cannot be overemphasized because materials technology has always profoundly affected the evolution of civilization. Historians characterize periods in this evolution by the materials that were dominant at that time. Thus terms such as "the Stone Age," "the Bronze Age," and "the Iron Age" have entered our vocabulary. Materials scientists and engineers are now creating new "smart" materials that contain intrinsic sensors, actuators, or microprocessors. Products made from these materials will be able to respond autonomously and intelligently to variable service conditions.

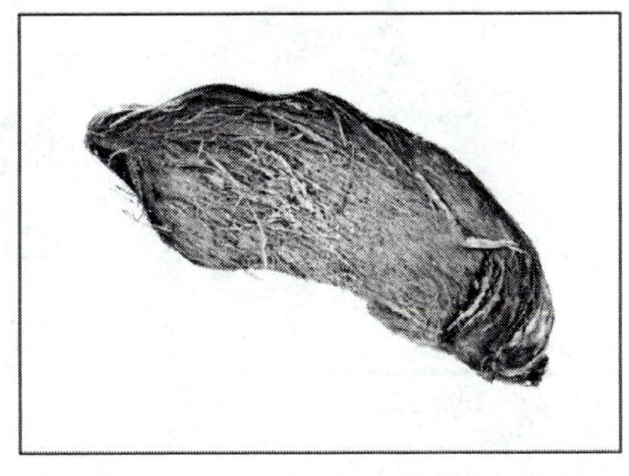

Figure 9.20

Husk of a coconut showing the fibrous structure.

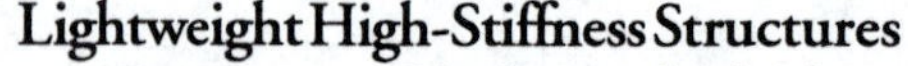

Lightweight High-Stiffness Structures

Designers of lightweight high-stiffness structures have frequently mimicked biological structures that contain hexagonal cells—the "honeycomb" structure. Wasp nests and bee honeycombs are characteristically lightweight and stiff. This structure can be seen clearly by the naked eye (Figure 9.21, right).

Hexagonal-cell structures occur in natural materials at the macroscopic and microscopic levels.

Hexagonal-cell structures also occur in biological systems at the microscopic level. With a microscope they can be seen, for example, in an insect cuticle. Hexagonal cell structures have been used to make stiff, lightweight structures of metal and other materials. These have appeared in the shells of space capsules, in components for space telescopes, in large tools for manufacturing parts, in racing cars, and in the airframes of planes such as the British Aerospace VTOL Harrier GR.3.

Another example of a lightweight high stiffness structure adapted from a biological system is the ski pole shown in Figure 9.22. This photograph also shows the fibrous composite nature of the material. Compare the hollow stemmed structure to that of a flower stalk in the same figure. Both

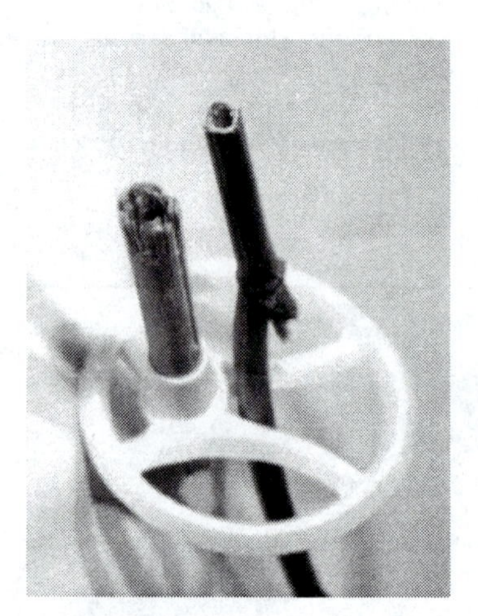

Figure 9.22

A broken ski pole showing its hollow stem and fibrous composite material mimicking the flower stalk of the Meadowrue on the right.

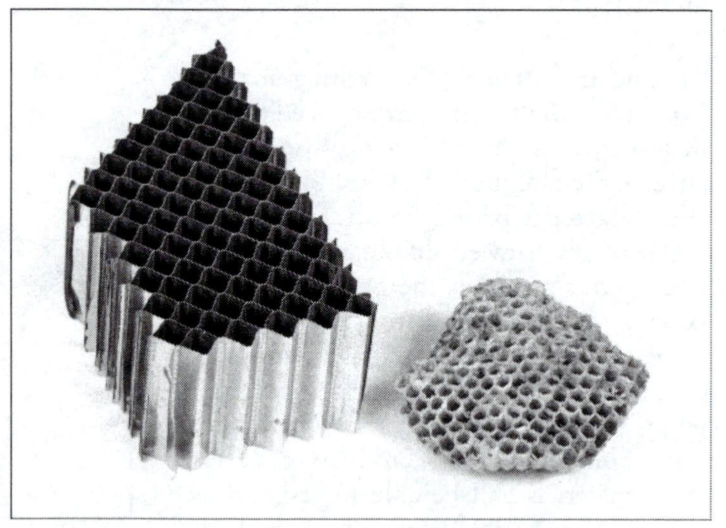

Figure 9.21

A wasp nest (right) and a metal hexagonal cell structural material (left).

structures are hollow, both feature long continuous fibers parallel to the longitudinal axis of the stem, and both show the same failure mechanism.

Hook-and-Loop Fasteners

Hook-and-loop fasteners—the primary product line of Velcro USA, Inc.—are the result of a serendipitous event involving the mimicking of a biological system. In the late 1940s, the Swiss inventor George de Mestral returned from a walk in the countryside and noticed that seeds from woodland cocklebur plants were attached to his trousers. The tenacity with which the burs adhered to his trousers motivated de Mestral to study the situation. Under a microscope he discovered that the cocklebur was covered with many small hooks, and the surface of his trousers contained tiny loops. These features are evident in Figure 9.23. The engagement of hooks and loops was the essential element in the attachment of the seed to the fur of passing animals for dispersion and propagation.

Figure 9.23

A photograph of a cocklebur (Xanthium). *Note the tiny hooks that hold the bur tightly to clothing.*

Cockleburs and woolen trousers inspired the hook-and-loop fastener.

Subsequently, de Mestral developed a hook-and-loop tape to mimic the cocklebur and trouser interface. The hook tape contained small stiff hooks while the loop tape had small soft loops (Figure 9.24). Velcro fasteners today are used in situations ranging from the space shuttle to the medical, aircraft, electronics, apparel, and automotive industries.

Figure 9.24

A photograph of a Velcro *hook-and-loop fastener.*

9.2 The Student Design-Build-Test Project

The student groups employed several different methods to create vehicle concepts. Many groups recorded their ideas randomly using sketches while others employed morphological diagrams to generate ideas. A collage of diverse ideas from a typical student group that adopted the first philosophy is presented in Figure 9.25.

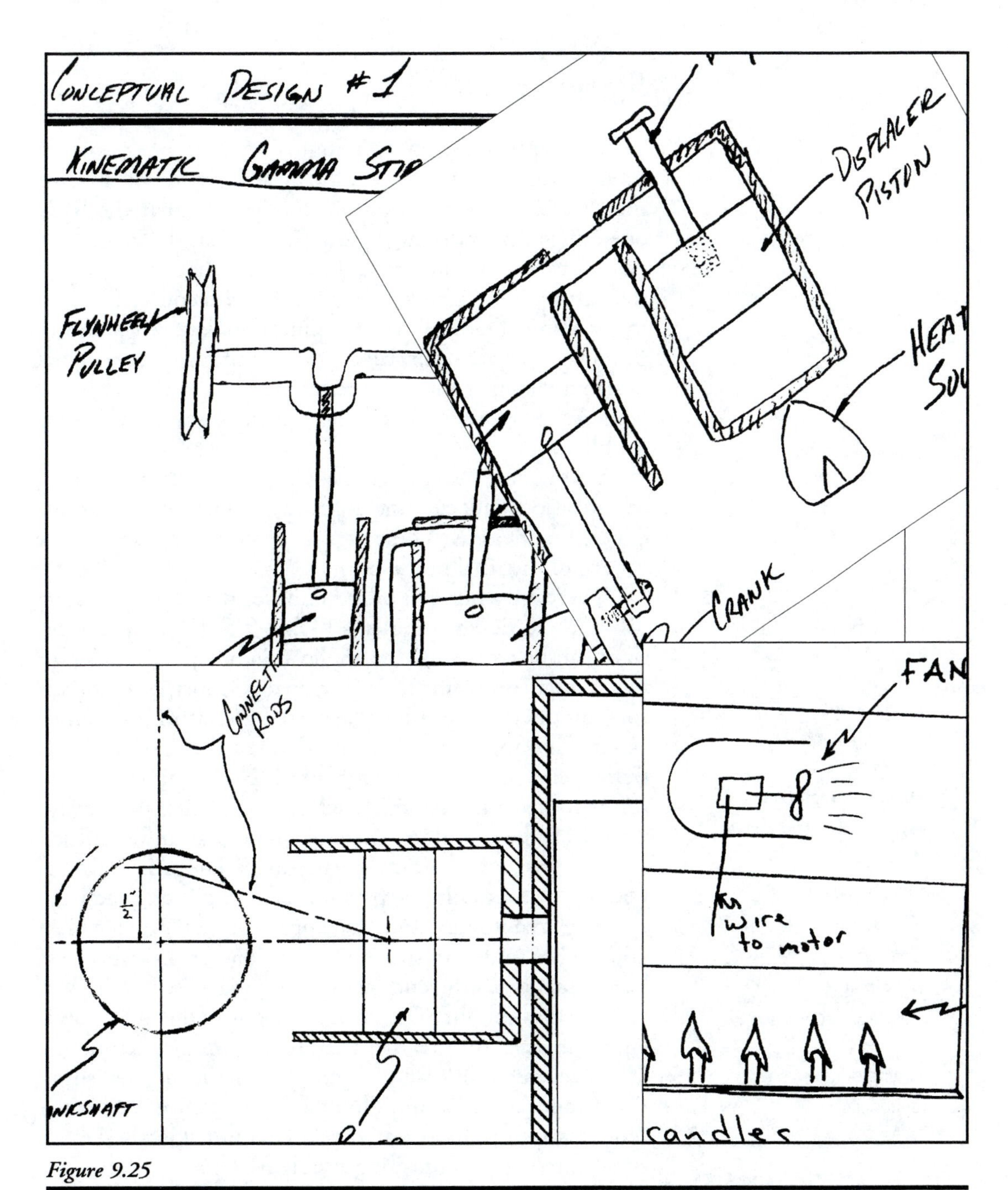

Figure 9.25

Some student sketches during concept development.

Summary

Conceptual design is a highly creative phase of the product realization process and is primarily responsible for the success of the final product. It is here that the basic operating principle of the product is established. Conceptual designs can be classified into three groups that depend upon the nature of the activity. *Original design* is concerned with the creation of radically new products; *variant design* creates products in which the size or configuration of an original design is changed but the function remains unchanged; and *adaptive design* uses an existing design to solve a different problem.

The conceptual design phase is initiated by a design specification and culminates in the evaluation of various conceptual designs. These designs typically are sketches or visions of the final product.

There are a number of methods for developing creative solutions to design problems (Table 9.5). Conceptual designs can be created by the application of a previous design to a similar problem. Indeed in many industries competitors' products are routinely analyzed and attributes duplicated. This approach is enhanced when the designer reads trade literature and visits trade shows. New combinations of standard machine elements such as gears, belts, and pulleys also yield new solutions. New technologies can be infused into old products to create new products. The prospecting of new opportunities in the patent literature can be beneficial.

The hypothesize-and-test approach can provide new approaches to design situations. Trial and error is the most basic and least efficient of these approaches, but it does provide the basis for brainstorming—one of the more popular techniques for creative thinking. Analogies frequently have been used to solve problems by relating a given situation to another situation. Thinking by analogy is used in the synectics approach to creativity and in mimicking biological systems to solve engineering problems.

Strategies for Individuals or Groups
Duplication
Combinations of elements
Historical evolution
Hypothesize and test
Analogy
Morphology
Empathy
Checklists
Biomimetics
Strategies Only for Groups
Brainstorming
Synectics

Table 9.5

Strategies for creative thinking.

Morphological approaches involve the decomposition of a problem into its major subproblems prior to systematically seeking different combinations of these solutions using a three-dimensional matrix. Empathy is an approach to creative thinking whereby the designer imagines that he or she is a part of the situation for which a solution is sought.

Checklists can also be used creatively to solve ill-defined problems. Questions such as "adapt?" or "reverse?" can have great utility as work is done on variants of original designs or on detail design tasks.

Conceptual designs created by these processes must be evaluated to determine which proposed solution offers the greatest potential for development into a successful product. Evaluation procedures are described in the next chapter.

Key Concepts

Adaptive design. Adaptive design involves changing an existing product to satisfy a new design specification. The term is applied either to the product resulting from such a design activity or to the design activity itself. Adaptive design typically requires the creation of new parts and assemblies but the principle of the product generally is not changed.

Analogy. A strategy for solving a design problem by recognizing some similarity between two things that are otherwise unlike. The designer may solve the problem by adapting or modifying ideas from other situations. The approach assumes that certain resemblances imply further similarities.

Biomimetics. A design strategy based upon a direct analogy between a biological system and an engineering system. A biological system is mimicked in engineering practice.

Brainstorming. A technique for developing creative solutions to open-ended problems. In brainstorming, a group of individuals with diverse backgrounds attempts to solve a specified problem by recording all ideas spontaneously suggested by the group members. After these ideas are recorded and evaluated, the most promising ones are developed.

Checklists. This method for creative thinking involves reviewing a list of questions that can stimulate new ideas. This approach is particularly useful in variant design and adaptive design. Typical questions are: *magnify? substitute? reverse? rearrange? combine? modify? adapt?* and *minify?*

Conceptual design. A conceptual design is created in an early stage of the product realization process. It is a vision—the basic concept—for a product in response to the design specification. Conceptual designs often are created in the form of freehand sketches. Subsequently they are evaluated and the best selected for further refinement, analysis, and optimization during the product design stage.

Combinations of elements. This method of creativity involves the recognition of a new relationship among conventional machine elements. A prerequisite is a wide knowledge of off-the-shelf elements that are variously regrouped until a potential solution is discovered.

Creative thinking. Creative thinking is associated with randomness, imagination, visualization, synthesis, nonjudgmental scrutiny, irregularity, and sensuality. It is the antithesis of deductive thinking, and it occurs in the right hemisphere of the brain. Creative thinking is sometimes called associative thinking, divergent thinking, or lateral thinking. It is, for example, employed to generate conceptual designs.

Deductive thinking. This thinking is associated with logic, analysis, judgment, and evaluation. It is also called convergent thinking or vertical thinking and it occurs in the left side of the brain. Deductive thinking is the antithesis of creative thinking. This type of thinking is employed during the evaluation of conceptual designs and also when computational techniques are used during the product design phase.

Divergent thinking. A synonym for creative thinking.

Duplication. A method for developing a conceptual design either by directly duplicating someone's concept with modifications to avoid patent infringements or else by recognizing a similarity between the problem to be solved and another solved problem.

Empathy. A method of creative thinking in which the individual imagines that he or she is a part of the situation for which a solution is sought.

Historical evolution. Old designs may have been constrained by scientific and technological limitations that are now resolved. Consequently, new concepts can arise as new technologies are infused into old designs.

Hypothesize-and-test. New designs can arise as prototype products are tested experimentally or by computer simulation.

Morphology. This method for developing new solutions uses a three-dimensional matrix or morphological box for a systematic search. The primary components of the problem are recognized and potential solutions to these subproblems are associated with cells of the three-dimensional matrix. The process can be employed with different combinations of subproblems.

Original design. This is either the product resulting from a first-generation conceptual design or it is the activity itself. Original design generally occurs at the highest level of creative design activity.

Reverse Engineering. The process of disassembling and analyzing a product to determine how it was designed and manufactured. Enhancements are then suggested or ideas are incorporated into other products.

Synectics. The study of creative processes, especially as applied to a group of diverse individuals

Variant design. A design activity—or the product itself—that results from a variation in the size or configuration of an existing product. The function and the solution principle remain unchanged.

Review Questions

1. Define conceptual design and explain why it is a crucial phase of the product realization process.
2. Define three different classes of conceptual design. How do they differ? Which is the most difficult?
3. List the different strategies presented in this chapter for creating conceptual designs and classify them as applying to groups of people or to individual designers.
4. Duplication of concepts is common. How is this approach implemented in practice?
5. If designers wish to extend their knowledge of the numerous conventional machine elements, what sources should they consult?
6. Discuss the broad array of possibilities associated with the strategy of hypothesize and test. What prolific inventor employed this approach with great success by using rudimentary forms of trial and error.
7. Why is trial and error the most inefficient method of creatively solving a design problem when it is used by only one individual?
8. What are serendipitous discoveries? What is the typical prerequisite for these discoveries.
9. List some important ingredients of a brainstorming session.
10. How is the strategy of analogy related to the first strategy of duplication?
11. The use of direct analogy is responsible for a strategy relating biological systems to engineering problems. What is it called?
12. List the primary steps in the solution of a problem by the strategy of morphology.
13. Describe three examples of the use of the checklist strategy.
14. Describe the group strategy of synectics and compare it with brainstorming.

15. Why do biological systems provide a rich source of ideas for conceptualization?
16. Describe three features of birds replicated in the wings of modern aircraft.
17. How has the mimicking of biological materials influenced the aeronautical industry? What are the main features of these materials?

Problems

1. Some 70 million American adults and schoolchildren spend part of their day working at a computer keyboard. Its an essential component of each day's activities. However, if you spend four or more hours at a computer keyboard each day, you risk repetitive strain syndrome. This injury, which causes pain and numbness in the fingers and wrists, has disabled hundreds of thousands of workers.

   Design a workstation that will minimize these injuries. The design should include the keyboard, the operator's chair, and the location and orientation of the computer screen relative to the chair. Specify exercises for the computer operator to minimize injuries.
2. Explore options for a family car that can travel 100 miles for each gallon of gasoline consumed. Your design should incorporate the same safety measures as in current vehicles, there should be no reduction in performance, and the car should have the same load-carrying capacity as current models. Explore a number of hybrid designs that use batteries, flywheels, fuel cells, and other energy-storage devices. Consider the minimization of losses by drag through a study of the vehicle's aerodynamic characteristics. The car must comply with exhaust-emissions legislation.

3. Propose a number of conceptual designs for a mechanism to activate a cast iron worktable of the type frequently found in machine shops. Evaluate these designs using a decision matrix and quality ratings.

   Assume that the cast iron table is rectangular with dimensions 1000 mm by 1500 mm and that the upper face contains T-slots. The actuation mechanism must provide a vertical rise-and-fall motion. There must not be any rotation about the vertical axis or any lateral motion. The design specification must satisfy the following requirements.

   a. The table must translate through a vertical distance of 150 mm.
   b. A cast iron tube support structure for the table must house the actuation mechanism. The internal diameter of the tubes is 350 mm.
   c. The table must be capable of being locked at any position of the vertical travel.
   d. The loading on the table may be represented by a single vertical load of 100 kg imposed at the center of its upper surface.
   e. The actuation mechanism must not interfere with use of the machined table top, which must support workpieces of different shapes and sizes.
   f. The actuation mechanism must be operated manually by a conventional handwheel and crank handle characteristic of many machine tools. This handwheel is to be keyed to a horizontal shaft protruding from the tubelike table support structure.
   g. The vertical position of the table must be adjustable within 0.10 mm.
   h. Twenty worktables are to be manufactured.

4. Develop a design specification for a wheelbarrow to be used in a residential garden. The design should be easy to use even if the ground is soft, must be storable outside without protection, and must withstand overloading

and abuse. The wheelbarrow must be designed for mass production and minimal maintenance. Develop and evaluate several conceptual designs for this product.

5. Develop a design specification for a manually operated kitchen aid that can slice a ripe tomato. Create and evaluate several conceptual designs for this product.
6. Design a machine to form office paper clips from a continuous strand of wire drawn from a drum. First develop a comprehensive specification by carefully studying the characteristics of paper clips from several different suppliers. Then create and evaluate several conceptual designs.
7. In Chapter 7 you developed a design specification for a collapsible umbrella that will fit inside a coat pocket. Now create several conceptual designs and evaluate them.
8. Create a machine to help bank tellers sort coins. The machine must accept an assortment of coins of differing thicknesses and diameters that have been deposited in a hopper. The design should sort quarters, dimes, nickels, and pennies. Develop the specification, create several conceptual designs, and select the best.
9. Design an adaptable fixture for a flexible manufacturing cell. This fixture must be capable of accommodating parts of different shapes and sizes weighing up to 5 kg. It is envisaged that a coordinate measuring machine and computer vision will be used with the fixture. Assume that the fixture is to be used for a variety of tasks such as automated assembly operations in which the fixtured parts must not move when subjected to a maximum force of 140 Newtons.
10. Design an energy dissipation device for translational motion. This device is to be incorporated into a mechanism for an electrical switchgear. When the device is triggered it must dissipate 200 Nm of kinetic energy during the final 15 to 20 millimeters of a total translational motion distance of 100 mm. The electrical re-

quirements of the switchgear mandate that the device have no bounce-back upon completion of the 100 mm travel.

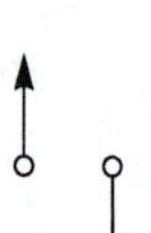

11. Develop conceptual designs for a device which must transmit a straight-line reciprocating motion from one machine member to a second member parallel to the first in order that the members travel equal distances but in opposite directions.
12. Design a device for use by impaired persons to help them turn the pages of a book without the use of fingers.
13. Develop a number of conceptual designs to help an arthritic person open a milk carton.
14. Generate a number of new conceptual designs for a product that will open a domestic tin can containing peaches in syrup.
15. Generate a number of conceptual designs for a product that will remove the cork from a wine bottle.
16. Develop conceptual designs for a device that will prevent an intoxicated person from driving an automobile.
17. Develop a morphological analysis of a machine for drying garments. Analyze an element of this machine.
18. Generate a number of conceptual designs to measure the velocity of a gas in a parallel-sided straight tube.
19. The domestic toilet accounts for almost 50 percent of the U.S. household water consumption. Because of the increasing cost of water, a better design is needed. Generate conceptual designs for a toilet flushing unit that will significantly reduce water consumption without affecting the product's efficiency.
20. Create conceptual designs for machine that will process onions for a company that markets pickled onions. The machine must be able to peel, top, and tail the

vegetable while minimizing waste from overpeeling or undesired cutting. The machine must accommodate onions of varying shape, size, and hardness.

21. Keyboards are an essential ingredient of our current lifestyle. They are a characteristic of telephones, typewriters, computer terminals, musical instruments, and the checkouts in stores. By reviewing the patent literature, chronicle the evolution of keyboards.

Further Reading

Alger, J. R. M. and C. V. Hayes. *Creative Synthesis in Design.* Englewood Cliffs, NJ: Prentice-Hall, 1964.

Avallone, E. A. and G. Baumeister, eds. *Mark's Standard Handbook of Mechanical Engineering.* New York: Reinhold, 1988.

Chironis, N. P. *Mechanisms and Mechanical Devices Sourcebook.* New York: McGraw-Hill, 1991.

Cross, N. *Engineering Design Methods.* New York: John Wiley, 1989.

French, M. J. *Invention and Evolution—Design in Nature and Engineering.* Cambridge: Cambridge University Press, 1988.

Gandhi, M. V. and B. S. Thompson. *Smart Materials and Structures.* London: Chapman and Hall, 1992.

Golley, J. *Whittle: The True Story.* Washington, D.C.: Smithsonian Institution Press, 1987.

Gordon, W. J. J. *Synectics: the Development of Creative Capacity.* New York: Harper and Brothers, 1961.

Hubka, V. and W. E. Ernst. *Theory of Technical Systems: A Total Concept Theory for Engineering Design.* New York: Springer Verlag, 1988.

Jones, F. D. *Ingenious Mechanisms for Designers and Inventors, Vols. 1, 2, and 3.* New York: Industrial Press, 1930, 1936, 1951.

Jones, J. C. *Design Methods.* New York: Van Nostrand Reinhold, 1992.

Laithwaite, E. *An Inventor in the Garden of Eden.* New York: Cambridge University Press, 1994.

Mayer, R. E. *Thinking, Problem Solving, Cognition.* New York: W. H. Freeman and Company, 1983.

Newell, J. A. and H. L. Horton. *Ingenious Mechanisms for Designers and Inventors, Vol. 4.* New York: Industrial Press, 1967.

Osborn, A. F. *Applied Imagination. Principles and Procedures of Creative Problem Solving.* New York: Scribner and Sons, 1953.

Pahl, G. and W. Beitz. *Engineering Design.* London: The Design Council, 1984.

Parnes, S. J. and H. F. A Harding. *Source Book for Creative Thinking.* New York: Scribner and Sons, 1962.

Poincaré, H. *Science and Method.* London: Nelson, 1914.

Roberts, R. M. *Serendipity: Accidental Discoveries in Science.* New York: John Wiley, 1989.

Van Gundy, A. B. *Techniques of Structured Problem Solving.* New York: Van Nostrand-Reinhold, 1988.

Weisberg, R. W. *Creativity: Beyond the Myth of Genius.* New York: W. H. Freeman and Company, 1993.

Evaluation of Designs

10

Fully to understand a grand and beautiful thought requires, perhaps, as much time [as] to conceive it.

Petrus J. Joubert
(1831–1900)

Objectives

- To emphasize the complexities of evaluating conceptual designs.
- To present several decision matrix methods for evaluating different designs.
- To introduce a quality rating factor for determining how successful the design process has been.

Contents

What If...

You're a student in a design projects course and you have been assigned the task of designing a better mousetrap. Your design team has created a large number of conceptual designs, but you are uncertain which is the best.

What criteria should you establish to evaluate these visions and what should be the target values for these criteria? How should you evaluate aesthetic aspects of your design? Since all design parameters have varying significance, should this variation be considered in the evaluation process?

How can you measure quality? How can you identify weaknesses? Will the designs work? Clearly there are numerous challenging issues to be addressed in this process of evaluation. What should you do?

10.0 Introduction

Design methodologies cycle through three classical activities: creativity, analysis, and evaluation. In this chapter we examine evaluation. Although our discussion here centers on evaluation in conceptual design, evaluation naturally plays a part in all design activities. It applies to the selection of parts and subassemblies for a product, to the acquisition of off-the-shelf items, and to the choice of manufacturing procedures.

Evaluation at the conceptual design stage has some additional challenges because of the embryonic nature of the ideas involved. Many of the psychological impediments to product creation were discussed in Chapter 8.

Evaluation is always a threat, always creates a need for defensiveness, always means that some portion of experience must be denied awareness.

Carl R. Rogers
(1902–1987)

Designs are evaluated by comparing each one to a set of design criteria. The summation of these data provides the basis for determining the strengths and weaknesses of individual designs and for selecting the best design. These data can be used to integrate perceived strengths of competing designs to create a new design and to rectify areas of weakness in the chosen design.

10.1 Evaluation of Conceptual Designs

The conceptual design phase of the product realization process has two primary ingredients: the conceptualization of a large number of potential solutions and the subsequent evaluation of these solutions. Generally only a small number of conceptual designs receive further development and continued scrutiny. One concept finally emerges as the best. This conceptual design then enters the product design phase of the product realization process.

The best design is the simplest one that works.

Albert Einstein
(1879–1955)

As discussed in Chapter 9, the first stage of the conceptual design phase involves creative thinking. However, evaluation requires deductive thinking. If evaluation is undertaken with the methodology of Figure 1.1—involving analysis,

Evaluation is an iterative process.

Methods for evaluating designs:

- Prototyping
- Authoritarian customer
- Dictatorial designer
- Decision matrices

It is with our judgements as with our watches; no two go just alike, yet each believes his own.

Alexander Pope (1688–1744)

Decision matrices provide:

- a framework for evaluation
- a systematic approach
- a basis for determining strengths and weaknesses

During the evaluation process, several decision matrices should be developed using more technical and quantitative criteria as the designs mature.

At the end of each evaluation cycle, the number of conceptual designs is reduced and the complexity of the evaluation process is increased.

evaluation, and synthesis—then it can also stimulate deductive thinking during scrutiny, comparison, and decision-making. By that means evaluation becomes an iterative process.

The evaluation process can assume several forms. Prototypes of the various concepts could be developed and tested; the customer could assume authority for the decision as to which is the best solution; or a dominant member of the design team may argue for a preferred approach. However, an approach employing decision theory and a decision matrix offers a number of advantages. This approach is even more effective if ideas are evaluated by a multidisciplinary team from product development, manufacturing, design, sales, and marketing.

Decision matrices for evaluating conceptual designs provide a framework for the review and evaluation process. They motivate a systematic comparison of the features of each conceptual design relative to the evaluative criteria. Such a systematic procedure is important because it minimizes the chance that conceptual designs with the greatest potential for success will not be identified.

Generally in a decision matrix the names of the conceptual designs appear along the top of the matrix and evaluative criteria are placed in the column at the left of the matrix. A decision matrix facilitates visual comparison of each design. Decision matrices should be developed several times during the evaluation process, and between each iteration both analysis and synthesis should be undertaken. The number of conceptual designs subject to evaluation can change and the evaluative criteria generally become more technical and quantitative as the iteration process evolves.

Evaluative criteria are derived from the design specification of the product and reflect requests of the customer through use of the quality function deployment (QFD) method described earlier. During initial evaluation cycles, the criteria are somewhat qualitative. As evaluation proceeds, however, knowledge of the product grows and the evaluative criteria change. Ultimately they become the target values of the design parameters. Thus the sophistication of the analysis increases as knowledge of the product grows.

It is difficult to decide when evaluation ends and early phases of product design begin. During the latter stages of evaluation, preliminary calculations are developed for the concept, although many simplifying assumptions may be required. The objective is to establish order-of-magnitude data. Rough sketches of layouts can be made, models can be built, and preliminary tests can determine some product characteristics. Economic models can be generated. A cost-benefit analysis at this time could increase the value of the product while reducing its manufacturing cost.

Use cost-effective models in the evaluation of concepts.

Evaluation involves risk and uncertainty.

Cautionary Notes on Evaluating Embryonic Ideas

The early phases of the evaluation of conceptual designs are challenging for several reasons:

1. Conceptual designs are, by their very nature, speculative concepts whose characteristics are not well understood. The characteristics of these concepts may change quite dramatically as the design matures, and they may differ substantially from those originally anticipated. Naturally as the concept evaluation process evolves and more detailed studies are completed, this uncertainty will gradually be alleviated.
2. In order to facilitate a comparison of the concepts, all designs should be developed to the same degree of maturity. Furthermore, all concepts should be evaluated in the same detail and relative to the same criteria.
3. Many evaluative criteria are subjective and qualitative. It is much easier to assign a numerical value to a quantifiable parameter such as product weight than to evaluate reliability or an aesthetic property. Consequently engineering intuition and judgment are at a premium.
4. The first few iterations of the decision matrix involve comparisons of tentative concepts. The

The maturity of concepts should be the same. Evaluative criteria often are subjective and qualitative.

information is generally not quantitative because these designs are too immature for numerical analysis. At that stage a ranking of +, 0, and – can be used to reflect notions of good, average, and inferior.

5. Even for embryonic products, costs should be considered during evaluation because they so affect the ultimate economic success of the product. All other things being equal, materials and manufacturing methods will determine the lowest price at which a product can be sold. In the early stages of evaluation, a part count could be used as a crude measure of cost.

10.2 Decision Matrices

If you can measure that of which you speak, and you can express it by a number, you know something about your subject; but if you cannot measure it, your knowledge is meager and unsatisfactory.

Lord Kelvin
(1824–1907)

A decision matrix can be established as shown in Table 10.1. Each evaluative criterion is assigned points and each conceptual design is evaluated relative to each criterion. In this way, points are assigned to each design and the score is recorded in the decision matrix.

This approach can be applied to the overall design, to subassemblies, and even to the design of individual parts.

Decision matrices apply to all open-ended problems from job offers to house purchases.

They apply to complete products, assemblies, subassemblies, and individual components.

	Conceptual Designs							
Evaluative Criteria	*a*	*b*	*c*	*d*	*e*	*f*	*C*	*I*
Action require	3	1	3	2	1	2	3	3
Simplicity	3	0	1	1	2	3	2	3
Product cost	2	...	1	3	1	2	2	3
Size	3	...	2	3	2	2	3	3
Energy consumption	3	...	1	2	1	2	2	3
Total	14	...	8	11	7	11	12	15

Table 10.1

Decision matrix for a set of conceptual designs.

Indeed it can be applied to any open-ended problem that requires a selection from among several alternatives—even to the purchase of a bicycle or a comparison of job offers.

Each conceptual design is assigned points depending on whether the potential solution is, for example:

Very suitable	3 points
Suitable	2 points
Marginally suitable	1 point
Unsuitable	0 points

If a conceptual design is awarded zero points relative to a particular evaluative criterion then it is unsuitable and the candidate is no longer considered in the evaluation process. Some aspects of such a design of course may have merit and can be incorporated into another concept.

By numerically evaluating each conceptual design relative to each evaluative criterion and summing the results, the design team can evaluate each concept. These numerical data can then be used to identify a potentially successful product. In addition, the data help to identify weak and strong areas in each design. Areas of weakness can be strengthened while areas of strength can be incorporated into sales literature for the marketing department.

Decision matrices help identify areas of weakness and strength.

Evaluation of conceptual designs and development of design improvements generally yield new conceptual designs. Consequently the size of the decision matrix often increases after removal of design candidates with the least potential. Throughout the evaluation process the trend is convergence toward a single conceptual design, but during this convergent process the size of the decision matrix often first decreases and then increases as poor conceptual designs are removed from the matrix and then new conceptual designs are added (Figure 10.1).

A typical evaluation scheme for six conceptual design candidates *a, b, c, d, e,* and *f* is presented in Table 10.1. For illustrative purposes they are evaluated relative to five criteria. Many more criteria would be used during an actual evaluation. In addition, they are evaluated relative to a competitor's product *C* and an ideal product *I*.

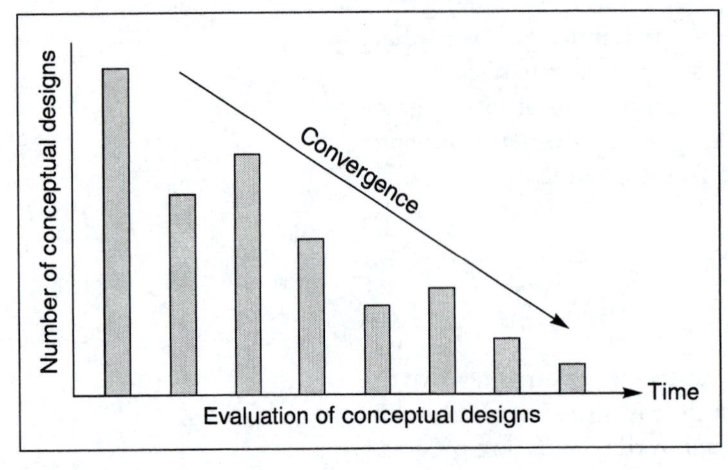

Figure 10.1

The number of conceptual designs decreases and increases as the evaluation proceeds. The overall trend is convergence to the best concept.

The best candidate is *a* because this design is assigned the most points. If after evaluation several designs have been assigned the same number of points, then a more detailed comparison could be undertaken. For example, each criterion could be assigned a range of 0–5 or 0–10 points in order to generate a broader range of numbers in the decision matrix. However, it is probably more appropriate to subject these design candidates to the next iteration of the evaluation process because during the early stages there is only tentative information on the different concepts. These numbers only provide guidance. Later, more comprehensive technical analyses are undertaken and the evaluation becomes more accurate.

Quality of Conceptual Designs

Measure the quality of conceptual designs by comparing them with an ideal design incorporating the best features of all designs, a competitor's product, or a previous generation product.

The effectiveness or quality of conceptual design *a* in the above example can be determined by comparing the total points assigned to this design with the total number of points assigned to the *ideal* design solution, *I*, which is listed in the extreme right hand column in Table 10.1. This can be accomplished by generating a *Conceptual Design Quality Rating* defined by the following quotient:

$$\text{Conceptual Design Quality Rating} = \frac{\text{Number of points awarded to the actual solution}}{\text{Number of points awarded to the ideal solution}}$$

The quality rating provides one measure of the success of the conceptual design process—a process that will typically be responsible for 75 percent of the product's total cost. Conceptual design is so important in the product realization process that it mandates comparison of designs with some reference that will stimulate refinements, enhancements, and the formation of superior concepts. If the company has completed some benchmarking activities, then the most competitive product, designated *C*, should be included in the decision matrix, as in Table 10.1. This competitive product can be used as the reference against which all other conceptual designs are evaluated.

Alternatively, the reference could be the most promising conceptual design, or a previous generation of the proposed product, or a concept that incorporates the best features of all the proposed concepts. With the latter approach, alternative conceptual designs can be evaluated as in Table 10.1 before comparison of each design with the reference design. The reference in the decision matrix can be assigned a set of zeros, and positive or negative points can be assigned to the other conceptual designs depending upon whether they are superior or inferior to the

. If a conceptual design satisfies an evaluative criterion with the same success as the reference design, then that design is awarded zero points or is assigned an *S* to indicate that the design is the *S*ame. The number of *S* (zero), positive, or negative points can be summed for each category to provide a measure of the quality of the conceptual designs relative to the reference design, as shown in Table 10.2. The results are presented in the lowest three rows of the matrix.

Compare your designs with a competitor's product.

Use a previous generation product as the reference for measuring the quality of a design.

Use a + or − points system to compare concepts to a reference design.

Table 10.2 presents a decision matrix for four conceptual designs relative to five evaluative criteria and a benchmarking reference design. The designs are evaluated relative to a scale (Table 10.3) with seven graduations.

Evaluation of conceptual designs by a decision matrix approach requires the designer to systematically evaluate the important aspects of all solutions. This triggers creative improvements. Numbers generated by a decision matrix are nothing more than pointers to the most promising conceptual designs. Each member of a design team should com-

Evaluative Criteria	Conceptual Designs a	b	c	d	Reference
Criterion 1	-2	+3	+2	+2	S
Criterion 2	S	-1	+2	+3	S
Criterion 3	-1	S	+1	+2	S
Criterion 4	+2	S	S	+1	S
Criterion 5	S	+1	-2	-3	S
$\Sigma(+)$	+2	+4	+5	+8	0
$\Sigma(-)$	-3	-1	-2	-3	0
$\Sigma(S)$	2	2	1	0	5

Table 10.2

Decision matrix: designs compared to a benchmark reference design.

+3	*criterion satisfied, very much superior to the reference*
+2	*criterion satisfied, much superior to the reference*
+1	*criterion satisfied, superior to the reference*
0	*criterion satisfied, the same as the reference*
−1	*criterion satisfied, but inferior to the reference*
−2	*criterion satisfied, but much inferior to the reference*
−3	*criterion satisfied, but very much inferior to the reference*

Table 10.3

Evaluation scale using the benchmark reference design approach.

Each team member should evaluate the conceptual designs independently before the group meets to compare results.

plete this evaluation independently prior to a meeting at which the results are discussed and consolidated. This ensures a more rigorous and effective evaluation.

A decision matrix often will stimulate thinking among members of the design group when the matrix is displayed as a wall chart. Visualization of the quality of the various design concepts stimulates discussion by colleagues who may provide different perspectives. New designs often are conceived.

Weighting the Evaluative Criteria

Actively evaluating conceptual designs through the interaction of creative and deductive thinking is one of several methods for producing the best product concept. It can be enhanced, at the expense of additional complexity, by weighting the evaluative criteria.

Weighting generates a matrix with greater distinctions between the numbers associated with the various designs. It recognizes that all evaluative criteria are not equally important because all design parameters from which they derive are not equally important. Some are essential for the product to function while others are only desirable if they are economically feasible. Thus in the latter stages of the evaluation process when all design parameters are invariably employed, the design parameter importance factors introduced in Chapter 7 can be used to enhance the procedure. The numerical value of the weighting factor for each evaluative criterion can be assigned on whatever scale seems

No human investigation can be called real science if it cannot be demonstrated mathematically.

Leonardo da Vinci (1452–1519)

Weight the evaluation criteria because they do not all have the same importance.

		Conceptual Designs									
		Rating (0–5)					Rating × Weighting Factor				
Evaluative Criteria	**Weighting Factor (0–5)**	a	b	c	d	Ideal	a	b	c	d	Ideal
Reliability	4	2	4	1	3	5	8	16	4	12	20
Product Cost	3	1	4	2	2	5	3	12	6	6	15
Maintenance Costs	3	3	3	4	1	5	9	9	12	3	15
Aesthetics	1	4	2	4	2	5	4	2	4	2	5
Weight	2	3	2	4	3	5	6	4	8	6	10
	Totals:	13	15	15	11	25	30	43	34	29	65

Table 10.4

A decision matrix in which the ratings for the evaluative criteria are multiplied by weighting factors.

QFD data can guide the assignment of weighting factors.

appropriate. The larger the weighting factor, the greater the significance of the evaluative criterion.

Generally these weighting factors are developed by the design team. However, if the customer's preferences are known through

, then the design team should use this information. For example, customers may rank reliability above operating costs, operating costs above product cost, and product cost above aesthetics. These data provide the basis for selecting the weighting factors.

To illustrate the mechanics of the process, Table 10.4 presents a decision matrix for a situation involving four conceptual designs *a, b, c,* and *d,* with five evaluative criteria *reliability, product cost, maintenance costs, aesthetics,* and *weight.* In the weighting factor column a 0–5 scale is used; 5 points are assigned to the most important criterion and 0 to the least important. Next, each conceptual design is rated on each criterion (Table 10.4, "Rating" block). A rating of 5 indicates a very suitable design whereas a rating of 0 indicates a design that is unsuitable. The numbers are multiplied by the weighting factors and are entered into the "Ratings × Weighting Factor" block. Numbers associated with each design can then be summed to arrive at the total value for each conceptual design.

The discriminatory characteristics of the weighting factors are evident by comparing the two ratings blocks of Table 10.4. Designs *b* and *c* have the same overall rating (15) but *b* is the better of the two designs because it has the higher ratings (4 *vs.* 1) and (4 *vs.* 2) in the more important criteria.

10.3 Quality of the Evaluation Process

Carefully consider the compromises and trade-offs involved in the final design choice. Were they justified?

Regardless of the method used to determine the concept with the greatest potential, designers should think carefully about the trade-offs and compromises made during evaluation. A period of incubation and reflection is useful; recall the psychological impediments described in Chapter 8. Careful consideration is important because all subsequent efforts will focus upon that single concept through the product design phase, and the cost of an inappropriate concept is high if it is ultimately found to be unworkable.

In addition to a period of incubation, the robustness of the decision-making process can be scrutinized if adequate documentation has been developed. For example, what would be the consequences of changing the number of points assigned to a particular evaluative criterion? Would the final choice be sensitive to small influences caused by changing the results of subjective criteria? If the decision matrix incorporates the weighting of different criteria, then this can be an important undertaking.

Change the weighting factors slightly to observe the robustness of the final design choice.

The analysis and evaluation of conceptual designs and the selection of the most promising concept for further development concludes this phase of the product realization process. The next phase—product design—transforms the concept into plans for the manufacture of a product.

10.4 The Student Design-Build-Test Project

The evaluation of conceptual designs is as important in designing a car for a student competition as for an industrial product. A poor selection can waste time and effort. This is illustrated in the following excerpt from an end-of-semester student-group report:

> Since a rather large number of conceptual designs were generated, an objective method of ranking each of the possibilities was employed. This method rated specific attributes of each of the designs on a scale from 0 to 4—four being the best and zero being the worst. Two ranking points of interest are starting and torque. Starting is particularly important because the competition rules state that the only input of energy can be from the candles. Torque is particularly important because most halls in the Engineering Building have a slight grade. Also, some of the attributes were considered more important than others and

were weighted relative to each other using a multiplier ranging from 2 to 8. Each of the entries in the concept rows were summed and the highest scores were considered the best concepts. Only those design parameters assigned a ranking of three or more-were deemed important to the success of the vehicle.

The two most promising solutions were the thermoelectric hybrid and the Rankine engine. The highest ranking conceptual design was the hybrid because of its simplicity and likely predictable functioning. The efficiency of this device, however, is rather poor (2 percent for the voltaics, 1.2 percent for the thermocouples, and 50 percent for the electric motor). The next best candidate was a variation of the Rankine engine. These engines are likely to be complicated mechanisms involving a pressurized fluid, so concerns about reliability and safety are justified. The efficiency of such a device can be reasonable, though.

While the merits of each design were made apparent from this list, it seemed important to familiarize ourselves with the fuel source to see if less efficient but easier designs might still be a good choice.

Note the observation of the students in the last sentence of the previous paragraph. The iterative nature of the design process is evident once again.

Summary

Conceptual designs must be analyzed and evaluated as part of the design cycle. Evaluation is particularly challenging because of psychological factors associated with conceptual designs. The selection process is a decisive step leading to a successful final product and sometimes must begin with limited information about a set of conceptual designs.

Alternative designs can be evaluated by a decision matrix approach in which the various attributes of the designs are transformed into numerical data—typically by comparing each attribute with parameters in the design specification. These data provide a measure of the strengths and weaknesses of alternate designs and can be summed for each design to determine the best one. This systematic procedure permits the strengths of one conceptual design to be identified and integrated with the strengths of another design to create a new concept. Thus the iterative design cycle is never ending.

A decision matrix evaluation can be applied to other open-ended problems in the development of a product. It can be used not only to evaluate a complete product but also to study subassemblies, individual parts, and manufacturing methods.

The decision matrix approach has several forms. Various designs can be compared with an ideal design. Alternatively, they can be compared with a reference design that can be one of the proposed concepts or a competitor's product. Finally, weighting factors can be introduced. This is useful because not all design parameters have equal significance. The degree of success of the design process can be measured by comparing the final design with an ideal design through a quotient called a design quality rating.

The evaluation process may require some detailed calculations of the conceptual designs in order to evaluate them effectively. These calculations will help the design team decide which concept should be subjected to much more detailed work in the product design phase of product development.

Key Concepts

Decision matrix. A matrix of rows and columns used to see how well each of several conceptual designs satisfies certain design criteria. The characteristics of each conceptual design are quantified by one of several approaches and the best design is frequently compared with an ideal design, a current design, or a competitor's product. This approach is relevant not only to the evaluation of conceptual designs but also to the evaluation of product components or the final design of a product.

Evaluation. Before a product is created, several conceptual designs must be analyzed and evaluated. Often a conceptual design is compared to design criteria using a decision matrix.

Period of reflection. Once the best concept has been identified a period of reflection should be initiated. This provides an opportunity to reevaluate compromises made during the evaluation and the appropriateness of the evaluative criteria and their numerical weighting.

Psychological Impediments. The evaluation of concepts is challenging because of the pyschological impediments discussed in Chapter 8. Evaluative criteria can be subjective and concepts change as they mature.

Quality rating. A comparison of the quality of a conceptual design with an ideal design. It is the ratio of the number of points assigned to a particular conceptual design divided by the number of points assigned to the ideal conceptual design. Quality rating can be applied to the complete concept or to subassemblies.

Review Questions

1. What type of thinking is primarily employed in the evaluation phase of the design process?
2. Can the evaluation phase be a creative phase of the design process?
3. What is the primary theory employed to evaluate designs and how is a matrix established?
4. How are the evaluation criteria determined for a decision matrix?
5. Why is the evaluation of conceptual designs a complex undertaking?
6. How are quantitative data established in the evaluative process? Why is this an important step?

7. How are numerical values employed in the evaluation process?
8. If several conceptual designs have the same number of points after completion of an iteration of the evaluative cycle, what should the next step be?
9. What is a conceptual design quality rating? How should it be employed?
10. Why is the evaluation of conceptual designs a crucial phase of the product realization process?
11. How should data from product benchmarking activities be incorporated into the evaluation matrix?
12. How could the best conceptual design be employed as the model for evaluation of the various design candidates?
13. How can design parameter importance factors be incorporated into the evaluation process? What advantages accrue from this approach?
14. What role can quality function deployment play in the evaluation of conceptual designs?
15. How can evaluating of designs with a decision matrix provide marketing data for a product?
16. It is important to identify areas of weakness in a conceptual design. How can this be done?
17. How can you ensure that you have evaluated the conceptual designs properly? Discuss the issues.
18. How can the quality of conceptual designs be determined?

Problems

1. Develop a specification for a computer for yourself. It might be desktop or laptop machine operating under Windows, or a Macintosh—or perhaps even a Unix workstation. Compare the attributes of four different computers that satisfy this product specification and select the best machine for you.

2. Evaluate the concepts proposed in Figure 9.3 (in the previous chapter) for changing the direction of translational motion in a straight line.
3. Numerous classes of products contain shafts that rotate in bearings. There are several categories of bearings. They include multi-element anti-friction bearings. These multi-element anti-friction bearings can be divided further into hydrodynamic bearings and plain journal bearings or bushings. Develop a list of all these types of bearings and then evaluate them relative to a set of general criteria. With these evaluations as a guide, develop a rule-based algorithm to assist a design engineer select the most appropriate type of bearing for any situation.
4. Develop several schemes to obtain variable load/deflection characteristics for coil springs. The load-deflection characteristics of each spring should vary nonlinearly during compression or extension. Using a decision matrix, evaluate these concepts.
5. Propose five different concepts for low-cost overrunning clutches. Such clutches are typically used when indexing, backstopping, and free-wheeling characteristics are required. Evaluate these concepts using a decision matrix.
6. Propose five different concepts for torque-limiting devices to protect light-duty shafts. These devices are to be based on the premise that a weak part breaks easily when the shaft is overloaded by a surge of torque. Use a decision matrix to evaluate these concepts.
7. A long slender lightweight structure is subject to a torque and a bending moment. If the mass per meter of the structure is 22 kg, evaluate the three proposed cross-sections shown below. All dimensions are in millimeters. Develop a table to evaluate the three cross-sections in terms of the maximum bending stress and the maximum shear stress, assuming that they are all manufactured in the same homogeneous material. Comment on the methods for manufacturing these shapes.

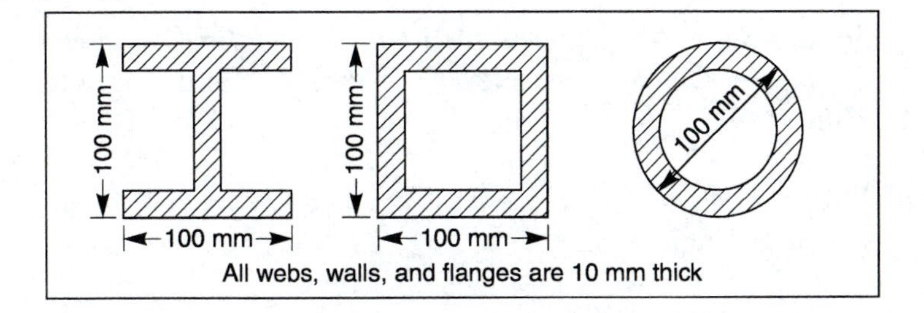

8. Threaded fasteners, featuring a nut and bolt, are frequently used to fasten two components. When the loads imposed upon these assemblies are dynamic, vibrations frequently loosen the nut and bolt combination. Propose several enhancements to this basic fastener system to inhibit this loosening. Analyze and evaluate these proposals relative to a set of design parameters using a decision matrix. The matrix should include maintenance costs, manufacturing cost, and other criteria introduced in this chapter.

Further Reading

Cross, N. *Engineering Design Methods*. New York: John Wiley, 1989.

Love, S. F. *Planning and Creating Successful Engineered Designs*. New York: Van Nostrand Reinhold, 1980.

Matousek, R. *Engineering Design: A Systematic Approach*. London: Blackie, 1963.

Pahl, G. and W. Beitz. *Engineering Design*. London: The Design Council, 1984.

Pugh, S. *Total Design: Integrated Methods for Successful Product Engineering*. Wokingham, England: Addison-Wesley, 1991.

Souder, W. E. *Management Decision Methods for Managers of Engineering and Research.* New York: Van Nostrand Reinhold, 1980.

Urban, G. L. and J. R. Hauser. *Design and Marketing of New Products.* Englewood Cliffs, N. J.: Prentice Hall, 1993.

Part Two

Product Design

Too much light often blinds gentlemen of this sort.
They cannot see the forest for the trees.

C. M. Wieland
(1733–1813)

Objectives

- To describe the complexity of product design.
- To illustrate the challenges of materials selection.
- To describe the selection of a manufacturing process.
- To reemphasize the practice of concurrent engineering in product design.

Contents

11.0 Introduction

The product design phase of the product realization process follows the conceptual design phase. Its primary aim is the creation of appropriate shapes for the product and its components. This requires both creative and deductive thinking in design, materials, and manufacturing. Product design uses concurrent engineering (Chapter 5), design-for-manufacture principles (Chapter 12) and design-for-assembly principles (Chapter 13) practiced by multidisciplinary teams of individuals not constrained by the barriers traditionally established between individual departments of a company. When these principles are applied, the products that evolve help companies compete because they reduce problems created when individuals cannot see the forest for the trees.

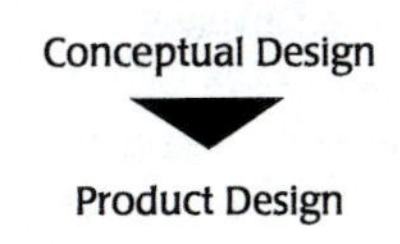

11.1 Product Design

Chapter 9 focused on conceptual design—a phase of the design process in which the designer has considerable freedom to establish the skeletal anatomy of a product. Ideally, the result of this design phase is a single conceptual design which is then refined into plans for a product. Sometimes a small number of conceptual designs must receive further refinement until the best design is identified with some degree of confidence. This refinement process is called *product design*, *form design*, or—in its early stages—*embodiment design*. It involves fleshing out the remainder of the anatomy—the shape or bodily form of the skeleton created in the conceptual design phase. This requires more detailed analyses than those undertaken at the conceptual design stage. Calculations are done to establish performance characteristics, costs, and sizes of important parts.

Generally several conceptual designs are carried forward into the early phases of product design before the best design emerges.

Product design transforms a selected conceptual design into plans for manufacturing a product. It typically involves detailed computational analysis and creativity.

Product design includes:

- embodiment design
- detail design

If preliminary product design activities generate a viable proposition, then the design group initiates the *detail design* phase of the product realization process. Comprehensive

The detail design phase involves using computers to perform analyses.

Detail design involves:

- shape and size
- engineering drawings
- tolerances
- heat treatment
- surface finish

Iteration is a component of all design activities.

analyses of all primary parts are undertaken. The analytical techniques involved in that work lie outside the scope of this text. (They involve CAD/CAM software packages that often use finite element methods and economic analyses.) Subsequently, detailed drawings and specifications are developed. They include tolerances, quality of surface finish, heat treatment, and other manufacturing information.

During product design, the conceptual design is transformed into assemblies and individual parts. At each level of design activity, the iterative method of Figure 4.2 in Chapter 4 is used. This involves a design specification, the creation of alternative potential solutions, the analysis of these solutions, and the choice of the best one on which to base a part, an assembly, or a product.

The iterative design process of Figure 4.2 is not restricted to the loop associated with an individual component; it is part of a process that includes the conceptual design of the complete product as represented by Figure 5.9. Indeed, this activity may result in the formation of new conceptual designs or, in extreme circumstances, to modification of the original design specification.

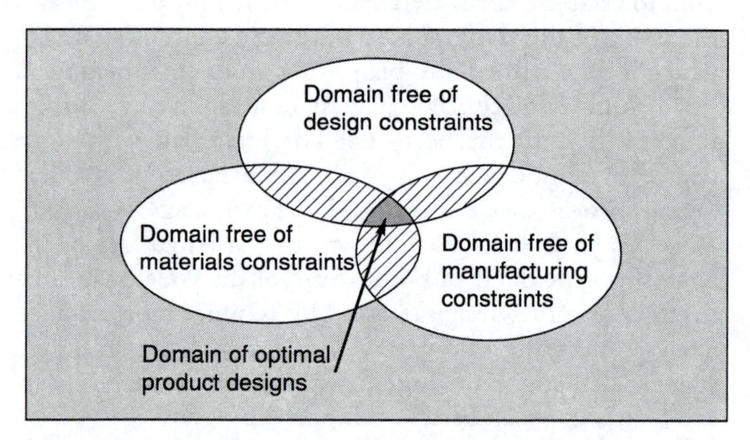

Figure 11.1

Products must be created to satisfy the constraints of the design, the materials, and the manufacturing processes.

At the heart of the product design phase is the complex set of relationships between design, materials, manufacturing, and economics. Figures 11.1 and 11.2 emphasize these four ingredients of product design in the development of appropriate forms for components of the product. Moreover, these ingredients are intertwined. The selection of an appropriate manufacturing process requires simultaneous consideration of the material to be used and the shape of the part to be made. The most appropriate material and the best manufacturing process both depend upon the design speci-

Product design involves:

- Conceptual design
- Materials selection
- Manufacturing processes
- Economics

Exploit the good attributes of a material or manufacturing process and minimize the bad.

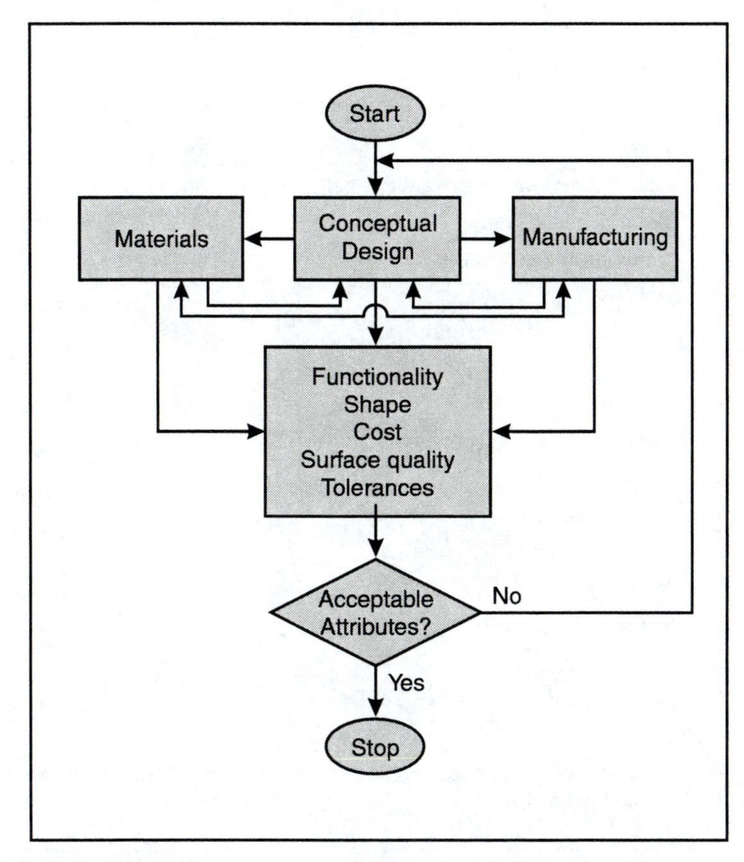

Figure 11.2

Product Design: the iterative cycling of concepts involving design, materials, and manufacturing attributes.

fication for a part—its form, surface finish, cost, and dimensional tolerances, for example.

Materials Selection

Over 50,000 commercial materials are available to the manufacturing industries.

Consider the material, the attributes of the part, and economics in the selection of a manufacturing process.

Figure 11.3 summarizes the principal commercially available materials used in manufacturing. It has been estimated that upwards of 50,000 materials are available to the engineering community and this number grows each year. These materials have diverse properties and vary enormously in the constraints they impose upon manufacturing methods. Some of the properties that must be considered during product design are listed in Table 11.1.

Cost
Constitutive behavior (elastic, plastic, viscoelastic, etc.)
Strength (yield, etc.)
Damping characteristics
Hardness
Modulus (tensile, flexural, etc.)
Electrical properties
Toughness
Fatigue characteristics
Fracture toughness
Wear resistance
Creep resistance
Melting point
Thermal conductivity
Thermal expansion
Specific heat
Density
Chemical properties (corrosion resistance, reactivity with acids and bases), etc.

Table 11.1

Some material properties which influence the design and manufacture of parts.

Figure 11.4 presents a way to retrieve information on this myriad of materials by first accessing the standard materials handbooks, literature from trade associations, and computer databases. If these sources are inadequate, then product data sheets can be obtained from materials suppli-

Information of materials:

- handbooks
- trade associations
- computer databases
- laboratory testing
- www internet

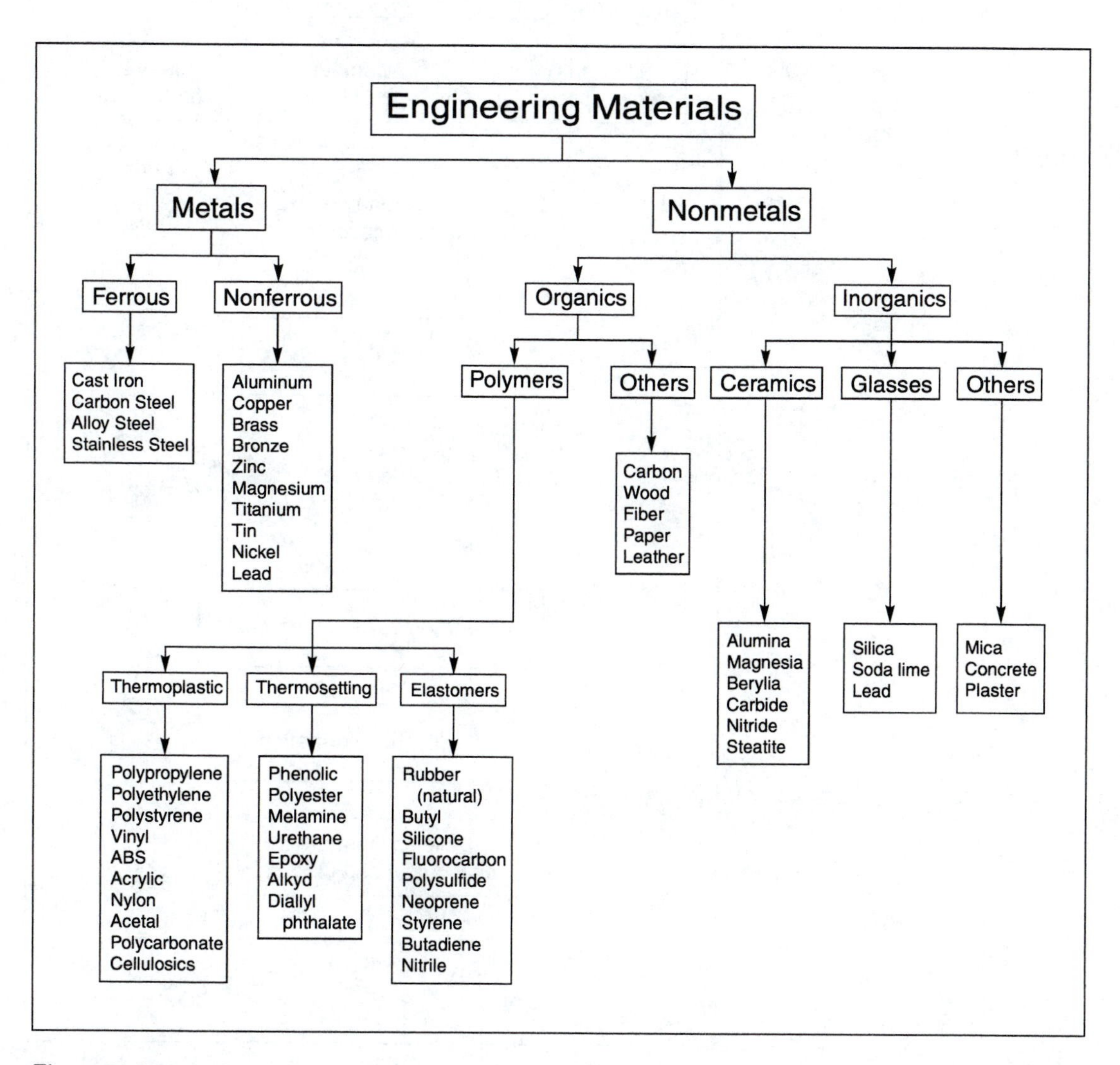

Figure 11.3

Industrial materials used to manufacture products.

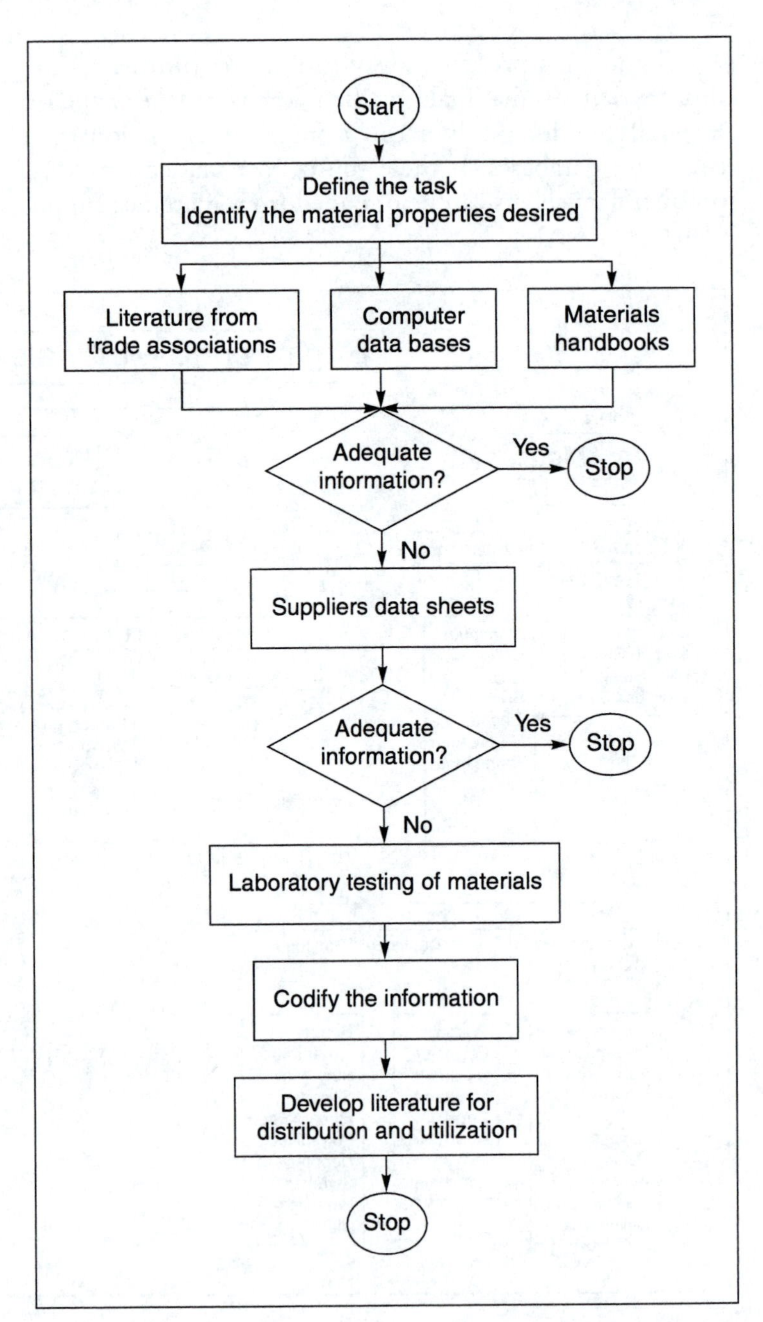

Figure 11.4

A scheme for retrieving information on commercially-available materials.

ers. Finally, if this does not provide the information required, then laboratory tests can be undertaken to determine the needed information. Laboratory testing should always be employed for materials whose properties depend upon the manufacturing process used to make a part. Fibrous composite materials are an example.

The selection of materials is aided by the approach proposed by *M. F. Ashby* in his book *Materials Selection in Mechanical Design.* He uses charts that present two-dimensional plots of the properties of common engineering materials. These charts condense a large data set but since only two properties are plotted on each chart, it is generally necessary to consult several.

Materials selection charts are useful design tools

An example of material selection for a lightweight part with a specific form can begin with an examination of charts similar to Figure 11.5. This chart describes the strength and density characteristics of a group of materials in addition to contours defining criteria for designs of minimum weight. Such charts can be used effectively in product and detail design to determine values of material properties.

Selection of Manufacturing Processes

The characteristics of a particular material and the design specification dictates the best and most cost-effective method for manufacturing a part. The most important characteristics of manufacturing processes which influence the selection of materials for a product are listed in Table 11.2. Figures 11.6 and 11.7 summarize common manufacturing processes.

The selection of a manufacturing process is governed by:

- materials
- economics
- design specification
- manufacturing processes

These manufacturing processes and the list of materials in Figure 11.3 suggest that the selection of a manufacturing process is a complex optimization problem. Since it belongs to the same class of problems as the design process, the solution to these problems clearly involves a cycling of creative and deductive thinking as represented in Figure 8.1. The method is illustrated in Figure 11.8. Form design and tolerances govern the material selection and manufacturing processes. These parameters provide the quantitative

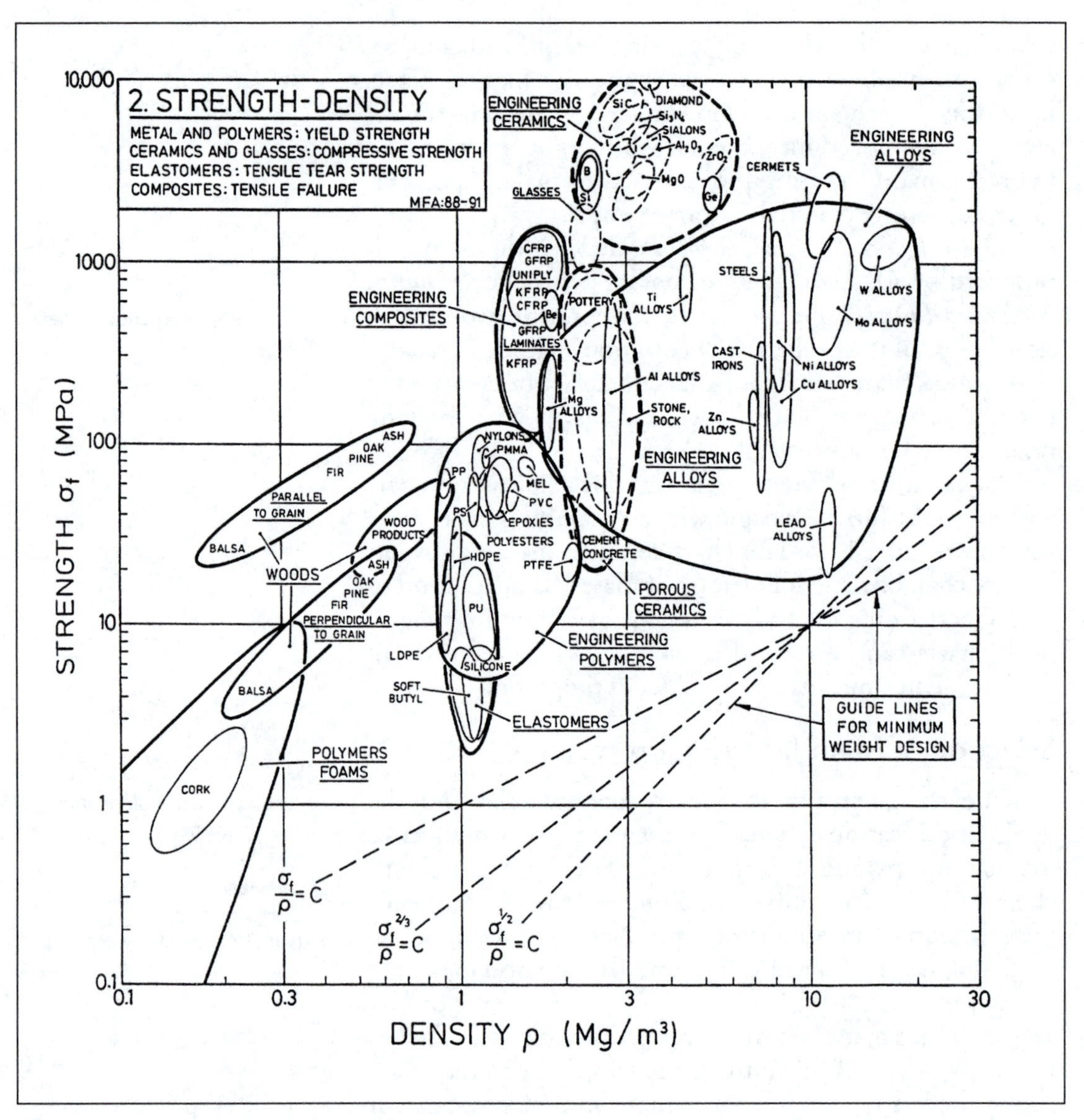

Figure 11.5

A materials selection chart characterizing strength and density. (Source: M. F. Ashby, Materials Selection in Mechanical Design.)

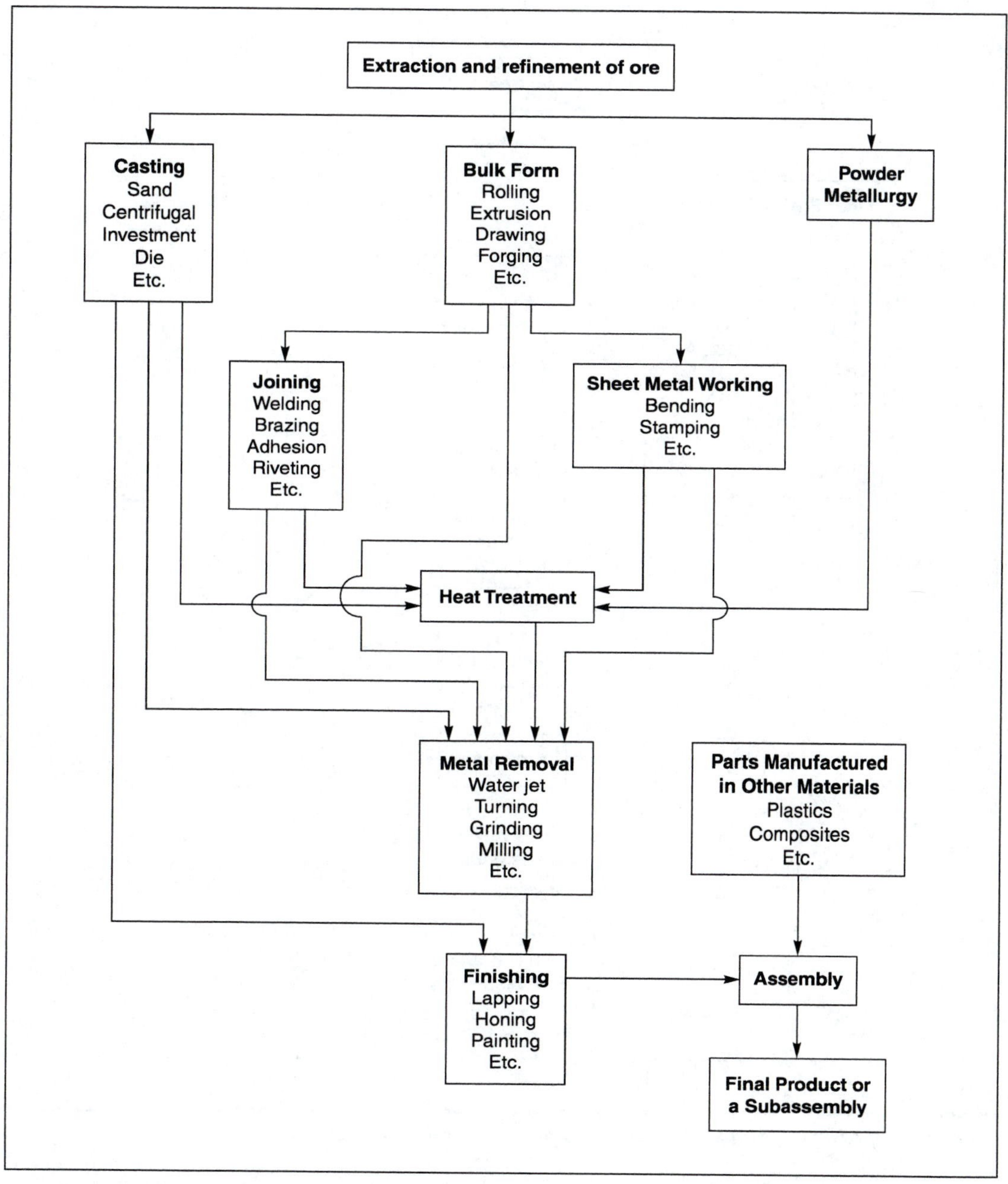

Figure 11.6

A summary of common manufacturing processes for metals.

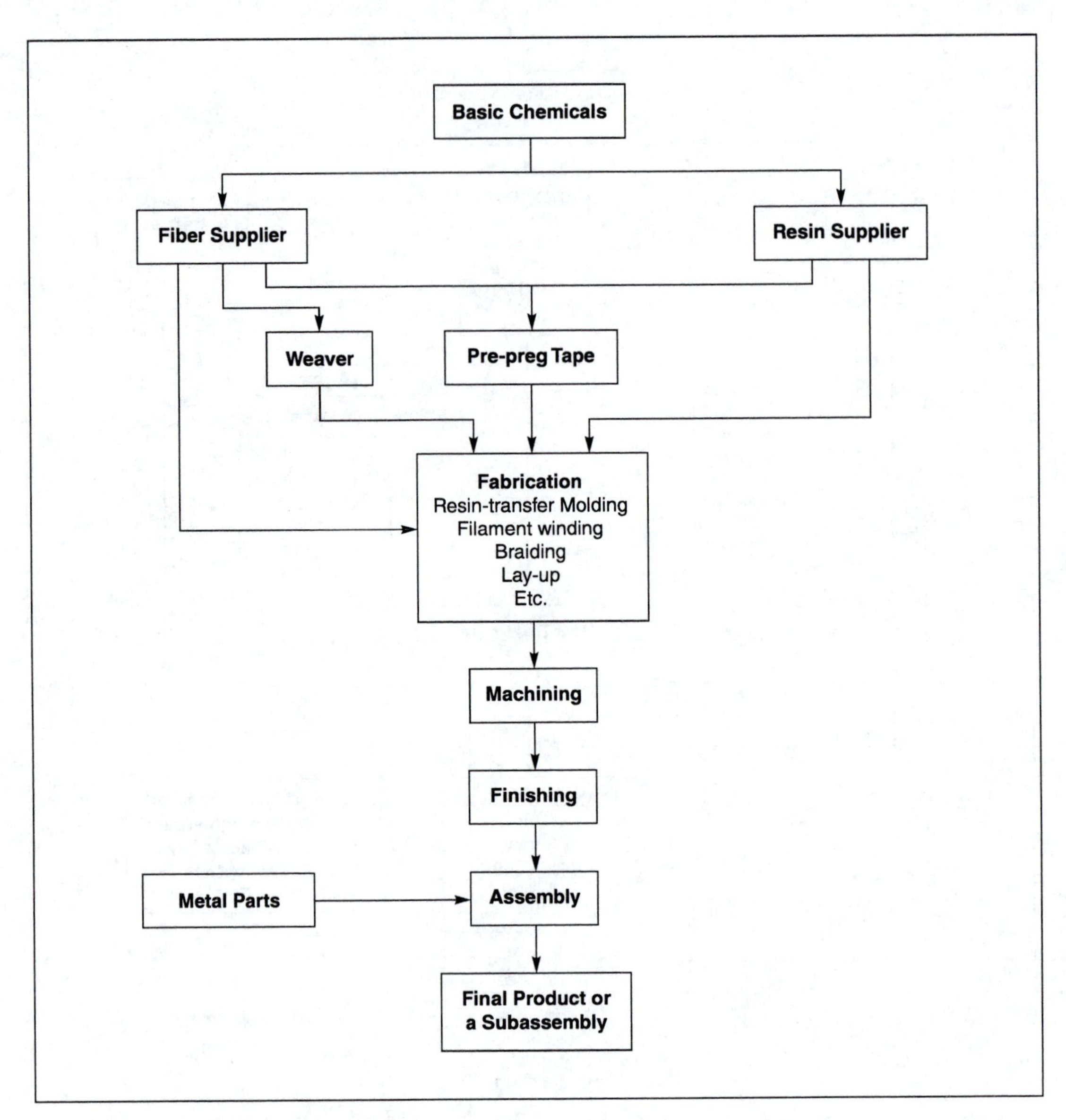

Figure 11.7

A summary of common manufacturing processes for fibrous composite materials.

Material characteristics and properties
Part shape
Surface finish
Part size
Tolerances
Batch size
Production rate
Capital cost
Cost of material wastage
Labor costs (skilled, unskilled etc..)

Table 11.2

Attributes of a manufacturing process which influence design and materials selection.

data on the characteristics of a part. Implicit in this is a specification for the manufacturing process such as the magnitude of the draft in a molding operation or the metal removal rate by a lathe. The final set of parameters for the part must be compared with the design specification for the part.

The use of Ashby's materials selection charts to link design and materials can be extended by a second group of charts that relate design attributes to processing characteristics. Each chart plots two manufacturing process attributes and each process occupies an area on the chart that characterizes the process relative to these two attributes. Generally several charts must be used during process selection. Figure 11.9 is a process selection chart showing the size attributes and information content of various manufacturing processes. The horizontal axis is a logarithmic scale in order to accommodate the large variation in size from microchips to castings weighing hundreds of tons. Similarly the vertical axis provides a logarithmic variation of the complexity of parts ranging from deformation processes typically characterized by 100 bits of information to microchips that may require megabits of information.

The selection of a manufacturing process involves iteration; it is an open-ended problem.

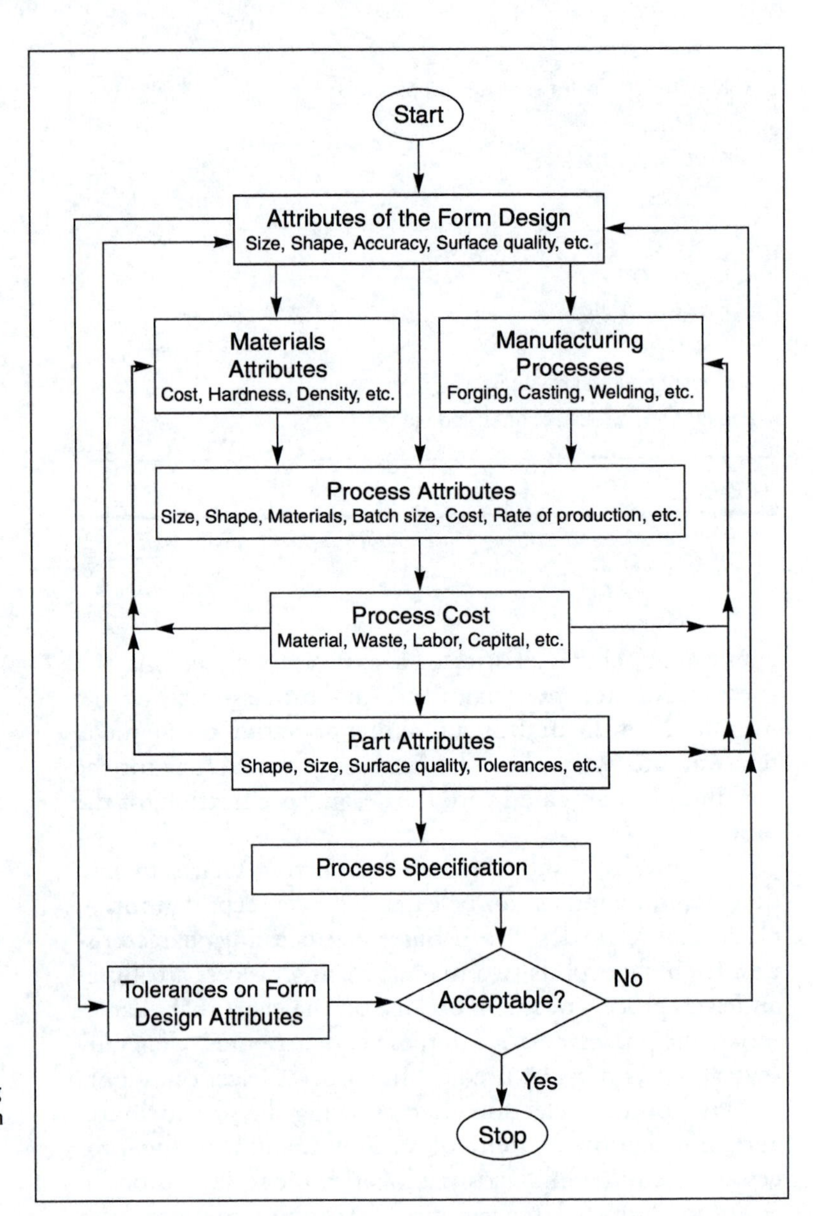

Figure 11.8

Iterative selection of a manufacturing process.

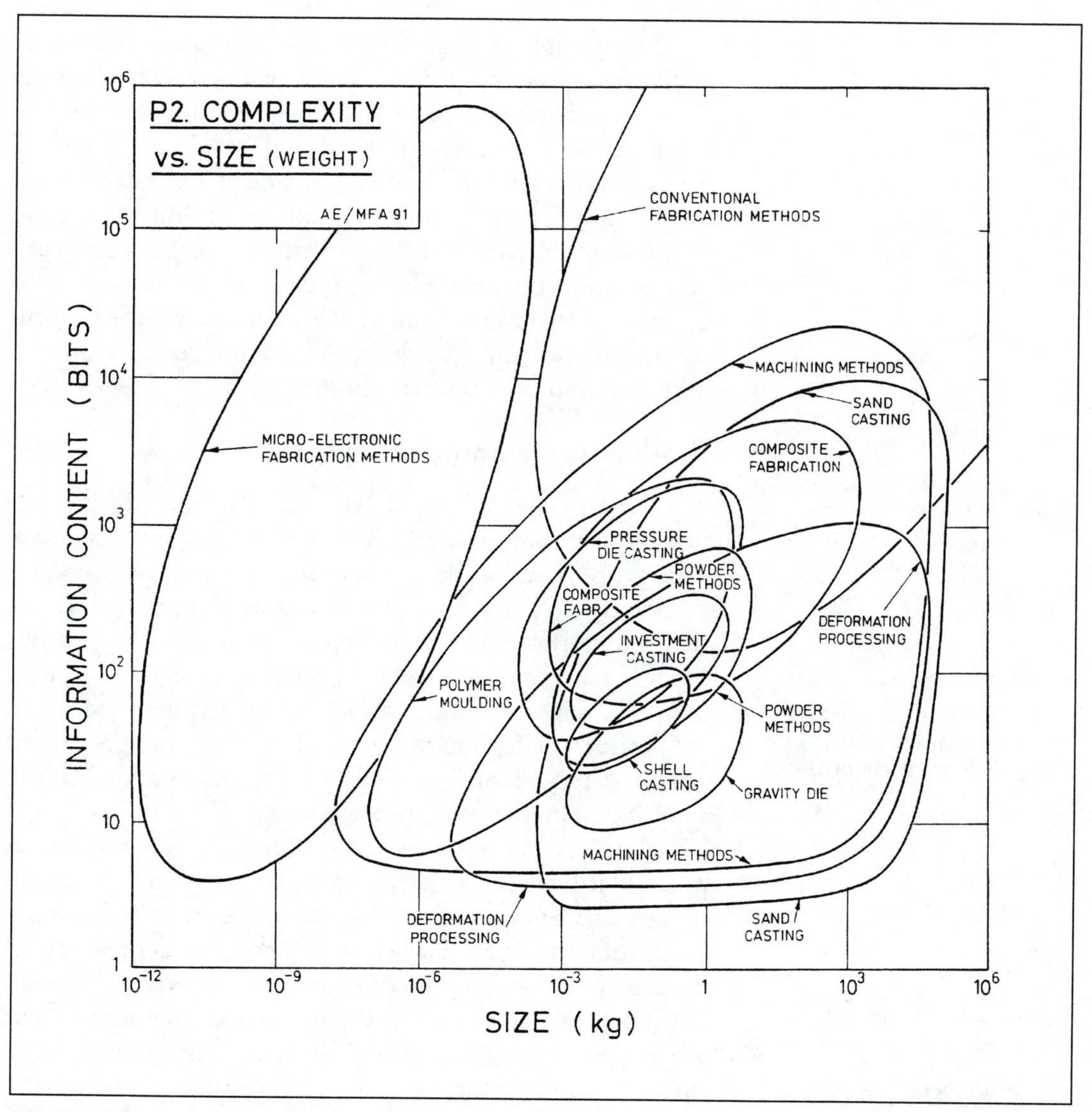

Figure 11.9

A process selection chart characterizing size and information content. (Source: M. F. Ashby, Materials Selection in Mechanical Design.)

This design tool is used by superimposing the design attributes upon the process selection chart. Thus the designer could impose upper and lower bounds that confine the search to a size between 0.001 and 1.0 kilogram with an information content between 10 and 100. Alternatively only an upper bound could be imposed by limiting the size to no more than 0.001 kilogram and the information content to 1000 bits. By relating the design attributes to the processing attributes through charts like these, the optimization process can quickly become more focused upon a narrow group of potential solutions.

Product Development Teams

Teamwork: Coming together is a beginning; keeping together is progress; working together is success.

Henry Ford
(1863–1947)

Product development teams are composed of multidepartmental, multidisciplinary groups of people.

The large number of materials available for the fabrication of products (Figure 11.2) as well as the many manufacturing processes available to produce a product (Figures 11.6 and 11.7) require that modern product design activities be undertaken by a team of people with diverse capabilities. These teams typically consist of design engineers, materials scientists, and manufacturing engineers who create products satisfying the constraints of the design specification, the appropriate engineering materials, and the available manufacturing processes.

Break down traditional artificial barriers among divisions of a company.

Functionality is the primary goal of traditional design departments.

Manufacturability is the primary goal of traditional manufacturing departments.

However, as discussed in Chapter 5, the traditional practice still employed by conservative companies is for the design and the manufacturing departments to work quite independently of each other during the product realization process. Moreover, a level of suspicion often exists between the design department and the manufacturing department when this traditional approach is used. The design department is primarily concerned with creating a functionally acceptable product. The manufacturing department primarily tries to design a product that is manufacturable. Thus there are two separate goals. These artificial barriers unfortunately lower the quality of the product and reduce the efficiency of the product realization process.

These barriers have largely been alleviated by the development of multidepartmental interdisciplinary teams that simultaneously address problems in the design and manufacture of products. An interdisciplinary team philosophy, exploiting concurrent engineering practices, avoids the conflicts and impediments associated with the traditional sequential approach discussed in Chapter 5. Table 5.2 presents the advantages of this approach.

The philosophy of defining a problem relative to the needs of the customer and the marketplace rather than to the internal structure of a company is a crucial observation represented in Figure 11.1. If the goal of a company is to develop a product of the best quality at the minimum cost, then usually only a few designs are satisfactory.

Employees should be focussed on product quality and customer loyalty. This is a primary objective of any company. Corporate policies should minimize artificial impediments and internal strife.

11.3 The Student Design-Build-Test Project

The design teams are now moving from paper studies to decisions about physical objects that must actually work. This is a new experience. The transformation of a conceptual design into a product involves a design process embracing the simultaneous consideration of several interrelated disciplines including materials selection, manufacturing, and the functionality of the design. Student reports include the following:

> Many exotic materials are available for fabricating the birthday candle powered vehicle but they are too expensive, or we don't have access to the small quantities that we need for our project, or we are unable to manufacture the part from the desired material because we don't have access to the necessary facilities. These constraints restrict us to conventional materials available in hardware stores and hobby shops.
>
> Every time a decision on a part or an assembly was made it was immediately necessary to think of how to manufacture it and from what material. Could it be made using the facilities in the College of Engineering's student shop and was the material available? All group members were involved with the simultaneous consideration of design and manufacturing decisions.

This concludes our series of snapshots describing the evolution of a student design-build-test project for the design, building, and testing, of a birthday-candle powered vehicle. During the past eleven chapters we have followed the evolution of the project from the original problem specification through the creation of conceptual designs, the evaluation of these designs, and the transformation of the chosen design into a product ready for competition. The result of one student group's work is illustrated in Figure 11.10.

Figure 11.10

A birthday candle-powered vehicle built by a student group.

Summary

Product design is the transformation of the best conceptual design into plans for its manufacture, support, and disposal. You have learned that this design phase is sometimes broken down into embodiment design in the early stages and detail design in the final stages. At the end of this latter phase, plans and engineering drawings for the manufacture of each part of the product will have been completed. Production of these plans typically requires an iterative process involving economics, conceptual designs, materials, and manufacturing processes.

Materials selection belongs to the triumvirate that underpins product design. This process can involve an enormous number of commercially available materials whose characteristics must be evaluated relative to manufacturing constraints and economic considerations. Charts have been developed to aid this process.

You have learned that the selection of an appropriate manufacturing process for a part is governed by the design specification for the part, including the batch size and production rate, attributes of the manufacturing process (including economics), and characteristics of the material being processed. Again, charts have been developed to facilitate this activity.

The complexity of the product design process often requires teams of people working together to address the many tasks of transforming a conceptual design into a product. This task is aided by design-for-manufacture and design-for-assembly principles that we shall consider in Chapters 12 and 13.

Key Concepts

Product Design. The simultaneous consideration of functionality, materials selection, manufacturing processes and economics to develop a product.

Materials Selection. The selection of an appropriate material for a part through the interaction of part shape, manufacturing processes, economics, materials and design functionality. Design charts enhance this complex iterative process.

Selection of a Manufacturing Process. An iterative process primarily dependent upon the part design specification and the characteristics of the material.

Product Development Teams. Groups of individuals with multidisciplinary skills, typically from different departments in a an organization, and also outside vendors. They generally include cost accountants, production personnel, manufacturing engineers and product designers.

Review Questions

1. Define product design and describe how it differs from conceptual design. What is another term for this phase of the product realization process?
2. Define embodiment design and describe how it relates to product design.
3. What is detail design? How does it relate to product design?
4. Product design is made up of three primary ingredients. What are they?
5. The general method for solving open-ended problems applies to product design activities and to selection of the manufacturing processes needed to make a part. What are the primary tasks in this method and what are the important thinking processes involved?
6. The product design phase is responsible for creating plans for the manufacture of each part. What attributes of the part govern this procedure?
7. The selection of material or a manufacturing process during the product design phase requires the determination of the good and the bad attributes of the material or process. What must the next step be?

8. Approximately how many different materials are commercially available for use in product design?
9. List at least eight different classes of engineering materials.
10. Describe how to access literature on materials for product design.
11. When should materials be laboratory tested to determine their properties?
12. What are the principal classes of material properties that dictate their use in design and manufacturing?
13. What are the attributes of a manufacturing process that influence design and manufacturing decisions?
14. What are product development teams?
15. List some of the disciplines typically represented on a team practicing concurrent engineering.

Problems

1. Develop several product designs for a keyless joint for connecting a hub and a shaft that is capable of transmitting power. Discuss the different attributes of these designs.
2. Seals are used in shaft-bearing assemblies to prevent the ingress of contaminants and to retain lubricants. Present five different designs of seals and discuss their advantages and disadvantages.
3. A coil spring is to fit inside a cylindrical space and be subjected to a static compressive force coincident with the principal longitudinal axis of the cylinder. Sketch the different classes of coil springs and other classes of mechanical springs suitable for this application. Discuss the advantages and disadvantages of these product designs.
4. Machine frames typically comprise an assemblage of structural members of uniform cross-section subjected to bending or torsional loads. Consider at least four

standard cross-sectional geometries and determine their relative attributes in terms of their weight, deflection, stresses, bending moments and torque.

5. A hinge is a simple mechanical joint with one degree of freedom. It permits rotational movement about one axis while permitting a load to be imposed along this axis of rotation and also a bending moment about an orthogonal axis. Neglecting minor details, propose a variety of mechanisms to perform this task and discuss their product design attributes.
6. A conceptual design sketch of a mechanical system requires a rod, subjected to a torque about its longitudinal axis, to move through a small angle relative to the supporting structure. An "elastic" element is proposed to accomplish this task. Propose several product designs and evaluate them.
7. The conceptual design of a hand-tool calls for the adjustment of two mating parts which translate relative to each other on a nominally smooth flat surface. Propose several indexing detentes for stopping, locating and facilitating this relative motion. Evaluate these different product designs and select the best relative to predetermined criteria.
8. A drive train of a machine includes a nut-and-screw combination that requires the elimination of the backlash at the interface of these two parts to ensure that the machine satisfies the performance specification. Transform the proposed concept into a product by developing and evaluating several concepts to achieve this task.
9. Discuss the options for manufacturing the fork-shaped part shown at the left. Present the advantages and disadvantages of each operating option in terms of the manufacturing process, the material selection, and the shape of the part. The principal dimensions are arbitrary.

10. Contrast several designs for a lap joint connecting two plates of equal thickness. The joint is subject to shear loading. The plates are to be connected by a series of nuts and bolts. A sketch of the concept is shown in the

figure at the right. Include sketches of the proposed product designs illustrating cross-sectional details of each assembly.

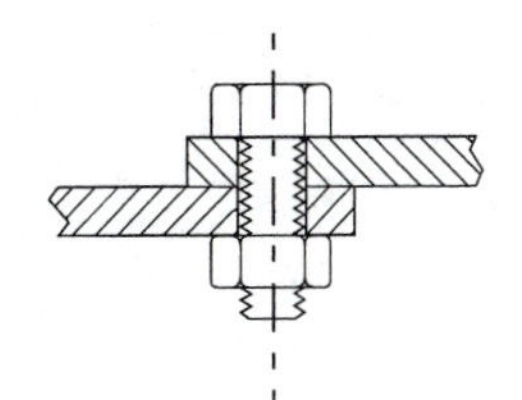

11. A set of steel studs connect two aluminum castings. This assembly is subject to a large range in temperature during operation. As a result, the integrity of the stud is compromised because of the difference in thermal expansions of steel and aluminum. Propose, and discuss, concepts for reducing the potential of failure of the stud.
12. Many product design situations involve the synthesis of a beam-like structure of low weight but high flexural rigidity. Hence the designer must typically create a lightweight structure which has a small deflection when subjected to a prescribed transverse load. A beam with a circular cross-sectional outer diameter of 47 mm has a second moment of area of 24 cm^4, a deflection of 0.66 cm, a stiffness of 0.15 kg/10^{-6} m and it weighs 13.6 kg/m. Propose alternative cross-sectional designs with superior structural characteristics and less weight fabricated from the same material. Evaluate them by considering both their structural properties and their manufacturing attributes.

Further Reading

Allen, C. W., ed. *Simultaneous Engineering*. Dearborn, MI: Society of Manufacturing Engineers, 1990.

Ashby, M. F., *Materials Selection in Mechanical Design*. Oxford, UK: Pergamon Press, 1992.

ASM Engineered Materials Reference Book. Metals Park, OH: ASM International, 1989.

ASM Metals Handbook. Metals Park, OH: ASM International, 1990.

Bralla, J., ed. *Handbook of Product Design for Manufacturing*. New York: McGraw-Hill, 1986.

Cornish, E. H. *Materials and the Designer*. Cambridge, UK: Cambridge University Press, 1987.

Crane, F. A. A. and J. A. Charles. *Selection and Use of Engineering Materials*. London, UK: Butterworths, 1984.

Dieter, G. E. *Engineering Design. A Materials and Processing Approach*. New York: McGraw Hill, 1983.

El Wakil, S. D. *Processes and Design for Manufacturing*. Englewood Cliffs, NJ: Prentice-Hall, 1989.

Engineers Guide to Composite Materials. Metals Park, OH: ASM International, 1987.

Farag, M. M. *Selection of Materials and Manufacturing Processes for Engineering Design*. Hemel Hempstead, UK: Prentice Hall International, 1989.

Hartley, J. R. *Concurrent Engineering*. Cambridge, MA: Productivity Press, 1992.

Design-for-Manufacture 12

When we mean to build,
We first survey the plot, then draw the model

William Shakespeare
(1564–1616)

Objectives

- To emphasize the crucial role of design-for-manufacture in product design.
- To provide rules for implementing the design-for-manufacture philosophy.
- To illustrate the application of design-for-manufacture in practice.

Contents

12.0 Introduction

Design-for-manufacture (DFM) is an approach that helps an engineer design a reliable product that can be manufactured easily and at low cost.

Design-for-manufacture (DFM) is an approach that leads to products that can be easily manufactured.

DFM is sometimes termed design-for-manufacturing or design-for-manufacturability.

When this philosophy is implemented, *all* product constraints and goals are considered early in the product realization process. If the problem is defined correctly, the process usually ensures an excellent solution. However, if DFM is not implemented, a manufacturer will generate one solution based upon the product design constraints and another dependent upon the manufacturing constraints. These two solutions are often different from the global solution created by the DFM approach where all constraints are considered simultaneously.

Design-for-manufacture assumes that:

Meticulously consider *all* the fine details.

- Design is the first step in the manufacture of a product.
- Every design decision, if not carefully scrutinized, can result in extra manufacturing costs and a loss in productivity.
- Product design takes into account a company's existing manufacturing facilities. Consideration must be given to flexible manufacturing, quality control, assembly operations, and material handling.

Design-for-manufacture, therefore, is the practice of taking manufacturing constraints into consideration as products are being designed. By doing this, products can:

Advantages of DFM:

- better quality
- better reliability
- lower cost
- faster concept-to-customer
- easy to manufacture
- easy to assemble
- easy to test

- Satisfy customers' requirements and still be competitive in the marketplace.
- Achieve the desired reliability and quality.
- Be created in the shortest time with the lowest development costs.
- Be transformed rapidly and successfully from the original concept into production.
- Be manufactured and tested quickly and at low cost.

Design-for-manufacture strategies have some subtle benefits as well as the obvious ones associated with quality, cost, and delivery. For example, assembly costs are lower for a product with fewer parts because assembly is easier and

Design-for-manufacture benefits cost, quality, and delivery.

Cost is minimized because of:

- easy assembly
- simple design
- ease of manufacture
- lower administrative burden

therefore labor costs are lower. Design for assembly is discussed in the next chapter. Simpler designs and easy assembly reduce the cost of achieving the desired product quality. Higher quality is associated with:

- Fewer parts
- Easy-to-inspect features
- Foolproof assembly techniques
- Specification of standardized parts of known quality
- Mature manufacturing processes

Similarly, higher reliability is achieved through the specification of proven processes and proven standardized parts; less time is spent solving unforeseen production problems.

Products with DFM attributes reach the market sooner because they are compatible with existing processes and therefore do not require specialized manufacturing equipment or procedures. The specification of standardized parts implies that they can be readily obtained. Because machine setup time is reduced, machine use is greater and capital equipment costs are lower. Furthermore, when manufacturing issues are considered early in the design process, new products usually are introduced smoothly with few production problems.

The advantages of DFM are summarized in Table 12.1. Data are from a survey by Boothroyd and Dewhurst, Inc., the primary developer of design-for-manufacturing and design-for-assembly procedures (Chapter 13).

- 50 percent reduction in the parts count
- 40 percent reduction in parts cost
- 50 percent faster time-to-market
- 70 percent increase in quality and reliability
- 60 percent reduction in assembly time
- 60 percent reduction in manufacturing cycle time

Table 12.1

Advantages of practicing design-for-manufacture.

12.1 Design-for-Manufacture Principles

Design-for-manufacture activities fall into four groups: general DFM principles, the subset of *Design-for-Assembly* (DFA) principles, facility-specific DFM principles, and process-specific DFM principles, as shown in Figure 12.1.

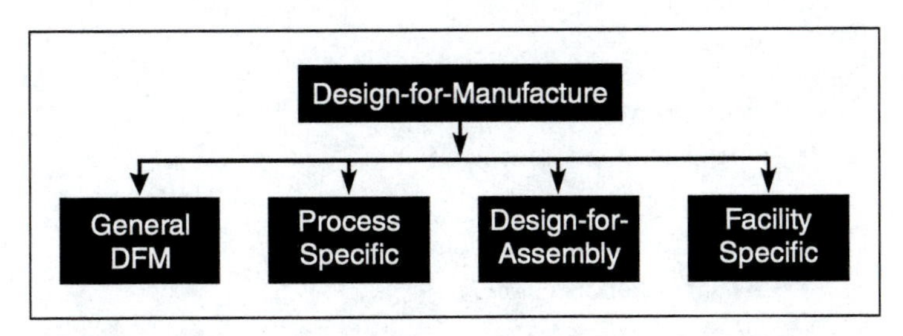

Figure 12.1

Classes of DFM principles.

Design-for-manufacture principles pertaining to assembly operations are called *design-for-assembly* principles.

The set of general DFM principles apply to all product design situations; they apply to both production and assembly processes. These principles can be presented as a complete unifying set, or they can be considered as two separate groups, one focused upon design-for-manufacture principles for *production* processes and the other focused upon design-for-manufacture principles for *assembly* processes. We will use the latter approach and call the second set design-for-assembly (DFA) principles. They are discussed in the next chapter.

Facility-specific DFM principles apply to products manufactured in a particular plant. The design of each part produced is restricted by the methods available in the plant.

Process-specific DFM principles pertain to the design of parts that will be manufactured by particular processes such as casting, forging, or welding. These will be discussed in Chapters 14–18. Each process has its own set of constraints and characteristics which influence the form design of the part.

In the remainder of this chapter we consider general design-for-manufacture principles (Table 12.2) along with the design consequences of adopting these principles. General DFM principles not only provide the basis for good manufacturing practices but they also stimulate creative thinking.

• Minimize the number of parts
• Specify standard components
• Create modular designs
• Create multifunctional parts
• Use the same part in many products
• Roughest surface, largest tolerance
• Avoid secondary operations
• Specify processible materials
• Design for production volume
• Minimize separate fasteners
• Minimize handling

Table 12.2

Eleven general DFM principles.

• Minimize the Number of Parts

Minimizing the product part count is the single biggest cost reducer in the DFM arena because:

- fewer manufacturing activities
- fewer engineering changes
- smaller inventory
- easier production control
- fewer purchase orders
- less accounting
- faster assembly
- less inspection

Minimizing the product part count is the single biggest cost reducer in the DFM arena. These cost savings occur both in the manufacture of an individual part and in the assembly of parts into a product. A product with fewer parts minimizes manufacturing activities. It requires fewer engineering drawings, a smaller inventory and less production-control. A smaller number of parts means fewer purchase orders and less accounting activities, less material-handling equipment, faster assembly, and fewer items to inspect.

By designing products with fewer parts, a company will interact with fewer vendors and may establish stronger vendor-partner relationships with less cost and fewer parts quality issues.

By combining two parts into one, the weight and complexity of the assembly often is reduced, expensive interface surface finishing and close tolerances are eliminated, and fewer fasteners are required. Stress concentrations are eliminated, and the structural integrity of the assembly is improved.

The parts count often can be reduced by specifying different materials or different manufacturing processes. Plastics or composite materials can be advantageous, and powder metallurgy can create parts with attractive attributes.

Reduce the number of parts by considering materials and manufacturing processes.

A part is a good candidate for elimination through combination with an adjacent part if the assembly satisfies one of the following criteria:

- There is no benefit from the relative motion of two adjacent parts.
- Adjacent parts need not be disassembled for servicing.
- Adjacent parts need not be made from materials with different properties.

As with most concepts, there is a point of diminishing return from a reduction in the parts count. That point will have been reached when a further part reduction adds cost or complexity. Under these circumstances, perhaps the best way to reduce the parts count is to find an alternative conceptual design with a different working principle and fewer parts.

Look for a different conceptual design to reduce the parts count.

• Specify Standard Components

Standard components mean:

- Lower parts costs
- Shorter lead time
- Simpler purchasing
- Simpler inventory control

Off-the-shelf bearings, valves, bolts and screws are always less expensive than custom-made parts (Figure 12.2 and also Figure 9.2). The use of common mass-production parts and standard materials reduces costs, shortens lead times, eliminates investment in tooling and equipment, and simplifies purchasing and inventory. If a company restricts the number of different fasteners and other components across its product lines, then economies of scale can be achieved.

Figure 12.2

Some off-the-shelp nuts. (Courtesy of Wyandotte Industries, Inc.)

• Create Modular Designs

A modular design allows a company to offer a variety of products that can be customized through different combinations of components built around a standard interface

For economies of scale, use common components across product lines.

(Figures 12.3 and 12.4). The connector of the Lego™ building block toy (Figure 12.3) illustrates this. A common interface simplifies process control and reduces tooling costs.

Modular designs permit:

- Diversity
- Standardized processes
- Lower tooling costs

Figure 12.3

A successful modular design: the Lego™ building block toy.

• Create Multifunctional Parts

Can the parts count be reduced by making each part serve several functions?

Parts should be designed to serve more than one function whenever possible. Replacement of several parts with one multifunctional part not only reduces the parts count but it can also simplify assembly. For instance, a part could be designed to serve as both a structural member and also a conductor of electricity. Alternatively, a part could be designed to aid in assembly by providing guiding, aligning, or self-fixturing attributes. Figure 12.5 shows the end carriage of the *IBM Proprinter.* Two parts were unified into a multifunctional part when a leaf spring was molded into the end carriage.

Figure 12.4

Modular designs of motors and gearboxes. (Courtesy of Sumitomo Machinery Corporation of America)

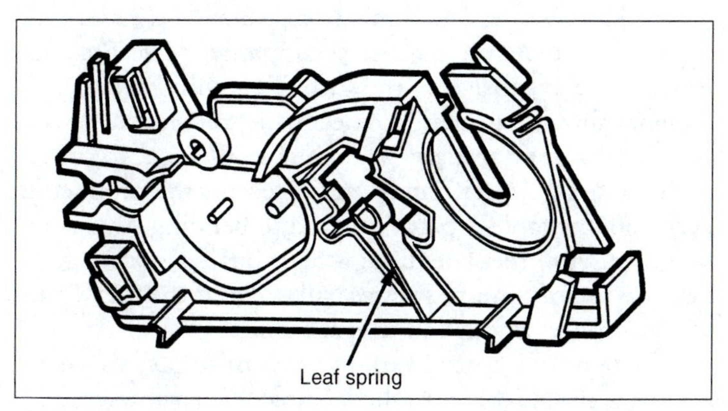

Figure 12.5

A multifunctional part for the IBM Proprinter. (Reprinted with permission of the Society of Manufacturing Engineers, Tool and Manufacturing Engineers Handbook: Volume 6, Design for Manufacturability, *copyright 1992.)*

• Use the Same Part in Many Products

Can a part be used in several different products?

Many parts can be designed for use in several products. A simple case is a standard stock item such as a bolt. This principle is a variation of the first principle; the number of parts in a company's product lines is reduced. If a company decides to adopt this DFM philosophy, then the first step is to study all parts used in the company's products—whether manufactured by the company or supplied by vendors. Parts should be classified as:

- Unique to a particular product
- Common to several products

Each group is then divided into categories of similar parts, and common attributes are identified. To recognize parts that can be used in several products, the designer should:

- Minimize the number of parts categories
- Minimize the number of variations within each category
- Minimize the number of design features within each variation

• Roughest Surface, Largest Tolerance

Dimensional precision and surface quality govern parts costs.

Very smooth surfaces and accurate dimensions are expensive.

Surface finishes and dimensional tolerances have a major impact upon the cost of a component because these features are expensive to produce. To minimize costs, the designer should be careful to specify the roughest acceptable surface finish and the largest acceptable tolerance to comply with the design specification and especially the product life cycle. For example, a part subjected to bending stresses in a fatigue environment requires a high quality surface finish because the maximum stresses will occur at the surface and surface imperfections can trigger failure.

Data on the cost of various levels of surface finish and accuracy should be consulted. Table 12.3 shows the approximate relative costs of surface finishes and the processes used to generate them. Figure 12.6 illustrates the range of surface finishes developed by a variety of common manufacturing processes and also shows how the process time (and

	Surface Roughness (Micrometers)	Approximate Relative Cost (Percent)
Case, rough-machining	6	100
Standard machining	3	200
Fine machining, rough-grinding	1.5	440
Very fine machining, ordinary grinding	0.8	720
Fine grinding, shaving, and honing	0.4	1400
Very fine grinding, shaving, honing, and lapping	0.2	2400
Lapping, burnishing, superhoning, and polishing	0.05	4500

Table 12.3

Relative cost of producing a specified surface finish by various processes.

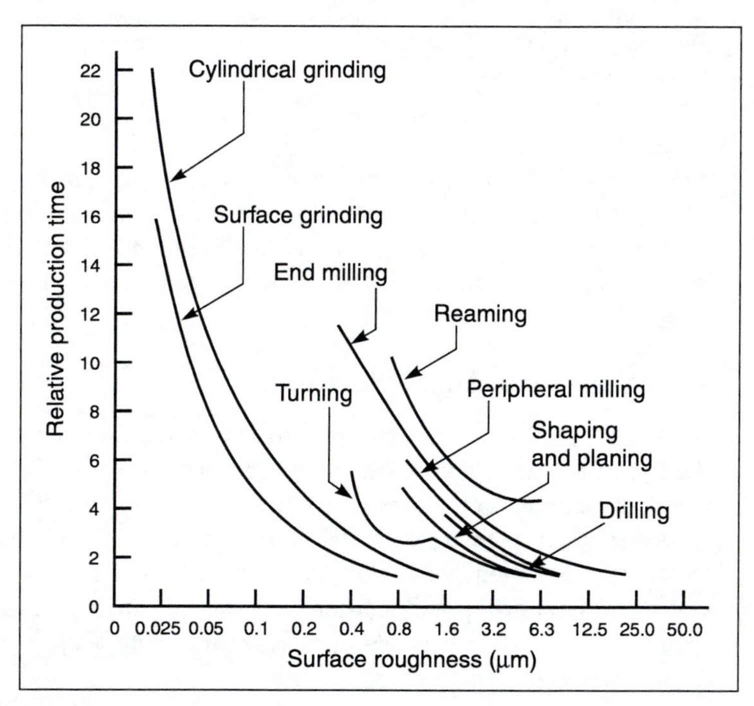

Figure 12.6

The range of surface finish generated by different processes and the relative time involved.

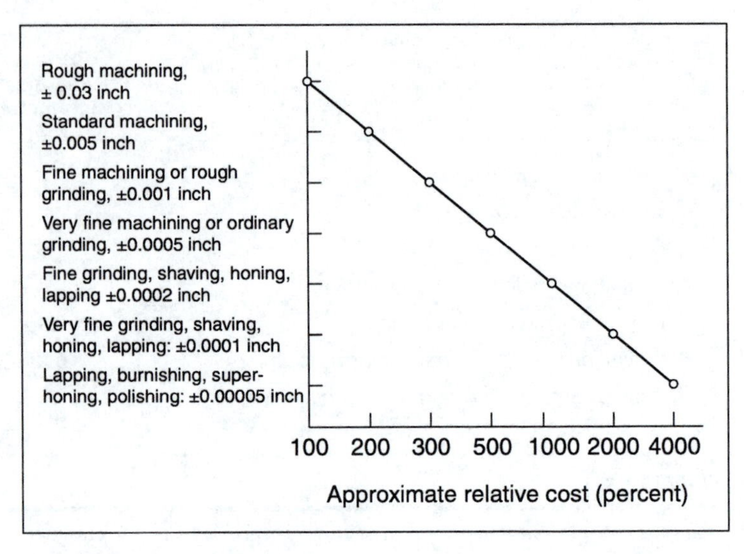

Figure 12.7

The relative cost of specifying tighter tolerances and the different processes used to generate them.

hence the cost) increases when extremely fine surface finishes are specified. Figure 12.7 illustrates the approximate relative costs of different tolerance levels, in addition to listing typical manufacturing processes for generating those levels.

• Avoid Secondary Operations

Secondary operations such as inspection, deburring, painting, etc., are expensive. Avoid them.

Each part should be designed to minimize the cost of the manufacturing operations necessary to produce it with the specified characteristics. Secondary operations such as inspection, heat treatment, deburring, and painting are frequently as expensive as the primary manufacturing operations. Whenever possible, therefore, design teams should specify near-net-shape processes such as forging and casting. Avoid secondary processes.

• Specify Processible Materials

Parts should be made from the cheapest material that satisfies the design specification and that minimizes costs associated with processing, scrap material, warranty costs, and service throughout the life cycle of the part. Consequently the most appropriate manufacturing process should be used to produce the part, and the part should be designed to exploit the characteristics of the. For example, mass production by powder metallurgy allows a single part to be made with a porous structure that will retain a liquid lubricant—thereby avoiding the manufacture of a part that incorporates a bushing at the interface with the rotating shaft. Injection-molded plastic parts can emerge from the mold with the desired color, surface qualities, and shape—thereby avoiding the secondary operations of painting and machining.

Parts should be made from the cheapest material that will satisfy the design specification.

Design the part to exploit the most appropriate manufacturing process.

• Design for Production Volume

The choice of an economical production process depends upon the total number of parts needed and the rate at which they are required. These considerations dictate whether money should be spent on special tooling or whether conventional methods should be used.

Consider the rate at which parts must be manufactured, and the size of the batch, when selecting the manufacturing process.

Processes for producing thousands of parts per year will be quite different from those used in a small shop making only a few hundred parts. The scale of production constrains the product design because of the limited shapes, tolerances, and materials associated with each manufacturing environment.

Small batch production mandates the use of general-purpose lathes and milling machines for metal cutting and the fabrication of welded structures from stock steel bars, rods, and sheets. Large production volumes afford the engineer the freedom to use casting or drop forging processes, and metal cutting operations can be automated with special-purpose machines. This situation is illustrated in Figure 12.8, where a bearing bracket is manufactured as a gray iron casting for large volume applications but as a steel weldment

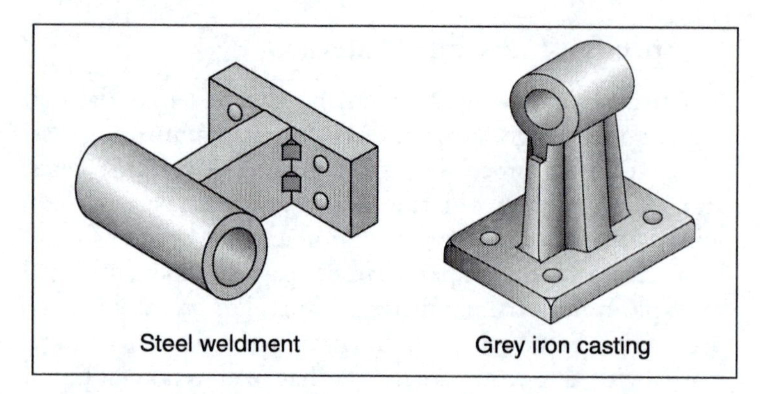

Figure 12.8

The production volume of a bracket determines the material selection, form of the product, and the manufacturing process.

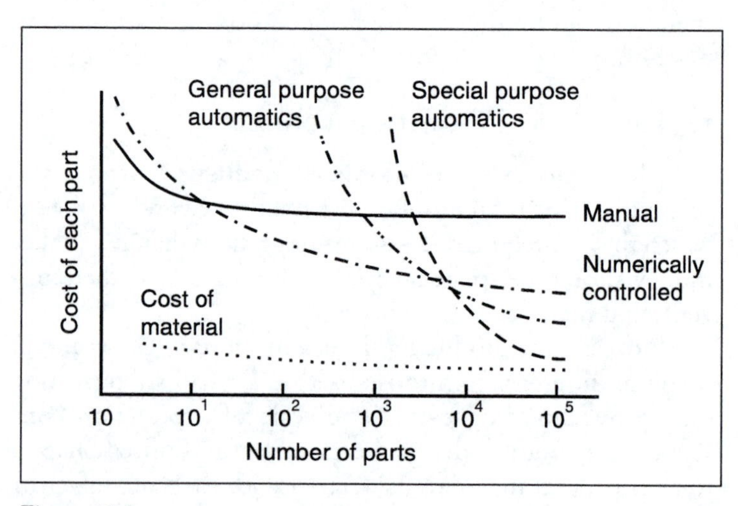

Figure 12.9

The relationships between production volume, economics of production, and machine tool selection.

for small batch applications. The economics of selecting appropriate metal cutting tools for batches of different sizes is illustrated in Figure 12.9.

There are more subtle aspects of process selection. Consider the choice of an appropriate casting process for an

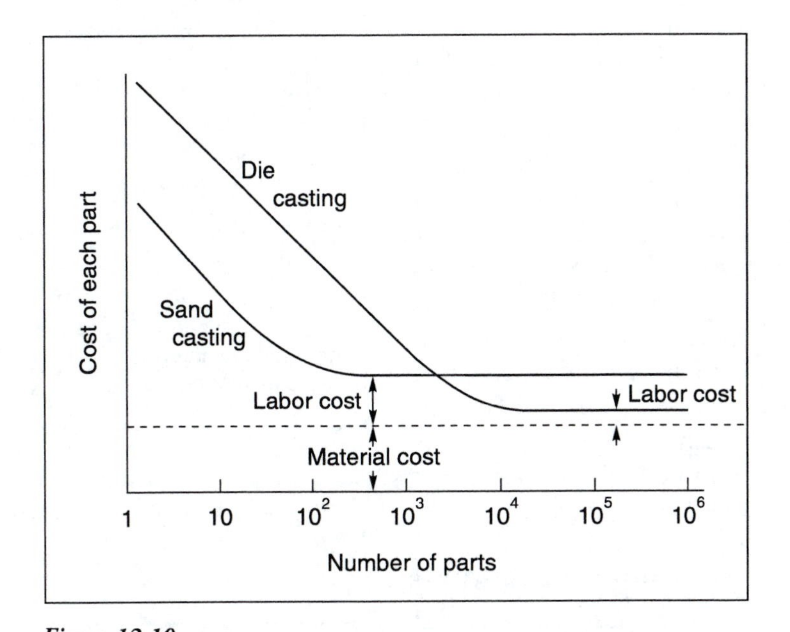

Figure 12.10

The most economical casting process depends upon production volume.

aluminum part to be manufactured in high volume. The designer may be able to use pressure die casting rather than gravity-fed sand casting as shown in Figure 12.10. Although it uses inexpensive equipment, sand casting is labor-intensive. Die casting requires expensive equipment but offers savings in material and labor costs at high production volumes.

Sand casting:

- inexpensive equipment
- labor intensive

Die casting:

- expensive equipment
- low-skilled labor

• Avoid Separate Fasteners

Parts and subassemblies should be created with as few separate fasteners as possible. Avoid bolts, screws, rivets, and pins. Naturally this is an extension of the DFM rule to minimize the number of parts in a product. Separate fasteners are associated with holes that create stresses; consequently products with many fasteners are more likely to fail unless they are carefully designed and manufactured.

Avoid separate fasteners. When they must be used, avoid loose washers. Use self-tapping screws.

This principle has ramifications for assembly operations. In automated assembly, separate fasteners are difficult to feed, they sometimes jam, and they require careful monitoring. In a manual assembly operation, the cost of driving a screw is typically six to ten times the cost of the screw itself. If fasteners must be used, use as few as possible. Use the least variation in size and type. Specify captured rather than loose washers (Figure 12.11) and choose self-tapping screws whenever possible.

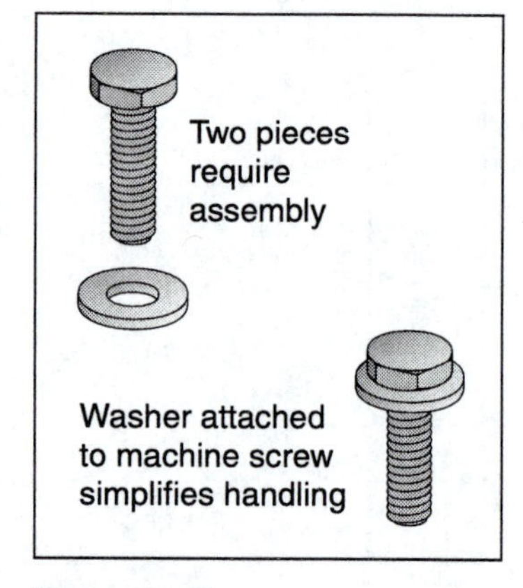

Figure 12.11

Fasteners with captured washers reduce assembly costs.

- ## Minimize Handling

Handle as few parts as possible.

The positioning and orientation of parts cost time and money.

The position and orientation of a part is defined by six degrees of freedom: three translational and three rotational (Figure 12.12). Because of this complexity, to accurately position a part costs both time and money. The associated handling operations require equipment, they increase the risk of low quality, and they slow feed rates and increase cycle times.

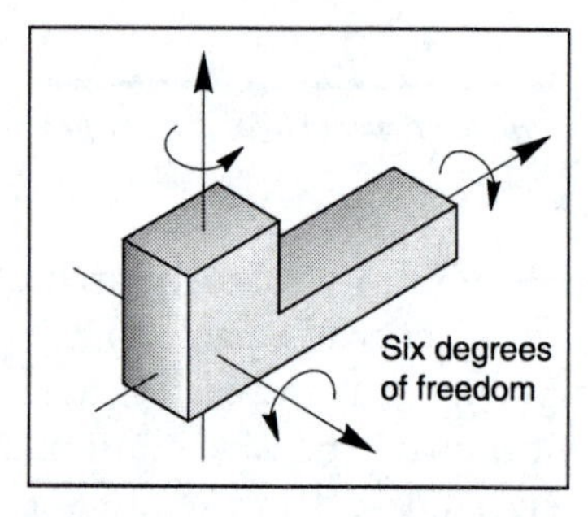

Figure 12.12

The six degrees of freedom of a rigid body.

Handling is a primary ingredient of assembly processes; therefore parts should incorporate features that assist in their positioning during assembly. Symmetry helps, for example, as do design features that locate and guide parts during an insertion process.

Summary

The eleven design-for-manufacture principles apply to the majority of manufacturing environments. The most important principle is to reduce the number of parts in a product. Standard parts should always be specified; modular designs offer advantages. Parts should serve several functions and, where possible, should be usable in several different products. Parts costs often are driven by the tolerances and quality of surface finish specified for a part. Therefore these attributes should be selected carefully. Secondary manufacturing operations should be avoided. Production volume will dictate the economic methods of manufacture, and this must be considered in conjunction with the selection of an easily processed material. Handling operations should be kept to a minimum during manufacture. Avoid separate fasteners.

- Reduce number of parts
- Use standard components
- Use modular designs
- Use multifunctional parts
- Use same part in many products
- Largest tolerance
- Fewer operations
- Use processible materials
- Design for production volume
- Fewer fasteners
- Less handling

Key Concepts

Commonality of parts. Economies of scale can be achieved by using the same part in several products.

Design-for-Manufacture (DFM). A methodology for ensuring that products can be manufactured cost-effectively and still satisfy the design specification.

Modular design. Design in which common interfaces provide for easy assembly, easy replacement of parts, easy servicing, easy upgrades or enhancements, and therefore reduced cost.

Multifunctional part. A part that has more than one function. When a multifunctional part is used, cost is reduced because of savings in manufacture, inventory, and assembly.

Part count. The number of parts in an assembly. This count should be minimized to reduce cost.

Processible material. A material that can be easily shaped and prepared for assembly.

Production volume. The number of items, products, assemblies, or parts manufactured relative to time. A larger production volume usually leads to reduced cost per item manufactured.

Secondary operation. A second operation required to finish a part and prepare it incorporation into the product. Examples are deburring, polishing, heat treatment, painting, plating, and other surface preparations. Every secondary operation increases cost.

Subgroups of DFM. Design for manufacture has the subgroups: process-specific DFM, facility-specific DFM, general DFM, and design for assembly (DFA).

Review Questions

1. What are the three primary assumptions embodied in the concept of DFM?
2. What are the advantages of DFM measured in terms of the product?
3. Design-for-manufacturing has three primary ingredients. What are they?
4. Define process-specific DFM principles. Provide illustrative examples.
5. List eleven DFM principles.
6. Why is the minimization of the number of parts in a product an important DFM principle? Provide illustrative examples.

7. Three criteria can be imposed upon a part to determine whether it can be eliminated from a product. What are they?
8. Define facility-specific DFM principles.
9. What advantages can be accrued by specifying off-the-shelf parts? List five of these products.
10. What is a modular design?
11. Define is a multifunctional part. Provide an illustrative example.
12. Define a multideployment part.
13. When a design team creates a multideployment part for product, what criteria should be considered to minimize three general attributes of these parts?
14. What role do tolerances and surface finish play in costing a part?
15. How much more expensive is it to fine-grind a cylindrical part instead of specifying a standard machined finish?
16. If a part were to be finished by cylindrical grinding and the surface finish was changed from 3.2 mm to 0.05 mm, how much longer would the production time typically be?
17. A DFM team decides to redesign an assembly so that a part with a surface finish that is fine-ground to an accuracy of ±0.0005 inch is replaced by a fine-machined surface with an accuracy of ±0.001 inch. What is the financial saving?
18. List two secondary manufacturing operations and explain why they should be avoided.
19. What are processible materials?
20. What is production volume and how does it influence product design?

21. Why should separate fasteners be minimized in the design of products?
22. Define handling operations. How do they influence the design of parts?

Problems

1. Select a kitchen appliance such as a coffee maker or toaster. Dismantle it and, by imposing a parts-consolidation philosophy on the product, develop a new product with fewer parts.
2. A 30 mm diameter steel shaft of length 150 mm must contain a shoulder to restrict the motion of the inner race of a deep-groove ball bearing at a point 75 mm from one end. Propose several methods for introducing this shoulder and evaluate the methods from a design-for-manufacturing perspective.
3. A spur gear and shaft combination must be designed for a low-power application for a packaging machine. Discuss the different options available to the design team confronted by this task. What are the governing parameters? When should a single part be manufactured combining the gear and the shaft in one part? If the shaft and gear are two separate parts, discuss the different methods for designing and manufacturing the assembly. What are the advantages and disadvantages?
4. A spur gear can be manufactured by a number of different processes such as extrusion, casting, powder metallurgy, drop forging, or machining from bar stock. Discuss the different options offered by these processes in the context of economy and production volume.

5. A pulley driven by a V-belt must be assembled on a long drive shaft. Explore the options for accomplishing this objective. Propose several different concepts and then analyze them from a manufacturing perspective before presenting their advantages and disadvantages.

6. A collar with a keyway on the inner diameter is illustrated in the margin. The keyway has sharp corners. Discuss the design. Consider the effect on the part shape if the collar must be quenched during heat treatment. Propose an alternative design.

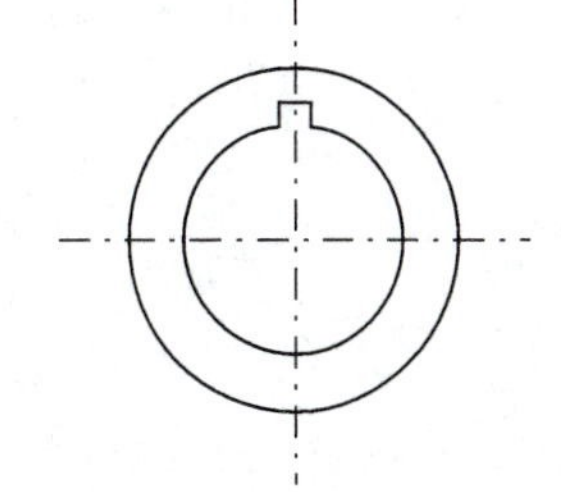

7. A lightly loaded valve body for a low-volume application is illustrated in the margin. It is manufactured by machining a thick plate of aluminum. The central hole is machined with a boring bar and four symmetrically located bolt holes are drilled. What are the consequences for the manufacturing processes if the part is used in a high-volume product line? How might you modify the shape of the part? Illustrate your solution.

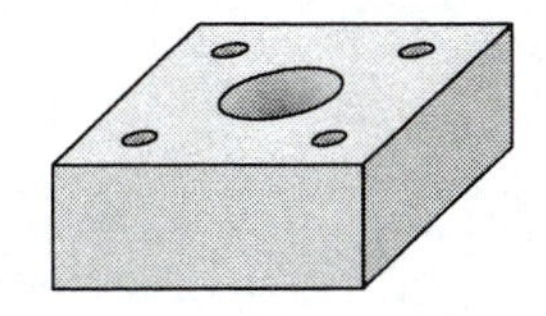

Further Reading

ASM Metals Handbook. Metals Park, OH: ASM International, 1990.

Bakerjian, R., ed. *Tool and Engineers Handbook, Volume 6, Design for Manufacturability*, Society of Manufacturing Engineers, Dearborn, MI, 1992.

Boothroyd, G. and P. Dewhurst. *Product Design for Assembly*, Boothroyd and Dewhurst, Inc., Wakefield, RI, 1987.

Bralla, J., ed. *Handbook of Product Design for Manufacturing*. New York: McGraw-Hill, 1986.

Corbett, J., M. Dooner, J. Meleka, and C. Pym. *Design for Manufacture*. Reading, MA: Addison-Wesley, 1991.

Dieter, G. E. *Engineering Design. A Materials and Processing Approach.* New York: McGraw Hill, 1983.

El Wakil, S. D. *Processes and Design for Manufacturing.* Englewood Cliffs, NJ: Prentice-Hall, 1989.

Farag, M. M. *Selection of Materials and Manufacturing Processes for Engineering Design.* Hemel Hempstead, UK: Prentice Hall International, 1989.

Miller, L. C. G. *Concurrent Engineering Design.* Dearborn, MI: Society of Manufacturing Engineers, 1993.

Stoll, H. W. "Design for Manufacture: an Overview," *ASME Applied Mechanics Reviews*, 39 (9, September 1986): 1356–1364.

Trucks, H. E. *Designing for Economical Production.* Dearborn, MI: Society of Manufacturing Engineers, 1987.

Design-for-Assembly

13

...parliament is a deliberative assembly of one nation, with one intent, that of the whole;...

Edmund Burke
(1729–1797)

Objectives

- To introduce design-for-assembly.
- To discuss assembly operations in manufacturing.
- To provide rules for reducing assembly costs.
- To illustrate the advantages of DFM and DFA in industry.

Contents

13.0 Introduction

In the previous chapter we discussed some general design-for-manufacture principles. Now we turn to assembly operations because the design of a product has a major influence upon how a product must be put together from its parts. Commonly 40 to 60 percent of the total time invested in the manufacture of a product is spent on its assembly.

Assembly operations take about half the time spent on the manufacture of a product.

13.1 Design-for-Assembly

Assembly is the putting together of a group of two or more parts into a product or some subassembly of the product. A collection of parts makes up a subassembly and subassemblies make up the product. The final product contains all necessary subassemblies and parts. Since the assembly process frequently involves components that contain machined surfaces that must interface accurately, it involves considerable expense (see Table 12.3 and Figure 12.7). The overall scheme of assembly operations for the manufacture of a product is shown in Figure 13.1.

Assembly is the fitting together of parts into a more complex whole.

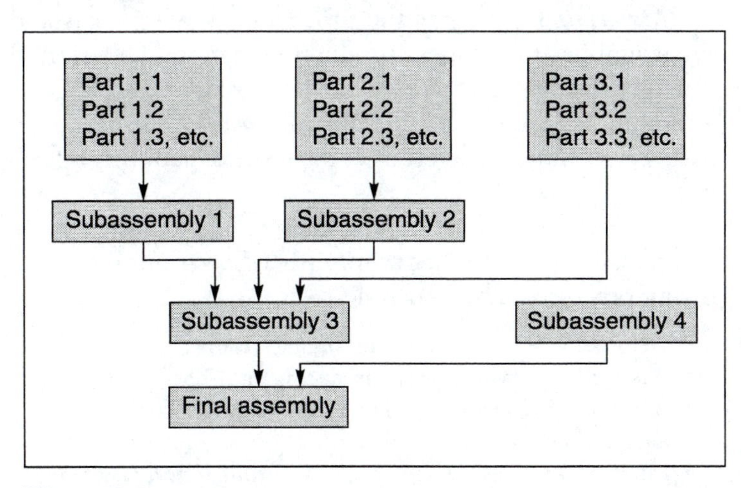

Figure 13.1

The primary stages of product assembly.

Assembly operations involve:

- Materials handling
- Assembly
- Quality control

Any assembly process has three primary attributes: materials handling, assembling, and quality control. These are shown in Figure 13.3.

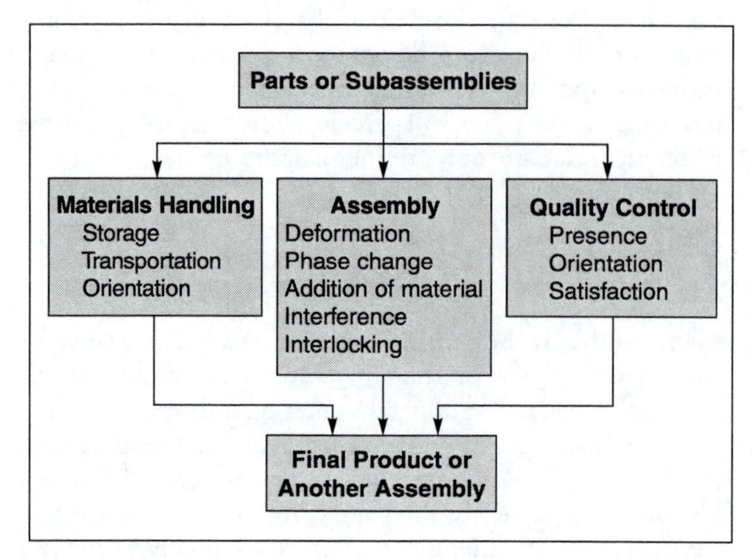

Figure 13.2

Primary attributes of an assembly operation.

Materials handling is the process of moving parts between manufacturing operations. It includes

- storage
- transportation
- orientation

Assembly operations can involve:

- deformation
- phase change
- adding material
- interference
- interlocking

Materials handling is the process of moving parts and subassemblies to the next production station. This station can be a buffer store, an automated parts feeder, a parts assembly station or a quality-control station. The manufacturing environment determines the types of handling operations required.

Assembly is the fitting together of parts into a more complex whole. It can be accomplished by various manufacturing processes. These include:

- **Deformation,** which occurs in forging, riveting, and swaging
- **Phase change,** which occurs in casting and adhesive bonding
- **Addition of material,** which occurs in welding, adhesive bonding, and metal spraying
- **Interference,** which occurs in shrink fitting, screw fastening, and nailing
- **Interlocking,** which occurs in sewing, braiding, and wrapping.

Quality control monitors whether a product has been assembled correctly. In addition, quality control may check for the presence of a part or the orientation of a part as the work enters an automated assembly machine.

Quality control monitors whether the product formed by one or more assembly operations has been generated correctly and whether it meets its performance criteria.

It is important to identify early in the assembly process the methods that will cost the least. If several million parts per year are to be manufactured, then special purpose equipment capable of high-speed automatic assembly can be used. However, if only a few hundred parts are to be made each year, then manual assembly is the logical choice. Between these two extremes there are a variety of options. The option selected has a major effect on the design of parts.

The number of parts to be made and their characteristics dictate the assembly process.

When are Assembly Processes Needed?

Assembly can be costly. Therefore the question arises as to why assembly processes are needed. Here are some reasons:

Assembly processes are needed in case of:

- different material properties
- relative motion
- different part functions
- manufacturing constraints
- accessibility requirements

- **Different material attributes.** For some products to operate correctly, the parts must perform functions that depend upon the properties of the materials from which the parts are made. For example, in a rigid piping system a soft gasket may be needed between two connected pipes.
- **Relative motion.** Many products involve relative motion between parts. Good examples are found in the mechanisms of machines—the piston in an automobile engine, for instance. These products depend upon an assembly of parts.
- **Different part functions.** A product must frequently perform many different functions, and these functions can mandate the use of various materials, shapes, and manufacturing processes. Consequently the product must often be assembled from a number of parts. An example is a bearing bracket, cantilevered from a wall, that has to support a rotating shaft (Figure 13.3). An iron casting can provide the stiffness and strength to support loads imposed by the shaft, but this material is unsuitable as a bearing for a rotating shaft. Good wear resistance calls for a ball bearing or a sintered metal bushing within the casting, and this requires assembly.
- **Manufacturing constraints.** Each manufacturing process has its own set of constraints and characteristics. The shapes of parts

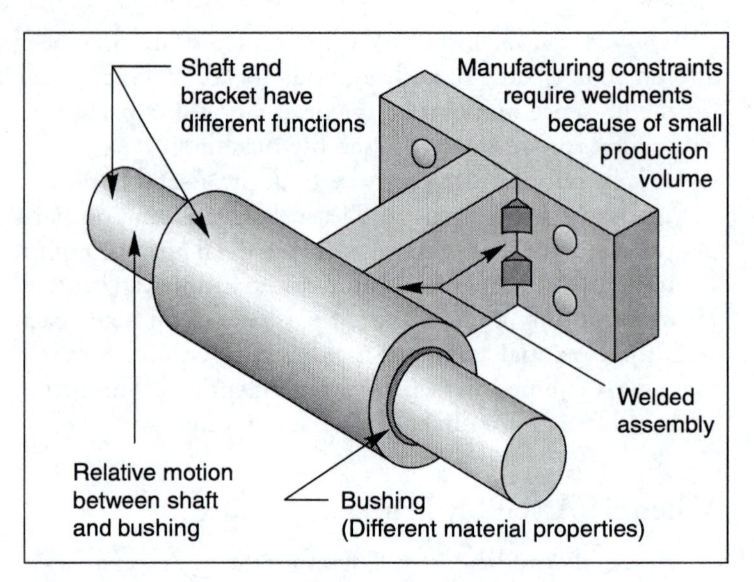

Figure 13.3

A product in which different parts have different functions.

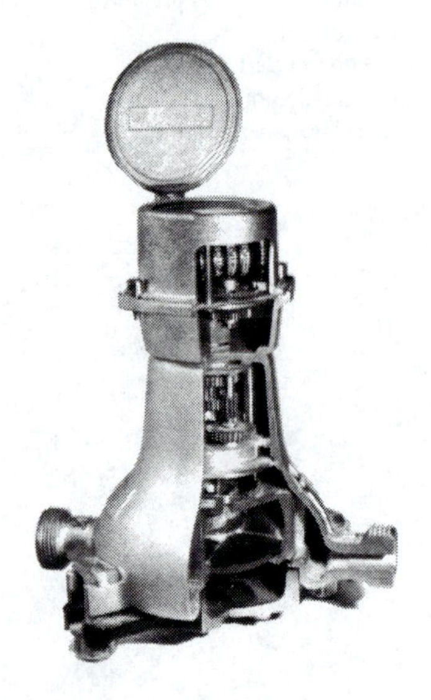

Figure 13.4

This water meter has been cut away to show the complex assembly required.

typically will require forming and finishing processes prior to assembly of the final product. If only a small number of products are to be made, then large parts can be made by welding together or bolting together pieces of rolled sections and plates because it is more economical than casting. This requires assembly.

- **Accessibility requirements.** Products often must be designed to facilitate dismantling for repair, inspection, and maintenance. This requires that the product be made up of separate parts. An example is the access cover for a gear box.

Some products are complex (Figure 13.4), but from the point of view of assembly, the best product has the fewest parts, the least expensive method of joining them together, and an overall assembly process that minimizes the cost of labor, parts, tools, and service. It is not surprising therefore that the pursuit of an easily assembled product has spawned the field of *design-for-assembly (DFA)*. We will treat design-for-assembly as a subset of design-for-manufacture.

In design-for-assembly the product structure and the structure of individual parts is the focus of attention. The primary goal is to simplify the structure so that assembly costs are minimized. A simplification of the structure of a product by use of design-for-assembly principles reduces the cost of manufacture because it leads to a relatively short manufacturing sequence and less intricate parts geometries. The consequences of DFA include increased reliability and easier servicing, while the economic consequences include a reduction in overhead costs.

The best design has the:

- fewest parts
- least expensive method of assembly

If a separate part can be eliminated from an assembly by parts consolidation or by creating a multifunctional part, then that part need not be tooled or assembled. The overhead for that part—storing, purchasing, and inspection—is also eliminated. These cost savings can be appreciable.

The elimination of a part implicitly eliminates:

- tooling
- assembly
- purchasing
- inspection
- storage

Design-for-assembly focuses upon the attributes of each and every part. Since DFA is part of DFM, and DFM is an important consideration in the design process, DFA principles should influence decisions made in the design process, the materials selection process, and the manufacturing process (see Figure 11.2). Some important parts attributes governing assembly costs are:

DFA is part of DFM and both influence product design.

- Form
- Material
- Dimensions
- Tolerances
- Surface quality
- Functionality

A number of general design-for-assembly principles can identified. These principles are concerned with avoiding or minimizing both assembly operations and materials handling. We shall discuss the nine general design-for-assembly principles listed in Table 13.1.

- Minimize the number of parts
- Minimize assembly operations
- Assemble parts from only one direction
- Allow for variations among parts
- Facilitate the transportation of parts
- Avoid orienting parts
- Facilitate the orientation of parts
- Select an appropriate method of assembling parts
- Facilitate the insertion of parts

Table 13.1

Nine general design-for-assembly principles.

• Minimize the Number of Parts

Eliminating a part simplifies assembly, reduces overhead, and eliminates opportunities for defects.

The first design-for-manufacture principle is repeated here for completeness and because it is the primary cost reducing principle in assembly operations. A reduction in the number of parts in a product saves overhead costs and reduces assembly operations. Savings occur because of reduced inventory, simplified factory layout, simplified assembly, fewer accounting activities, fewer purchasing activities, less materials handling, and less quality control. Furthermore the quality of the product increases because each additional part in a product is another opportunity for the introduction of a defect. Methods for reducing the number of parts in a product have been addressed already in the design-for-manufacture principles presented in Chapter 12. Others will appear within some of the following design-for-assembly principles.

• Minimize Assembly Operations

Combine two parts into one by redesigning the assembly.

Reducing the number of assembly operations is closely tied to reducing the number of parts. Assembly typically involves many direct and indirect costs and should be minimized in the manufacture of a product. Sometimes this can be achieved by combining two parts into one, as shown in Figure 13.5 where the welding operation to unite the snap-in device with the rolled steel stock of the base is avoided by implementing an alternative design. A single sheet metal part is manufactured by stamping and bending the plate stock. Alternatively, the total number of parts can be reduced through exploitation of the properties of different materials. For example, a different material can provide the functional characteristics of two separate parts which previously have been made of two different materials.

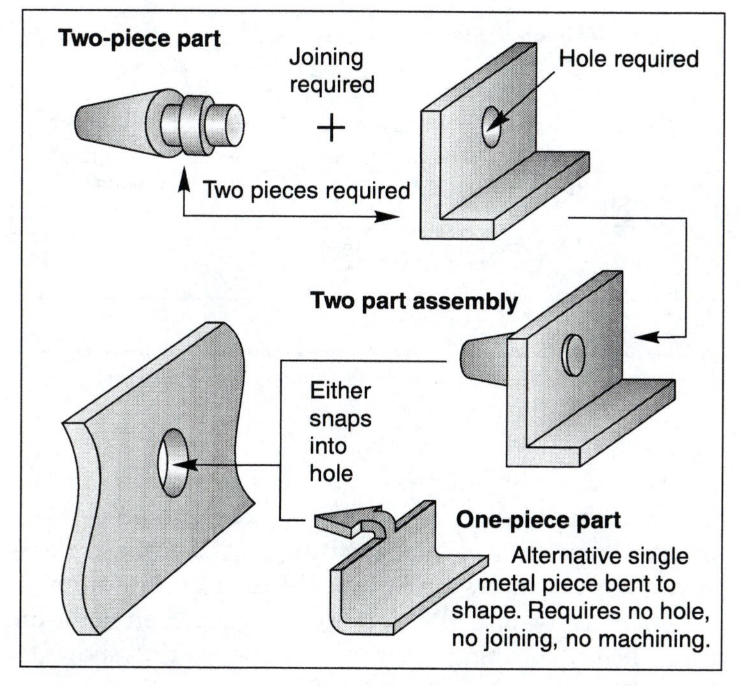

Figure 13.5

Minimize assembly operations by minimizing the number of parts.

Figure 13.6 shows an example in which the rotation of a relatively rigid low carbon steel lever is controlled by a leaf spring made from a strip of medium carbon steel. These two parts can be replaced by a multifunctional injection-molded plastic part of variable stiffness.

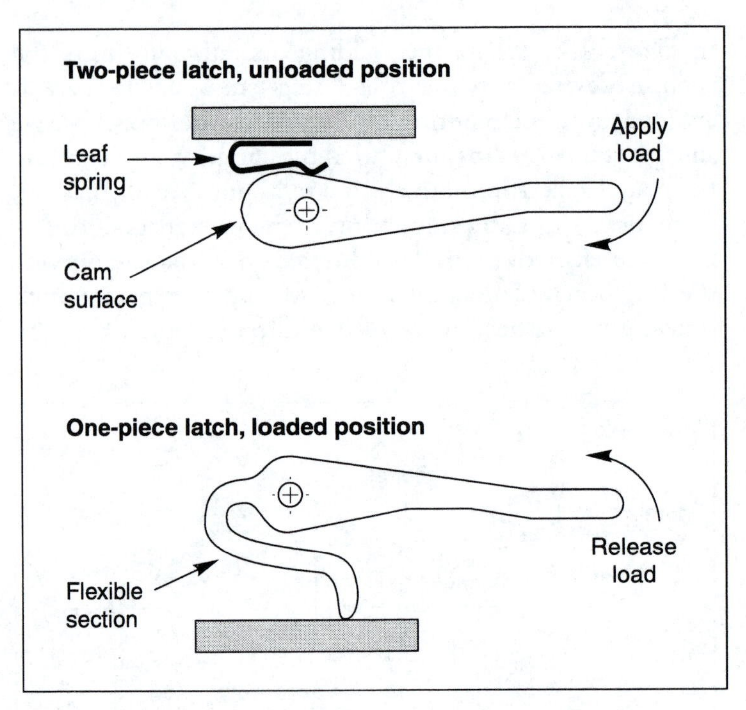

Figure 13.6

Minimize assembly operations through the creation of multifunctional parts.

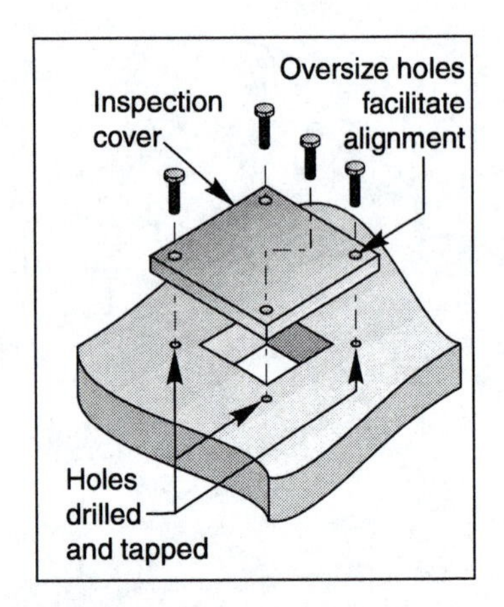

Figure 13.7

A product assembled downward.

Use large, heavy parts as assembly fixtures for other components.

Some assembly operations can be eliminated by specifying production processes which promote the joining of two or more parts during manufacture. For example, a high-strength metal insert can be assembled with a low-strength pressure die cast zinc alloy by positioning the insert in the die cavity before injecting the zinc alloy. Figure 12.5 shows an example of this type of assembly process.

If a product cannot be manufactured without assembly operations, then every effort should be made to design the part for easy assembly. Some ways to do this follow.

• Assemble Parts From Only One Direction

All parts should be assembled from only one direction (Figure 13.7). Whenever possible, use a large component as the primary object to which others are attached. Take advantage of gravity. If pieces logically hang from a superstructure, work top down; if parts are logically layered upon a base, work bottom up. Each part should be positively located and should have locating and guiding elements for other parts. This avoids assembly fixtures and offers a number of advantages. For example, a fixtureless process reduces the need for subsequent adjustments.

Let gravity help as parts are assembled.

If the designer specifies assembly from multiple directions, it will generally be more expensive because additional equipment, transfer stations, and inspections will be needed. All assembly operations must provide access for tools such as screwdrivers and socket wrenches.

Allow access for tools during assembly.

• Allow For Variations Among Parts

A part from one batch does not have exactly the same dimensions as a part from another batch, nor is it probable that one part will have exactly the same surface finish as another. During assembly, these variations can cause misalignments that require excessive assembly forces, or excessive clearance in assemblies, or parts that cannot be assembled because of a tolerance stack-up. The result is an unreliable product and undesirable downtime of automated assembly equipment.

Anticipate variations among parts during assembly.

These problems can be minimized by the manufacture of parts with designed-in tolerance for variation. Designed-in characteristics include the specification of generous radii, tapers, and chamfers as well as guiding elements to facilitate insertion. Sometimes the elastic properties of plastic parts can be used to absorb variations in size or shape.

• Facilitate the Transportation of Parts

Parts often are transported in chutes, rails, or channels. In these systems parts can tip over and they will require reorientation before assembly. The behavior of parts during transportation should be carefully considered. If necessary the shape should be modified to prevent transportation problems. Examples are shown in Figure 13.8.

Design parts with a low center of mass to avoid tipping.

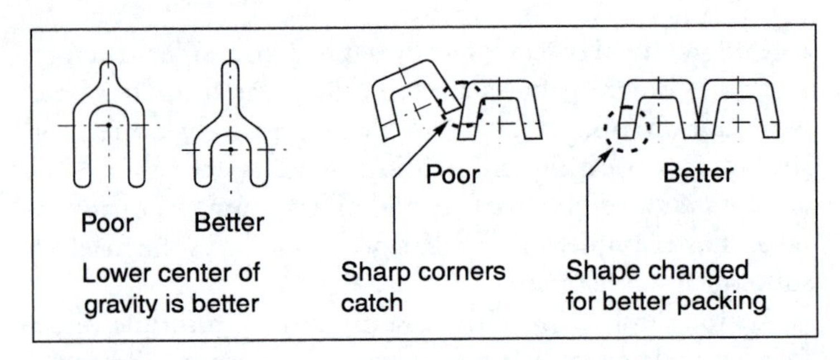

Figure 13.8

Avoiding transportation problems by selecting appropriate shapes for parts.

• Avoid Orienting Parts

Carefully consider the geometrical attributes of the part in the context of the static and dynamical response when designing automated handling systems.

If parts arrive at an assembly station in random orientation, the position of a part must be identified before it can be oriented correctly. That involves time and money. One way to avoid this is to assemble parts as they are being manufactured. Another way is to use magazines to hold parts. They keep parts in a desired orientation during handling and transportation before their final assembly.

The orientation of a part is generally necessary whenever there are several surfaces to which other parts must be mated. Sometimes this can be avoided by properly stacking parts as they are manufactured.

• Facilitate the Orientation of Parts

Avoid parts shapes that tangle easily.

Parts with hook-like appendages or holes and slots easily become tangled during storage or transportation. This frustrating situation can be avoided by careful detail design. Examples of good and bad parts are shown in Figure 13.9.

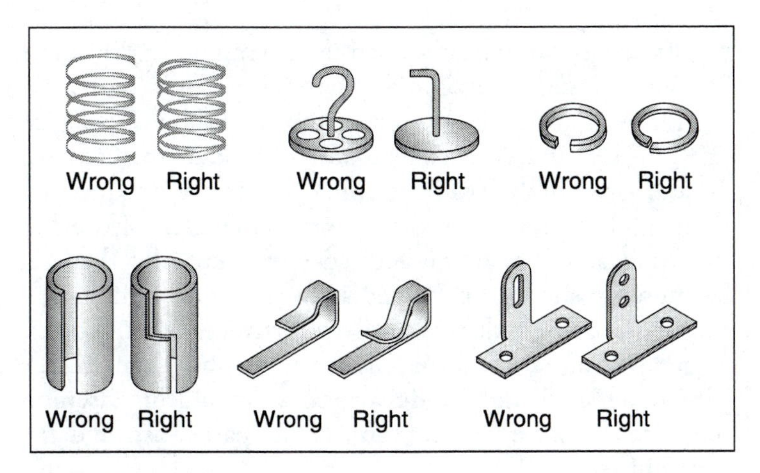

Figure 13.9

Avoid parts that tangle.

A part is more easily oriented if it has many planes of symmetry. A spherical part is ideal. Parts should, where possible, require the least rotation during assembly. Examples of good and bad practice are shown in Figure 13.10.

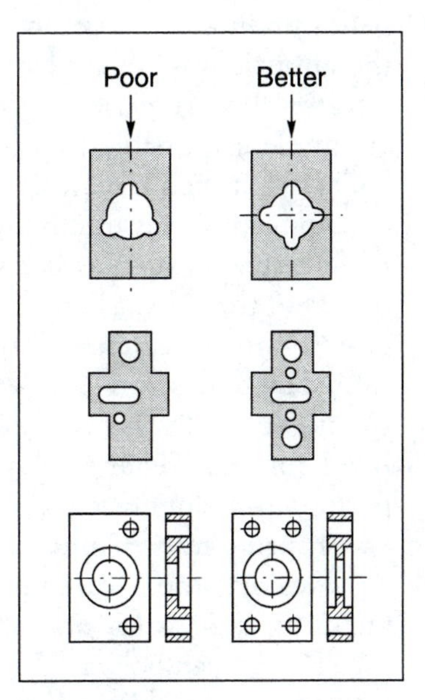

Figure 13.10

Use symmetrical parts whenever possible.

Design parts with many planes of symmetry if possible.

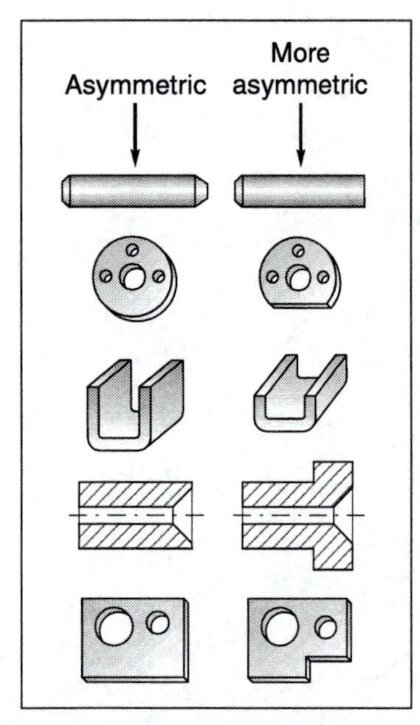

Figure 13.11

If parts must be asymmetric, make it obvious.

If a part can be assembled incorrectly, then it will be. Design parts so that they can be assembled in only one way.

If a part cannot be symmetrical, consider asymmetry to facilitate orientation.

Specify self-fastening parts.

If a part cannot be designed to be symmetrical, the other extreme should be explored. Asymmetry can be in length, width, weight, or shape, for example, as shown in Figure 13.11. This can be especially useful for product maintenance as well as assembly if different pin diameters or unequal hole spacings can prevent products from being assembled incorrectly. If a part is manufactured asymmetrically then the asymmetry should be immediately clear to a human or a mechanical assembler. Asymmetric attributes should be exaggerated so there is absolutely no doubt about the orientation of a part. Failure to do so wastes time during assembly because the correct orientation of the part must be determined first.

• Select an Appropriate Method of Assembling Parts

Selection of the best method of assembling parts depends upon what assembly is needed, and the latter reflects the application—or lack of application—of other design-for-assembly principles. When final assembly has been made as simple as possible by careful application of DFA principles, what then remains?

Appropriate assembly methods should avoid small element fasteners such as bolts and screws even though they represent the classical approach to joining. This reduces the part count, lowers assembly costs, and eliminates failures caused by high stress levels at bolt and screw locations. Parts should be self-fastening unless they must be regularly separated for maintenance. Use snap-fit fasteners on plastic parts, tab-in-slot devices in sheet metal, and push-on fasteners or spring nuts, as illustrated in Figure 13.12.

Two or more parts can be assembled during the manufacturing process for one of the two parts. For example, a metal insert can be placed in a mold cavity before the mold is filled with a polymer (Figure 12.5). This was described in the section on minimizing assembly operations.

Figure 13.12

Various fasteners. (Courtesy of Arden Fasteners.)

An alternative approach to replacing a multipart assembly is to reevaluate the original assembly and select a different material and a different process to manufacture it as a single part. Figure 13.13 illustrates a situation where an assembly consisting of two bronze bushes, a steel shaft, and a steel gear are replaced by a single multifunctional plastic part.

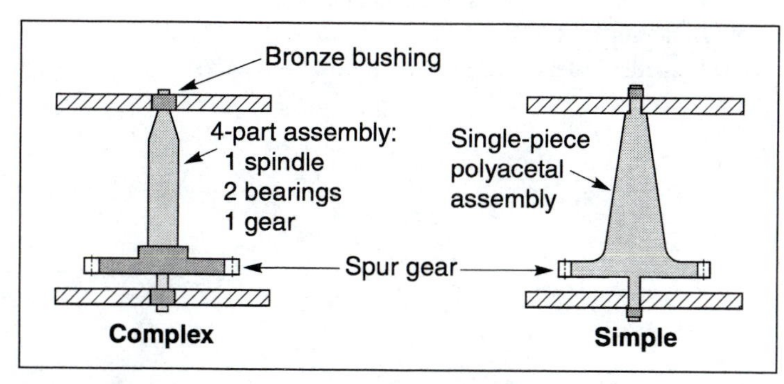

Figure 13.13

Replacing a multipart assembly with a single part.

• Facilitate the Insertion of Parts

The insertion of one part into another is a basic element of many assembly operations. This is more easily accomplished if only simple movements in a straight line are required. Avoid processes that require a change in direction of motion during insertion. This is illustrated in Figure 13.14.

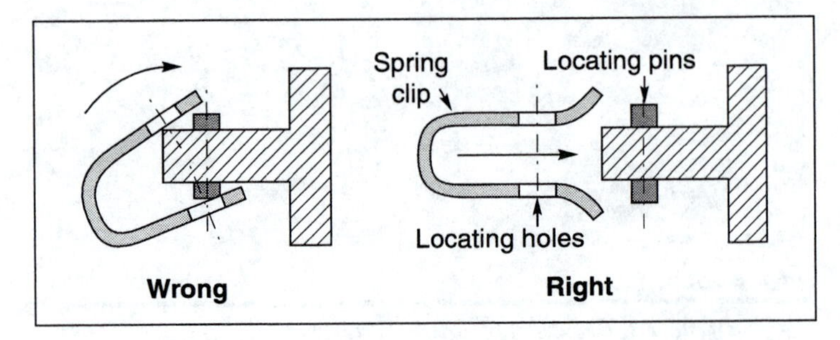

Figure 13.14

Use simple straight-line motions for assembly operations.

Another approach is to design parts to be self-locating. Figure 13.15 shows some ways to do this. Assembly processes can be assisted by providing:

- A centering spigot
- Widely separated dowel pins
- Chamfers on the edges of holes and pins
- Guide surfaces
- Elongated holes to relax dimensional tolerances

Use simple assembly motions and design parts with self-locating attributes.

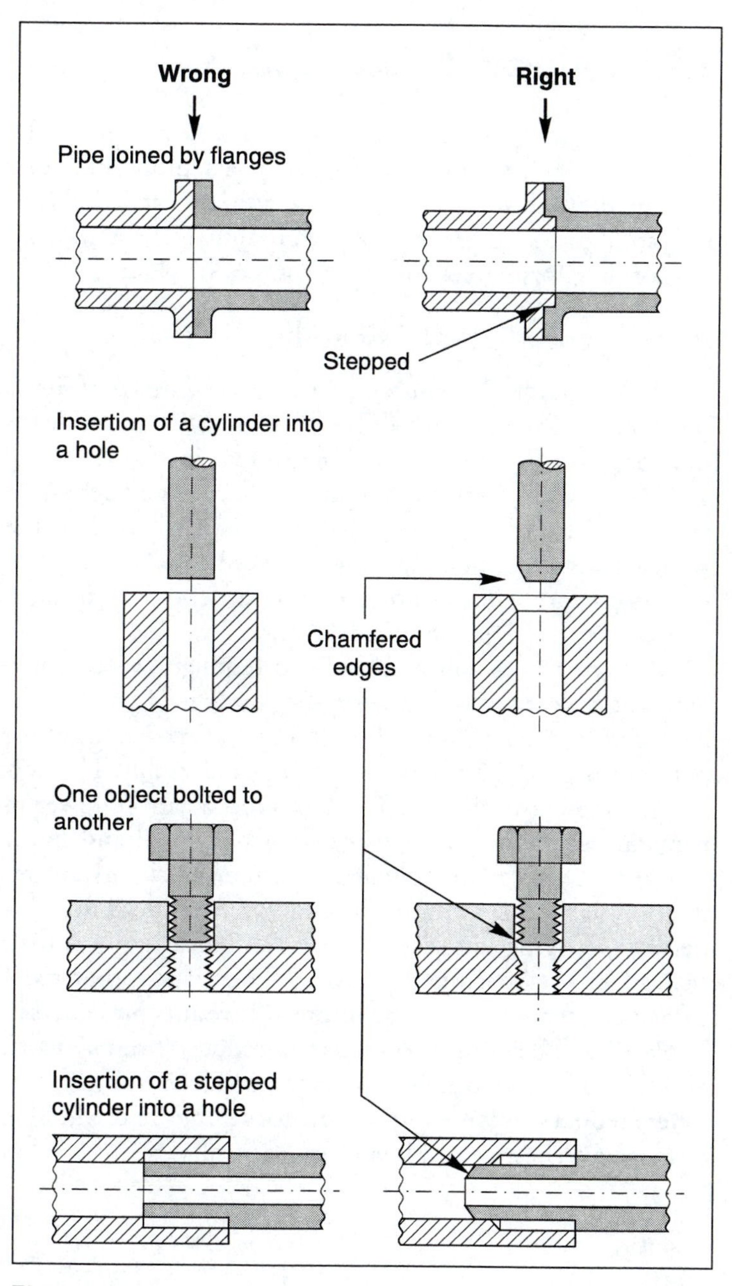

Figure 13.15

Strategies to facilitate the insertion of parts.

13.2 Industrial Case Studies

The objective of this section is to show how effective the design-for-manufacture and design-for-assembly principles presented in this and Chapter 12 are in practice. Three industrial case studies are presented to illustrate the application of these principles to three classes of products.

Redesign of a Reticle Assembly

The Defense Systems and Electronics Group of Texas Instruments Incorporated (TI) of Dallas was responsible for the design of the reticle assembly shown in Figure 13.16. This lightweight assembly supports a thermal gunsight used by ground-based armored vehicles. It is used to align the vehicle's weapon optically while the vehicle is subjected to vibrations and ballistic shocks under battlefield conditions. The original design contained 47 parts.

A team was established by TI to redesign the assembly with the objectives of reducing the number of parts, standardizing the parts, and eliminating the reorientation of parts during the 2.15 hours required for assembly. The new design, shown in Figure 13.17, contains a cam-follower to provide the connection between the rotational and linear motions. This kinematic pair permitted the significant reduction in the parts count. Fasteners were reduced in number by specifying self-securing parts and a casting was used instead of the machined housing. These and other redesigned elements allowed the impressive results presented in Table 13.2. The number of parts was reduced from 47 to 12 and the number of different parts was reduced from 24 to 8. These reductions were responsible for savings in the hidden costs of purchasing and inventory control.

The redesign of the reticle assembly reduced the number of parts from 47 to 12 and the number of different parts from 24 to 8.

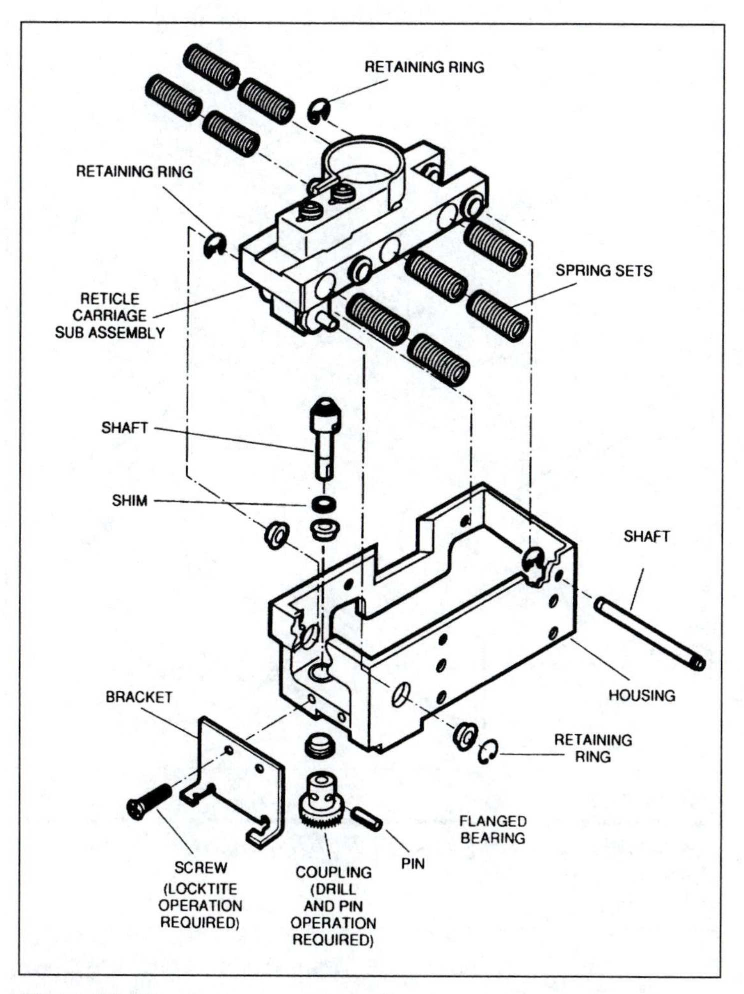

Figure 13.16

Original design of a reticle assembly. (Source: Texas Instruments, Inc.)

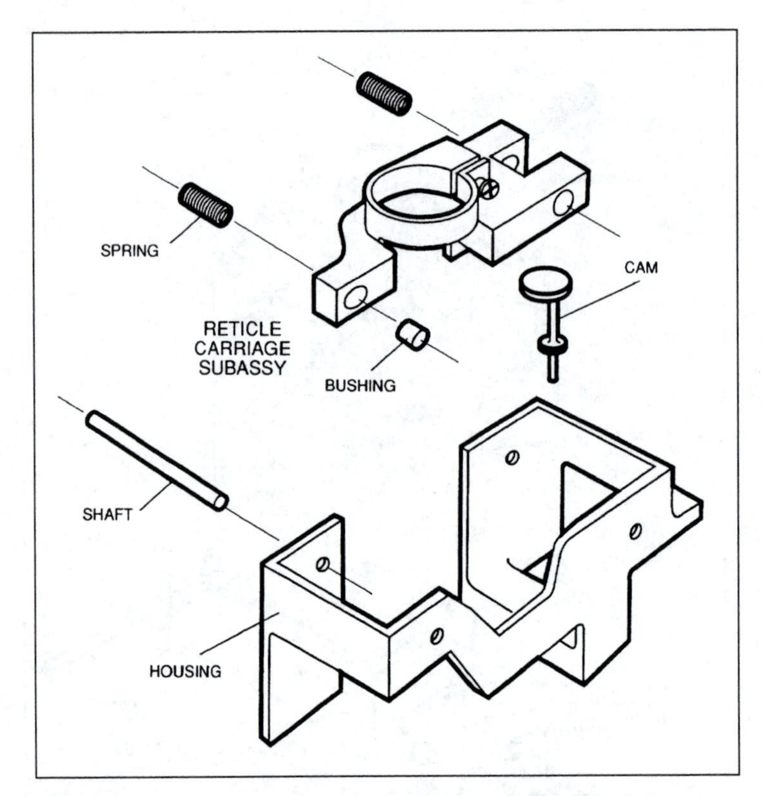

Figure 13.17

Redesigned reticle assembly using DFM/DFA principles. (Source: Texas Instruments, Inc.)

	Original design	Redesign	Improvement (%)
Assembly time (hours)	2.15	0.33	84.7
Number of different parts	24	8	66.7
Total number of parts	47	12	74.5
Total number of operations	58	13	77.6
Metal fabrication time (hours)	12.63	3.65	71.1
Weight (lbs)	0.48	0.26	45.8

Table 13.2

Comparison of original and new designs of a reticle assembly (Source: Texas Instruments, Inc.).

Redesign of a Spindle-Housing Assembly

Figure 13.18 presents two views of an inefficient spindle-housing assembly. The assembly is inefficient because it contains ten separate parts: a sheet metal housing, two nylon bearing inserts, six screws, and a steel spindle. An analysis of this assembly reveals that the housing and the shaft must be separate parts because there must be relative motion between these two bodies. However, none of the other parts move; consequently it is plausible that they can all be made from the same material and they do not need to be separate entities.

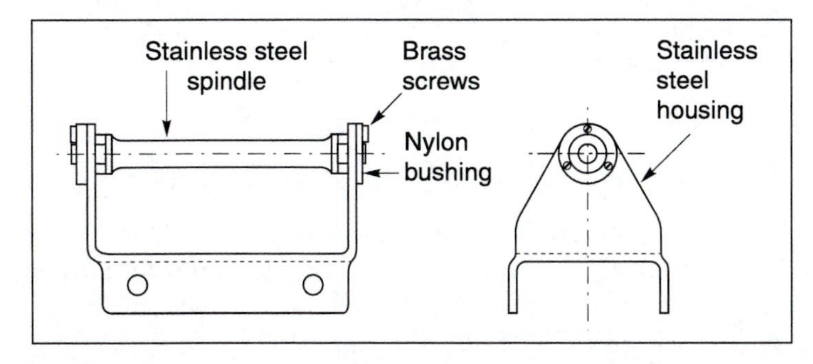

The original housing had 10 parts: sheet metal housing, two nylon bearing; six screws, and a steel spindle.

Figure 13.18

Original spindle housing design with ten separate parts (Reprinted with permission of the Society of Manufacturing Engineers, Simultaneous Engineering: Integrated Manufacturing and Design, *copyright 1990 and* Manufacturing Engineering Magazine, *copyright April 1988).*

A superior two-part design was created with nylon bearing surfaces and a one-piece injection-molded nylon housing to support the unchanged steel spindle. Two views of this design are shown in Figure 13.19. This redesigned spindle housing contains 80 percent fewer parts, a simplified assembly process, and a reduced assembly time.

The redesigned housing has 2 parts: an injection-molded housing and a steel spindle.

- 80% fewer parts
- simplified assembly
- faster assembly

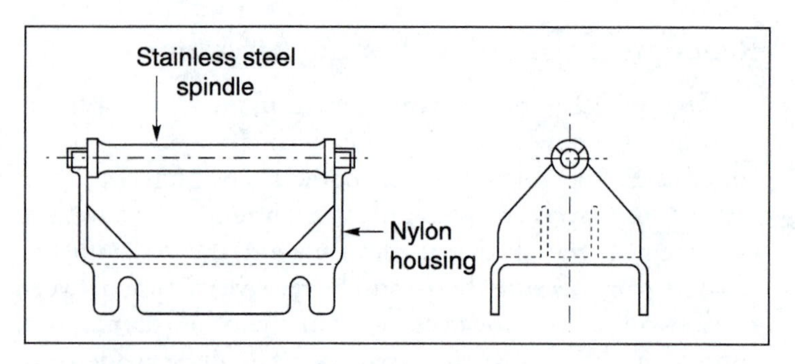

Figure 13.19

Improved two-part spindle housing with the same steel spindle and an injection-molded nylon housing (Reprinted with permission of the Society of Manufacturing Engineers, Simultaneous Engineering: Integrated Manufacturing and Design, *copyright 1990 and* Manufacturing Engineering Magazine, *copyright April 1988).*

A Comparison of Two Computer Printers

The *Epson MX80* was the most popular dot matrix printer in the 1980s. It is compared here with one of the next generation of printers, the *IBM Proprinter*, which was designed using a DFM-DFA philosophy.

Figure 13.20 is an exploded view of the *Epson MX80* showing the principal subassemblies and parts. The *final* assembly involves 57 separate operations that take an estimated 552 seconds to complete. The chief reason for this is that the parts being assembled require positioning to facilitate the insertion of fasteners, and they are often difficult to locate and align. Furthermore, the number of fasteners is greater than the number of functional parts in the product. The elimination of fasteners was a primary design goal for the next generation of products. The 552 seconds of final assembly consists of three groups of activities:

211 seconds to insert 21 parts or subassemblies
305 seconds to insert and secure 34 separate fasteners
36 seconds to adjust mating parts

One step in the final assembly process is the installation of the main printer subassembly shown in Figure 13.21

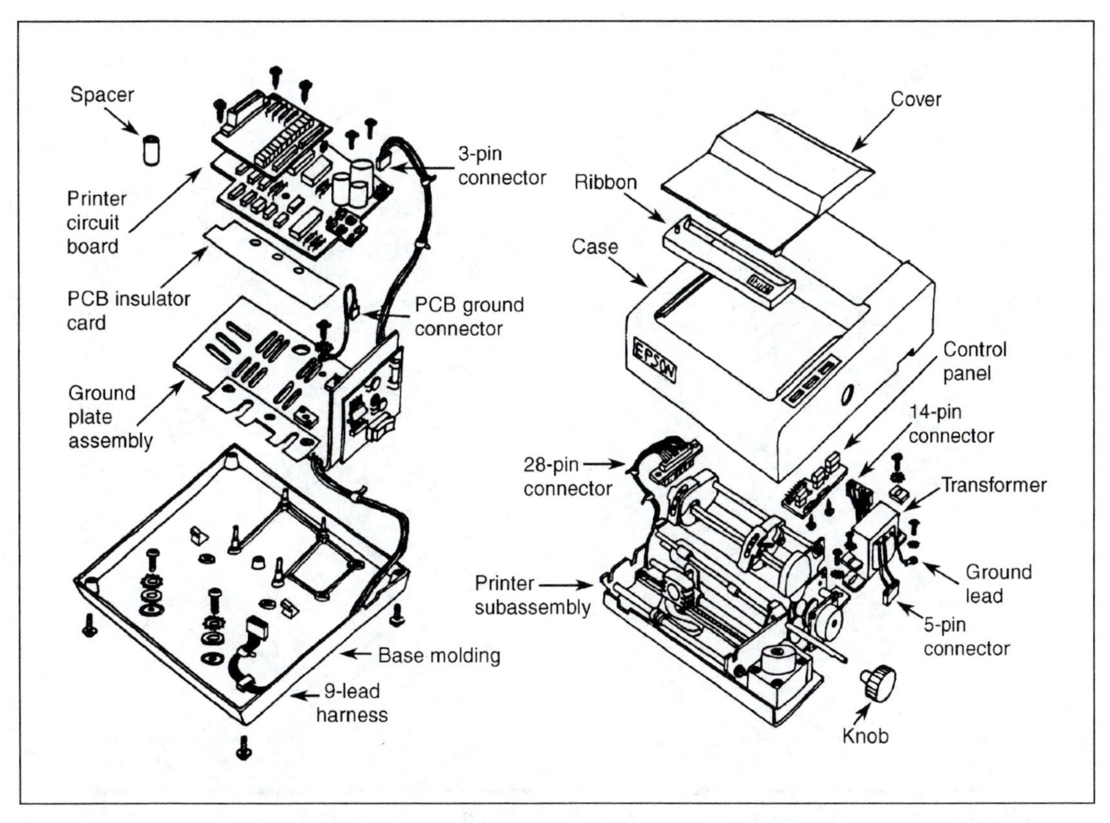

Figure 13.20

An exploded view of the Epson MX80 *dot matrix printer which contains 49 parts or subassemblies and requires 57 assembly operations (Source: Assembly Engineering, January 1987).*

which incorporates the paper and print-head drive units. This subassembly suffers from an absence of design-for-assembly practices. For example most of the functional parts either are located on the side brackets or are hung between these two supports. Furthermore, the parts are not firmly secured until the upper front rod and the upper rear rod are secured by hexagonal nuts with the lower right bracket and the lower left bracket. This practice is not recommended. Parts should be added to the base plate in a layer-by-layer approach in which each part is self-locating and immediately secured in place.

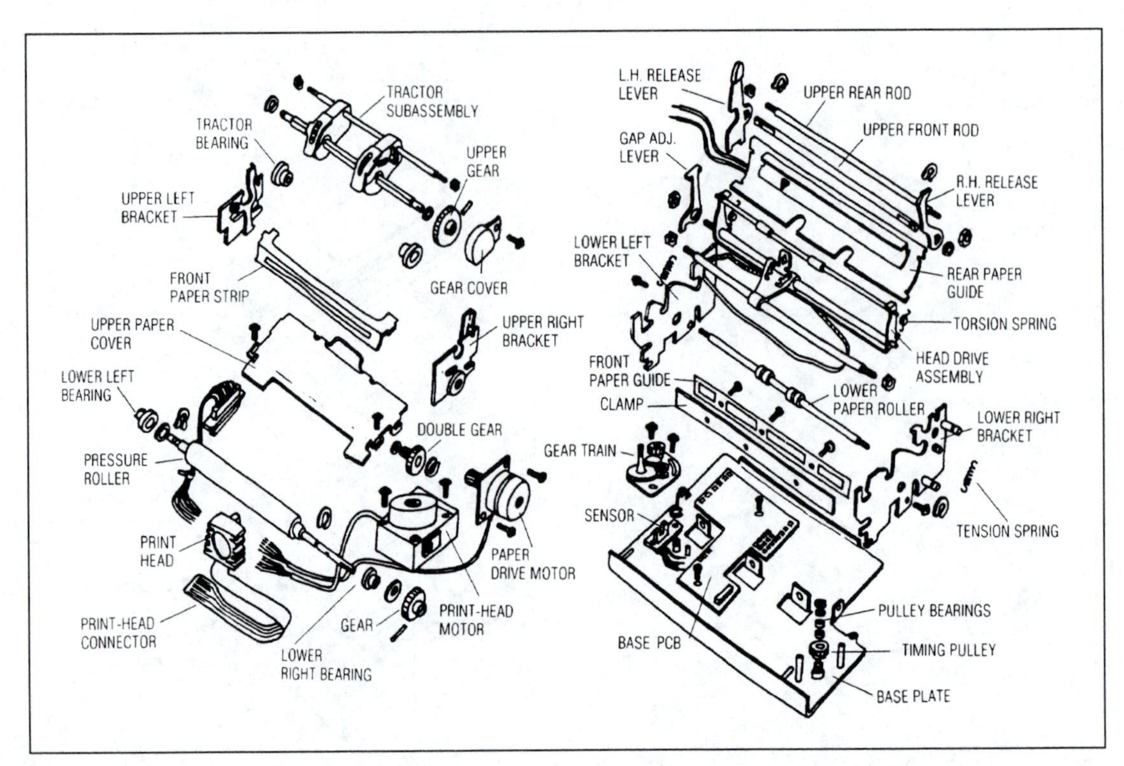

Figure 13.21

An exploded view of the main printer subassembly for an Epson MX80 *dot matrix printer (Source: Assembly Engineering, January 1987).*

The printer subassembly requires 128 separate operations during the assembly process, and these take 1314 seconds to perform. It is evident from Figure 13.22 and from the following data that the design has too many separate fasteners:

708 seconds for inserting 65 parts or subassemblies
454 seconds for inserting and securing 40 separate fasteners
152 seconds for parts adjustments and soldering operations

The *IBM Proprinter,* designed to supersede the *Epson MX80,* is shown in an exploded view in Figure 13.23. This product was designed with three primary DFM-DFA objectives. The printer has fewer parts and is devoid of separate fasteners.

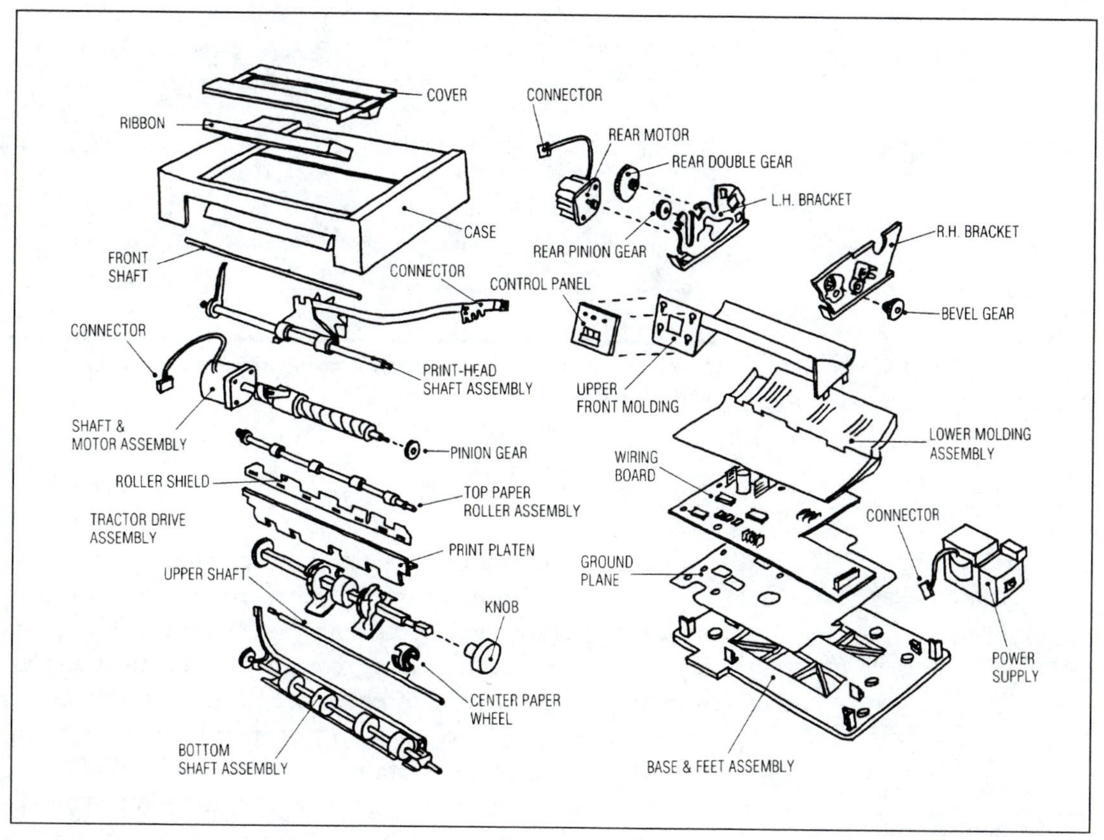

Figure 13.22

An exploded view of an IBM Proprinter *which contains 32 parts or subassemblies and requires 32 assembly operations (Source: Assembly Engineering, January 1987).*

The *Proprinter* requires only 32 assembly operations which can be completed manually in 170 seconds. The *Epson MX80* requires 185 assembly operations which can be completed manually in 1866 seconds. Thus the *IBM Proprinter* is assembled in less than one-tenth the time it takes to assemble the *Epson MX80*. If labor rates are assumed to be $25 per hour, this time difference translates into a saving of $13.78 for each product assembled.

The IBM Proprinter is assembled in one-tenth the time required to assemble the Epson MX80 printer.

A summary of the design and manufacturing attributes of both products is presented in Table 13.3, assuming that both products are assembled manually.

	Epson MX80 Final Assembly	Proprinter Assembly
Total assembly time (seconds)	552	170
Total labor cost (cents)	383	118
Total number of operations	57	32
Number of parts or subassemblies	49	32

Table 13.3

Summary of the design and manufacturing attributes of the Epson MX80 *and the* IBM Proprinter.

Summary

- Minimize number of parts
- Minimize assembly operations
- Minimize assembly directions
- Maximize compliance
- Facilitate transportation
- Avoid orienting parts
- Facilitate orientation of parts
- Select appropriate assembly method
- Facilitate insertion

Design-for-assembly principles are important because typically 50 percent of the time spent manufacturing a product is consumed by assembly operations. DFA principles reduce manufacturing costs by reducing materials handling, assembly, and inspection. Selection of protocols is governed by the batch size and properties such as form, material, dimensions, tolerances, and surface quality.

Assembly operations are necessary if different materials are required for the product, relative motion of parts is necessary, parts must perform different functions, constraints are imposed by manufacturing processes, or if access for maintenance is required. Once these criteria are satisfied then the number of parts—and hence the number of assembly operations—should be minimized. Assembly processes should be designed to exploit gravity and they should accommodate dimensional variation.

The shapes and symmetries of parts govern their behavior during transportation and assembly. During assembly, parts should not be reoriented but if this is unavoidable then reorientation should be facilitated by symmetry or asymmetry in the part. The insertion of parts should be facilitated by self-locating features. Finally the most appropriate method of assembly should be selected.

The chapter concludes with several examples of how these principles have been used in practice to reduce manufacturing costs. Attention will now focus upon design-for-manufacture principles for some specific manufacturing processes. The next chapter will examine gravity-fed sand casting.

Key Concepts

Assembly. The fitting together of discrete parts to make a more complex whole. This can be accomplished by deformation, phase change, addition of material, interference, or interlocking.

Choice of assembly method. If assembly is required then the best method should be selected. Consider snap-fit fasteners, tab-in-slot interfaces, and self-tapping screws.

Cost of assemblies. Assembly consumes 40% to 60% of the time taken to manufacture products.

Design-for-assembly (DFA). A method for designing products that are easily assembled. It is essential to reduce the number of parts and to ensure that they are easy to assemble.

Facilitate insertion. Parts should be easily inserted. Facilitate insertion by detail design features such as self-locating fits, chamfers, and single direction motion.

Facilitate orientation. Parts are easily assembled if there are several planes of symmetry. If the part is not symmetrical, its asymmetry should be exaggerated.

Facilitate transportation. Ensure that parts can be transported in chutes or on rails without loss of orientation. Parts should not require reorientation at the assembly station.

Justification for assembly. Assembly is justified if adjacent parts: are manufactured from different materials, need relative motion, perform different functions, are manufactured by different methods, or require separation for access.

Materials handling. This is an essential ingredient of all assembly process because, in the true sense of the phrase, it describes the movement of parts and assemblies.

Minimize the parts count. Numerous cascading savings can be achieved if the parts count is reduced.

Minimize the number of assembly operations. The fewer assembly operations required, the less the cost.

One-direction assembly. Costs are reduced if all the parts are assembled in the same direction.

Quality control. A discipline dedicated to determining whether products comply with the performance specification.

Review Questions

1. What percentage of the total time invested in the manufacture of a product is typically concerned with assembly?
2. Describe the structure of a complex product in terms of parts of assemblies.
3. What are three primary attributes of any assembly process?
4. List four basic methods of assembly.
5. How are assembly methods influenced by production volume?
6. In which phases of the design process are DFA principles applied?
7. If the number of parts in an assembly is minimized, how does this affect the manufacturing process?

8. When the number of assembly operations is minimized, how does this reduce the cost of an assembly? Provide illustrative examples.
9. Contrast the methods for assembling a spur gear on a steel shaft. Discuss the advantages and disadvantages of these different designs.
10. What is fixtureless assembly?
11. How can tolerance for variations among parts be designed into assembly operations?
12. What are some important considerations for ensuring that parts can be easily transported?
13. Why should the reorientation of parts be avoided? How can this be accomplished?
14. How do rules of symmetry and asymmetry influence the orientation of parts during the assembly process?
15. How do materials and parts attributes affect the selection of assembly methods for a product?
16. How can the insertion of parts be facilitated during assembly?

Problems

1. Select a domestic appliance, such as a vacuum cleaner or an electric iron, and dismantle it. Identify the subassemblies or parts that are difficult to disassemble and devise a superior design
2. Consider the task of employing a threaded fastener to hold together two steel plates of thickness 10 mm each. Discuss the different options when the loading has static characteristics and is imposed in the same direction as the longitudinal axis of the bolt.
3. A shoulder must be created in the bore of a steel tube to restrain the outer race of an antifriction multielement bearing. The tube has an internal diameter of 15 mm and the shoulder is to sited near the end. Present four different methods to accomplish this. Discuss each one.

4. Two pieces of flat sheet metal are fastened together with bolts. Redesign this assembly using snap-fit or press-fit fasteners. Sketch your solution.
5. A pin and a washer must be assembled so that the washer cannot rotate on the pin. Devise several designs to prevent washer rotation.

Further Reading

Allen, C. W., ed. *Simultaneous Engineering*. Dearborn, MI: Society of Manufacturing Engineers, 1990.

Andreasen, M. M., S. Kahler, T. Lund, and K. G. Swift. *Design for Assembly*. Kempston, UK: IFS Publications Ltd., 1988.

Bakerjian, R., ed. *Tool and Engineers Handbook, Volume 6, Design for Manufacturability*, Society of Manufacturing Engineers, Dearborn, MI, 1992.

Boothroyd, G., Dewhurst, P., and Knight, W. *Product Design for Manufacture and Assembly*, Marcel Dekker, Inc., New York, 1994.

Bralla, J., ed. *Handbook of Product Design for Manufacturing*. New York: McGraw-Hill, 1986.

Dieter, G. E. *Engineering Design. A Materials and Processing Approach*. New York: McGraw Hill, 1983.

El Wakil, S. D. *Processes and Design for Manufacturing*. Englewood Cliffs, NJ: Prentice-Hall, 1989.

Farag, M. M. *Selection of Materials and Manufacturing Processes for Engineering Design*. Hemel Hempstead, UK: Prentice Hall International, 1989.

Redford, A. and J. Chal. *Design for Assembly: Principles and Practice.* New York: McGraw-Hill, 1991.

Sand Castings

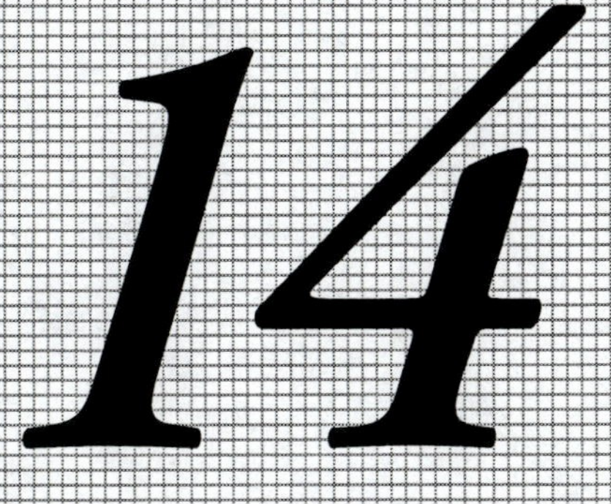

...the cast is equal in merit to the original...

Leonardo da Vinci
(1452–1519)

Objectives

- To present a procedure for selecting an appropriate casting process.
- To develop an appreciation for the characteristics of sand castings.
- To present rules for designing gray iron sand castings

Contents

14.0 Introduction

Casting is a class of common production processes capable of generating parts in a broad range of sizes, qualities, surface finishes, and accuracies. However, because sand casting is the most common casting process (accounting for approximately ninety percent by weight of metal poured) it merits special attention. In this chapter we examine the process itself, identify its advantages, and list guidelines for selecting an appropriate casting process. Finally we present a set of rules for designing gray iron sand castings.

Sand casting is the most widely used casting process.

14.1 The Sand Casting Process

Casting is the first stage in the manufacture of many metal parts. It begins with the melting of metal in a furnace in a foundry. The melt is poured into a carefully prepared cavity in sand—a *mold* (Figure 14.1)—whose shape and dimensions mirror the shape of the desired part. After cooling and solidification, the resulting solid *casting* is then removed from the mold, cleaned, and risers and runners are cut off. Sometimes parts of the casting must be machined before use to provide dimensional accuracy or a superior finish.

A *mold* is a hollow form or cavity that shapes a material in a molten or plastic state before it solidifies.

The word *casting* describes both a manufacturing process and the solid object emerging from this process.

14.2 Guidelines for Selecting a Casting Process

Many kinds of casting processes can be specified by a design team. They include sand casting, investment casting, die casting, and centrifugal casting. Each has advantages and limitations in surface finish, size, minimum production quantities, production rates, part geometry, and the material from which a part is to be made. These attributes must be considered not only as a particular casting process is

Casting processes include sand casting, investment casting, die casting, and centrifugal casting.

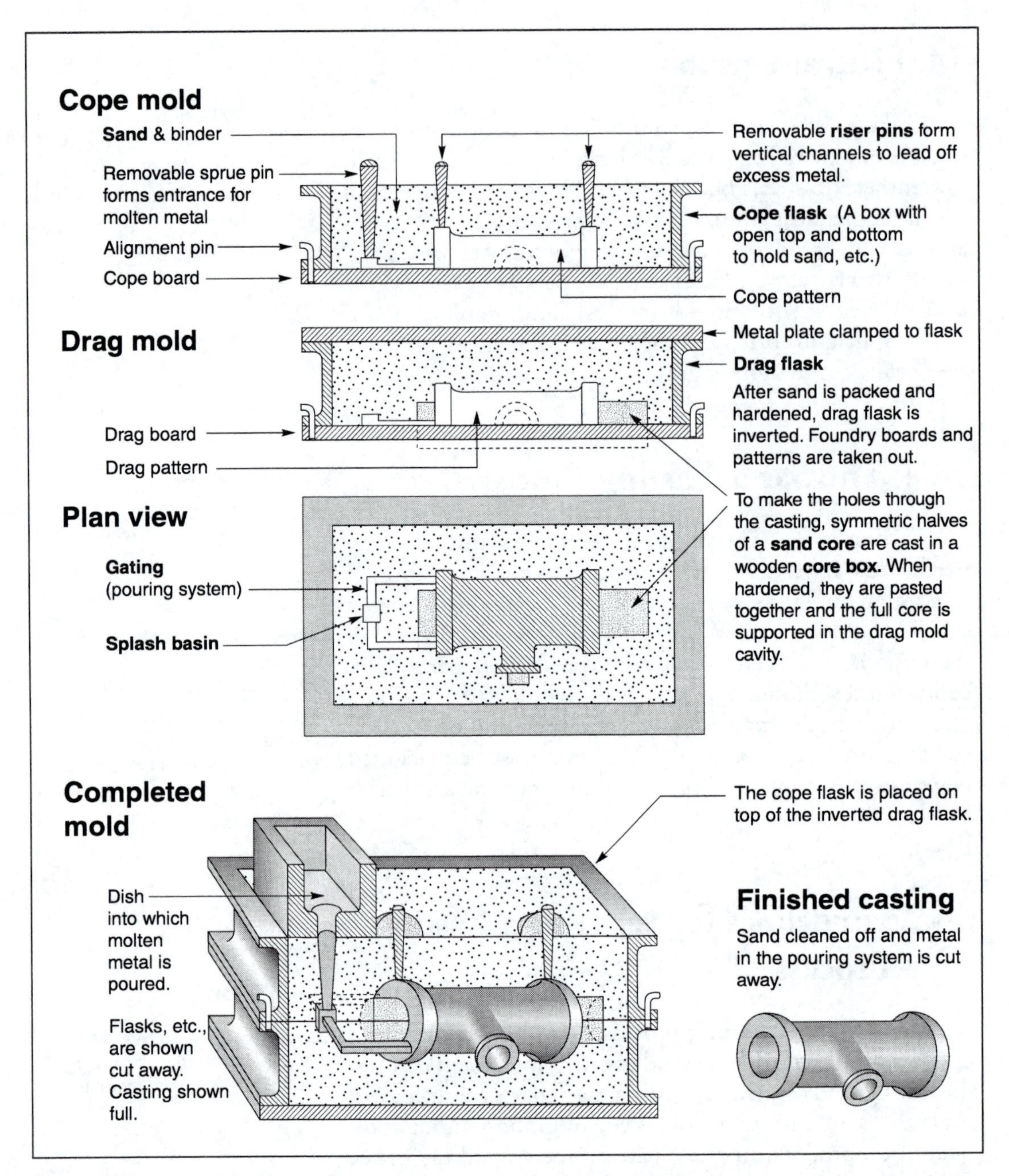

Figure 14.1

A mold for sand casting.

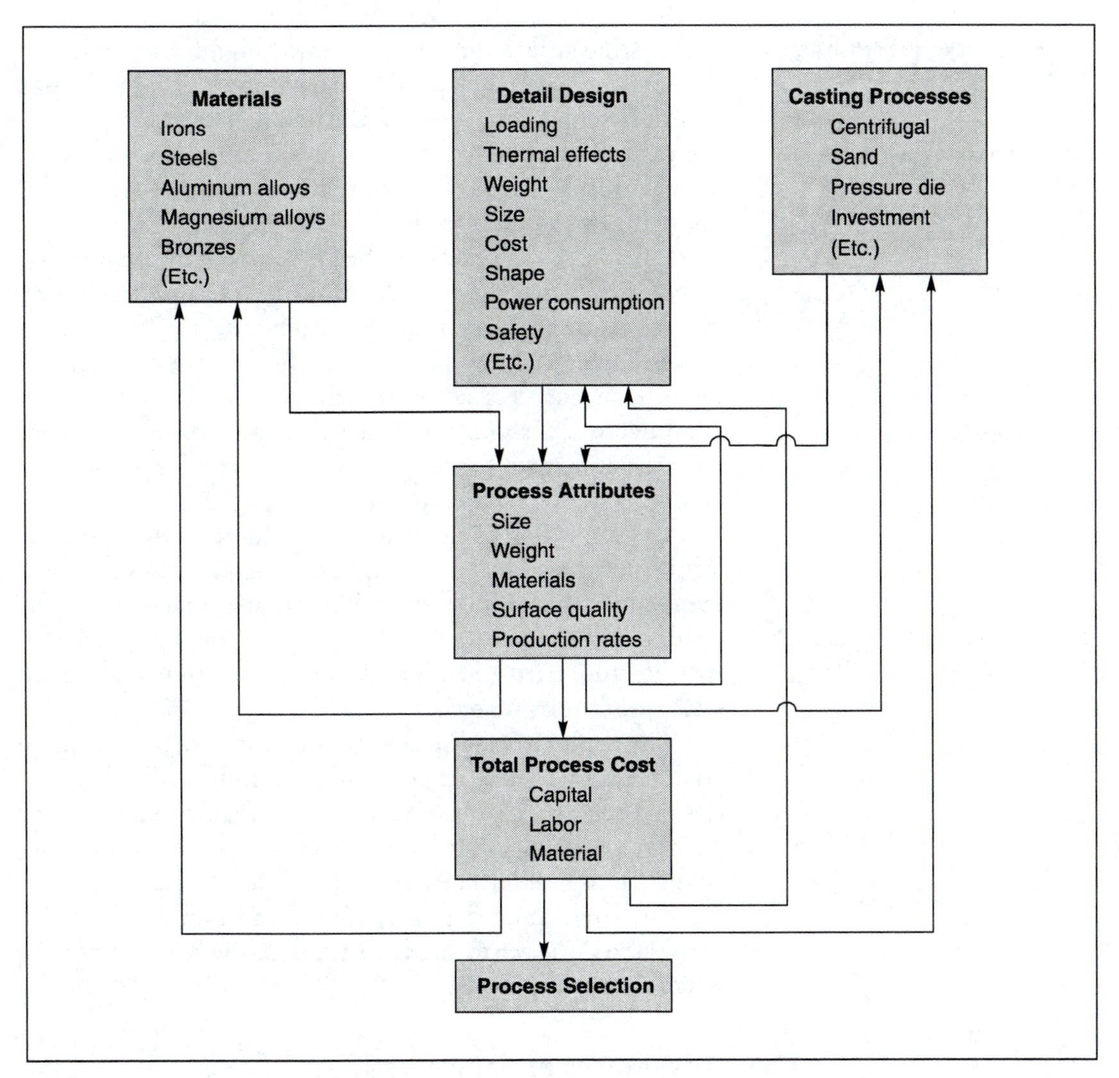

Figure 14.2

Selection of an appropriate casting process.

chosen but also as the casting itself is designed. In seeking the most cost-effective process, the design team must consider the attributes of the various casting processes, the materials available for casting, the design of the part, and the costs involved (Figure 14.2). Here design-for-manufacture is again an iterative selection process.

Selection of a casting process requires consideration of:

- processes
- materials
- part design
- costs

Castings are only specified in high volume situations.

Near-net-shape manufacturing: the minimization of secondary operations by generating parts of high accuracy and surface quality.

Iron, steel, and nonferrous castings can be manufactured.

Casting offers the designer some significant near-net-shape advantages when production volumes are large enough to amortize the initial costs of the metal dies used in pressure die casting or the patterns for sand casting. *Near-net-shape manufacturing* is a method of reducing costs by using only enough raw material to make a part as close to the desired final shape as is practical. Secondary machining operations—and hence costs—are minimized by specifying precision castings which afford high surface finish and dimensional accuracy rather than the relatively rough surface finish and less accurate dimensions of sand casting. Manufacturers should not have to pay once for excess material and again for its removal by machining in order to achieve a desired shape or surface quality.

Casting is one of the primary manufacturing processes. While perhaps a dozen common casting processes are used, some ninety percent of all castings are manufactured by the sand casting process. This process is used primarily to manufacture parts of iron and steel, although it is also used with nonferrous materials.

The ability to readily change the geometry of a part is an advantage of casting that should be fully utilized in the design process. Casting permits the mechanical stiffness of a part to be changed by varying cross sections and other shapes. This freedom relieves the designer of the constraints imposed by standard plates, tubes, and bar stock. It also allows the designer to make castings aesthetically appealing when that is important.

Advantages of Sand Castings

1. Castings can be almost any shape.
2. Castings can be almost any size—approximately 100 g to 250 metric tons.
3. Typical surface finishes are in the range of 10 to 25 micrometers.
4. Typical accuracy ranges from approximately ±0.5 mm for a 25 mm dimension to ±2 mm for a 600 mm part.

5. Castings can be cheaper than forgings, depending on the quantity of parts required, the properties of metals used, and the cost of forging dies relative to the cost of patterns for castings.
6. Castings may be cheaper than welded structures if a large number of parts must be manufactured. The cost of welding jigs and fixtures relative to the cost of patterns for castings must be considered.
7. The material properties of castings are generally isotropic; consequently the strength is the same in all directions.
8. Molten iron has good flow properties. Therefore if the mold and the associated gating system are designed correctly, the liquid metal can flow into thin sections of the mold cavity and generate parts with complex shapes.
9. Cast iron has excellent energy dissipation characteristics. Consequently it is often used for parts of machine tools, engines, and other applications in which vibration must be absorbed.
10. The sand casting process is very flexible because it accommodates many different materials, part sizes, and part shapes. Figure 14.3 illustrates some of this diversity. The photographs show a General Motors Corporation V-8 cylinder block for an automobile engine. It was manufactured in a gray iron by the green sand casting process. The section of the casting shows the complex internal passages of the part.

In the context of a casting, a design engineer considers loads, failure modes, aesthetics, and the physical properties of a material. In contrast to this, a foundry engineer considers fluid flow, heat transfer characteristics, shapes, cores, parting lines, and hot spots. The special knowledge of these two kinds of engineers must be integrated through effective communication to provide a good DFM framework. The two primary considerations are the physical properties of a metal as it changes from a liquid to a solid and the foundry requirements dictated by the particular casting process.

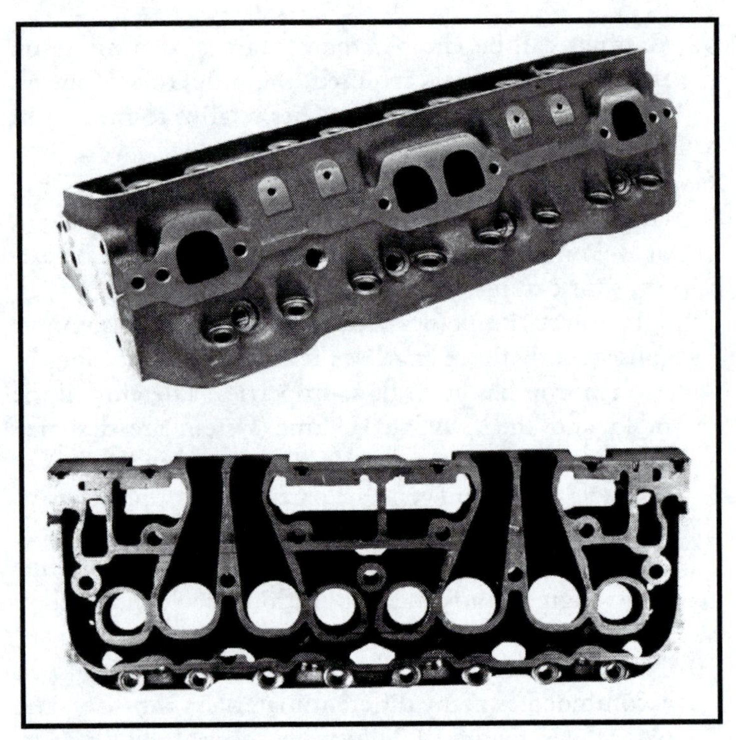

Figure 14.3

Two views of a V-8 cylinder block (courtesy of General Motors Powertrain).

14.3 Rules for Designing Sand Castings

The design of a high quality casting requires a multi-disciplinary approach. It involves consideration not only of the structural properties of the machine element but also consideration of the flow of molten metal into the mold cavity, finishing operations, assembly, testing, and the use of the casting. The following rules can serve as guides.

• Molten metal characteristics influence castability

Three physical characteristics govern the performance and castability of most alloys used in foundries—commonly cast irons, bronzes, and low-carbon steels. These three characteristics are the life of the molten metal prior to solidification, the shrinkage of the metal as it solidifies, and contamination of the melt by slag.

The most common alloys cast are the cast irons, the low-carbon steels, and the bronzes.

Castability is influenced by:

- life of the molten metal
- contraction during solidification
- contamination by slag

The time a metal remains liquid is an important consideration in the design of a mold because it determines the metal's ability to fill the mold cavity. This affects the minimum wall thickness that can be cast by a specific metal and also the maximum length of thin-walled sections. The fluid life also governs how readily the material conforms to fine surface details. If a material does not have a long fluid life, this attribute does not adversely affect the economic viability of the casting design but it does require that the mold accommodate this property.

A liquid metal has three distinct shrinkage properties as it cools to a solid state. The first is liquid shrinkage before solidification begins. Liquid shrinkage is of no consequence to the design engineer. The second phenomenon is the phase change shrinkage as liquid metal transforms to the crystalline structure of the solid phase. This is of great importance for the designer because this shrinkage causes voids in castings that are responsible for weak regions. The third kind of shrinkage occurs after the metal has solidified in the mold. Solid shrinkage changes the dimensions of the casting and thus affects the final dimensions of the part. It affects decisions about the location of the parting line and about other characteristics of the cores and mold.

Kinds of shrinkage:

- Liquid (not important)
- Liquid-to-solid (very important)
- Solid (moderately important)

Formation of slag or dross during the casting process affects the properties of a part because slag can introduce small quantities of impurities into the microstructure of the metal. The formation of these impurities is metal dependent.

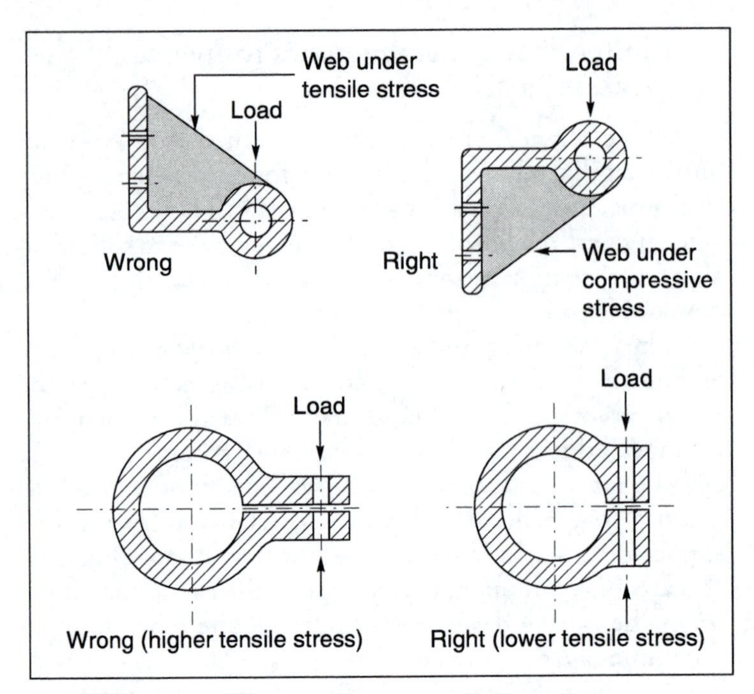

Figure 14.4

Iron castings should be designed to avoid tensile stresses.

- ## Material properties influence the geometry of the casting

Cast iron is stronger in compression than tension.

The physical properties of a cast metal affect the design of cast metal parts. For example, the compressive strength of cast iron is approximately four times greater than its tensile strength; consequently iron castings should be designed so that the primary stresses are compressive rather than tensile. Examples are shown in Figure 14.4 where the intent is to avoid tensile stresses.

Naturally this limitation of cast iron also manifests itself in parts subject to bending loads where one face of the member in subjected to tensile stresses and the opposite face experiences compressive stresses. By creating an unsymmetrical cross-section in which the neutral plane of bending is closer to the face experiencing tensile stresses, the magni-

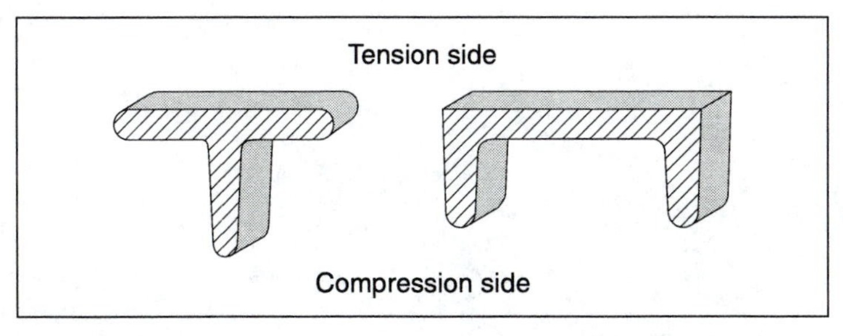

Figure 14.5

Unsymmetrical cross-section geometries reduce tensile bending stresses in cast iron parts subjected to bending.

tude of the tensile stress on the part can be reduced. Figure 14.5 shows sections of parts designed in this way. Curved and doubly-curved sections are structurally superior to planar surfaces. Therefore, whenever possible, castings should have cylindrical, spherical, conical, or other curvilinear shapes.

Design castings with curved sections to enhance stiffness and control stresses.

The elastic modulus of gray cast iron has a wide range of values that are considerably less than that of some common steels. Therefore, sections of iron castings (and hence their mass) should be increased to provide the same rigidity and the same deformations as steel structures.

• Avoid concentrations of metal

Concentrations of metal should be avoided in gray iron castings because the resulting uneven rates of cooling can cause undesirable shrinkage cavities. These voids occur in the more highly stressed regions and cause weakness. Figure 14.6 illustrates portions of two castings, one poor and the other better.

Concentrations of metals cause shrinkage cavities and structural deformation.

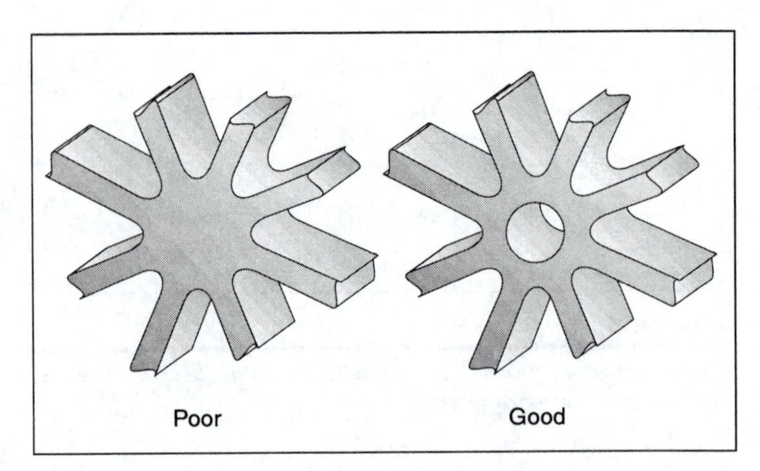

Figure 14.6

Avoid concentrations of metal.

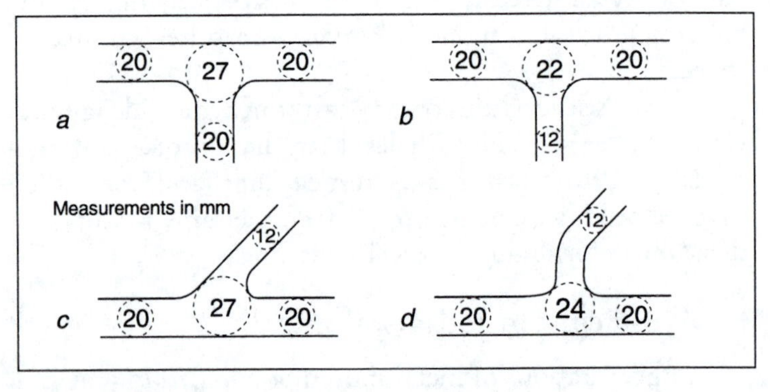

Figure 14.7

A method for evaluating metal concentrations at wall intersections. Designs *b* and *d* are better.

Design walls to intersect at 90° and ensure that ribs are thinner than the walls.

A *wall* is an enclosing barrier that defines the outer limits of a cavity.

A *rib* is a stiffening element that strengthens or supports a part, often by dividing a surface into panels.

A *web* is a stiffening or strengthening element that connects heavier sections such as walls, ribs, flanges, beams, rails, truss components, etc. A web forms a continuous, flat, rigid connection between other members.

Castings often involve intersections of walls with *ribs* or *webs*. Concentrations of metal always occur where two walls of *equal* thickness intersect, as in Figure 14.7*a*. This concentration of metal, illustrated by the circles, can be reduced by designing one wall thinner than the other, as shown in Figure 14.7*b*. When one of the walls is a rib in a casting this strategy can be used very effectively. Figure 14.7*c* illustrates the problems that arise when a rib joins a wall at an acute angle. The associated concentration of metal can be reduced by designing the two walls to intersect at 90 degrees as shown in Figure 14.7*d*.

A *beam* is a rigid structural member supported at the ends and subject to bending stresses from a direction perpendicular to its length.

A *truss* is a straight element that strengthens a structure by virtue of the rigidity of a triangle. It is subject to longitudinal compression, tension, or both.

A *rail* is a horizontal bar used for support, as a barrier, etc.

- **Avoid porosity in casting**

During the casting process, gases are generated in the liquid metal in the mold cavity. These bubbles must be allowed to escape by encouraging them to rise to the surface of the casting. Relevant sections of the casting should have sloping walls and additional ribs to facilitate the escape of these bubbles. If these bubbles are trapped in the metal, porous regions will be created which are weaker than solid metal. Figure 14.8 presents an example of both good and bad casting design. The better gray iron casting in Figure 14.8 has superior porosity characteristics because the escape of gas bubbles is encouraged and is stronger because of the removal of the stress concentrations at the sharp corners.

Encourage the escape of gas bubbles from the molten metal in the mold by incorporating slopping walls and additional ribs.

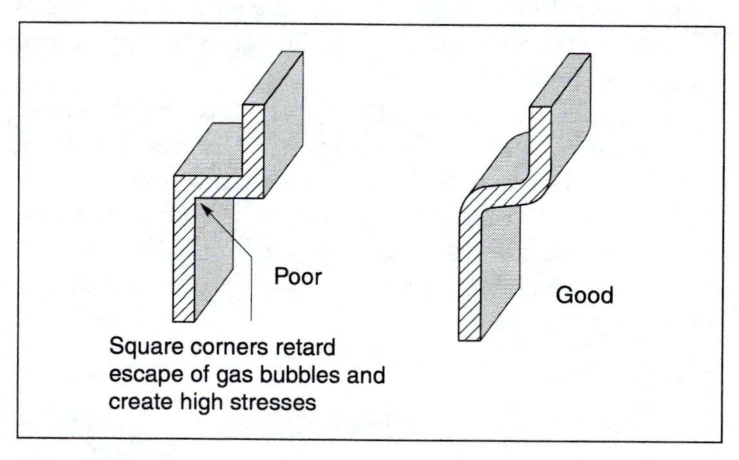

Figure 14.8

Designing for gas emission.

• Round edges of the mold cavity

Provide a generous radius at all internal edges to promote:

- desired shape
- better cooling
- lower stresses

All sharp edges protruding into the mold cavity should be given a generous radius. Otherwise the edge of the sand will be eroded by the impinging molten metal as it fills the mold cavity. While outside corners radiate heat from the two adjacent faces and cool quickly, inside corners have different geometries and cool slowly, causing thermal gradients and distortion. This thermal problem can be alleviated by specifying generous radii. These larger radii also reduce stress concentrations in these regions. Thus castings are characterized by smooth contours with no sharp edges. The cylinder block shown in Figure 14.3 illustrates this.

• Avoid residual stresses and distortion

Unequal rates of cooling in different sections of a casting will cause unequal thermal contraction of the structure. This causes deformation and permanent warping of the casting. This can be avoided by designing castings with uniform wall thicknesses without concentrations of metal. Where this is not possible, gradual changes of cross section should be specified.

Thick sections cool more slowly than thin sections.

Consider the cooling of the T-shaped section of a casting designated in Figure 14.9 as design *a*. As the metal cools, rib *y* will cool faster than flange *x* and will solidify first. As flange *x* cools further, rib *y* will prevent it from contracting, and this constraint will be responsible for the final shape *b*. Design *c* is superior because its thicker rib *z* reduces the rate of cooling.

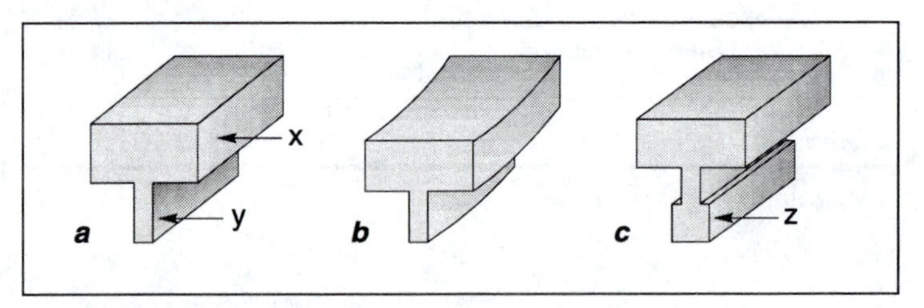

Figure 14.9

Avoid thermal distortion. Design c *is recommended.*

• Avoid shrinkage-cavity formation

A cold casting has a smaller volume than a hot casting and phase-change phenomena cause additional shrinkage as the molten metal solidifies. When the solidification is not uniform, the development of residual stresses cause deformation of the casting and the formation of shrinkage cavities in critical regions. These cavities are particularly prevalent in castings that have large variations in wall thickness because outer and thinner regions of the casting solidify before interior and thicker sections do.

Nonuniform rates of cooling cause thermal distortion, residual stress, and shrinkage cavities.

To minimize shrinkage cavity formation, a difference of more than 2:1 in section thicknesses should be avoided. If this difference must be exceeded, then either two separate castings should be manufactured and subsequently bolted together, or a tapered section with a slender taper of no more than 4:1 should be specified.

Shrinkage cavity formation can be avoided by providing gradual changes of cross section and uniform wall thicknesses.

Figure 14.10 illustrates castings with both good and bad shrinkage characteristics. These principles are evident in the sectioned cylinder block shown in Figure 14.3.

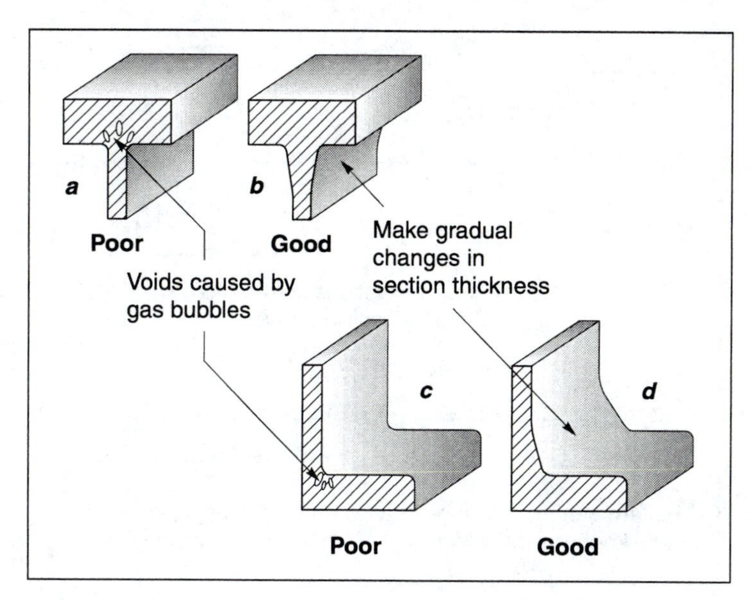

Figure 14.10

Avoid shrinkage cavity formation.

• Facilitate pattern removal

A pattern (usually wooden) is used to develop the desired shape of mold cavity in the sand within the cope and drag. Once this has been accomplished, the pattern must be removed from the sand to generate the mold cavity. To facilitate pattern removal from the compacted sand without damaging the shape in the sand, the sides of the pattern should have a taper, called a *draft*. This amounts to only a few degrees in the direction in which the pattern is removed. This is shown in Figure 14.11 where the parting line for the mold is the horizontal line with the chevrons. While this is a trivial design detail for castings with simple geometry, it may be more difficult for complex parts with projections such as lugs and bosses.

Taper casting walls so that the pattern can be removed easily from the molding sand.

The orientation of the part relative to the mold parting line dictates which faces of the pattern will have a draft. Figure 14.12 presents two parts with nominally the same shape which have been cast in two different configurations offset by 90 degrees. Thus they have the draft on different

The orientation of the pattern relative to the parting plane dictates which faces have draft.

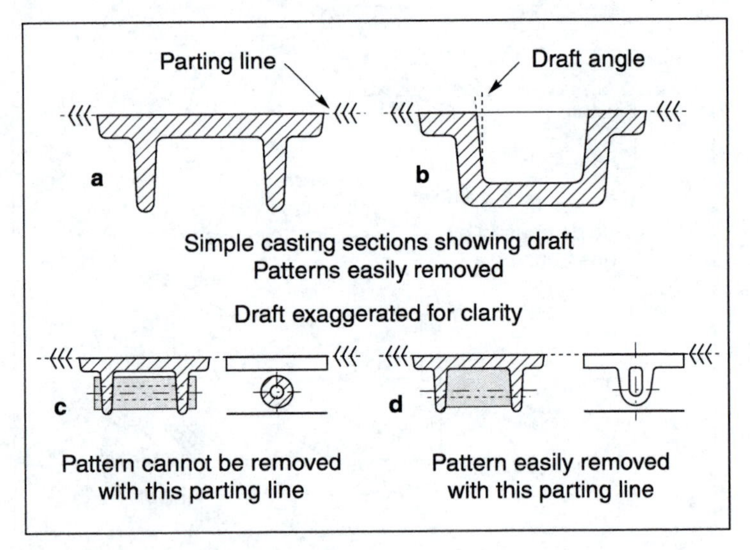

Figure 14.11

Features to ensure pattern removal.

faces. The orientation of the part in the mold is often dictated by the function of the part. For example if a dense structure is required at a particular face, then that face generally will be located at the lowest part of the mold.

- **Pattern orientation influences casting properties**

The position of a pattern in the sand within the cope and drag flasks of a mold affects the properties and the cost of the casting. For example, whether a part is cast in a horizontal or a vertical configuration can affect the foundry cost because one configuration may require more flasks, and hence cost more, than an alternative configuration. This is illustrated in Figure 14.13 which shows the horizontal and

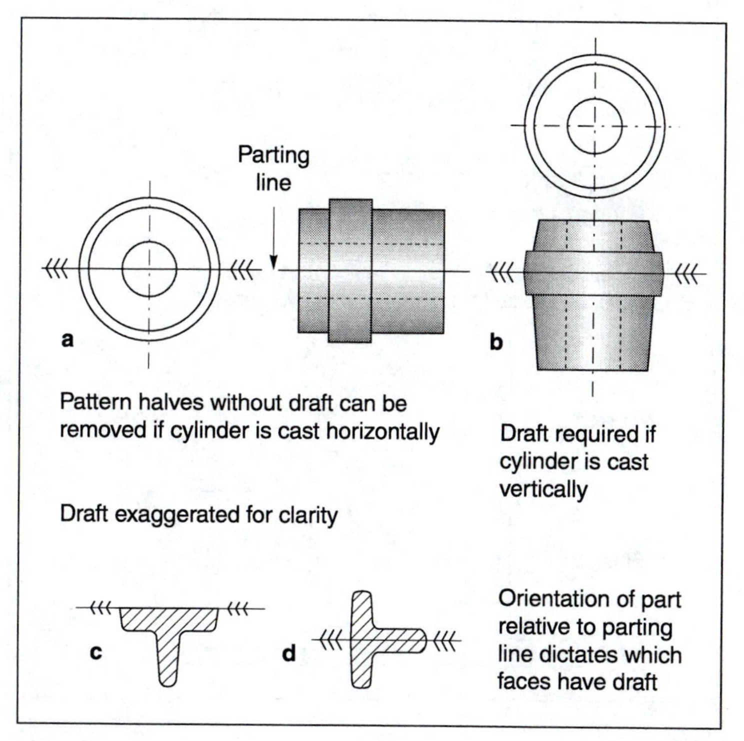

Figure 14.12

Draft and part orientation.

vertical casting of a cylindrical cover plate. A vertical casting configuration requires three flasks and a central core, while a horizontal design requires only two flasks and a central core.

Castings should be designed so that the parting line is on a flat plane.

A parting line is a continuous line on a casting at the interface between the cope and the drag.

Castings should be designed so that the parting line is on a flat plane. (The parting line is a continuous line on a casting that defines the interface between the cope and drag.) If contoured parting lines are required, then a number of nontrivial costs arise because more highly skilled labor is

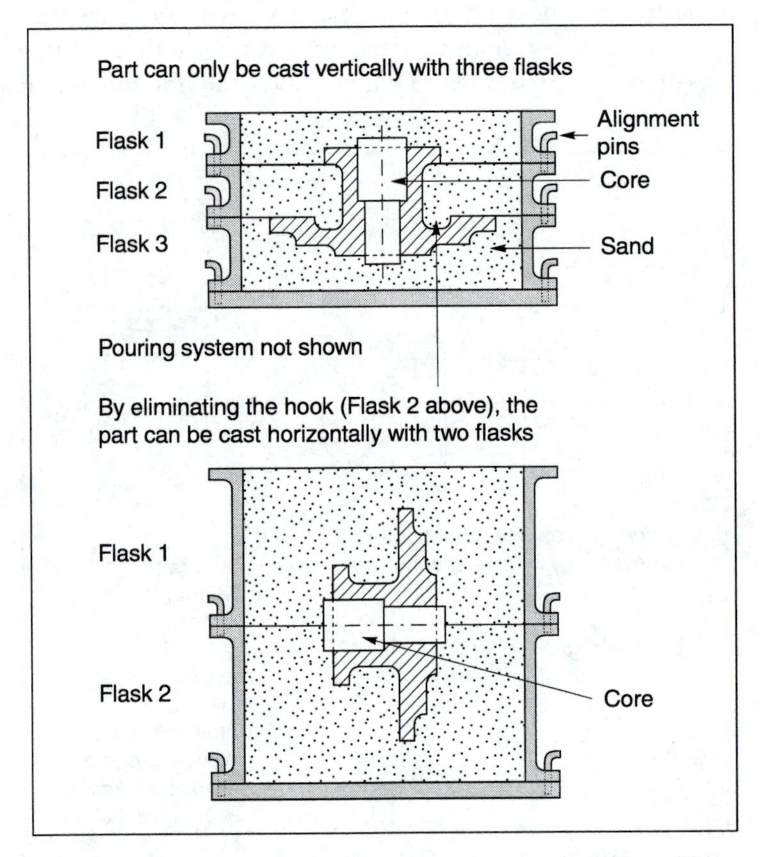

Figure 14.13

Horizontal and vertical casting configurations influence the properties of the casting.

required, and more time will be spent troubleshooting problems.

The orientation of a pattern relative to the parting line dictates the density of the material at the lower faces of the casting, which may or may not be desirable. Figure 14.14 presents two alternatives for casting a cover plate. In Figure 14.14*a*, the lower face of the of the bearing housing will possess the greatest density; while in Figure 14.14*b*, the large face of the cover will possess the greatest density.

The lowest surfaces of the casting will be the densest.

• Avoid cores

Holes in castings can be created using a *core*. Cores typically are rigid pieces of a composite material made from a mixture of sand and binders which are positioned in the mold once the shape of the mold cavity has been formed by the pattern. Cores should always be avoided whenever possible because of the expense associated with their design and manufacture. If they must be used, then they should always be as simple as possible. Cores generally are not used to form holes smaller than about 20 mm because of poor economics. Smaller holes are formed by drilling. Stress concentrations are associated with holes, but the freedom offered by the casting process permits additional material to be used to reinforce the structure to reduce these effects.

Cores are rigid pieces of material made from a mixture of sand and binders. They are positioned in the mold once the mold cavity has been formed by the pattern.

Cores:

- Avoid them
- Provide core support
- Keep them simple
- Position multiple cores on the same parting plane

Create holes smaller than 20 mm diameter by drilling.

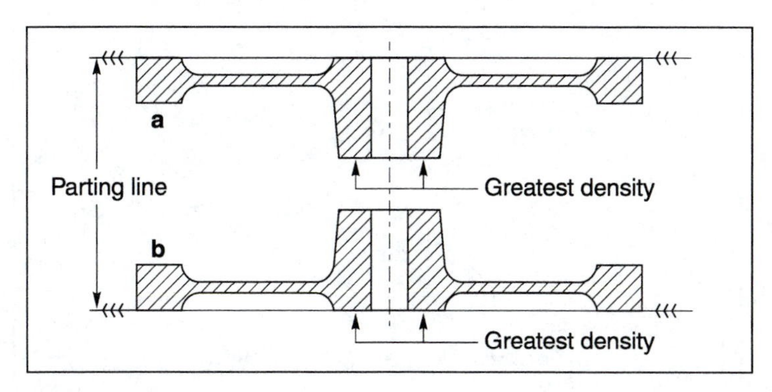

Figure 14.14

Developing a dense structure in castings requires consideration of the pattern position relative to the parting plane.

Figure 14.15 presents two classes of hollow casting designs requiring the use of cores. Designs *b* and *d* are recommended because they use simpler, cheaper cores.

During casting, cores can easily be displaced by the stream of molten metal as it enters the mold cavity unless the cores have adequate support. This support is typically provided by recesses in the mold cavity, generated by the pattern, which accommodate the cores (see Figure 14.1). Figure 14.16 presents several designs of parts that require cores during casting. Design *a* is inferior because the core is only self-supporting as a cantilever beam before the mold is filled. Design *b* is better because the core is supported at both ends. If a mold is to have several cores, then they should all be positioned in the same parting plane, as shown in design *c*.

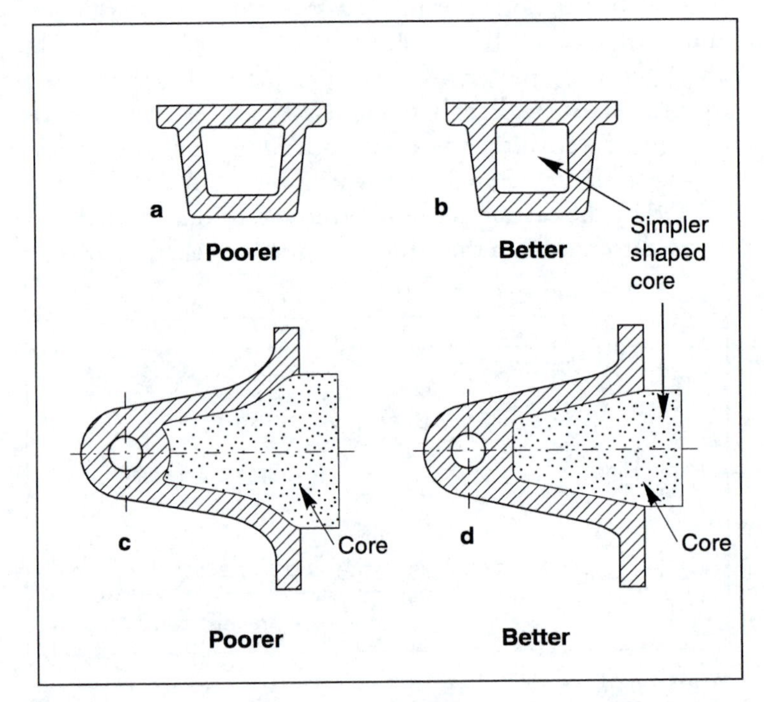

Figure 14.15

Design cores with simple shapes: select shapes b *and* d.

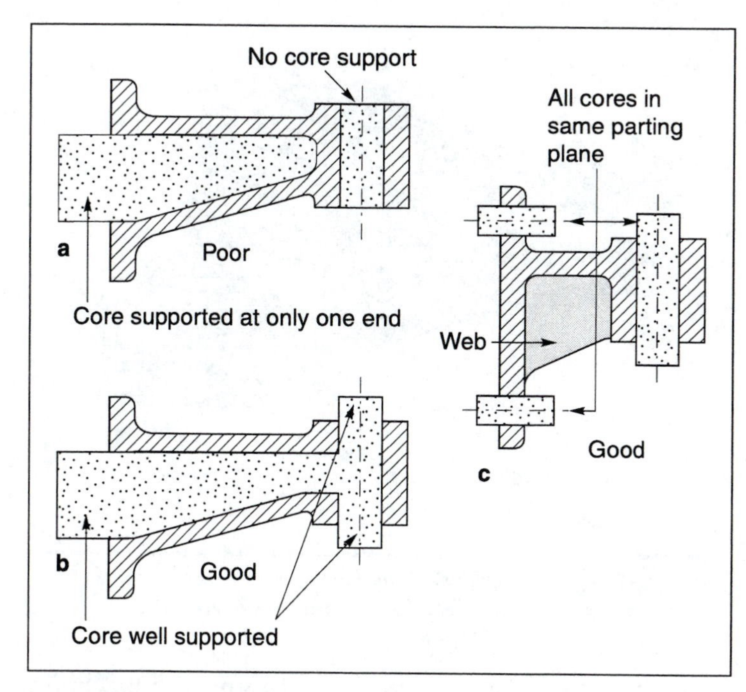

Figure 14.16

Provide adequate core support.

Machining is used to create a superior surface quality and dimensional accuracy.

Provide a machining allowance but minimize the area to be machined.

• Provide a machining allowance

The casting process alone is often not sufficient to manufacture a gray iron part with the desired attributes. Secondary operations may be required, and these typically involve machining to develop the surface qualities desired in specific regions of the part and to create the dimensional tolerances needed to interface with other parts. Machining allowances must, therefore, be provided in the pattern. The casting is made larger in regions where machining must be performed although the casting should be designed to minimize the area to be machined (Figure 14.17).

Figure 14.17

Photograph of pillow block bearing showing minimum area to be machined at the left and right box sections of the base.

Figure 14.18 shows a variety of designs with machining allowances. Designs *a* and *c* are expensive propositions because the intricate details specified in corners designated by the arrow cannot be formed with standard grades of sand. Designs *b* and *d* are recommended.

Walls that define hollow recesses in a casting should not enter another surface at an acute angle when that interfacing surface must be machined later. Figure 14.18*e* illustrates this case, where the top face of the casting must be machined and diameter *D* is subject to close tolerances. If the top face is to be machined, then the diameter of the hole with diameter *D* subsequently will be increased. If this dimension is crucial, then the situation can be avoided by employing design *f* in which the walls defining the hole intersect the wall at the top of the casting at 90 degrees.

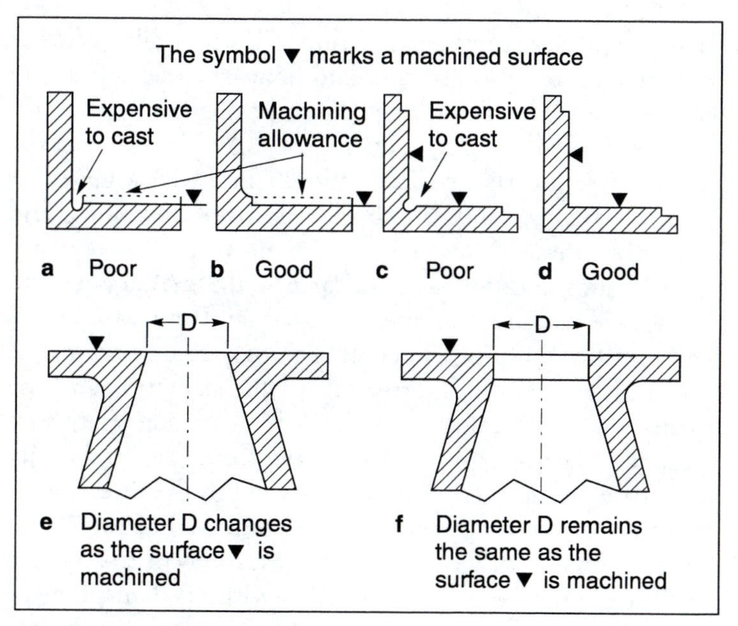

Figure 14.18

Machining allowances must be provided.

Summary

You have learned some of the basic attributes of the gravity-fed sand casting process. This is the most common casting process. Its basic steps involve first making a pattern (a slightly enlarged replica of the final part) which is used to form a cavity in molding sand. The cavity is filled with molten metal that is allowed to cool before the cast object is removed.

Because there are many casting processes, a design team must develop a scheme to select an appropriate process. This scheme should consider the various attributes associated with materials, product design, and the casting processes. Once the best process has been selected, DFM rules should be used to design the casting.

Concentrations of metal should be avoided in the design of castings because they cause formation of shrinkage

- Ensure uniformity of wall thickness.
- Ensure gradual changes of section.
- Avoid concentrations of metal.
- Avoid sharp edges at the intersection of surfaces.
- Provide draft.
- Minimize the use of cores.
- Ensure that the shape reflects the strength of the material.
- Incorporate a machining allowance.
- Minimize area to be machined

cavities and structural deformation. This typically occurs at the intersection of walls, ribs, and flanges. Casting porosity can be avoided by facilitating the escape of gases by prescribing sloping walls and introducing ribs. All internal edges at the intersection of two faces should be given a generous radius to reduce sand crumbling, minimize hot spots, and reduce stresses.

Castings should have uniform wall thicknesses, and changes of cross-section must be gradual. Failure to do this will result in nonuniform cooling rates that are responsible for residual stresses, deformation of the structure, and the formation of shrinkage cavities which create regions of weakness. The removal of the pattern from the mold cavity is made easier if a draft (slope) oriented in the direction of pattern removal is applied to the pattern walls. Protrusions from the main body of the pattern can prevent the pattern from being removed from the mold without damaging the sand contours. These should be avoided by appropriate design otherwise multipart patterns must be developed.

The orientation of the pattern in the mold influences the characteristics of the casting. Material in the lower regions assumes a greater density than elsewhere. The location of the parting line can influence the cost of producing the part. Cores (of sand and binding agent) are used to manufacture castings with holes. They are expensive and should be avoided whenever possible. If cores must be used, they should be simple and well-supported. If several cores are used, they should be positioned on the same parting plane. Cast parts that require machining must incorporate a machining allowance.

Key Concepts

Concentrations of metal. Thickened regions where walls, flanges, and ribs intersect in complex castings. They cause warping and porosity in a casting.

Cores. Used to form large holes in castings. These separate pieces of a mold are made from sand and a binding material. They are expensive and should be avoided whenever possible.

Draft. A small taper introduced into those walls of a pattern that are parallel to the direction of pattern removal from the mold cavity. Drafts ensure that the pattern can be removed from the mold cavity without damage to the sand surfaces.

Gradual changes of wall thickness. A desirable part characteristic when uniform sections are impossible. A gradual change of cross section minimizes shrinkage cavity formation and thermal deformation.

Internal edges. The internal edges of a casting should have generous radii to reduce stress concentrations.

Machining allowance. The casting geometry should provide sufficient material for subsequent machining if this is necessary.

Material properties. The inherent properties of the material of the casting influences the casting geometry. For example, the design of cast irons should reflect the fact that the compressive strength is about four times greater than the tensile strength.

Pattern orientation. The orientation of the pattern can influence foundry costs.

Pattern removal. Facilitated by introducing draft into pattern walls and by avoidance of protrusions from the pattern.

Phase change process. The change is state from liquid to solid in the casting process. The process influences the design of castings (for example, the gating system and the wall thickness).

Porosity. Small voids in a casting that create weak regions. They are caused by nonuniform cooling and failure to permit gases to escape from the mold cavity. Avoid it by good web design and by sloping walls.

Residual stresses. Stress and distortion produced as a result of uneven cooling. The geometry of the casting should ensure even cooling.

Sand casting. Liquid metal is poured into a cavity in the sand called a mold. The metal solidifies in the mold to produce a casting of a desired shape. A broad range of part sizes, shapes, and materials can be cast.

Shrinkage cavities. Voids, and hence areas of weakness, caused by uneven cooling rates in castings at changes of wall thickness and at discontinuities.

Uniformity of wall thickness. Ensures uniform cooling rates and thus avoids thermal distortions and porosity.

Review Questions

1. Describe the gravity-fed sand casting process.
2. What part attributes need consideration in the selection of a casting process?
3. List three classes of attributes that must considered in selecting a casting process.
4. Define and discuss near-net-shape manufacturing.
5. List six advantages of sand casting.
6. List six attributes of a mold that concern a foundry engineer.
7. What are three important areas that must be considered in the design of sand castings?
8. How do the characteristics of the material being poured influence the design of a sand casting?

9. What is fluid life and what influence does it have on mold design?
10. When a liquid metal cools to form a solid object, it contracts. What are the consequences of this for the foundry engineer designing a mold?
11. How does the formation of slag influence a casting?
12. How does the strength of gray cast iron influence the geometry of a casting?
13. Discuss the design of a cast gray iron part subject to bending loads.
14. Why should castings be designed to avoid concentrations of metal? How are these concentrations reduced?
15. What causes porosity in castings and how can it be avoided?
16. Why should all edges in a mold cavity be given a generous radius and why should a rounded surface be generated at the intersection of two faces? What are the advantages of doing so?
17. How can the distortion of a casting be prevented by selection of an appropriate geometry? Explain.
18. What is shrinkage cavity formation and how should it be avoided in castings?
19. Why is the easy removal of a pattern important in the design of a mold and how can this be accomplished? What is a typical magnitude for the draft of the wall of a casting?
20. How does the orientation of a mold cavity influence the properties of a casting?
21. What is a parting line? How does it influence the design of a casting?
22. What is a core and why should they be avoided?

23. If cores must be used in a mold, what rules should be followed for their use?

24. If a sand casting must have a machined surface, how is a casting design affected?

25. List the primary principles for designing castings from gray cast iron.

Problems

1. The strength of a cast iron beam of uniform section and prescribed length is a function of its cross-sectional shape. Propose alternative shapes of superior transverse strength and compare them to an I-beam. Comment on the results.

2. The torsional strength of a parallel-sided rodlike structure is a function of its cross-sectional geometry. Plot the strengths of different shapes and compare them to a solid circular section. Comment upon the results.

3. The strength of cast iron parts subjected to bending loads is different in tension and compression. Propose some sectional configurations of parts with uniform cross-sections that are subjected to bending loads. Indicate the regions subjected to tensile and compressive stresses.

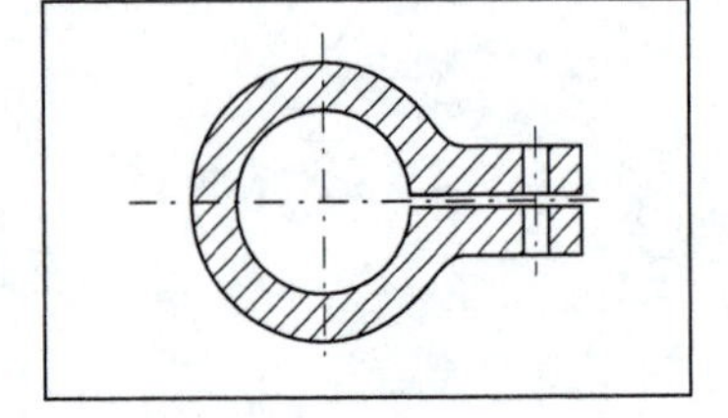

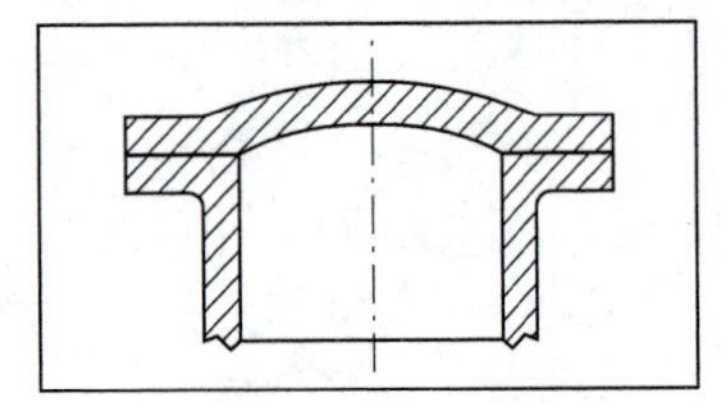

4. Portions of two products made from gray iron castings are shown in the accompanying figures. The first product is used to clamp a tube that fits through the larger hole, and the smaller hole accommodates a bolt that clamps the tube when tightened. What is wrong with this design? Propose alternatives made from gray iron.

   The second drawing shows a section of a pressure vessel with a shaped cover plate subject to high internal pressure. Analyze the product, identify the problem with this design, and propose a superior design in gray cast iron.

5. Drawings of three ribbed castings are shown below. Develop alternative ribbed designs of similar stiffness that offer economic advantages by avoiding the use of cores.

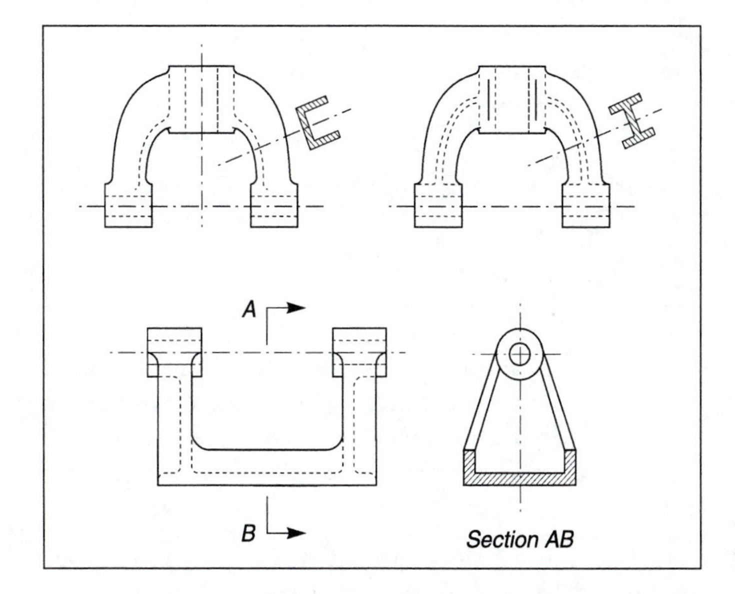

6. The figure on the right presents sectional drawings of parts of two castings that contain undesirable features. The first casting shows a partially enclosed cavity, and the second casting shows a core to generate a loose-fitting bolt hole. Redesign these castings to avoid using cores.

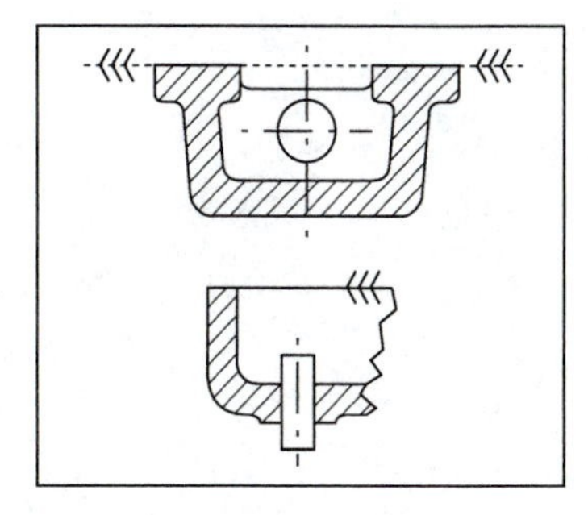

7. Some detail design characteristics of castings are shown in the accompanying figure. Note that the thickness of the flange is greater than the wall thickness, t, to increase strength and to provide a machining allowance. Note also that the location of the bolt hole center line relative to the edge of the flange should be greater than the hole diameter, D. However, both of these designs contain a fundamental error of casting design. What is it? Propose alternatives.

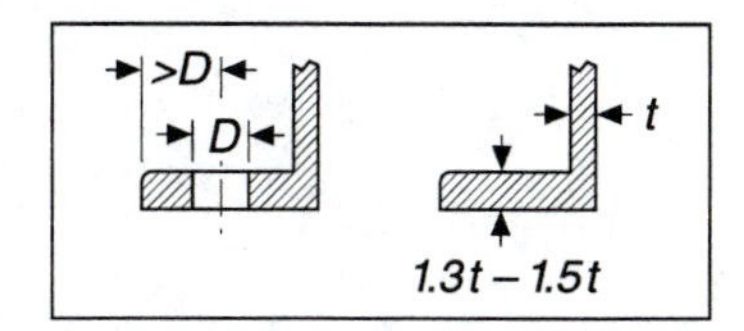

8. Castings should be designed with compact shapes to avoid generating high stresses. This can invariably be accomplished by specifying ribs to increase the stiffness of the structure. The figure below shows drawings of castings that are not compact. Redesign them to increase their functionality.

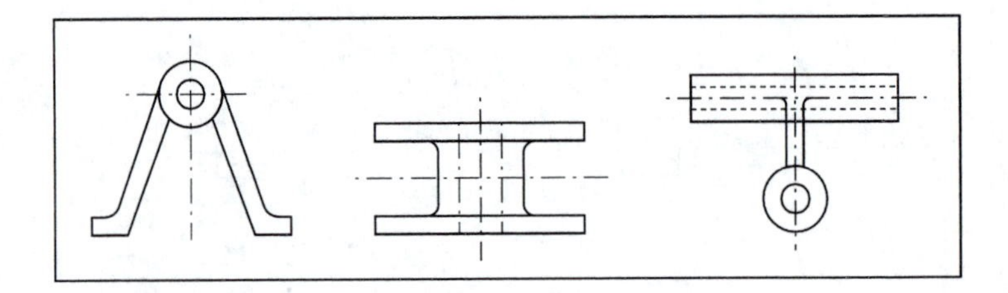

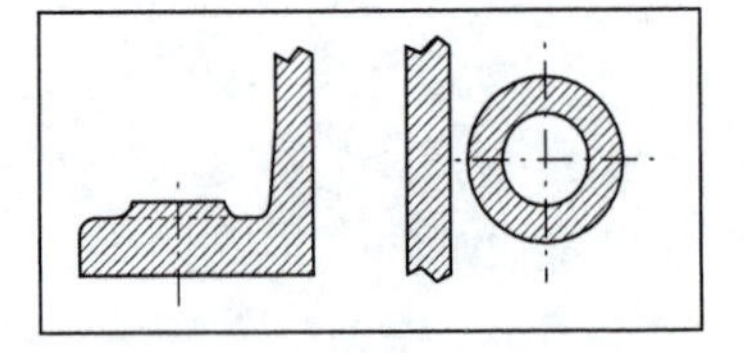

9. Thin sections should always be avoided in sand molds because they invariably break when metal is poured into the mold. Two unsatisfactory designs are presented in the figure on the right. Redesign them to avoid this error.

10. The figure below shows a welded fabrication to support a sintered metal bushing for a shaft. Redesign it for fabrication as a casting in gray iron.

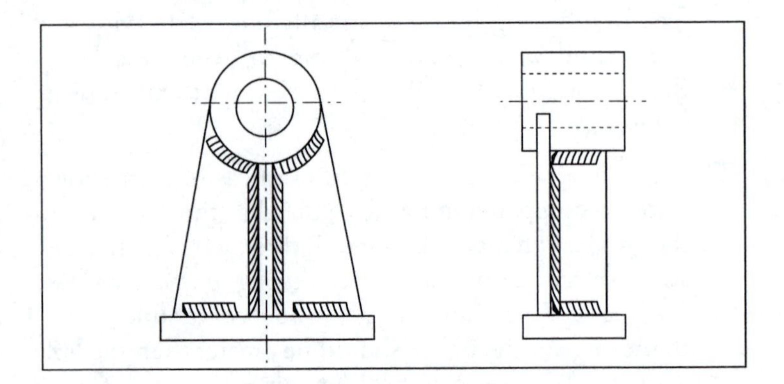

11. A specification for the loading imposed upon a pin that must be housed in a bracket fixed to the wall is presented on the left.

Assuming that the loads are constant, propose designs for castings that can fulfill this specification. Assume that the face of the casting in contact with the wall must be machined.

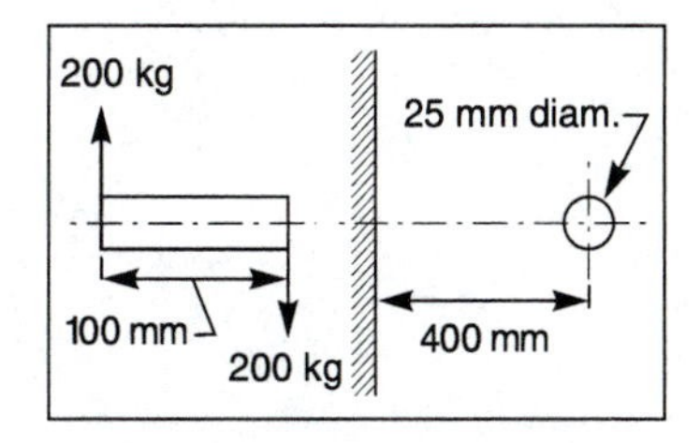

12. Redesign the structure in the figure on the right so that it can be manufactured by the casting process in gray iron without using a core.

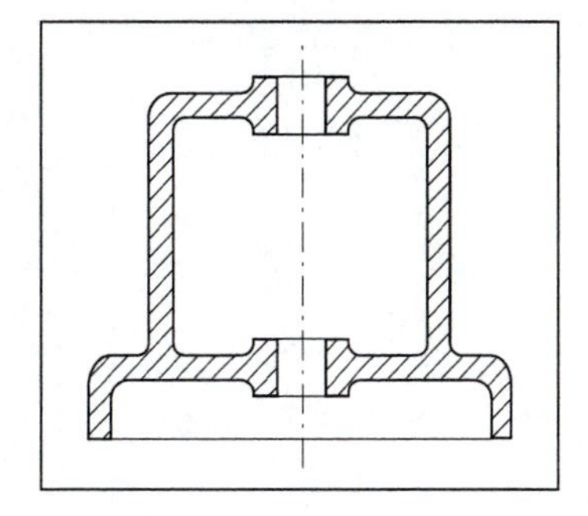

13. Two orthogonal views of a welded steel cover are illustrated below. Redesign this cover for fabrication as a gray iron sand casting.

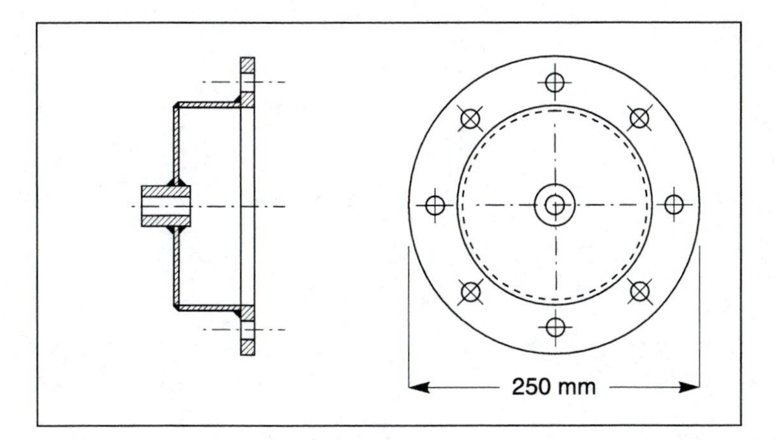

Further Reading

Bakerjian, R., ed. *Tool and Manufacturing Engineers Handbook: Volume 6, Design for Manufacturability.* Dearborn, MI: Society of Manufacturing Engineers, 1992.

Bralla, J. G., ed. *Handbook of Product Design for Manufacturing*, New York: McGraw-Hill, 1986.

El Wakil, S. D. *Processes and Design for Manufacturing.* Englewood Cliffs: Prentice-Hall, 1989.

Matousek, R. *Engineering Design.* London: Blackie, 1969.

Forgings

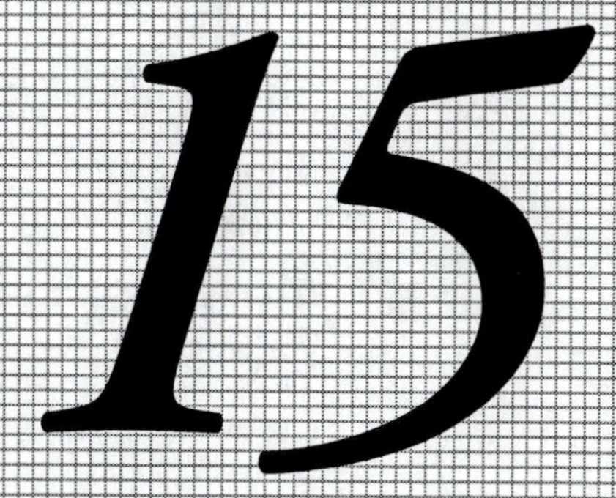

Thus at the flaming forge of life
Our fortunes must be wrought;
Thus on its sounding anvil shaped
Each burning deed and thought.

Henry Wadsworth Longfellow
(1807–1882)

Objectives

- To describe the forging process
- To delineate the advantages of the forging process
- To document rules for designing drop forgings

Contents

15.0 Introduction

Forging is a manufacturing process that shapes metals in the plastic state. This can be accomplished at ambient temperature or at elevated temperatures depending upon the shape and the desired properties of the final part. Parts manufactured by forging are characterized by superior durability and mechanical properties relative to similarly shaped parts manufactured by other processes such as casting, welding, powder metallurgy, or machining. These enhanced properties are a result of the macroscopic grain flow imposed upon the metallic structure of the part by the plastic deformation of the material during the forging process.

Metallic parts can be shaped by forging at ambient or elevated temperatures. The choice is governed by the desired attributes of the part.

The forging process is typically employed to manufacture critical parts that are subjected to high shear or tensile stresses and impact loads. Examples include hand tools such as box-end wrenches, parts of automobile steering systems, automotive connecting rods and crank shafts, crane hooks, parts for aircraft landing gear, and heavy forgings for the drive shafts of ships and power plants. Thus, forgings can weigh from a few hundred grams up to several metric tons.

15.1 An Overview of the Forging Process

There are several types of forging processes that are classified by the method used to shape the metal and by the temperature of the blank during the process. Cold forging imparts great precision and a high quality finish to the part, although the pressures imposed on the blank are higher than those imposed during hot forging, and the change in shape is limited by work-hardening of the material. Hot forging permits much greater changes of shape, but the accuracy and surface quality is somewhat inferior to cold forged parts because of warpage and oxidation.

Characteristics of cold-forged parts:

- Great precision
- High quality finish
- High working pressures
- Limited change in shape of the workpiece

Hot forging or impression-die forging—sometimes called drop forging—is used in mass production manufac-

Hot forging:

- Considerable shape changes possible
- Inferior precision
- Inferior surface finish

Drop forging dies are very expensive, so many parts must be produced from one die set for the process to be economical.

turing. Only large numbers of parts can be made this way because the metal dies used to shape a hot metal blank are very expensive. As the lot size increases, this high initial cost becomes less significant relative to the costs of raw material, labor, and overhead. Die costs naturally depend upon the complexity of the part, but something weighing approximately 100 grams might require a minimum lot size of 5000 parts before drop forging becomes economical.

Figure 15.1 shows the basics of the drop forging process. A hot metal blank is placed between the upper and lower metal dies that shape the part. As the upper die is moved

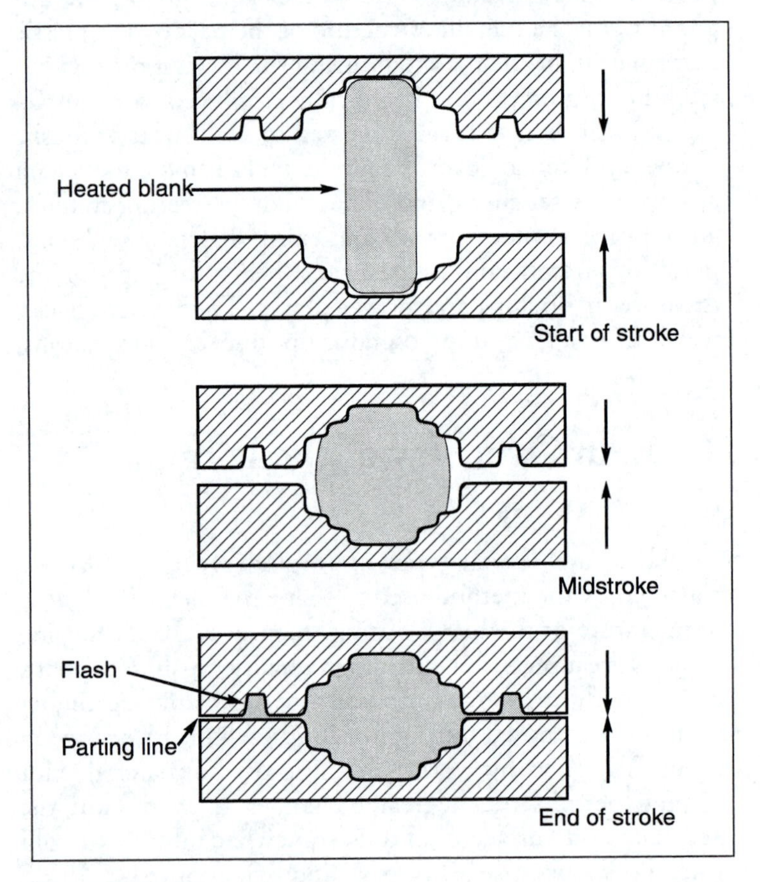

Figure 15.1

Basics of the drop forging process.

down onto the fixed lower die, the blank is squeezed into the shape of the cavity to form the part. When the shape of a blank must be changed a great deal to create the shape of the final product, several intermediate forgings will be created using a series of dies. Semifinished parts may require reheating during this process. The procedure is illustrated in Figure 15.2 showing the manufacture of an automobile crankshaft. The initial blank is shown at the left and, after passing through four die sets, the final part emerges with the separated flash. A photograph of an automobile crankshaft is shown in Figure 15.3.

Large shape changes may require several intermediate forgings, possibly with reheatings.

As dies close, the high pressure imposed on the metal blank refines the metal's grain structure and improves its physical properties. Most metals can be forged at elevated temperatures but some are much easier to forge than others. Specification of easily-forged metals whenever possible can

By refining the grain structure of the metal, forging improves the properties of a part.

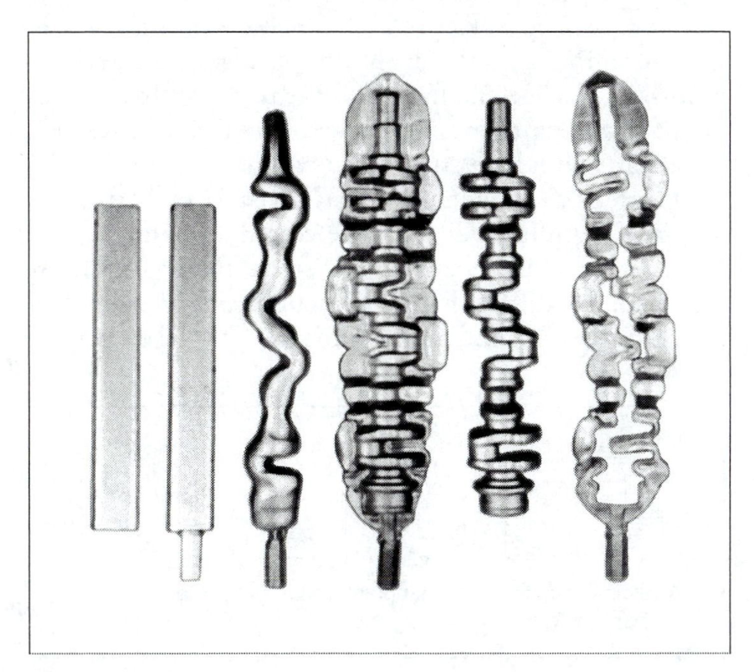

Figure 15.2

Multistep forging of an automobile crankshaft (courtesy of the American Forging Association).

Figure 15.3

An automobile crankshaft.

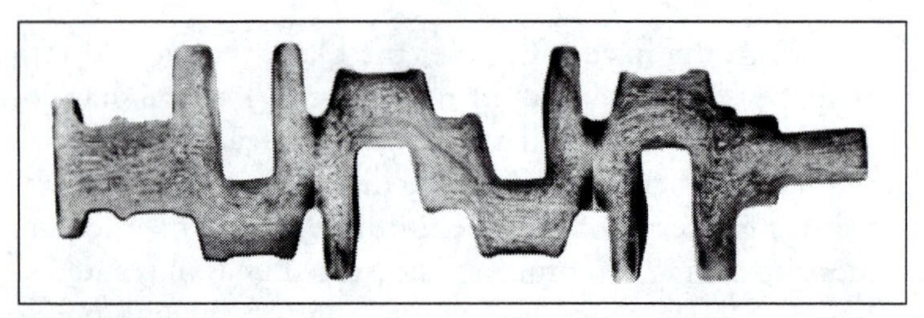

Figure 15.4

Microscopic grain flow of a forged automobile crankshaft. (Courtesy of the American Forging Association)

reduce costs through use of fewer dies and smaller forging machines. Easily forged metals include aluminum alloys, low-carbon steels, magnesium alloys, and copper.

It is possible to strengthen a part by aligning grain flow with the principal stresses on the part.

The primary advantages of forged parts derive from the controlled grain structure of the part which is established during the plastic deformation of the material. This grain flow is illustrated in Figure 15.4 which presents the section of an automobile crankshaft. With appropriate die design, it is possible to align the flow with the principal stresses that the part will experience in service. Doing so can result in metal parts with high strength-to-weight ratios.

Figure 15.5 provides a comparison of the grain structure of a part manufactured by three common manufacturing processes: machining from rolled stock, forging, and casting. Note the different macrostructural characteristics imparted by each manufacturing process. The forged part has

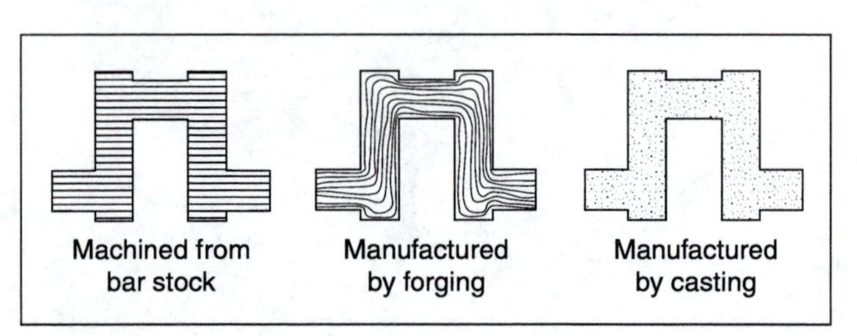

Figure 15.5

A comparison of the macrostructure of a crankshaft made by machining from solid stock, by forging, and by casting.

an unbroken grain flow parallel to the boundary of the part. (*Grain flow* is the directional pattern assumed by metallic crystals during forging.) This flow is responsible for the development of a part with superior ductility, impact resistance, and strength because the characteristics in the direction of the longitudinal axis of the crystals are superior to those of a material with the random grain structure that occurs in a cast part.

A forged part has a continuous grain flow parallel to the boundary of the part. This imparts superior toughness and impact resistance.

The intense pressures imposed upon metals during forging are responsible also for a part that seldom has any internal flaws and is less prone to failure. Thus critical parts such as those in aircraft engines are invariably forged if their failure could be catastrophic. Figure 15.6 shows some typical forged parts.

Figure 15.6

Some typical forged parts. (Courtesy of the American Forging Association)

15.2 Advantages of Drop Forging

The manufacture of parts by drop forging offers the designer a number of advantages relative to other manufacturing processes, provided that production is high enough to amortize the high cost of the dies or the part is a crucial component of a system that will be subjected to extreme conditions. These are some of the advantages:

- Plastically deforming a metal reduces its grain size so that the strength, impact resistance, and toughness are increased.
- Higher allowable strengths than parts manufactured by other processes.
- Unmatched toughness among monolithic materials.
- The absence of internal flaws, voids and defects in forgings are responsible for parts with higher structural integrity.
- Forged parts have a longer service life than parts manufactured by other processes.
- Forged parts have high safety margins because characteristics such as ductility and toughness vary less.
- High quality and repeatability among different lots permits the designer to offer extended warranties with less exposure to product liability. Less quality control testing is required.
- The drawing action during the forging process causes the grains to flow parallel to the boundary of the part. This can be made to coincide with the directions of the principal stresses in the part. A forging therefore can be stronger and have a higher strength-to-weight ratio than a part of comparable geometry manufactured by machining or by casting.
- A forged part requires less material than a machined part.

15.3 DFM Rules for Drop Forgings

Design-for-manufacture rules for drop forging follow. They are summarized in Table 15.1.

- Position the starting plane so that the impression in the die is as shallow as possible. Make the principal plane of the part parallel to the parting plane.
- Give draft to the side walls to facilitate removal of the forging from the die.
- Avoid multipart dies; they are expensive.
- Avoid sharp edges; they cause high wear in the dies.
- Provide machining allowances where appropriate.
- Perform an economic analysis to determine whether the part can be manufactured by another process, such as welding, drawing, or stamping.
- Avoid high narrow ribs and thin webs.
- Avoid undercuts that prevent the part from being removed from the die.
- Avoid redesigning forgings. It requires scrapping or modifying the dies. Both are expensive.
- Ensure that changes in cross section are gradual.

Table 15.1

Design-for-manufacture rules for forgings.

• Parting plane placement

The parting plane is the interface between the two halves of a closed die set. This interface creates a parting line around the perimeter of the part as shown in Figure 15.1. The location of the part relative to this plane is important because it dictates the flow of metal in the dies, some properties of the part, the cost of trimming operations to remove the flash, as well as the cost of the dies and their wear characteristics. As die halves come together in a forging operation, the primary direction of metal flow is parallel to

The parting plane is the interface between the two dies.

Make the principal plane of the part parallel to the parting plane.

the parting plane causing grain flow in the part parallel to the parting plane. This should be the direction of primary stresses when the part is in service. The part also should be positioned to produce a low profile in the die to facilitate this flow (Figure 15.7). The parting line can be a horizontal or inclined straight line or it can be developed on several planes. In any case it must be shown on the detailed drawings used to manufacture a part. Chevrons are often used to indicate the parting plane.

Chevrons indicate the parting plane.

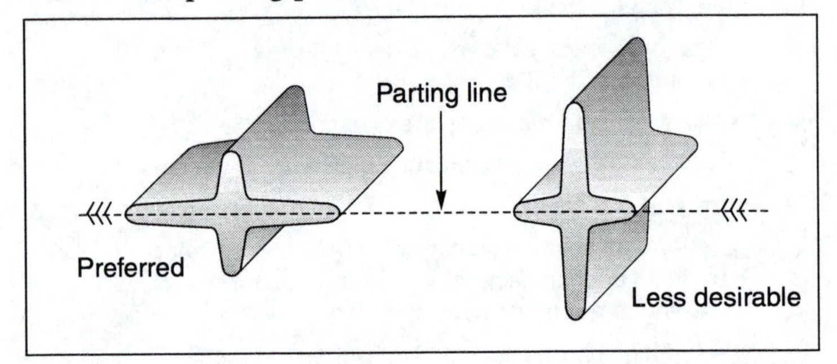

Figure 15.7

Position the part in the die to create a low profile.

Ideally the parting plane should be perpendicular to the direction of die motion and coincident with the axis of symmetry if the part has a single *axis* of symmetry (Figure 15.8). Figure 15.9 shows a part with a single *plane* of symmetry. Options for positioning it relative to the parting plane are shown in Figure 15.10.

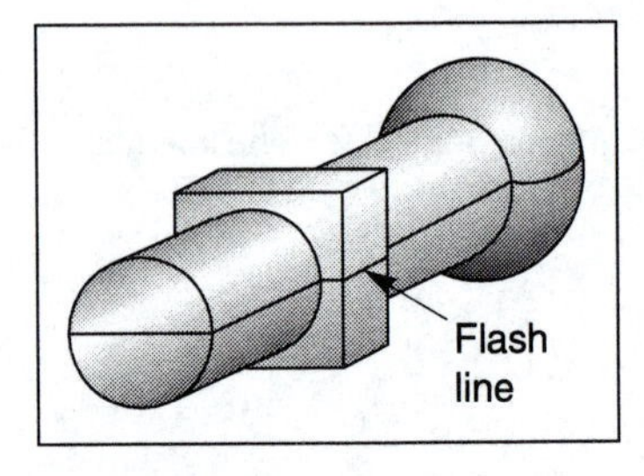

Figure 15.8

Parting plane coincident with a single axis of symmetry.

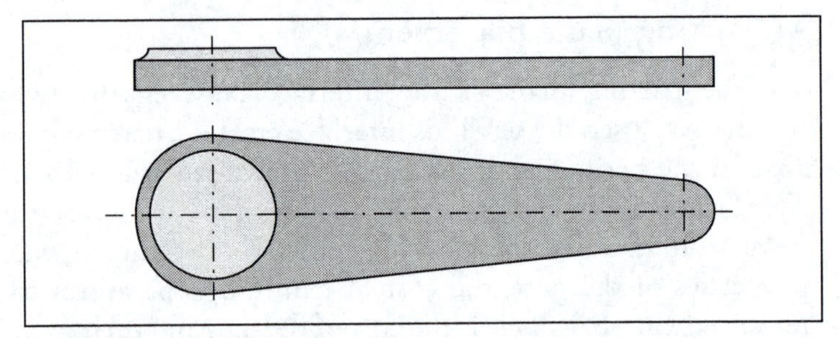

Figure 15.9

A forging with a single plane of symmetry.

While a first thought might be to position the part shown symmetrically in the die (option *a*), a deep die impression would be required and the part would need to be forged sequentially using a series of dies. This would be an expensive choice. It is better to position the part as flat as possible in the die in order to avoid a deep die impression. Of the options presented in Figure 15.10, option *c* is best because it avoids the more expensive die shapes of options *b* and *d*, while option *e* requires the part to be formed primarily in the upper die.

Ensure a shallow impression in the die set.

Selection of the part configuration relative to the parting plane is critical.

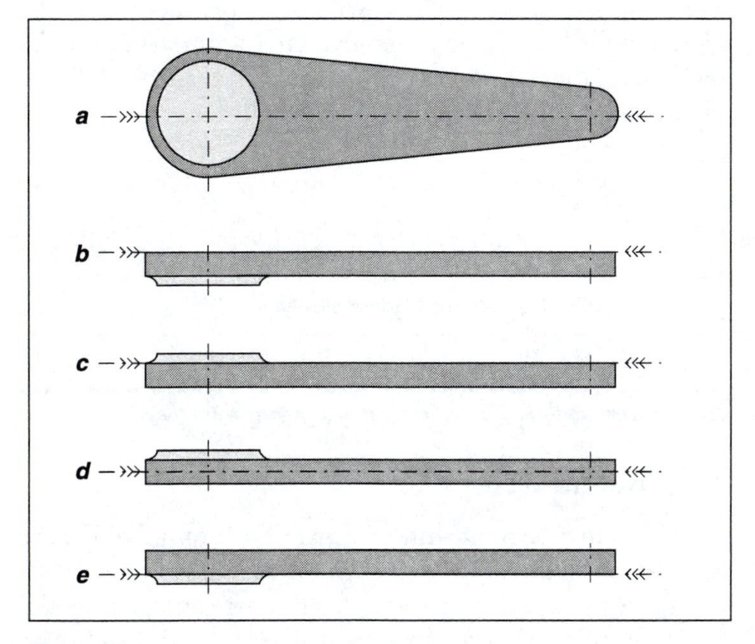

Figure 15.10

Possible orientations in a die for a forging with a single plane of symmetry.

• Draft

Ensure that side walls are tapered to ease removal of a forging from the die.

The sides of a die parallel to the direction of travel of the moving die must have a sloped face (a draft) in order that the forging can be removed easily from the die cavity. Furthermore the draft should assist the flow of plastic metal during the forging operation. The draft angle is specified relative to the axis of motion of the upper die. Its size depends upon the material being forged and whether the face is in the upper or lower die, but the draft angle commonly ranges from 0 to 8 degrees. Figure 15.11 illustrates some forging cross sections with draft on the faces. The necessity for being able to remove the forging from the die set dictates that protrusions from the forged shapes perpendicular to the die motion are not permitted.

Avoid lateral protrusions from the part because they prevent the forging from being removed from the die.

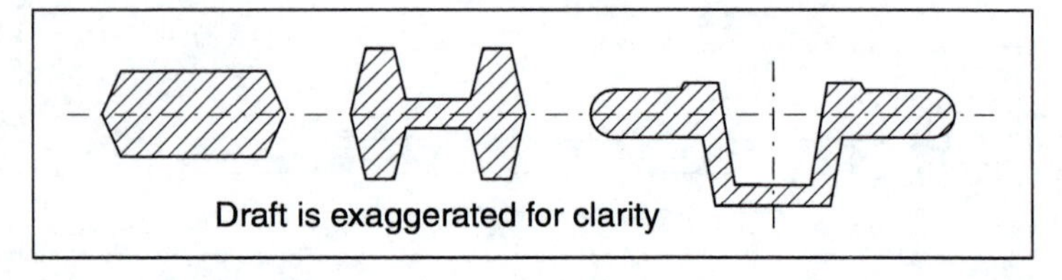

Figure 15.11

Draft on side walls facilitates removal of a forging from the die.

• Ribs and webs

A *rib* is a wall parallel to the die motion.

A *web* is a wall perpendicular to the die motion.

Ribs are relatively thin sections of a forging which are parallel to the direction of motion of the moving die. Webs are sections of a forging which are perpendicular to the direction of motion of the moving die. Ribs and webs can be forged when there is a satisfactory flow of metal. This requires ribs that are not too high and narrow, as well as webs of uniform and appropriate thickness relative to other parts of the forging. Webs are usually thicker than ribs. Metal flow can be enhanced by ensuring gradual changes of cross-section, and high wear rates in the dies can be avoided by ensuring that there are no sharp edges. Figure 15.8 shows two designs of a part, one that is easy to forge and one that is difficult to forge.

Summary

You have learned in this chapter that drop forgings are made by deforming a hot or cold blank in a closed cavity between two metal die halves. Hot and cold processes produce parts with quite different attributes. Parts made by cold forging have great precision and a high quality surface finish, but each forging operation achieves only a limited change in the shape of the workpiece. Parts manufactured by hot forging have inferior precision and inferior surface finish but can be produced in fewer steps.

The high pressures imposed upon the workpiece by metal dies is responsible for a void-free macrostructure and a refined grain structure. Because of this grain structure, forgings possess several distinct advantages over cast or machined parts. These include high strength-to-weight ratios and superior ductility and toughness. This is the reason for the specification of forgings for critical parts such as those in aircraft engines.

The attributes of drop forgings include superior physical properties and the absence of internal flaws. This permits specification of higher safety margins, longer service life, and longer warranties. Drop forging also uses less material than casting or machining.

The design of a drop forged part begins with deciding the location of the parting plane. This decision influences the cost of manufacturing the dies, the cost of manufacturing the part, the direction of grain flow, and the shape of the forged part. Ideally the parting plane should be perpendicular to the direction of die motion and it should be positioned to avoid deep die cavities. The impression should ensure lateral flow of metal perpendicular to die motion. Lateral metal flow affects web formation while flow parallel to the die motion affects rib creation. Draft on rib walls eases removal of the forging from the die cavity.

Die forging produces parts with unmatched toughness and reliability. It is typically used when many parts are to be manufactured and when specific material properties are required.

- Make the die as shallow as possible.
- Make the principal plane parallel to the parting plane.
- Give draft to the side walls.
- Avoid multipart dies.
- Avoid sharp edges; they cause high wear.
- Provide machining allowances.
- Perform an economic analysis.
- Avoid high narrow ribs and thin webs.
- Avoid undercuts.
- Avoid redesigning forgings.
- Make changes in cross section gradual.

Key Concepts

Draft. The taper introduced on walls the are parallel to the die motion. Draft ensures that forgings can be easily removed from the die. It promotes flow in deep cavities.

Drop forging. A mass production forging process that produces high quality parts.

Flash. A fringe of excess metal extruded into a gutter in the parting plane as the two dies close. This excess metal ensures that the die is filled completely. It is subsequently trimmed off.

Forging. A manufacturing process that plastically deforms metal to create a part. Also the name given to the part that results. The process can be performed with the blank either heated or at room temperature.

Grain flow. The elongated arrangement of the structure of a metal due to mechanical working during the forging process. Proper grain flow adds toughness to a forged part.

Parting plane. The plane of separation of the die halves. It must be indicated on a drawing of the forging. Selection of this plane is an important DFM consideration.

Ribs and webs. A rib is any wall created by metal flowing parallel to the die motion. A web is any wall created by metal flowing perpendicular to the die motion and parallel to the parting plane.

Superior physical properties. The forging process can produce parts with a controlled macrostructure. Hot working refines the grain structure. In a forging the strength, ductility, and impact resistance parallel to grain flow are significantly higher than corresponding properties in a cast metal part where crystals of the metal are randomly oriented.

Review Questions

1. Describe some important attributes of the forging process. Which metals are frequently forged?
2. How do the attributes of parts manufactured by hot and cold forging differ?
3. What is the range of weights of parts manufactured by forging? List typical parts manufactured by this method.
4. What other name is often given to drop forging? Why is this process usually limited to a large batch size?
5. In forging, high pressures are imposed upon a metal blank as the dies close. What metallurgical phenomenon occurs during this phase of the forging operation? What properties are developed in a part by this phenomenon?
6. Explain why forgings are often used for critical parts in a product.
7. List some geometric attributes of forged parts that are often used to classify them.
8. List six advantages of forgings.
9. List three aspects of the forging process that require consideration during the design process.
10. How does the parting plane influence the design of a part?
11. Under what conditions should the axis of symmetry of a part not coincide with the parting plane? Illustrate your answer with sketches of a forging.
12. What is draft and what role does it play in the forging process?
13. How do ribs differ from webs?
14. Explain why drop forging is typically a production process involving a large number of parts.
15. List ten design-for-manufacture rules for forged parts.

Problems

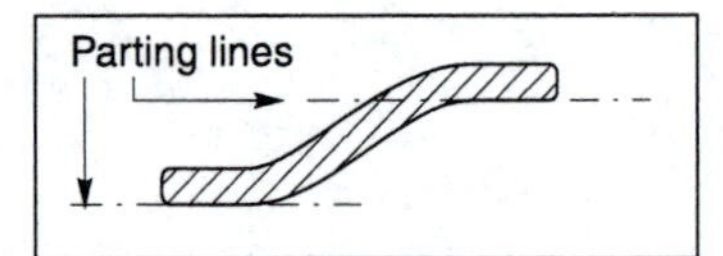

1. Considerable pressure is generated during a forging operation. The situation is exacerbated when the parting line does not lie in a plane perpendicular to the motion of the die. Large lateral forces can develop. The figure at the left shows a section of a forging that will be subjected to large lateral forces during manufacture. Reconfigure the forging to avoid this.

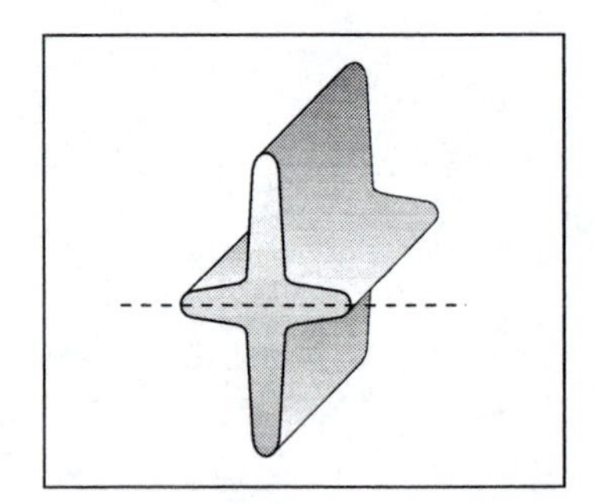

2. The figure at the left shows a section of a forging. The dotted line represents the parting plane. Determine what is wrong with this situation and propose an alternative.

3. The figure below shows a section of a forging. Reproduce the drawing and label the ribs, webs, and parting line.

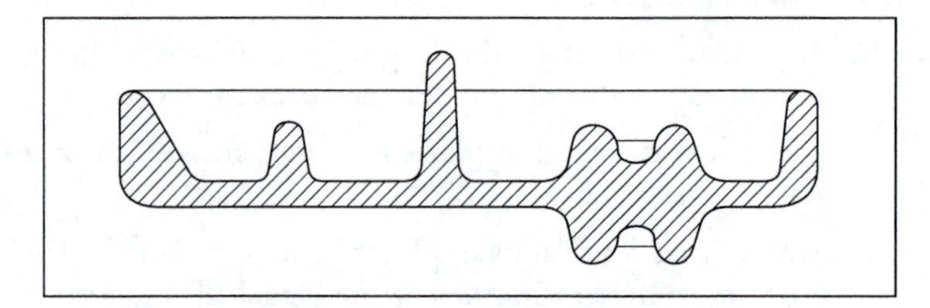

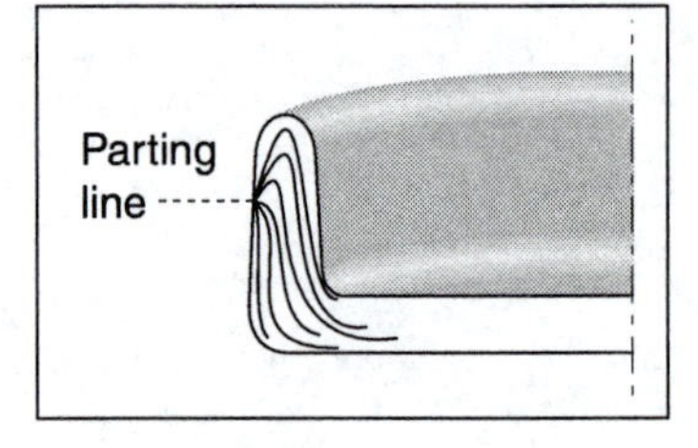

4. Deep impressions in dies should be avoided. This is often accomplished by positioning the parting line so that each die half has the same impression depth. However, as shown by the discontinuous grain flow in the dished part shown in the figure at the left, this is sometimes undesirable. Redesign the forging operation for a superior location of the parting plane and better grain flow.

Further reading

ASM Metals Handbook: Volume 14, Forming and Forging. Metals Park, OH: ASM International, 1988.

Bakerjian, R., ed. *Tool and Manufacturing Engineers Handbook: Volume 6, Design for Manufacturability.* Dearborn, MI: Society of Manufacturing Engineers, 1992

Bralla, J. G., ed., *Handbook of Product Design for Manufacturing.* New York: McGraw-Hill, 1986

Dieter, G. *Engineering Design: A Materials and Processing Approach.* New York: McGraw-Hill, 1983

El Wakil, S. D. *Processes and Design for Manufacturing.* Englewood Cliffs, NJ: Prentice-Hall, 1989.

Welded Fabrications

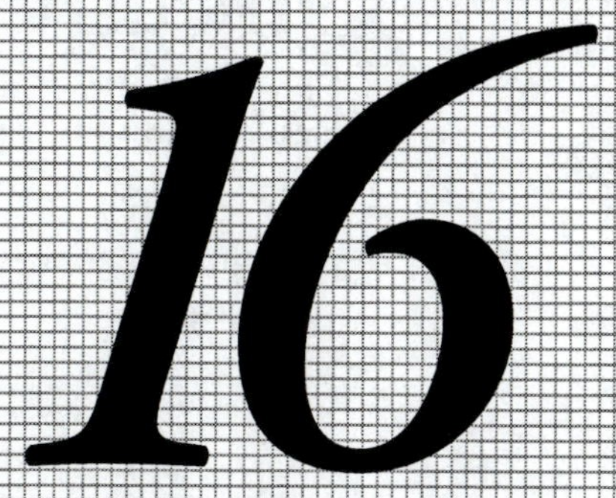

So let us melt.

John Donne
1573–1631

Objectives

- To develop an appreciation for design and manufacturing attributes of weldments
- To present some rules for designing welded structures

Contents

16.0 Introduction

Welding is a manufacturing process in which two metals are joined by the intermixing of atoms at the metal interface. The resulting structure is called a *weldment*. This manufacturing process offers maximum homogeneity since it creates bonding between the adjacent components at the atomic level. Welding permanently unifies components with a homogeneous joint having a strength approximately equal to that of the parent metal. Unification is usually accomplished by joining the parts in either a plastic or a molten state, depending upon which welding process is employed.

Welding is a manufacturing process in which two metals are joined by the intermixing of atoms at the metal interface.

A weldment is a structure formed by welding parts together.

An advantage of welded joints is that the stress pattern in the joint is much simpler than the pattern in riveted or bolted joints that have stress concentrations around rivet or bolt holes. This is a distinct advantage when parts must operate in fatigue environments or when lightweight parts are needed as in the automotive and aerospace industries. Additional advantages of weldments include great savings in material, low manufacturing costs, and dependability of the joint. Furthermore, the designer has considerable latitude in designing the structure.

Weldments are often lightweight structures.

Many ferrous and nonferrous metals can be welded to create weldments. These welded structures are diverse, ranging from buildings and bridges in civil engineering to electric motors, the hulls of ocean-going vessels, mining machinery, agricultural equipment, furniture, automotive components, and aerospace parts. Typically the more complex the structure and the bigger it is, the more advantageous it becomes to combine semifinished products and off-the-shelf hardware to create a welded structure.

Typical weldments:

- steel building frameworks
- road and rail bridges
- ocean vessels

Semifinished products fabricated into weldments are usually made by primary production processes such as rolling and extrusion. Hence components of weldments are often extrusions, plates, sheets, bar stock, tubes, machined parts that have been turned, milled, and drilled, and rolled sections of uniform cross section such as I, T, and L cross sections (Figure 16.1).

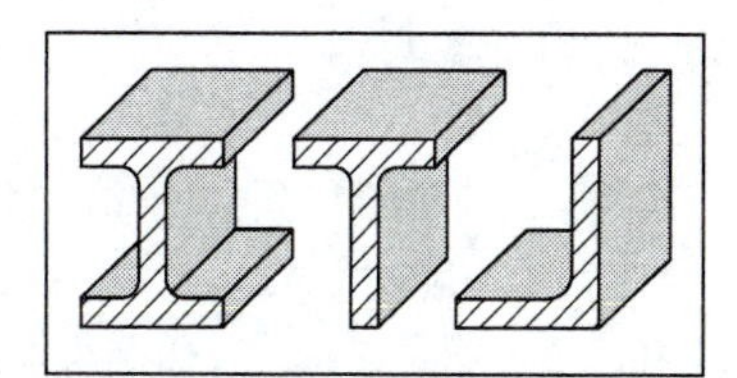

Figure 16.1

The primary components of weldments are extrusions and rolled sections.

16.1 A Summary of Welding Processes

The selection of the appropriate welding process involves consideration of:

- welding processes
- materials
- design
- cost

There are a number of welding processes available to the designer, each with advantages and disadvantages. Thus when a welding process is to be used, the design team must select an appropriate technique. This will dictate the size, quantity, quality, and type of welds that must be formed.

Welding processes can be classified by the state of the metal in the weld zone during the welding operation. Two categories of welding processes exist:

Pressure welding generates plastic flow at the interface of the two parts.

- *Pressure welding,* in which external pressure is applied to create a plastic state at the interface. Some of these processes are listed in Table 16.1.

Fusion welding generates a molten pool at the interface of the two parts.

- *Fusion welding,* in which localized heating and melting of the metal in the interface creates a weld pool that unifies the two metallic parts as they cool and solidify. Some of these processes are listed in Table 16.2.

Pressure welding processes
Explosive welding
Ultrasonic welding
Cold-pressure welding
Hot-pressure welding
Molten-metal bonding
Thermit welding
Resistance flash welding
Resistance seam welding
Resistance spot welding
Resistance projection welding
Diffusion bonding
Friction welding
Induction welding

Table 16.1

Pressure welding processes.

Shielded metal arc welding
Submerged metal arc welding
Metal-inert-gas welding (MIG)
Tungsten-inert-gas welding (TIG)
Plasma arc welding
Gas welding
Electron beam welding
Laser beam welding

Table 16.2

Fusion welding processes.

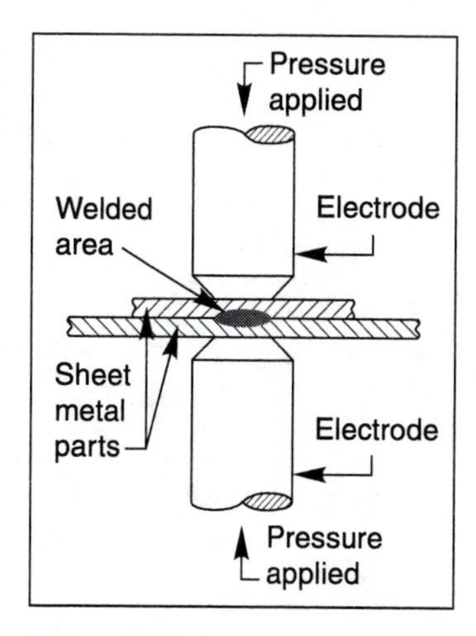

Figure 16.2

Resistance spot welding.

Figure 16.2 presents a sketch of resistance spot welding, a pressure welding process that has wide application in industries that join sheet metal parts. Resistance spot welding equipment is characterized by two water-cooled electrodes that press metal sheets together. When an electrical potential is applied between the electrodes, electrical resistance at the contact between the two sheets causes local heating until a weld pool is formed. Usually only partial melting of the metal sheets occurs. When the current is turned off, the weld pool solidifies to unite the two sheets, and pressure on the electrodes is removed. The surface of the sheets typically has a small depression and some localized discoloration of the metal. A spot welded part is shown in Figure 16.3.

Figure 16.4 presents a schematic of tungsten inert gas welding (TIG), a fusion welding process sometimes called gas tungsten-arc welding. In this process an arc is maintained between the workpiece and a nonconsumable tungsten electrode. The intense heat from the arc melts the adjacent metal in the weld region and the two parts coalesce at the weld pool. Since most metals become chemically reactive when heated to welding temperatures, the molten weld metal and the hot zone in the base metal must be protected from the atmosphere. This protection must prevent oxygen and nitrogen in the air from combining with molten metal to form metal oxides and nitrides which lower

Figure 16.3

A spot welded part in thin metal sheet.

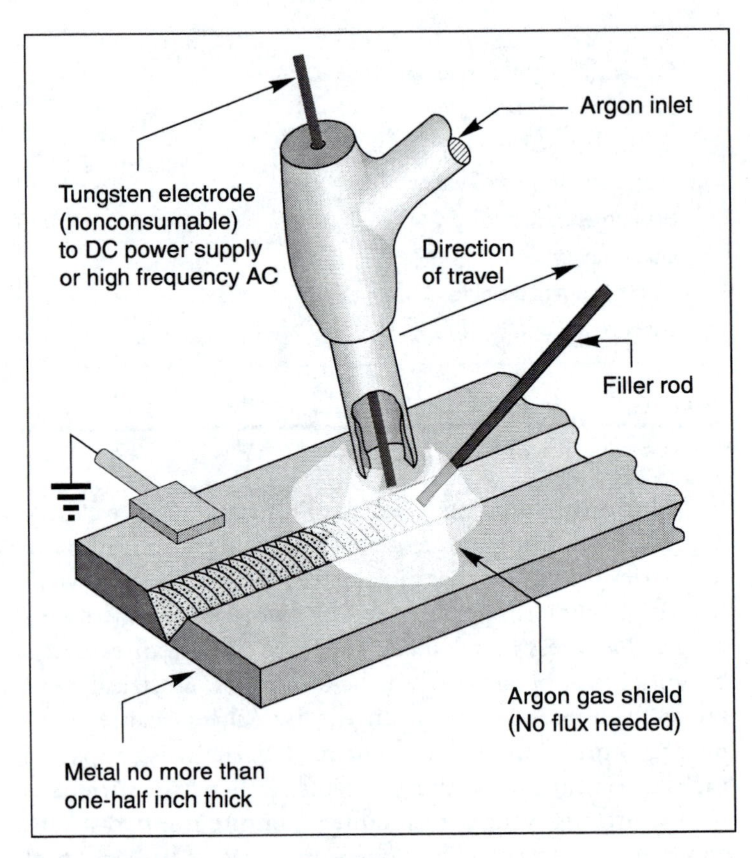

Figure 16.4

Tungsten inert gas welding.

Some processes for welding different metals:

- low carbon steel: any arc welding method
- medium carbon steel: generally any arc welding process
- high carbon steel: TIG or MIG
- alloy steels: generally TIG or MIG
- stainless steels: generally TIG or MIG
- aluminum alloys: TIG or MIG
- magnesium alloys: TIG or MIG

TIB: Tungsten-inert-gas welding

MIG: Metal-inert-gas welding

the ductility, toughness, and strength of the weld. In TIG welding this protection is provided by a shield of inert gas such as argon or helium.

The reader interested in these manufacturing processes should consult relevant literature because the details of the processes lie outside the domain of this text. Some literature is listed at the end of this chapter.

The characteristics of each welding process impose constraints upon the design of a weldment. Other constraints are imposed by properties of the metals being welded. Parts being welded usually are made of the same metal since dissimilar metals are rarely joined by welding. Most metals can be welded by arc welding, but the engineer should check for metallurgical problems such as a reduction in strength or brittleness. Figure 16.5 shows a fusion welded aluminum bracket.

Low carbon steels are the most common metals welded commercially. All of the arc-welding processes can be used to weld these metals, but as the carbon content increases, weldability decreases. Medium carbon steels are generally easy to process but extra precautions may be necessary to avoid stress cracking. High carbon steels are also prone to stress cracking and are generally welded by TIG or MIG processes. These processes are also recommended for welding alloy steels where weldability varies greatly depending upon the alloying elements. The weldability of stainless steels depends upon their composition. Ferritic and martensitic stainless steels are less weldable than austenitic stainless steels. Generally these steels are welded by shielded-metal arc, MIG, and TIG processes. Cast iron parts generally are not joined by welding in a production environment because the heat-affected zone suffers from embrittlement. Magnesium and aluminum alloys are both readily welded, generally by TIG or MIG processes.

Figure 16.5

A fusion welded aluminum bracket.

16.2 Weld Forms

A wide variety of welding processes and weld forms can be used to fabricate a weldment. The principal weld forms are the *butt weld* and the *fillet weld.* Each of these has a number of variations that depend upon the edge preparation of the parent material. A variety of butt welds are illustrated in Figure 16.6. It is evident that they join two members that are aligned edge to edge and of substantially the same sectional area and shape.

Butt welds can be created on plates up to about 4 mm thick without edge preparation. When thicker plates must be welded together, edges must be beveled as shown in Figure 16.6.

Butt welds join parts aligned edge to edge.

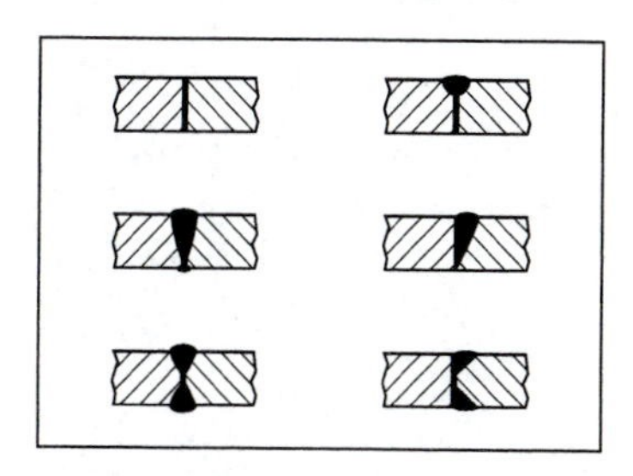

Figure 16.6

Forms of butt welds when the two members are of equal thickness.

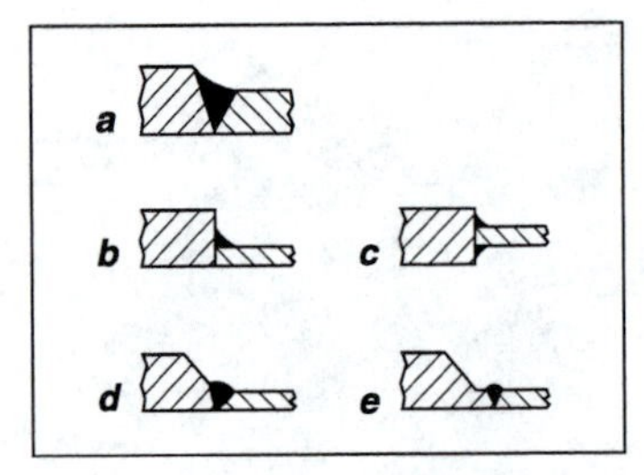

Figure 16.7

Forms of butt welds when the two members are of unequal thickness.

Butt welds should join two planar sections of the same thickness.

Concave fillet welds are the strongest type of fillet weld.

Fillet welds join parts oriented at non-colinear angles. Often the edge of one part contacts the face of the second part.

Sometimes it is necessary to weld members of unequal thickness. Figure 16.7 presents a number of different options for this task. Option *a* can be employed when there is not a great difference in thickness of the plates being joined. Options *b* and *c* can be used when the joint is only subject to static loading and options *d* and *e*, which are both characterized by a gradual change of cross section, can be used when the weldment is subjected to dynamical loading.

Figure 16.8 shows a variety of fillet welds. The forms of fillet welds comprise the concave fillet weld *(a)*, the miter fillet weld *(b)*, and the convex fillet weld *(c)*. The most desirable weld is concave because the smooth contour is responsible for a smaller stress concentration than the other welds. A convex weld has more weld metal but has larger stress levels at the notched interface between the weld metal and the parent metal.

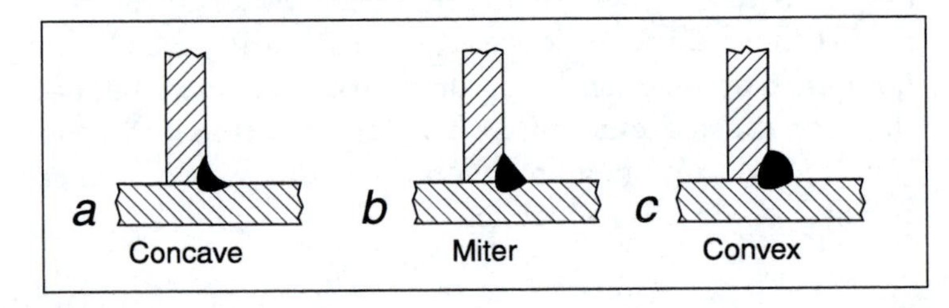

Figure 16.8

Forms of fillet welds.

Butt welds are stronger than fillet welds

Fillet welds are inherently weaker than butt welds and are subject to higher stresses. Consequently, designers should specify butt welds for situations requiring high strength. However by appropriate edge preparation involving the bevelling of the edge of the plate to be welded, this situation can be enhanced to provide a joint with superior strength. Figure 16.9 shows several options. Option *a* is the weakest and option *d* is the best. Options *a* and *b* involve no edge preparation whereas options *c* and *d* do.

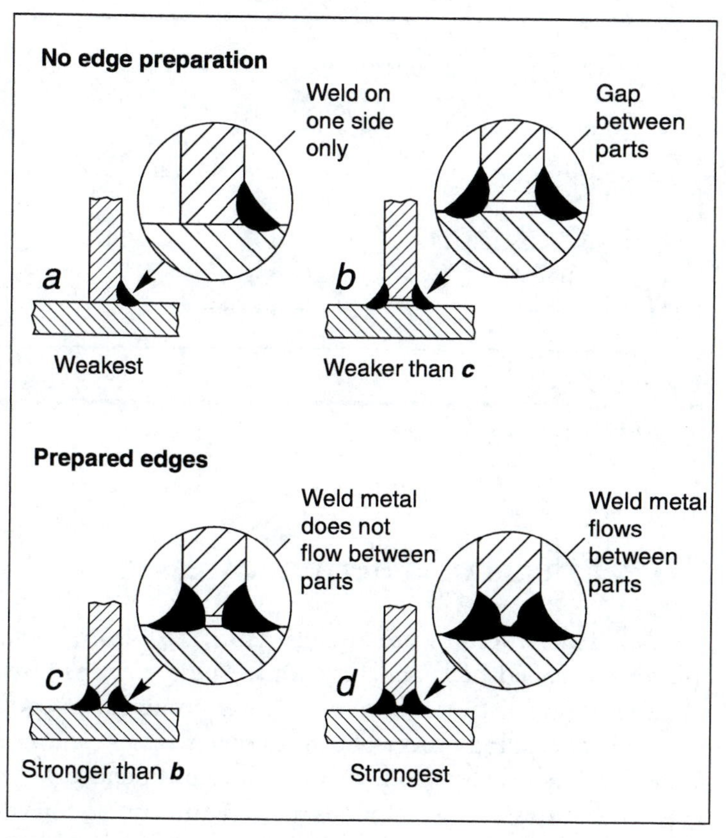

Figure 16.9

Types of fillet welds.

Butt and fillet weld forms provide a variety of joints useful in welded structures. Some common joints are presented in Figure 16.10, which illustrates the butt joint (a), the lap joint (b), the corner joint (d), the edge joint for thin sheet material no more than about 2 mm thick (d), and the T-joint (e). There is a variety of each of these joints; consult welding handbooks to learn the advantages and disadvantages of each type.

Consult welding handbooks for details on the design of different types of welded joints.

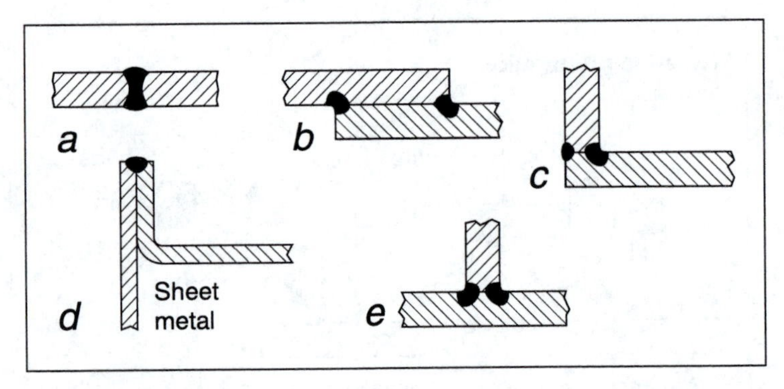

Figure 16.10

Some basic types of welded joints.

16.3 Effects of Thermal Stresses

Create weldments with:

- plates
- extrusions
- I, L, and T sections

Residual stresses can cause warping and permanent distortion of the weldment.

Welded structures are usually fabricated from metal plates and rolled L, T, and I sections. These sections often contain residual stresses because of their nonuniform cooling after the rolling process. Such residual stresses can cause unacceptable structural deformations if recommended welding procedures are not followed. In addition, the heat-affected zone in the immediate neighborhood of a weld can cause contractions that result in distortion, as shown in Figure 16.11. The microstructure of the heat-affected zone is generally quite different from that of the base metal because of the elevated temperature developed during welding.

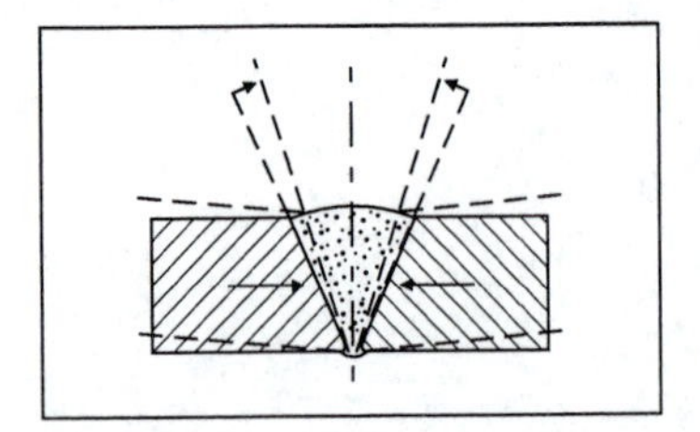

Figure 16.11

Consequences of residual stresses.

Remedies for residual stresses include following recommended weld-sequencing protocols, post-weld annealing, hammering the weld after welding, and heating the parts prior to welding. Strategies to minimize the undesirable consequences of thermal distortion include the following:

Minimize thermal effects by:

- post weld annealing
- pre-weld heating
- post weld hammering
- recommended weld sequence

- Use heavier, more rigid sections that are less prone to thermal distortion.
- Ensure maximum contact at the interface between two welded members.

- Use symmetrical weld configurations to generate a balanced thermal load.

These and other DFM rules are discussed in the next section.

16.4 DFM Rules for Welded Structures

The design of welded structures is primarily dictated by the materials to be used, the functionality of the weldment, the types of welding processes available for the manufacturing operations, the type of semifinished products to be employed (rolled sections or extrusions, for example), the types of joints to be used, and the location and distribution of the welds. These characteristics are reflected in the following rules:

DFM rules:

- appropriate welding process
- interfacial conditions
- use semi-finished sections
- symmetrical layouts
- loading and weld design
- fewest parts and welds
- access for welding equipment
- horizontal joints

• Select an appropriate welding process

Assess the characteristics of the material from which the structure is to be fabricated, relative to the welding processes and functionality of the design. For example, alloy steels which become brittle in the region of the weld when conventional arc welding is used may be welded successfully by electron-beam welding.

• Ensure good interfacial conditions

Design the interface between two welded members for maximum contact. This will reduce welding time because less filler is required in the gap and will reduce thermal distortion. The cost of the joint will be reduced. This rule is illustrated in Figure 16.12.

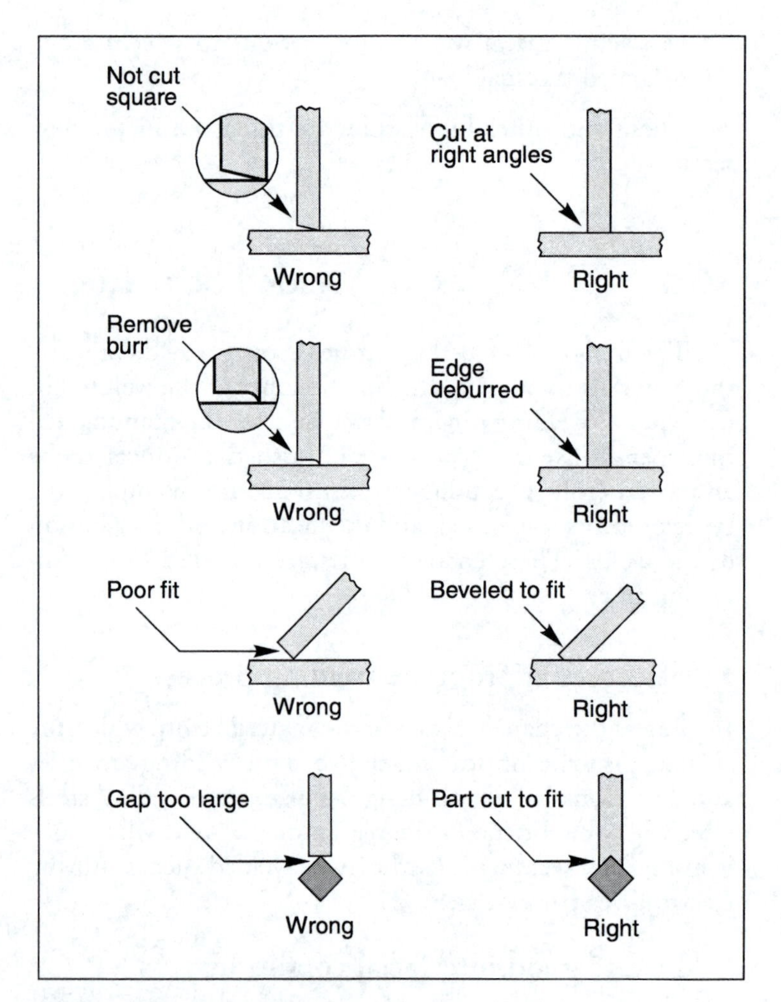

Figure 16.12

Ensure good interfacial conditions.

• Use rolled sections and other semifinished parts

Make welded structures from semifinished products because their variety gives the designer great freedom. Semifinished products include plates, sheets, strips, extrusions, and rolled sections. Structures can be made by welding together plates of different thicknesses, but additional processes such as plate bending may be needed. A design is shown in Figure 16.13 for a bracket made from plate material, while Figure 16.14 shows a design for the same bracket made from a combination of a rolled section and plates. This latter design can be cheaper and more accurate than designs that only use plate stock.

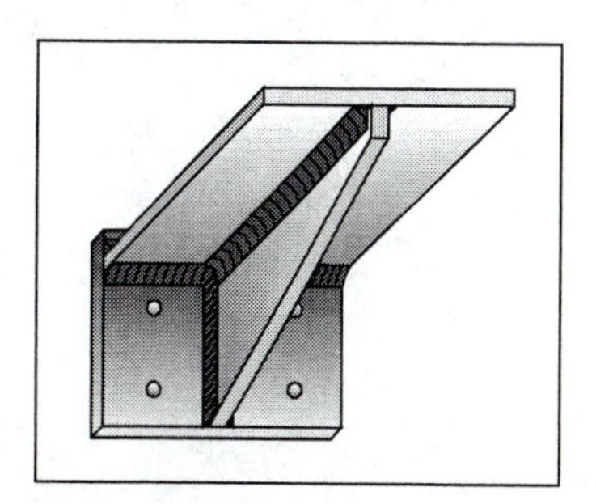

Figure 16.13

A weldment made from plate.

- **Create symmetric weld layouts**

Design structures with symmetric weld configurations in order to generate a balanced thermal load on each weld and avoid thermal distortions.

- **Consider loading and weld thickness**

Consider the thickness of the members being welded together as well as the load characteristics on the weld. Is the load constant, impactive, or dynamic? The various butt welds shown in Figure 16.7 are all appropriate for loads of constant magnitude. However, for more severe loads, the weld in option *e* is recommended.

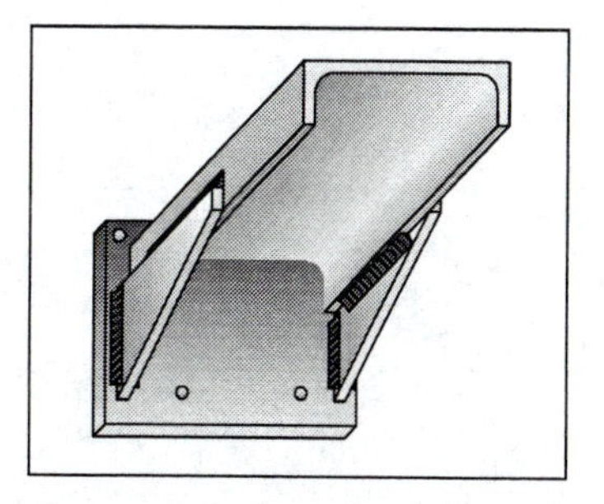

Figure 16.14

A weldment made from a rolled section and plate.

- **Perform a stress analysis**

When possible, make a computer-aided analysis of the proposed structure to determine stress levels, deflections, and the reliability of the weldment. This can help determine the best locations for welds and the characteristics of each welded joint.

- **Avoid welds in highly stressed regions**

Because welds are usually weaker than the parent metal, don't locate welds at high stress points. See Figure 16.15 which shows a pressure vessel composed of a tubelike central section welded to two domed end pieces.

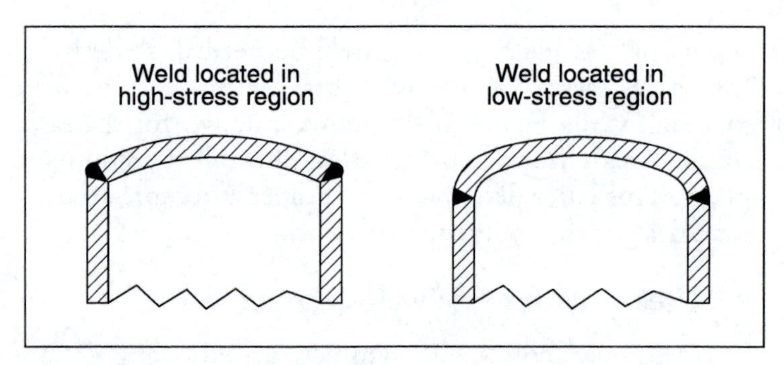

Figure 16.15

Avoid locating welds in highly stressed regions.

- ## Create welds perpendicular to maximum stress

Configure welds carefully relative to the principal stresses in a structure. The longitudinal axis of a weld should be perpendicular to the maximum stress. Consider the simple case of two lap joints (Figure 16.16). Option *a* is stronger than option *b*. The lap joint in option *a* has a weld along the narrow end of the strip material at right angles to the applied uniaxial load, while option *b* has two welds along the edges of the strip material parallel to the applied load.

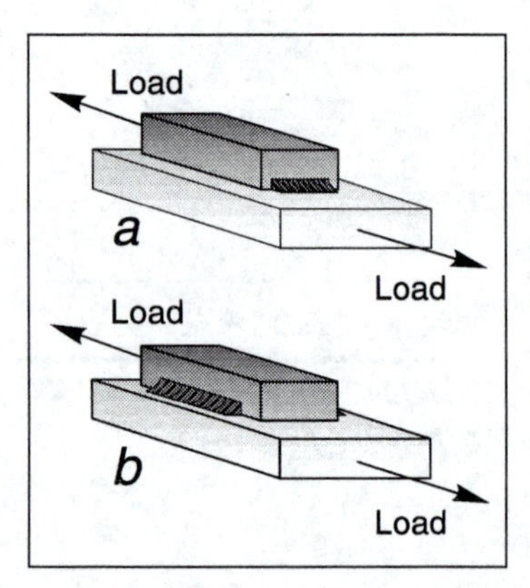

Figure 16.16

Weld locations relative to applied loads.

- ## Specify the fewest parts

Minimize costs by using the fewest parts in weldments. Use semifinished material, bent plates and machined parts when possible.

- ## Specify the cheapest form of welds

Minimize costs by using the fewest welds, the shortest preparation time, and the shortest possible arc time. Don't specify a continuous weld unless it is mandated by a stress analysis or unless a continuous seam is required for the weldment to function effectively. Intermittent welds with a large amount of weld metal are cheaper than continuous welds with a smaller bead.

- **Provide access for welding equipment**

Joints to be welded should have adequate access for the welding equipment. For example, if a semiautomatic process is specified, then the wire feed assembly requires more space than single stick welding. When a welding process has a stream of gas providing a shield, the nozzle must naturally be close to the weld zone to provide adequate protection.

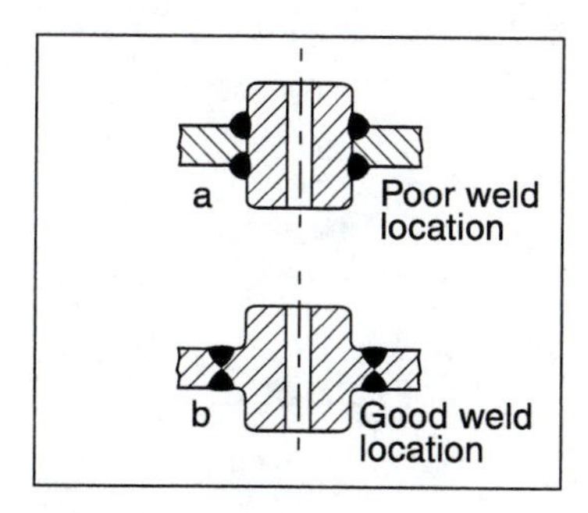

Figure 16.17

Ensure equal wall thicknesses of semifinished parts to be welded together.

- **Design the weld interface carefully**

If a weldment combines forged, cast, and semifinished parts, then the interfacial conditions should not only be good (Figure 16.12) but the wall thicknesses of the parent metal in the region of the weld should be the same (Figure 16.17). This will minimize thermal distortion and the welding time. Option *a* should not be selected because the welds are located in regions of high stresses between the bearing housing and the adjacent plates. In addition, the wall thicknesses at the weld are not equal. This undesirable situation is rectified in option *b*. A part failure in a lawn mower (Figure 16.18) illustrates this principle; the parts welded together are of unequal thickness. The wire has a diameter of 8 mm and the sheet metal is 1 mm thick.

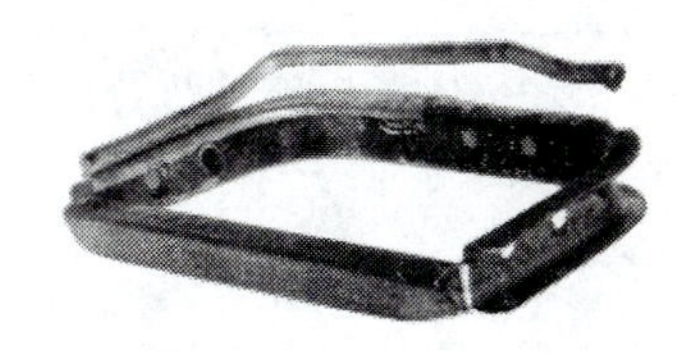

Figure 16.18

Part failure due to unequal wall thickness.

- **Consider post-weld machining**

Weldments should be designed with enough removable stock to allow subsequent machining operations to bring the product into compliance with the design specification. Thus a thicker plate may need to be specified for a weldment so that machining can provide the desired parallelism, surface quality, and dimensional tolerances. Welds should be located where they will not be subsequently machined. Machining can weaken a welded joint by removing material from the weld region, and hard regions in the heat-affected zone can blunt cutting tools. Figure 16.19 illustrates how to avoid this problem by good design.

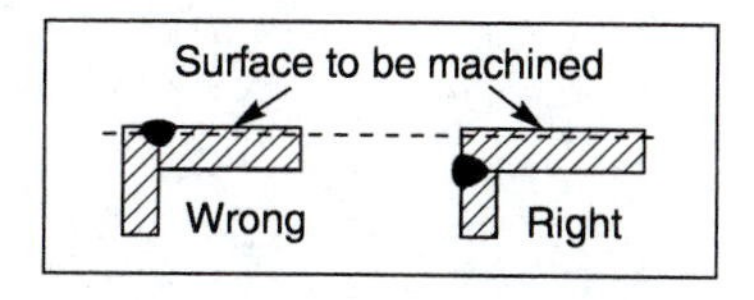

Figure 16.19

Avoid machining weld regions.

- **Weld joints horizontally**

Design weldments to ensure that the welding process can be done with the joint horizontal and the electrode or torch above the parts being joined. This configuration is the most convenient and cheapest. With large weldments it may be necessary to manufacture the structure by developing a number of subassemblies prior to final assembly.

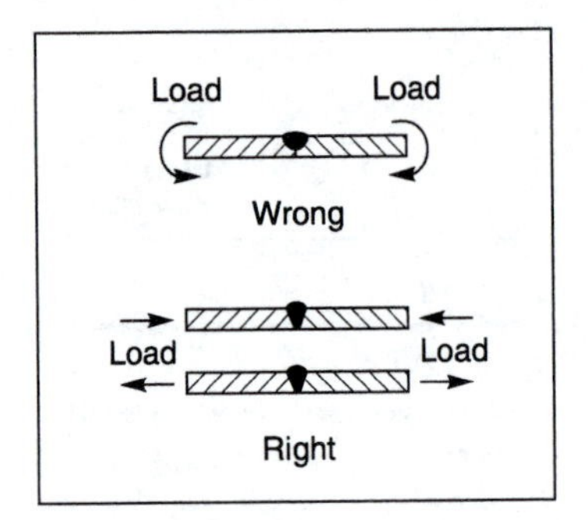

Figure 16.20

Weld-strength considerations.

- **Consider loading and joint design**

The strength of a welded joint depends upon the loading and the type of weld. Butt welds are the most efficient. However, they should be subjected to purely tensile or compressive loading rather than bending loads. Fillet welds should only be subjected to shear loads. Figure 16.20 illustrates these principles.

Summary

In this chapter you have learned about the joining of parts by welding. The resulting weldment can be created by a variety of welding processes broadly classified as pressure welding processes and fusion welding processes. Selection of the appropriate process requires consideration of materials selection, cost, the design attributes and functionality of the final weldment, and the characteristics of the different welding processes.

Welding processes enable a diverse group of ferrous and nonferrous metals to be welded together to create lightweight structures by capitalizing on a variety of butt joints and fillet joints. This is typically accomplished by using semifinished products like plates and L-, T-, and I-shaped rolled sections.

DFM rules should be employed to design weldments (Table 16.1). Special attention should be given to thermal stresses through consideration of pre- and postweld sequencing and the design of the joint and the individual

- Fabricate weldments from semifinished products such as extrusions, rolled sections, and plates.
- Minimize thermal distortion by various procedures including the specification of equal wall thicknesses at a joint.
- Use butt welds because they are strongest.
- Do not specify welds at locations with stress concentrations or at critical cross sections.
- Reduce the cost of the weldment by limiting the number of welds.
- Avoid designing welded structures that utilize welding fixtures because of the engineering, time, cost, storage, and handling consequences of this approach.
- Ensure easy access to the weld region so that the welding equipment can operate easily.
- Consider the sequence in which parts are to be welded together and prepare the manufacturing instructions. Ensure that production drawings contain all necessary information to manufacture the assembly (weld form, weld quality, and weld length, for example).

Table 16.1

Design-for-manufacture rules for weldments.

welds. Design the joint configuration and welds by considering the thicknesses of the parts to be welded and the characteristics of the loading imposed upon the weldment. Perform a computer-based design of the weldment. This analysis should permit welds to be specified in noncritical regions of the structure and permit them to be oriented correctly relative to the maximum principal stresses.

Weldments should be designed to use the fewest parts and the least expensive weld formats. These decisions are influenced by the edge preparation prior to welding and the necessity for access to the weld region by welding equipment. Other factors include the possible requirement for postweld machining. Joints are generally cheaper to weld when they are horizontal.

Key Concepts

Butt weld. This weld joins the edges of two metal parts.

Fillet weld. A weld joining the edge of a metal part with the face of a second metal part.

Fusion welding. A welding process in which the welded joint is created by the localized heating of the edges of the base metal above the melting temperature.

Heat-affected zone. The region of the base metal, adjacent to the weld zone, where the elevated temperature created by the welding process has caused the microstructure to change from the original structure.

Pressure welding. A welding process in which the application of external pressure is required to create the unification of the two parts.

Residual stresses. Stresses caused by the restrained expansion or contraction of a heated metal part. This can occur during localized heating and cooling in the region of a welded joint.

Thermal stresses. Stresses in welded joints attributed to the heating and uneven cooling.

Weld. A permanent unification of two pieces of metal at their interface by the application of heat, pressure, or both.

Welding process. A sequence of operations for creating a weld.

Weldment. A structure fabricated from two or more parts by one or more welding processes.

Weld zone. That region of a welded joint comprising the heat affected zone and the weld metal zone.

Review Questions

1. What is a weld and how does it differ from a weldment?
2. List some of the disadvantages and advantages of a welded joint.
3. Weldments usually consist of semifinished products. List these products and discuss their advantages and disadvantages in the context of welded structures.
4. How can welding processes be classified? Define the two common classifications.
5. Describe a welding process that is typically used to join sheet metal parts.
6. Describe the TIG process. What is the role of the gaseous stream of argon or helium?
7. How does the MIG process differ from the TIG process?
8. List five types of pressure welding processes.
9. How does the carbon content affect the weldability of ferrous materials? What particular phenomenon occurs with the medium carbon and high carbon steels?
10. Describe the consequences of not protecting the weld region from the atmosphere when welding steel parts.
11. How does a butt weld differ from a fillet weld? Which is stronger?
12. Describe three forms of fillet weld? Which form is the strongest? Why?
13. List five types of fusion welding processes.
14. Sketch three types of welded joint.
15. What are thermal stresses and why are they important in weldments?

16. What is the heat-affected zone and why is it important in a welded joint?
17. Define residual stresses. How can the undesirable consequences of these stresses be minimized in weldments?
18. How does the interface region of two parts prior to welding affect the design of the welded joint?
19. Contrast the design of weldments using rolled sections rather than plates.
20. How do the characteristics of the loading imposed upon a weldment affect its design?
21. Describe methods for reducing the cost of a weldment by changing its design.
22. How does the equipment used to create a welded joint influence the design of that joint?
23. Weldments often interface accurately with other parts to create a final product. This frequently requires the imposition of close tolerances on surface finish and spatial dimensions. How do these constraints influence the design of the weldment?
24. What is the preferred configuration for performing a welding operation?
25. Describe the primary parameters governing the strength of a welded joint.

Problems

1. Weldability is a term used to denote the relative ease of fabricating a welded joint free from defects which is able to perform satisfactorily during service. This term is governed by five primary parameters. List these factors and discuss their interrelationships and affect on both the design and manufacture of weldments.

2. Discuss the different methods of determining the structural integrity of welded joints. Discuss both destructive and nondestructive testing.
3. Propose a variety of T-joints featuring fillet welds. Subsequently analyze and evaluate the relative strength of these joints.
4. Propose a number of alternative designs for a joint connecting a smooth circular pipe to a flange. Analyze and evaluate these joints from a strength perspective.
5. Numerous classes of parts can be either cast or fabricated by welding. The debate typically focuses upon the manufacture of the part as a gray iron casting or as a steel weldment. Discuss this situation contrasting the two alternatives in terms of a variety of design parameters. These could include their relative manufacturing cost, stiffness, part shape, stresses, weight, cost of material, material properties, vibration, and lot size.
6. Two plates of equal thickness are to be joined at their common edge by a weld to produce a joint that does not leak. Propose a weld sequence for this task which minimizes distortion.

Further Reading

ASM Handbook Volume 6: Welding Brazing, and Soldering. Materials Park, Ohio: ASM International, 1994.

Bralla, J. G., ed. *Handbook of Product Design for Manufacturing.* New York: McGraw-Hill, 1986.

Doyle, L. E., C. A. Keyser, J. L. Leach, G. F. Schrader, and M. B. Singer. *Manufacturing Processes and Materials for Engineers.* Englewood Cliffs, NJ: Prentice-Hall, 1985.

El Wakil, S. D. *Processes and Design for Manufacturing.* Englewood Cliffs, NJ: Prentice-Hall, 1989.

Funk, E. R., and L. J. Rieber. *Handbook of Welding*. Boston, MA: Breton Publishers, 1985.

Gray, T. G. F., J. Spence, and T. H. North. *Rational Welding Design*, Butterworths, London 1975.

Hine, C. R. *Machine Tools and Processes for Engineers*. New York: McGraw-Hill, 1971.

Lindberg, R. A. and N. R. Braton. *Welding and other Joining Processes*. Boston: Allyn and Bacon, 1976.

Niebel, B. W., A. B. Draper, and R. A. Wysk. *Modern Manufacturing Process Engineering*. New York: McGraw-Hill, 1989.

Romans, D. and E. N. Simons. *Welding Processes and Technology*. London: Pitman, 1974.

Schey, J. A. *Introduction to Manufacturing Processes*. New York: McGraw-Hill, 1987.

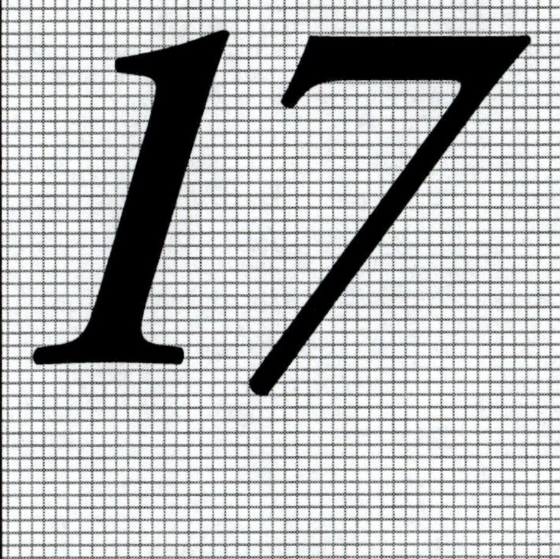

Machined Parts

No, Honey, you can cut that right out.

W. H. Auden
1907–1973

Objectives

- To discuss common methods of machining plane and curved surfaces
- To provide a set of rules for designing parts that can be easily machined

Contents

17.0 Introduction

When a primary forming process such as casting cannot directly create a part with the desired shape, dimensional accuracy, or surface quality, secondary operations must be used to finish the part. *Machining* is an important secondary operation. It involves cutting away relatively small amounts of material with a sharpened wedge-shaped tool of harder material by the processes of turning, milling, planing, drilling, grinding, etc. Machining can be used to manufacture extremely large parts, such as the huge drive shafts of ocean-going ships and it can also be employed to make parts as small as the miniature screws and shafts of mechanical wrist watches.

Typical primary forming processes are:

- Casting
- Rolling
- Forging
- Extruding

Machining is used to manufacture parts from a very small size to a very large size.

Precision parts that are shaped by metal-cutting processes are found in almost all commercial and industrial products because machining is an economical and versatile procedure. Furthermore, machinable materials are not just ferrous and nonferrous metals; composite materials, ceramics and plastics also can be machined. Machining can create precision parts with attributes that are economically unattainable by other methods. While there are indeed specialized primary manufacturing processes such as powder metallurgy or investment casting that can produce high quality parts directly, nevertheless parts that must exhibit precise roundness, flatness, parallelism, or finish, generally are made by machining. Parts formed to a rough shape by sand casting or drop forging may require certain surfaces to be machined to specified geometrical tolerances or surface quality. For example, the engine blocks of most automobile engines are cast and then cylinders, bearing surfaces, and faces are machined to size.

Precision parts often are manufactured by machining a semifinished product such as a casting.

Machinable materials include:

- metals
- plastics
- composites

Many manufacturing processes are available to designers when precision parts must be machined economically. Engineers must consider the sequence in which operations will be performed to ensure that the product design allows machining operations to be done easily. They must be able to visualize the machining of a semifinished part through to

Machining operations:

- tool–workpiece interaction
- fixturing
- inspection

A workpiece is the object that is worked upon by a tool and that ultimately becomes a finished part.

the finished product and must understand the characteristics and constraints of each process. These include tool–workpiece interaction, fixturing, assembly protocols, and inspection procedures—all of which pertain to all batch sizes and most machining processes.

17.1 Common Machining Processes

A machine tool is a powered device that removes material from the workpiece in the form of chips to generate a specified shape and surface finish.

A *lathe* is a machine tool used to produce surfaces of revolution.

Machine tools can generate a variety of surfaces depending upon the shape of the tool used and the path it traverses through the material. Some examples of machined parts are shown in Figure 17.1. If a workpiece is rotated in a lathe and a tool passes through the workpiece in a prescribed path, then a surface of revolution is generated. This surface can assume many different forms. If the tool moves parallel to the rotational axis of the workpiece, then a cylindrical surface is generated (Figure 17.2). This process is called *turning* if an external cylindrical surface is produced and it is called *boring* if the cylindrical surface is internal. If the tool

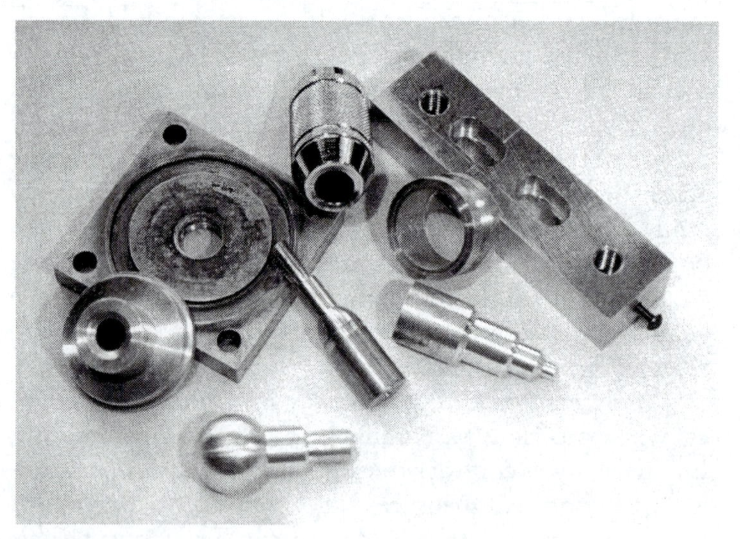

Figure 17.1

Machined parts.

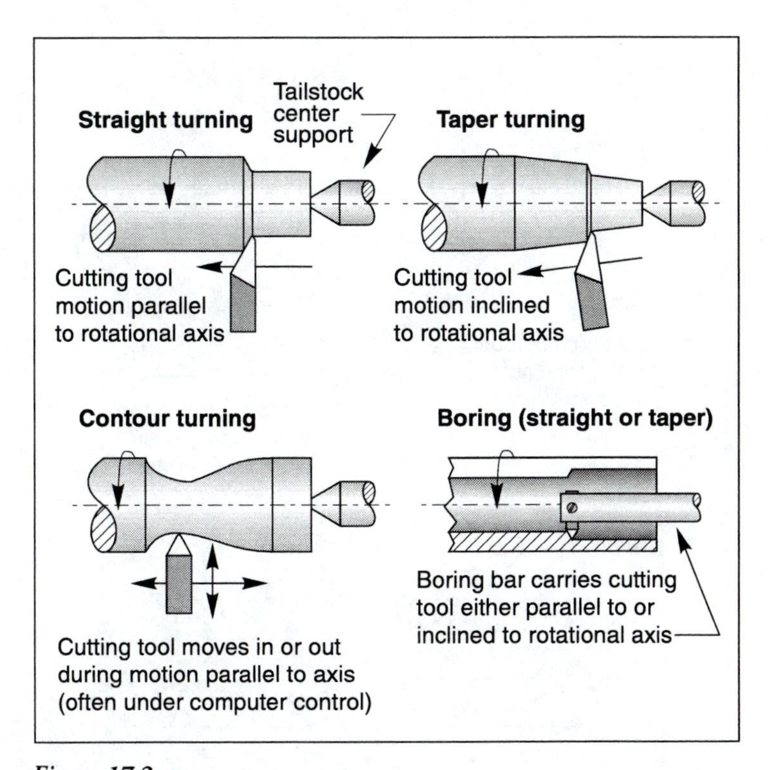

Figure 17.2

Generating surfaces of revolution using a lathe.

moves in a straight-line path that is inclined to the rotational axis of the workpiece, then a conical face is developed; this is *taper turning*. If the path of the tool curves toward or away from the rotational axis of the workpiece, then a curved shape is generated. This is *contour turning*.

A flat surface perpendicular to the longitudinal axis of a workpiece can be generated in a lathe by a facing operation in which the tool remains stationary at a point along the rotational axis but moves toward or away from the rotational center. (Figure 17.3). This is *face-turning*.

Traditional machining processes:

- Turning
- Milling
- Planing
- Drilling
- Grinding

In addition to turning in a lathe, typical machining operations include drilling holes, grinding flat and cylindrical surfaces with abrasive wheels, planing flat surfaces with a planing machine in which there is a relative translational motion between a single-point cutting tool and the work-

A single-point cutting tool is one with only a single cutting edge, such as a lathe tool.

A shaping machine generates flat surfaces by reciprocating a tool relative to the workpiece.

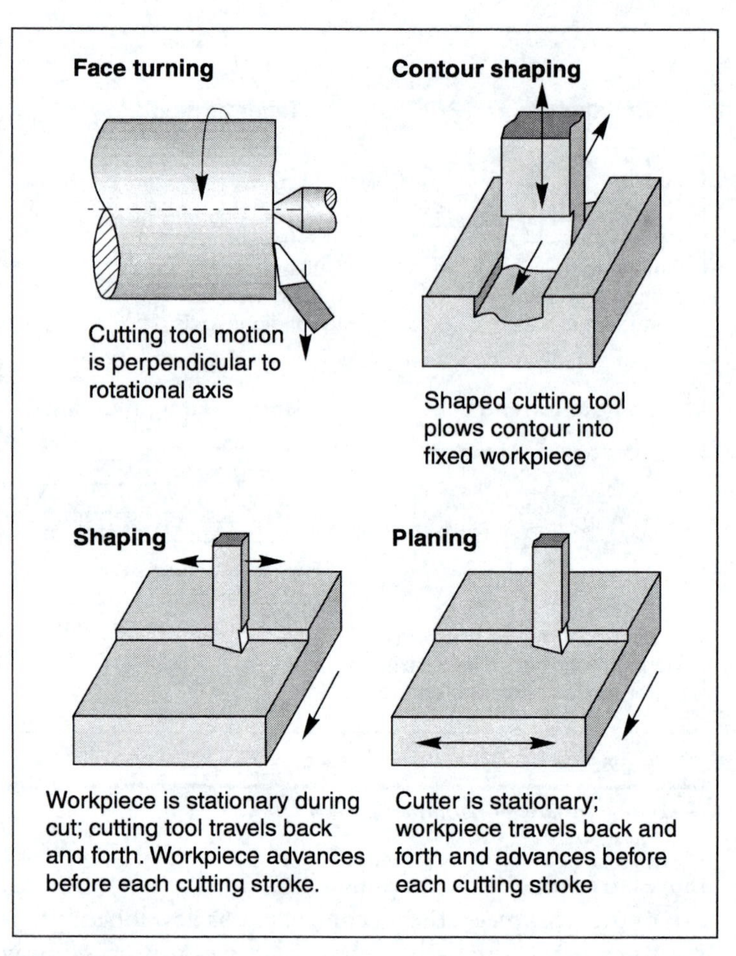

Figure 17.3

Generating planar surfaces using single-point cutting tools.

piece, and the production of various surfaces with shaped rotating cutters using milling machines. Grinding with abrasive wheels can produce surfaces with very high precision. Rotating cutting tools can achieve high rates of metal removal. Some of these processes are illustrated in Figures 17.3 and 17.4. The literature at the end of this chapter should be consulted for more information on machining processes.

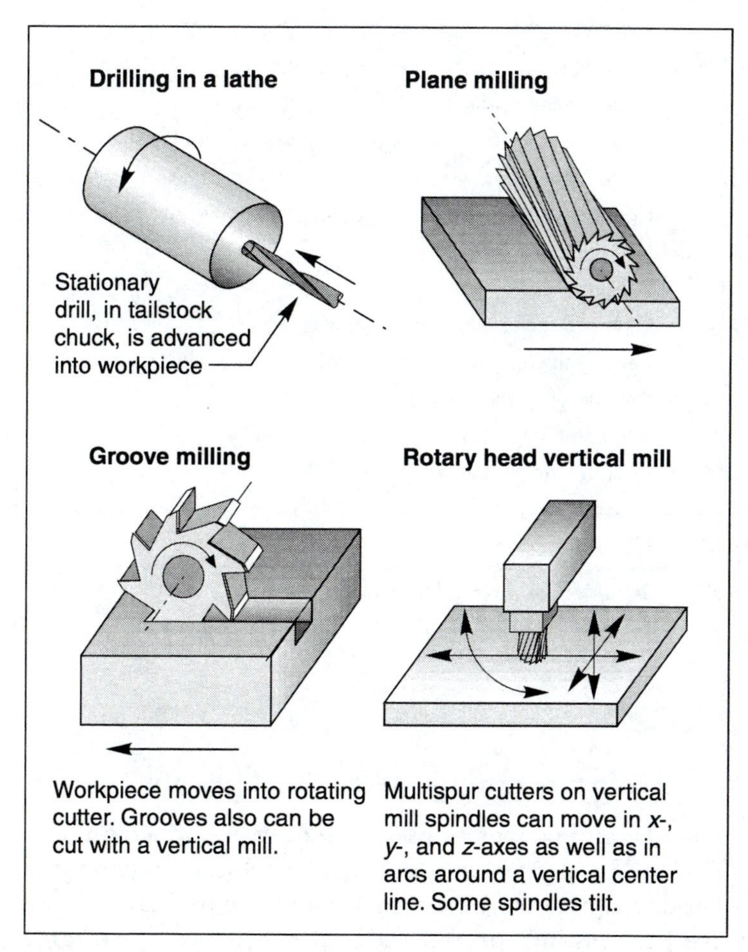

Figure 17.4

Generating planar surfaces using multipoint cutting tools.

Multipoint cutting tools include grinding wheels, milling cutters, and drill bits. They can achieve high metal removal rates.

17.2 DFM of Machined Parts

The more common machine tools are lathes, milling machines, drill presses, and grinding machines. These come in various types, each with its own DFM rules. From these rules, the following general machining principles have been developed (Table 17.1).

Parts are machined to achieve:

- A quality surface finish
- Roundness
- Flatness
- Parallelism

- Avoid machining operations if possible
- Minimize the number of machining operations
- Specify the worst acceptable surface characteristics
- Make fixturing easy
- Hold the workpiece securely during machining
- Use standard tools
- Facilitate tool access
- Minimize reclamping the workpiece
- Keep machining operations simple
- Provide good clamping surfaces
- Accommodate burrs
- Ensure machinability

Table 17.1

Principles of design-for-machinability.

• Avoid machining operations if possible

Costs of machining operations:

- Financial burden of the machine
- Setup time
- Labor
- Tooling
- Quality control
- Fixturing
- Materials handling
- Storage

Machining operations are typically performed on semi-finished products such as standard metal sections or on parts produced by one of the primary forming processes such as forging, casting, or extrusion. If the desired features or surface qualities can be generated by one of these primary forming processes without undertaking secondary machining operations, then generally it will be more economical to do so. Costs associated with machining operations include the financial burden of the machine tool, setup time, labor, quality control checks, tooling, fixturing, materials handling, and storage. These costs often can be avoided by selecting an appropriate forming process or by relaxing the requirements for the shape, tolerances, or surface finish of the part.

• Minimize the number of machining operations

If it is not possible to avoid machining operations completely and the worst acceptable surface has been specified, then the next way to minimize costs is to minimize the number of machining operations and the area to be machined. The area of a part that requires machining should be minimized to reduce the volume of scrap material, the rate of tool replacement, and the time spent machining.

Minimize the number of machining operations and the area to be machined. This reduces machine time, tool wear, and the volume of scrap to be handled.

Machining operations can be reduced in various ways that depend upon the situation. If a shaft with shoulders is to be manufactured, it could be machined from solid bar stock as in option *a* of Figure 17.5. However, there are several alternatives that involve less machining. These include welding or shrink-fitting a collar to the shaft, as in option *b*, or putting a small shoulder on the shaft with a Circlip or a socket screw as in option *c*.

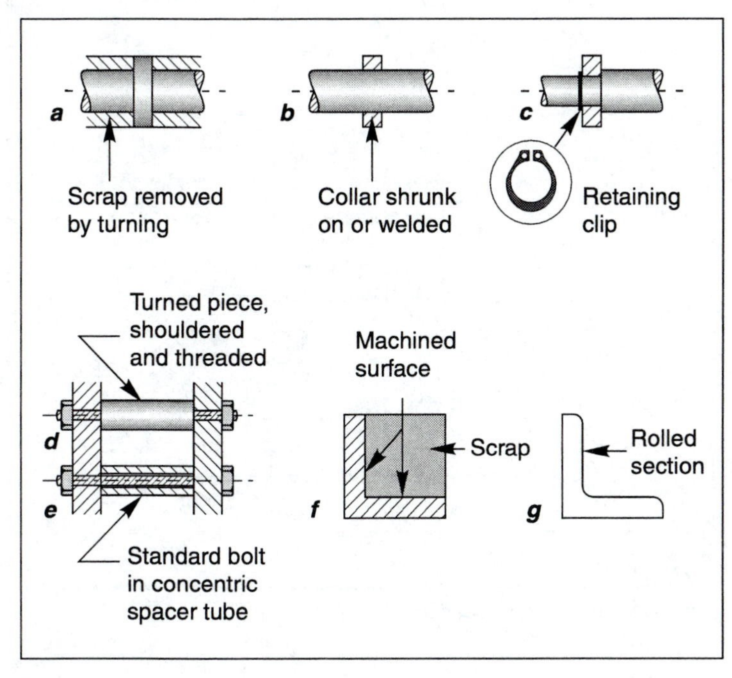

Figure 17.5

Minimize machining operations.

Specify standard off-the-shelf items to minimize machining operations.

Similarly, the turned part separating the two vertical platelike members in option *d* can be replaced by a standard off-the-shelf bolt housed within a tube, as shown in option *e*. The specification of standard stock items is generally advantageous in minimizing the number of surfaces to be machined or the area to be machined. Option *f*, involving the removal of a large volume of material from rectangular stock, can be effectively replaced by a rolled section as shown in option *g*.

Figure 17.6 presents two scenarios involving a large pedestal manufactured as a gray iron casting and a part with a hole bored through it. Options *a* and *c* are unsatisfactory compared to option *b* with machined pads at the perimeter of the base for seating the pedestal (see Figure 14.17) and option *d* has a recess near the middle of the bore that is not machined.

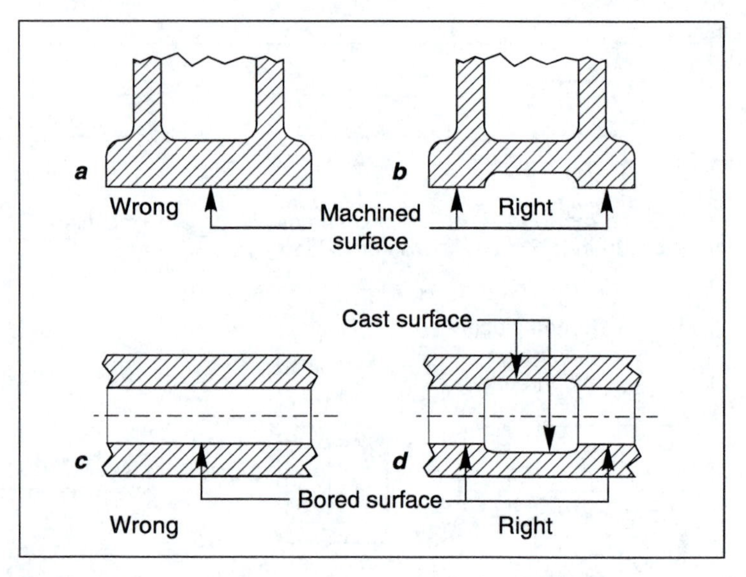

Figure 17.6

Reduce the size of machined areas.

Fixtures are tools that facilitate the accurate manufacture of duplicate parts by locating, holding, and supporting the workpiece securely while providing a reference for the cutting tool.

Fixturing of a workpiece is enhanced by large, rigid, parallel-sided external and internal surfaces.

• Specify the worst acceptable surfaces and tolerances

Machining operations can sometimes be avoided by specifying the largest acceptable tolerances and the roughest acceptable surface finish commensurate with the function of the part. Under these conditions it may be possible to use a forged or cast part without further expensive operations. This approach is illustrated in Figure 17.7. Alternatively, a milling operation might be used to finish a part rather than a more costly process such as lapping or reaming.

Always specify the roughest acceptable surface finish and the largest acceptable tolerance.

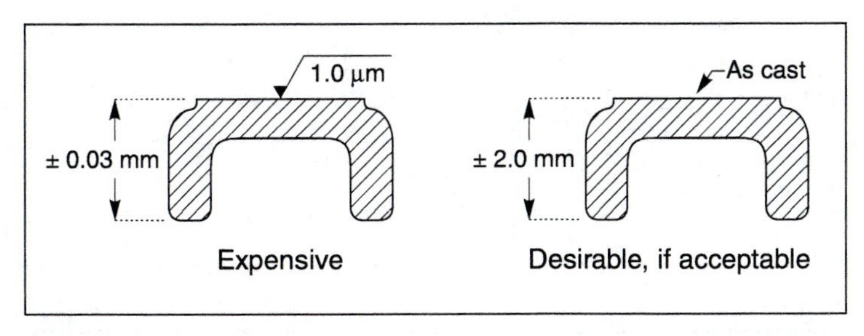

Figure 17.7

Specify the worst acceptable surfaces and tolerances.

• Make fixturing easy

Where possible, parts should be designed to eliminate holding fixtures. Otherwise, the cost of fixturing should be minimized by facilitating workpiece location, setting up, and holding. Parts to be fixtured for machining must have a reference surface that permits the workpiece to be securely clamped. This permits the imposition of the large forces associated with metal-cutting operations and ensures the manufacture of an accurate part. Generally this requires the part to have relatively large, rigid, parallel-sided external or internal surfaces to interface with a fixturing device. These surfaces can be such things as lugs, bosses, registration faces, or cylinders that will fit into lathe chucks. Examples of both good and bad product designs are presented in Figure 17.8.

Setup time is the time taken to configure and adjust a machine tool to perform a prescribed set of operations.

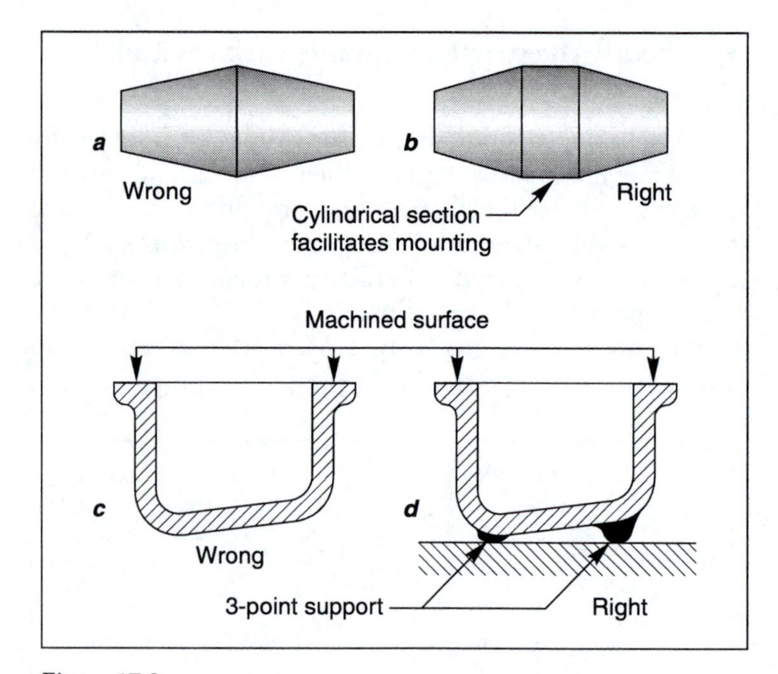

Figure 17.8

Design for clampability.

Option *a*, which is a part with a surface of revolution, cannot be machined readily on the tapered face because, without a cylindrical section, the piece is difficult to hold in a lathe chuck. Option *b*, with a small cylindrical section in the middle, is preferable. Options *c* and *d* illustrate how clamping the part to machine its top face can be accomplished by incorporating three small feet in the base of the vessel to provide stability and ensure that its upper face and the supporting plane are parallel. A three-point contact is preferable to clamping a workpiece on a large flat surface because it is cheaper to generate. If the flat surface is not accurately machined, then the workpiece can be unstable.

If the desired tolerances and surface finish are to be achieved, then the workpiece must be rigid.

Workpiece deflection is governed by:
- Shape
- Material
- Surface being machined
- Metal removal rate

• Hold the workpiece securely during machining

Machining imposes large forces upon workpieces, and high clamping forces are necessary to secure parts during metal-cutting operations. If the desired tolerances are to be achieved then parts must not move during machining operations. The shape of a part, the material from which it is manufactured, the types of surfaces being machined, and the metal removal rates govern this task. Workpieces that cause problems are those that require the drilling or machining of deep holes or the machining of thin walls.

If long, slender workpieces must be machined in a lathe; they must be supported. That costs time and money.

Such problems are illustrated in Figure 17.9. The end milling operation in option *b* is much more desirable than the machining shown in option *a* because of the thin and flexible wall of the part. Similarly the long hole with a small bore in option *c* will be less accurate than the larger hole in option *d* because the boring bar used to make the larger hole is stiffer.

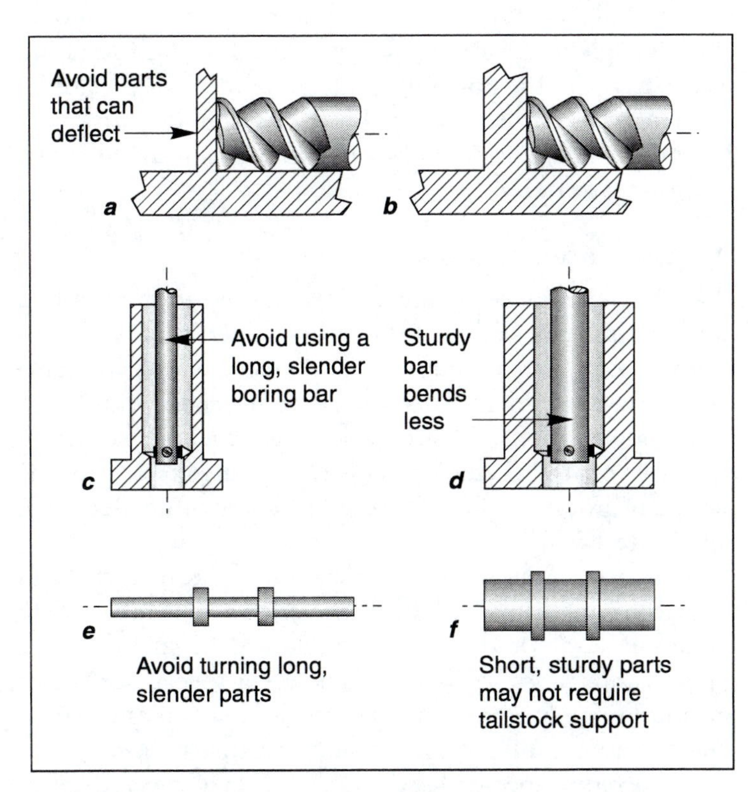

Figure 17.9

Design so that tools don't bend the workpiece.

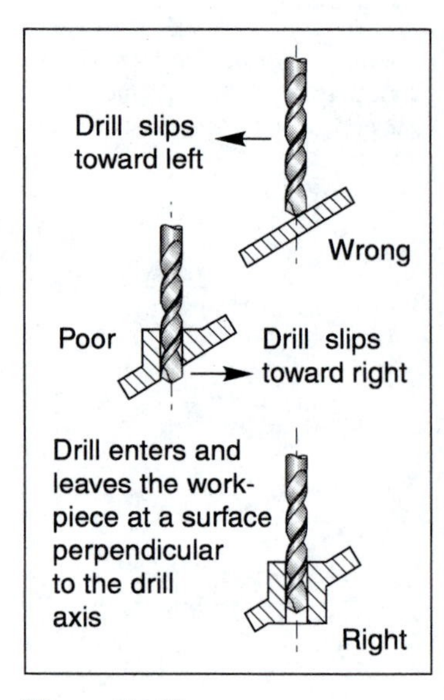

Figure 17.10

Placement of holes made with twist drills.

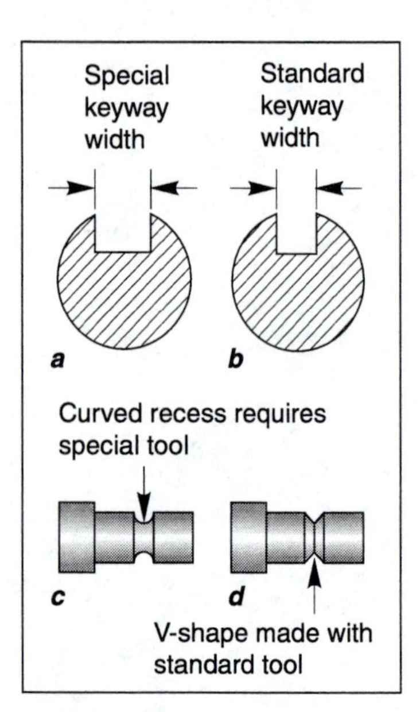

Figure 17.11

Use standard tooling.

Tolerances can be difficult to attain in shaping a flexible workpiece by metal cutting operations. For example, if workpieces must be turned in a lathe, unsupported slender parts should be avoided because they deflect under normal loading. Furthermore they typically require support from the lathe tail stock which involves extra setup time. Parts should be kept relatively short with a large diameter-to-length ratio. Options *e* and *f* in Figure 17.9 illustrate this situation. Option *e* is undesirable; option *f* is preferable.

Twist drills have two cutting edges that must have the same cutting conditions for the drill to function correctly. If this is not the case, the hole may not be circular and the part may then have to be scrapped. This can happen especially when holes are specified too close to the edge of a part or when a hole must be drilled into a sloping face that is not at a right angle to the rotational axis of the drill. This situation is illustrated in the upper drawing in Figure 17.10. A similar situation arises when the drill bit erupts from the workpiece as shown in the middle drawing. The design in the lower drawing is better.

• Use standard tools

Parts should be designed to use conventional tooling and equipment. Failure to do so will increase the cost of parts. Designers should know the characteristics of available tools and equipment to correctly size dimensions for machining operations. Wherever possible, machine parts with standard off-the-shelf cutters instead of cutters that must be specially ground to size. These considerations are illustrated in Figure 17.11.

Situation *b* shows the specification of a keyway that can be created with a standard diameter end mill rather than with a nonstandard tool in situation *a*. In situation *c*, a special tool must be used to generate a special surface of revolution on one part while a conventional tool can be used to generate a similar shape in the alternative part *d*.

The radius specified at the intersection of two surfaces at right angles—such as at the interface between a shoulder on a shaft and the cylindrical portion—should permit the most appropriate cutting tool radius to be used to shape the part. The only time this guideline should not be followed is when special radii are specified because of stress concentrations. This consideration applies to the edges of grinding wheels and milling cutters as well as to the radius of the single-point tool of a shaping machine.

Flanges, lugs, bushes, and other major shape irregularities can inhibit tool access to surfaces that must be machined.

- **Facilitate tool access**

Cutting tools should interact easily with the workpiece. Avoid awkward locations for fixturing elements, bushes, flanges, and other elements of a part that must be machined. Furthermore, most cutting tools require run-out (tool travel beyond the machined area) in order to generate a uniform surface finish. This has some consequences for determining an appropriate shape for a part. Additional clearance at the end of the cut can generally be created during primary part formation, and this surface usually does not require machining. Figure 17.12 presents some design-for-machinability scenarios. Options *a* and *c* require holes to be drilled in the walls of castings, but access for the drill bit is limited; options *b* and *d* are recommended.

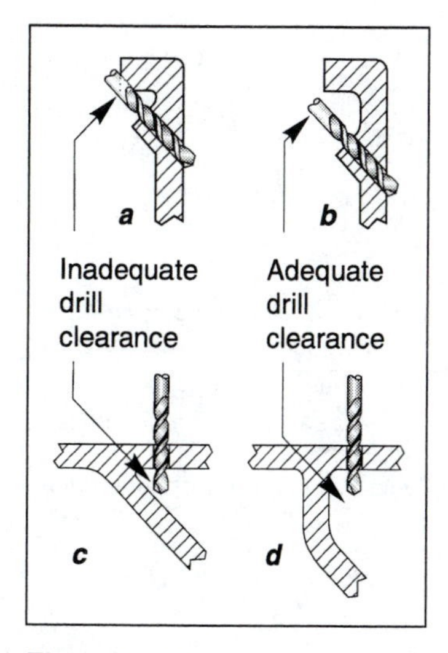

Figure 17.12

Design for drill access.

The designs presented as options *a, c, e, g, i,* and *k* in Figure 17.13 limit tool access in the generation of both internal and external flat surfaces by a variety of machining operations. Options *b, d, f, h, j,* and *l* are recommended.

- **Minimize reclamping the workpiece**

Many parts require several machining operations to generate the final shape, dimensional tolerances, and surface finishes. Resetting tools, reclamping the workpiece, and rechucking must be minimized to save time and money. The cost of machining can be minimized if all surfaces to be machined are in the same plane. If this is impossible, then the part should be designed so that all the faces can be machined with the same set up. In some situations it is possible to machine several surfaces with one pass of a cutter, such as in a straddle milling operation.

Reduce machining costs by ensuring that all surfaces to be machined:

- are on the same plane
- can be shaped with the same setup
- can be in one pass of the cutter

Provide room for tool run-out to obtain a uniform surface finish.

To reduce machining costs, minimize:

- Resetting of tools
- Reclamping of the workpiece

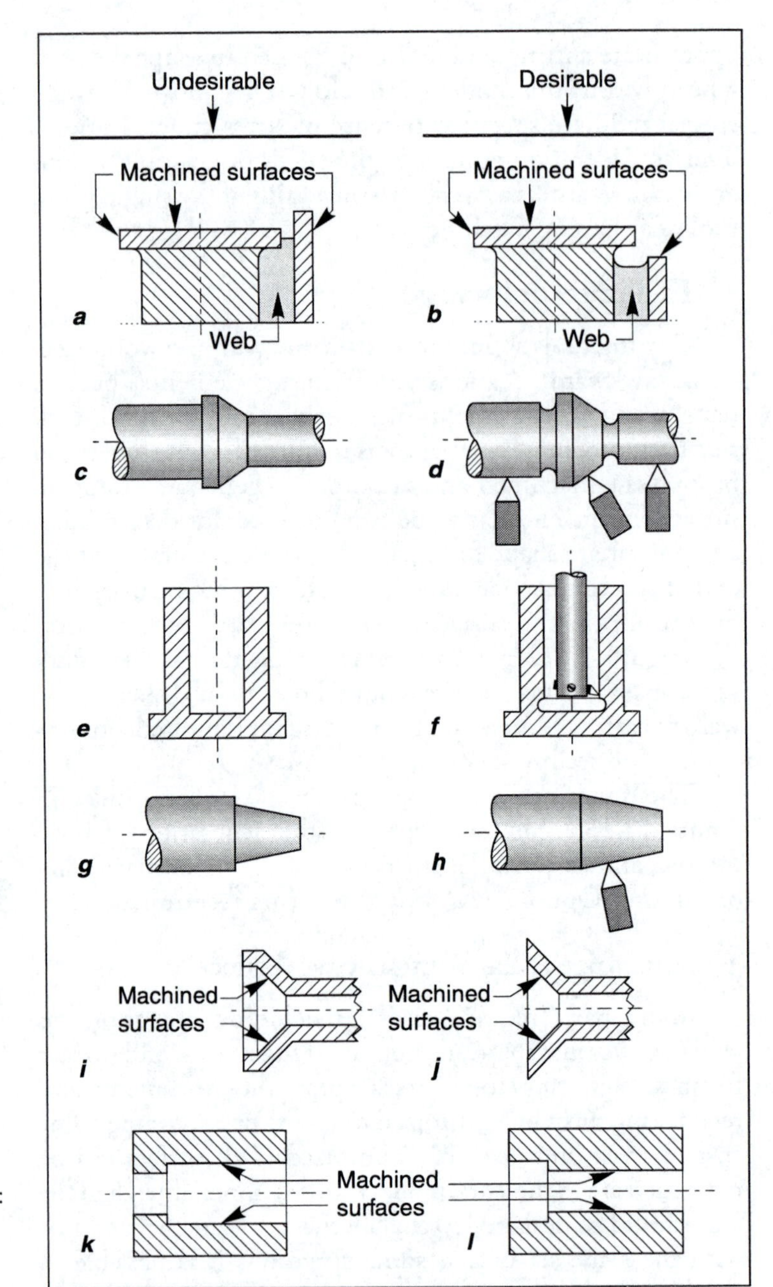

Figure 17.13

Design for easy machining operations.

Figure 17.14 illustrates both good and bad product design. Option *a* contains a tubelike member with several machined internal faces. This expensive design can be replaced by an alternative, option *b*, that includes a tube whose external surface is machined. Option *c* suffers from the necessity to machine the sloping face of a lug on the casting in addition to a horizontal face on the boss on the upper section. This time-consuming activity is avoided in option *d*. Option *e* requires two flat faces to be machined at different heights, while option *f* presents a less expensive alternative.

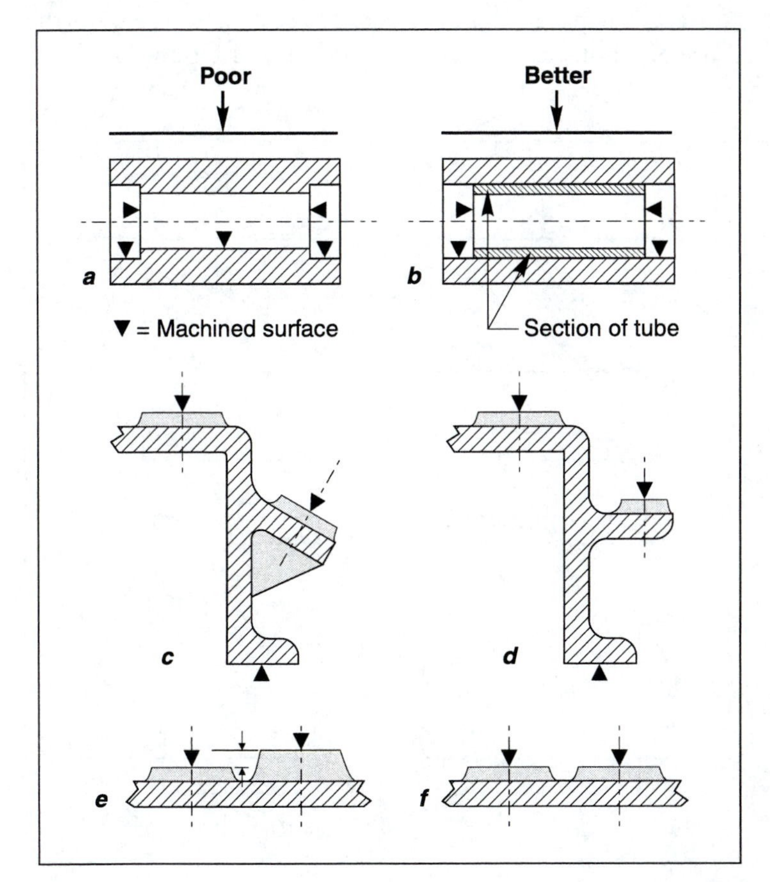

Figure 17.14

Design for minimal setup costs.

- ## Use simple machining operations

Design parts to use the simplest and least expensive machining operations.

Machining operations can be simple or more complex and hence more expensive. Therefore parts should be designed to utilize the simplest possible machining operations. Avoid undercutting operations that require extra machining with specially ground tools. Figure 17.15 illustrates this aspect of design for machinability. The undercutting in options *a* and *c* should be avoided. Options *b* and *d* are preferable. If parts of complex geometry must be assembled, then it is generally much easier to machine features on the outer surface of a piece of circular stock in a lathe where the surface is exposed than to machine similar features on the bore of a hole, as illustrated in option *f* of Figure 17.15.

It is easier and cheaper to machine complex features on exterior surfaces.

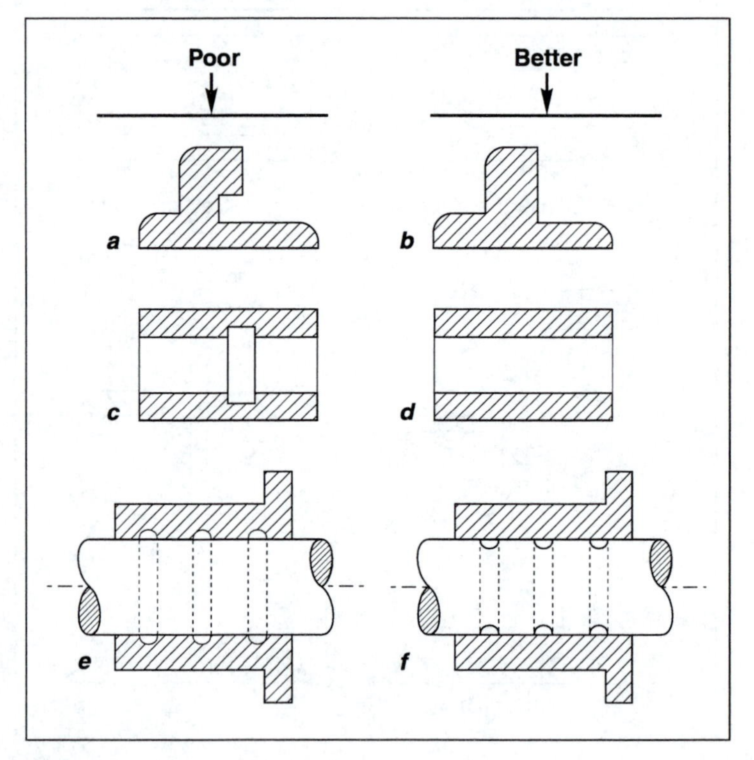

Figure 17.15

Design for easy machining operations.

• Provide good quality clamping surfaces

Machining requires the workpiece to be clamped while material is being cut. In order to facilitate clamping, the surface interfacing with the machine tool should be smooth. Therefore the part should be designed so that relatively large surfaces are available for locating or clamping the workpiece. These surfaces should not have the flash asperities created by drop forging, the draft taper associated with casting processes, or parting line irregularities found on castings. Figure 17.16 illustrates this situation. Option *b*, which shows an additional cylindrical face, is preferable to option *a* which must be clamped in the lathe chuck by the large-diameter surface that contains a flash or parting line.

Provide smooth clamping surface. Avoid clamping on forging flash, casting draft, or on casting parting lines.

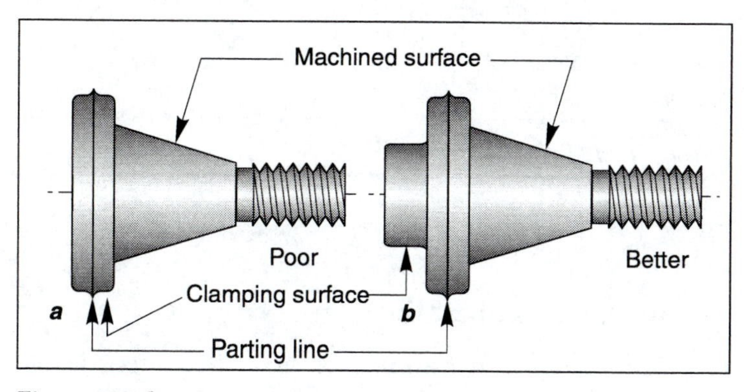

Figure 17.16

Avoid clamping surfaces that contain flash or parting lines.

• Accommodate burrs

Burrs form on the edges of all parts subjected to a metal-cutting process, especially when the faces intersect at right angles. They are an inherent characteristic of this manufacturing process. Burrs are small, flexible, sharp, plastically deformed protrusions from the edge of a workpiece and they should be removed in order to prevent human injury and interference with assembly operations. Parts can be deburred by a variety of methods but these secondary operations should be avoided whenever possible since they can be

A burr is a sharp ribbon or ridge formed at the intersection of two orthogonal surfaces after one surface has been machined.

Remove burrs by:

- Providing space to accommodate the burr
- Ensuring that faces do not intersect at right angles

relatively expensive. Deburring can be accomplished by designing the part to provide space to accommodate the burrs or, alternatively, they can be avoided altogether by chamfering faces that intersect at an acute angle. Figure 17.17 presents two parts with surfaces of revolution that must be machined. Option *a* shows a collar with edges at right angles to the longitudinal axis of the part. This edge will develop burrs after machining the outer diameter of the shoulder. Option *b* alleviates this situation with a curved surface between the two diameters of the part so that an acute angle is developed at the intersection of the two faces.

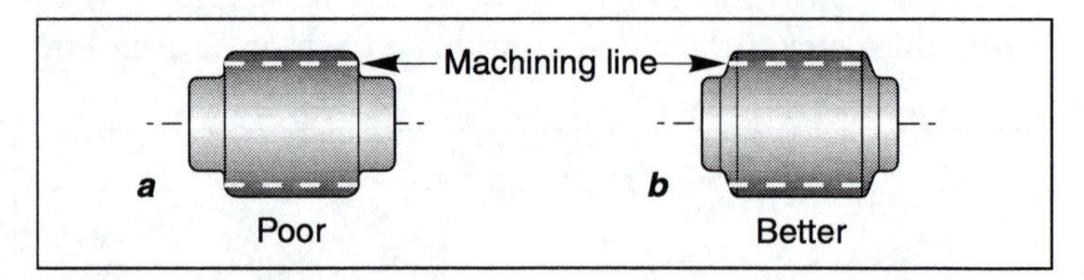

Figure 17.17

Avoid deburring operations.

• Ensure machinability

Machinability is the relative ease with which a material can be machined.

Machinability is assumed to be a characteristic of a material, but there are several different measures of it. Machinability is often defined as the relative ease with which a material can be cut, but that depends upon what process is used for the cutting. A material is said to be highly machinable when parts of acceptable quality can be manufactured easily with a specified machining process.

Machinability is generally characterized by the following criteria:

- The surface finish produced at standard feeds and cutting speeds.
- The wear of the cutting tool, which manifests itself as a change in dimension of machined parts as quantities are produced.

- The life of the cutting tool, which manifests itself as the chipping or cracking of the cutting edge under specified cutting conditions.
- The maximum rate of metal removal through chip production.

Materials with a high machinability may manifest this characteristic by requiring lower cutting forces and lower power consumption to generate a specific surface and shape.

High machinability:
- lower cutting forces
- lower power consumption
- high surface quality

Similarly this characteristic may manifest itself by a series of short chips from the metal cutting process rather than long string-like cuttings from a material with poor machinability. These latter chips tend to clog the machine tool and can be a hazard for machine tool operators. Alternatively, some materials are inherently more difficult to finish to a desired surface quality than others because of microstructural properties imparted by a prior manufacturing process. All of these attributes must be considered in the material selection process.

Summary

Interchangeable precision parts for a broad range of commercial and industrial products are manufactured by machining operations. These operations can generate both surfaces of revolution and flat surfaces using various machine tools such as lathes, milling machines, drill presses, shapers, and grinding machines. These machine tools can machine parts in materials ranging from the ferrous and nonferrous metals to ceramics and reinforced plastics. Each machine tool has advantages and disadvantages that influence the design of the product; consequently to achieve desired tolerances and surface finishes, a set of design-for-manufacturability rules should be employed.

- Avoid machining operations
- Minimize the number of machining operations
- Specify the worst acceptable surface characteristics
- Make fixturing easy
- Ensure that the workpiece is held securely
- Use standard tools
- Facilitate tool access
- Minimize reclamping the workpiece
- Keep machining operations simple
- Provide good clamping surfaces
- Accommodate burrs
- Ensure machinability

Key concepts

Burr. The rough sharp ribbon or ridge of metal formed at the intersection of two nearly orthogonal surfaces, one of which has been machined.

Boring. The metal cutting process of enlarging a hole that has already been drilled. The process is generally performed with a single-point cutting tool on a lathe or specialized boring machine. The accuracy of the cut depends upon the rigidity of the boring bar spindle and the cutting tool rather than the guiding effect of the drilled hole. A bored hole is typically of good accuracy and surface finish.

Cutting tool. A piece of wedge-shaped material used to cut the workpiece. The cutting tool is harder than the workpiece and is sharpened to a shape that makes cutting easy. Tools are usually made from high speed steels or cemented carbides.

Drilling. Use of a twist drill to produce a round hole or to enlarge a hole. Drilling can be done with various kinds of drill presses or in a lathe with the drill bit held stationary in the tailstock.

Fixturing. A manufacturing procedure involving a structure, called a fixture, to locate, hold, and support a workpiece during machining and to provide a reference surface or device for the cutting tool. Fixturing permits duplicate parts to be manufactured accurately.

Grinding. A process used to produce high quality surfaces on precision parts with a rotating abrasive wheel. Each abrasive grain acts as a cutting tool. It can be used on both conventional materials and also on very hard materials. Grinding can generate flat surfaces, internal and external cylindrical surfaces, and surfaces of revolution.

Lapping. A microfinishing process in which a master form, called a *lap*, and a fine abrasive are moved continuously over the workpiece surface at low velocity and with a light pressure. The resulting surface finish and dimensional accuracy are superior to those attainable by grinding.

Lathe. A machine tool used to produce surfaces of revolution and flat edges. The workpiece is rotated about an axis while a cutting tool moves along the workpiece.

Machinability. The relative ease with which a material can be machined. A material with high machinability permits parts to be machined easily in an acceptable quality at low cost.

Machine tool. A powered device used to produce a part of specified size, shape, and surface finish by the removal of material in the form of chips. This is accomplished by rigidly supporting the tool and the workpiece, by providing relative motion between the tool and the workpiece, and by providing a range of cutting speeds.

Milling. A manufacturing process in which a workpiece is fed against a revolving multipoint milling cutter to generate various kinds of flat or curved surfaces. Alternatively the rotating cutter can be fed into the workpiece. There are many kinds of milling operations performed using a variety of milling machines.

Multipoint cutting tool. Cutting tools with at least two cutting edges that sequentially interact with the workpiece. These tools are capable of high rates of material removal. A face milling cutter and the common twist drill are examples of multipoint cutting tools.

Planing machine. A machine tool used to shave a surface flat with a single-point cutting tool. In a planing machine, the tool and the workpiece move relative to each other in a translational motion. The workpiece is clamped to a long horizontal reciprocating machine table while the tool is attached to a rigid stationary structure.

Primary forming processes. Manufacturing processes in which the cast structure of a metal is destroyed by successive deformation processes. The resulting product is said to be semifinished because it must be processed further to create a part with the desired shape, dimensional tolerances, and surface qualities. Typical primary forming processes are extrusion, forging, rolling, and drawing.

Reaming. This metal-cutting operation is a sizing process in which a drilled hole is slightly enlarged to the desired size and surface finish with a long cylindrical fluted tool called a *reamer.*

Secondary forming processes. Manufacturing processes in which semifinished products of primary forming processes are transformed into finished products with more accurate dimensions and higher surface quality. Typical secondary processes are the sheet-metal working processes, machining, and finishing operations.

Setup time. The time taken to configure and adjust a machine tool to perform a prescribed set of operations. It typically involves installation of cutting tools, selection of spindle speeds and tool feed rates, and the adjustment of stops to limit travel of parts of the machine tool such as the cross-slide of a lathe.

Shaping machine. A machine tool used to create flat surfaces on horizontal, vertical, and angled planes. The basis of the process is the relative motion between a single-point cutting tool and a stationary workpiece. A chip forms during the forward stoke of the tool, and as the tool returns, the machine table on which the workpiece is clamped moves laterally a preset increment prior to the next cutting stoke.

Single-point cutting tool. A tool with only one cutting edge, such as a lathe tool.

Surface finish. A parameter associated with the quality of a part and measured in terms of the microscopic character of the surface such as waviness or roughness. Typical measures include the centerline average, the height of the waviness, the roughness width and the waviness width. Each manufacturing process can be characterized by a range of surface finishes.

Tolerances. Limits imposed upon dimensions to ensure that a part will fulfill its specification. Tolerances are necessary because no manufacturing process can routinely manufacture a part to exact dimensions. Tolerances should be set as large as possible within the function of the part because close tolerances are synonymous with higher costs. Tolerances are typically imposed upon the location of geometric features and upon geometric properties such as perpendicularity, parallelism, concentricity, flatness, and straightness.

Turning. A machining process by which cylindrical or irregularly shaped internal and external surfaces of revolution are generated on a rotating workpiece by one or more single-point cutting tools. Turning is done on a *lathe*.

Twist drill. A tool, sometimes called a drill bit, with two cutting edges and two helical flutes that continue along the length of the drill. It is used to produce through holes or blind holes in a workpiece. This is generally accomplished by rotating the drill about its axis and forcing the tool against the workpiece with a drill press. Other approaches involve rotation of the workpiece relative to a stationary drill bit when the drill is held in the tailstock of a lathe and the workpiece is held in the chuck.

Workpiece. The object, part, or work that interacts with a tool and subsequently becomes a finished or semifinished part.

Review Questions

1. In manufacturing, how is the final shape of most parts generated?
2. List five traditional metal cutting processes.
3. List five classes of materials that can be shaped by machining operations.
4. What are typical features of parts shaped by machining processes?
5. Describe the basic features of a metal removal process by metal cutting.
6. List the primary financial considerations in metal cutting operations.
7. Define a lathe and describe the shapes that this machine tool can generate.
8. Define a shaping machine and describe the types of surfaces that it can produce.
9. List some examples and advantages of common multipoint cutting tools.
10. List five common types of machine tools.
11. Why and how can machining operations be avoided?
12. What are the financial burdens associated with machining operations?
13. What are the advantages of specifying the roughest acceptable surface quality and the largest tolerances?
14. Describe some examples of the minimization of the area to be machined and of the minimization of the number of machining operations.
15. What are the economic advantages of minimizing the total area to be machined?

16. What are fixtures? What role do they play in manufacturing? Provide examples of surfaces that interface between a part and a fixture.

17. Why should a workpiece be rigidly supported during metal cutting operations? What factors govern this situation?

18. Discuss the drilling of a circular hole in a sloping surface. What are the concerns with this situation and how can they be avoided?

19. Discuss the advantages of using standard tools and equipment.

20. Develop a series of sketches to illustrate how part geometry hinders a cutting tool from reaching a surface of a workpiece. Illustrate how a part can be redesigned to avoid this problem.

21. What is tool run-out and why is it important in the design-for-manufacturability of machined parts? Use illustrative examples and sketches.

22. Most parts require several machining operations to generate the desired dimensions and surface finishes. How can the cost of these tasks be reduced?

23. Discuss the primary rules of selecting simple and hence cheaper machining operations. Provide examples and sketches to illustrate these principals.

24. How can a designer ensure that good clamping surfaces are provided for machining operations? Why are these surfaces necessary?

25. What are burrs and why should they be avoided in the manufacture of parts? Describe two rules that permit this to be accomplished. Provide examples and sketches of redesigns.

26. Material selection influences the cost and quality of a part. Discuss this process in the context of the four primary ingredients of the selection process.

27. What is machinability? What criteria are typically used to characterize this term? If a material has a low machinability, what does this mean? Illustrate your answer.
28. Tolerances and surface finishes are attributes of detailed engineering drawings of a part. Define these terms and describe their role in the manufacture of a part.
29. How does a machine tool differ from a cutting tool?
30. Define the term *workpiece*.

Problems

1. The figure below presents three cases of manufacturing parts by machining. Each figure shows an instance of poor practice. Identify the poor practice, briefly describe it, and propose alternative designs that incorporate design-for-manufacturability rules. Design *a* shows a threaded portion of a shaft formed with a single-point cutting tool on a lathe. Design *b* shows a single-point cutting tool of a shaping machine forming a flat external surface. Design *c* shows a single-point cutting tool of a shaping machine generating a keyway on an internal surface.

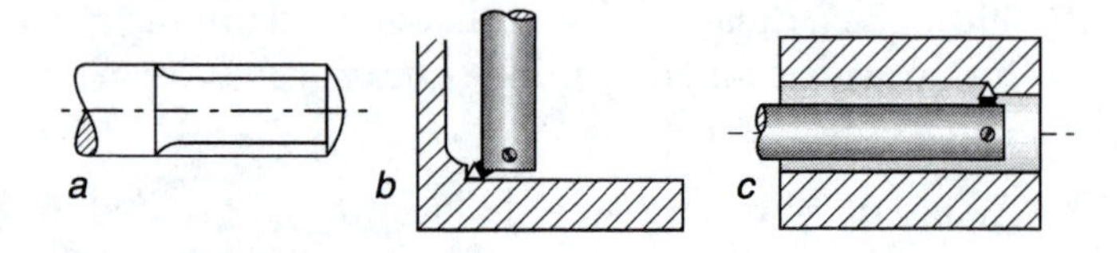

2. Some poorly designed assemblies are shown in the following figure. They contain redundant fits—namely fits that are duplicated in the same assembly. Identify the faults with each design, document these faults, and propose improved designs.

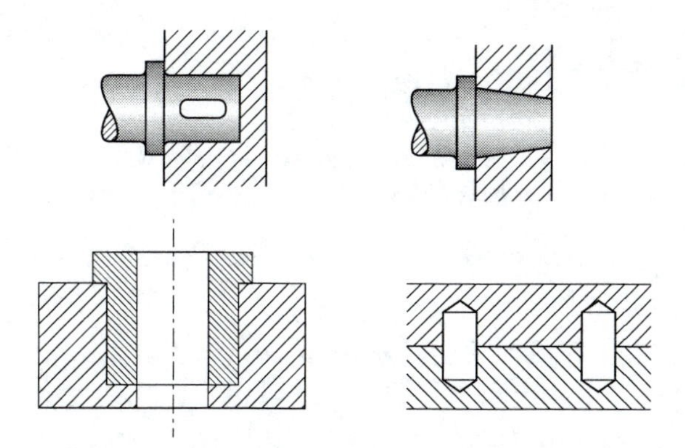

3. The figure at the right presents two assemblies containing two separate types of errors. Identify these errors and redesign the assemblies to provide superior designs.

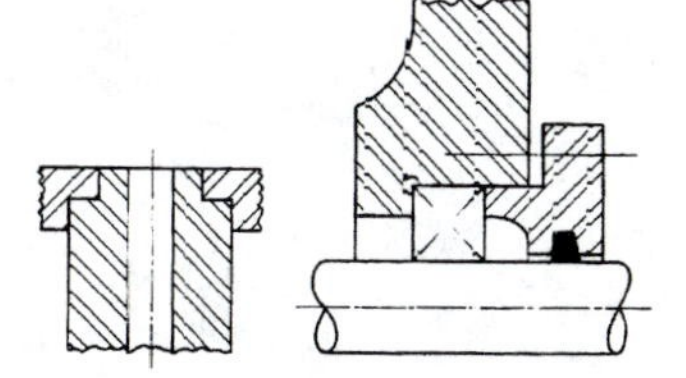

4. Several product designs that have not considered the manufacture of the part are illustrated in the figure below. All involve milling operations. Identify the erroneous aspect of each detail design feature and propose superior designs.

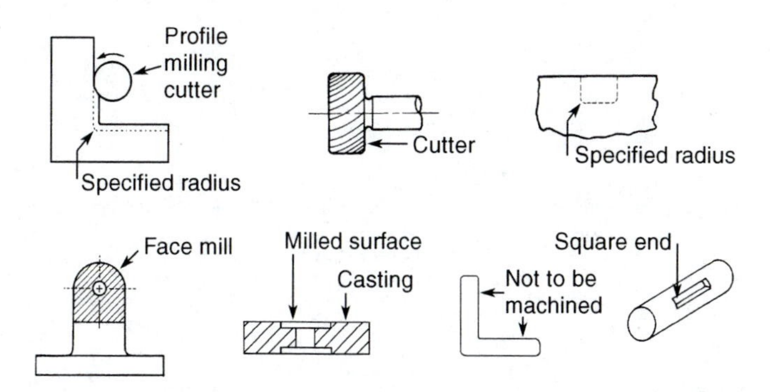

Further Reading

Bralla, J. G., ed. *Handbook of Product Design for Manufacturing*. New York: McGraw-Hill, 1986.

Dieter, G., *Engineering Design*. New York: McGraw-Hill, 1983.

El Wakil, S. D. *Processes and Design For Manufacturing*. Englewood Cliffs, NJ: Prentice-Hall, 1989.

Farag, M. M. *Selection of Materials and Manufacturing Processes for Engineering Design*. Englewood Cliffs, NJ: Prentice Hall, 1989.

Niebel, B. W., A. B. Draper, and R. A. Wysk. *Modern Manufacturing Process Engineering*. New York: McGraw-Hill, 1989.

Matousek, R. *Engineering Design*. London: Blackie & Son, 1963.

Schey, J. A. *Introduction to Manufacturing Processes*. New York: McGraw-Hill, 1987.

Yankee, H. W. *Manufacturing Processes*. Englewood Cliffs, NJ: Prentice-Hall, 1979.

Polymeric Composite Parts

18

So the same day Pharaoh commanded the taskmasters over the people and their foremen, saying, "you are no longer to give the people straw to make bricks as previously..."

The Bible, Exodus 5, Verses 6, 7
450 B.C.

Objectives

- To introduce polymeric fibrous composite materials
- To classify composite materials
- To identify parameters for designing fibrous polymeric composite parts
- To discuss advantages of polymeric composite materials
- To discuss disadvantages of polymeric composite materials
- To present a summary of the common methods of manufacturing polymeric composites
- To identify some design advantages and disadvantages of methods for manufacturing polymeric composites
- To introduce a method for designing and manufacturing polymeric fibrous composite parts

Contents

18.0 Introduction

Fibrous composites are nontraditional materials that can be created with properties that satisfy a particular need. These properties can be tailored to meet complex material specifications. The freedom to design the macromechanical characteristics of the material from which a part is to be made does not exist with traditional monolithic materials such as low carbon steels or aluminum alloys because their macromechanical properties are inherent characteristics of these homogeneous material. While the advantages of using heterogeneous materials has been known for centuries, it is only during the latter half of the twentieth century that the technology has matured enough for an engineer to design a broad range of unusual materials by selecting the appropriate form and distribution of ingredients in the macrostructure.

The properties of composite materials can be adjusted to suit the situation.

The properties of the commercial metals cannot be changed significantly.

One of the differences between the design and manufacture of parts in monolithic materials and the design and manufacture of composite parts is that these two activities are inextricably intertwined with the latter class of materials. Until recently, this has not been the case when products are created in commercial metals because design and manufacturing processes have typically been two quite independent activities. The part is first designed, and almost as an afterthought the manufacturing process is then selected.

The design of fibrous composite parts is intertwined with the manufacture of the material to be used because the design team is responsible not only for creating a part of specified shape, tolerances, and surface characteristics but also for devising the macrostructure of a material with properties that may be unique to the part. The manufacturing process creates not only a part but also its structural material. Properties of composite materials depend upon the placement, orientation, and distribution of fibers throughout the binder or *matrix material*—characteristics that result from the manufacturing process. Thus the integration of both design and manufacturing is crucial in the development of composite parts.

Processes used to manufacture composite parts are responsible for creating both the attributes of the part and the attributes of the material.

18.1 Polymeric Composite Materials

Artificially created fiber-reinforced composites have been recognized as a superior class of materials for over 2000 years. In 450 B.C. the Israelites were compelled to manufacture bricks for Pharaoh in Egypt using a combination of clay and straw to create a composite building material. This is perhaps the first recording of such an activity in the embryonic field of materials science. However, the field lay largely dormant until the latter half of the twentieth century, with the exception of the addition of animal hair—initially to clay and later to plaster for walls of dwellings.

All biological materials are composites.

As the field of biomimetics began to mature, especially during the past fifty years or so, people began to develop a more insightful appreciation for the innovative features and superior structural properties of biological materials. All of these are composite materials. The emulation of these naturally occurring composite materials by the scientific and engineering communities was stimulated by studies of the chitin fibers and proteinaceous matrix of insect cuticles, the flexible cellulose fibers in a rigid lignin matrix of trees, and the characteristics of other flora and fauna. The quest for materials with superior properties was motivated by the inherent limitations of commercial metals and the insatiable demand for superior weapons systems and commercial products in the international marketplace.

Composites are a combination of two or more constituents differing in form or composition on a macroscopic level.

Macroscopic characteristics are revealed by visual examination without magnification.

Composite materials are a combination of two or more materials differing in form or composition on a macroscopic scale. Generally the constituents of the material can be physically identified and they exhibit an interface between them. They do not merge or dissolve and the result is a heterogeneous structure. This contrasts with most alloys in which two or more metals combine homogeneously. One of the cornerstones of designing with composite materials is the philosophy that composites will generally exhibit the most desirable qualities of the constituents and often some desirable properties that none of the constituents alone possess.

Thus composite materials differ from conventional monolithic materials in that they comprise two or more distinct phases. One phase, the reinforcement, provides the desired strength, toughness, or thermal characteristics, while the other phase—the matrix—provides the adhesive properties that unite and protect the reinforcement.

Composites contain reinforcements and an adhesive (the matrix).

Composite materials are usually made by mixing the matrix and the reinforcement in such a way that the dispersion of one material in the other provides the desired macromechanical properties. For example, toughness can be created in a ceramic composite by addition of ceramic whiskers. Thermal expansion characteristics of some metal–matrix composites can be tailored to comply with a material specification by addition of silicon carbide particles. Strength and stiffness can be created in a polymeric matrix by the addition of carbon fibers.

Typically metallic, polymeric, or ceramic matrix materials are employed in composites, and they are responsible for terms such as metal–matrix composites, polymer–matrix composites, and ceramic–matrix composites. Terms like these are used to classify nontraditional materials. This diversity is illustrated by Figure 18.1. The classification can become more detailed in each of the three primary categories of materials. For example there are two major classes of polymeric matrices, the thermosets and the thermoplastics. Thermosets are plastics that become an insoluble and infusible material upon curing during the manufacturing process, while thermoplastics are materials capable of being repeatedly softened or hardened by heating or cooling.

Types of composites:

- Metal matrix
- Polymeric matrix
- Ceramic matrix

Types of polymeric matrices:

- Thermosets
- Thermoplastics

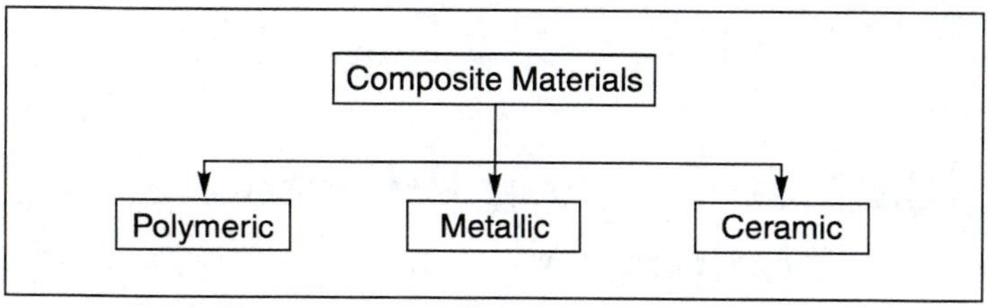

Figure 18.1

A classification of composite materials by matrices.

Types of fibrous reinforcements:

- Continuous
- Discontinuous (chopped)

Reinforcements for composite materials typically are continuous or discontinuous fibers, whiskers, flakes, thin platelets, or spheres. Composites can be classified by the reinforcements used, as shown in Figure 18.2. Naturally each class of reinforcements and matrices has its own advantages and disadvantages, and the design team must make trade-offs to create a part with the desired properties. This is a much more challenging task than with a monolithic material. Fiber-reinforced polymeric materials currently dominate the composites industry, and the most common reinforcements are glass, graphite, and aramid fibers.

Common fibrous reinforcements:

- Glasses
- Graphites
- Aramids

Fiber reinforced composites dominate the composites industry.

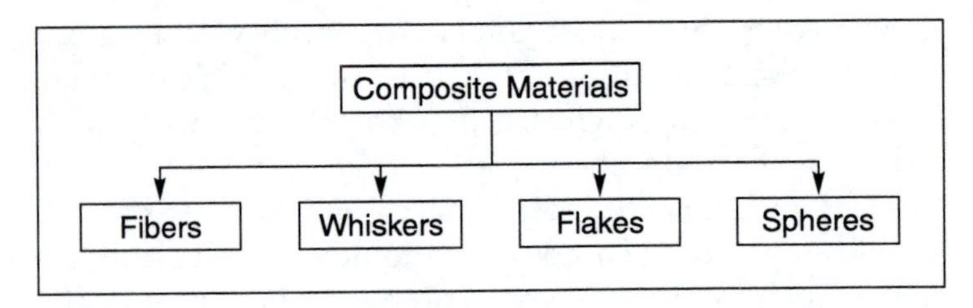

Figure 18.2

A classification of composite materials by reinforcements.

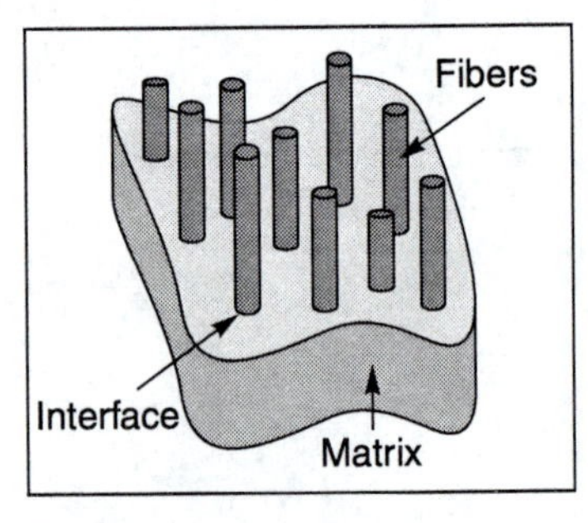

Figure 18.3

The principal macroscopic features of a fibrous composite material.

Figure 18.3 illustrates the basic structure of fibrous composite materials—namely a matrix with embedded fibers. The development of design tools for predicting the static and dynamic behavior of these materials, which is typically a component of the product design phase of the product realization process, involves the development of mathematical models for the fibers, the matrix medium, and the interfacial properties between the fibers and the matrix.

18.2 Advantages of Polymeric Composites

A product development team practicing concurrent engineering must carefully evaluate the advantages offered by polymeric composite materials relative to commercial metals at the product design phase of the product realization process. The attractive properties and characteristics of polymeric materials have been utilized in a diverse group of products for various industries including sporting goods, automotive, aerospace, high-speed production machinery, and medical devices. Some of these advantages are described on the following pages and summarized in Table 18.1.

Industries exploiting composite materials to produce new generations of products:

- sporting goods
- automotive
- aerospace
- medical
- defense
- production machinery
- boating
- industrial plant

Macrostructural design
Hybridization
Parts consolidation
High specific strength
High specific stiffness
Lightweight products
Superior fatigue properties
Superior damping
Acoustical insulation
Superior corrosion resistance
Reduction in material wastage during production
Superior impact properties
Advantageous design cascading consequences
Superior tribological properties
Superior thermal properties

Table 18.1

The advantages of fibrous composite materials.

The advantages of designing parts in polymeric composite materials have been exploited by product development teams in many industries. These advantages have been responsible for new products with performance characteristics superior to the previous generations of products that used the commercial metals. Some of these products are listed in Table 18.2.

Aircraft structures
- Control surfaces
- Radar domes
- Wings
- Fuselage
- Floor panels
- Doors and hatches
- Helicopter blades

Energy storage and conversion
- Solar collectors
- Flywheels

Ground transportation
- Bodies for high-speed trains
- Train pantographs
- Carriages for commuter trains
- Bodies for mobile homes
- Bicycle frames and other recreational equipment
- Truck, car and bus bodies
- Bumpers, drive-shafts, leaf springs

Marine structures
- Hulls of small craft and canoes
- Sailing masts
- Spinnaker booms

Industrial plant
- Pipelines
- Pressure vessels

Medical applications
- X-ray analysis and treatment devices
- Implants, prostheses

Sporting goods
- Golf clubs, hockey sticks, tennis rackets
- Archery bows, gun stocks
- Skis and ski poles
- Fishing poles

Industrial machinery
- Robotic arms
- Parts for textile looms
- Telescopic cranes
- Wire bunching bows
- Vanes for pumps and compressors

Table 18.2

Some applications of fibrous composite materials.

Macrostructural design

Fibrous composites can be designed to tailor properties to suit applications:

- Toughness
- Density
- Strength
- Thermal
- Electrical

Fibrous composite materials offer a design team the ability to tailor the properties of a material to comply with a specification. These characteristics can include toughness, stiffness, strength, density, orthotropy, thermal insulation, and electrical conductivity, for example. Such versatility is based upon the large number of design parameters that influence the macromechanical properties of composites. Thus, by appropriate selection of ingredients, a material with unique properties can be designed. Generally these properties cannot be developed in any other class of engineering materials.

Uniqueness can extend to situations where some regions of parts have isotropic properties because of randomly-oriented short chopped fibers while other regions are anisotropic because long, continuous fibers have been specified. Furthermore different kinds of fibers with different properties—glass and aramid, for example—can be combined in a common matrix to create a material with unique properties superior to those of the individual constituents.

Hybridization

The large number of parameters available for synthesizing a set of material properties permits a macrostructure to be designed from two or more composite material systems. This kind of hybridization can manifest itself as the specification of a variety of different *forms* of reinforcement such as a composite part containing chopped fibers, continuous fibers, and woven fabric. Another primary class of hybrid designs combines two or more different *types* of fibrous materials in a component. For example, the part may contain both aramid and carbon fibers.

Hybrid macroscopic design:

- Different forms of reinforcement
- Different types of fiber

Hybrid designs often are used in parts that must comply with stringent cost constraints. Thus a low-strength, relatively cheap material such as E-glass can be used in low stress regions of a part where a high-strength and typically expensive material such as a graphite fiber is not needed. Stronger reinforcements can be used only where they are necessary.

Parts consolidation

Hybridization permits parts consolidation.

The ability to design composite materials with the desired properties in conjunction with designing the associated manufacturing processes enables a significant consolidation of parts to be achieved in a product. Instead of designing a part for an automotive suspension in several separate pieces using a combination of steel components welded together prior to incorporating rubber bushings, for example, a single composite part can be created using a single manufacturing process to fulfill the same role. This achievement is significant in design-for-manufacturability, as discussed in Chapter Twelve. The parts count is reduced, an integrated design is developed, both structural weakness and weight are reduced, and the cost of designing joints is lowered.

High specific strength

Unidirectional carbon–epoxy materials are six times stronger than steel on a weight basis.

Aramid-epoxy composites are approximately four times stronger than the commercial metals per unit mass.

Parts fabricated with continuous fibers are stronger than parts fabricated with discontinuous fibers.

The ratio of the strength to density, or the specific strength, of unidirectional carbon fiber epoxy materials are approximately four to six times greater than the specific strengths of aluminum alloys or the steels. Furthermore, composites with an epoxy matrix and aramid fibers, such as the DuPont Kevlar, can achieve specific strengths approximately four times greater than those of commercial metals. This is illustrated in Figure 18.4. All these data depend upon the orientation of the fibers of the composite relative to the direction of the loading. These characteristics can be exploited in numerous aerospace applications, where high strength lightweight parts are mandatory. Parts fabricated from continuous fibers are stronger than parts fabricated from discontinuous fiber, other parameters being the same.

High specific stiffness

The specific modulus of unidirectional carbon-epoxy materials is 3 to 6 times greater than for the commercial metals.

The ratio of stiffness to density, or the specific modulus, of unidirectional carbon fiber epoxy materials is approximately three to six times greater than the specific moduli of the steels and aluminum alloys. These properties are an order of magnitude greater than the aramids and the glass composites for the same fiber configuration and matrix, as shown in Figure 18.4. High specific stiffnesses such as these can be exploited in the design of high-speed linkage machinery, where inertial loading dictates the operating speed, and hence the productivity, of the machine.

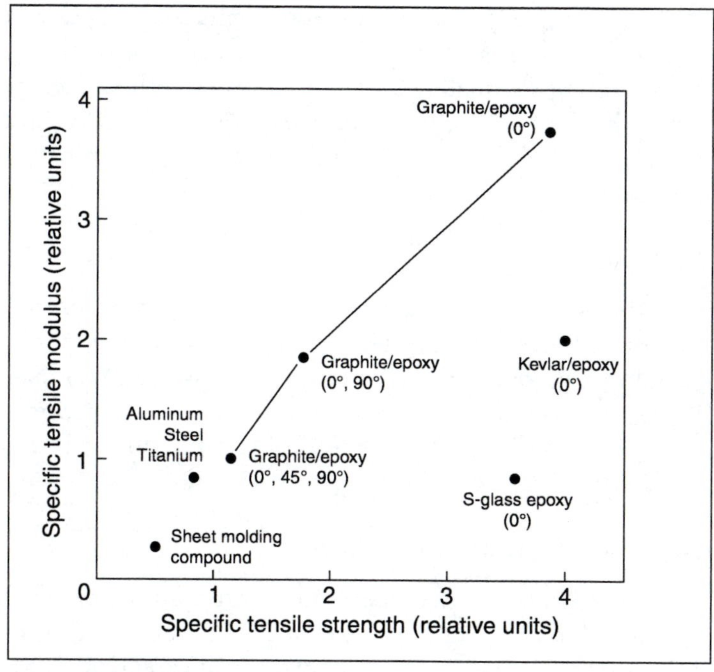

Figure 18.4

A comparison of the specific strength and specific stiffness properties of polymeric fibrous composite materials and the commercial metals using different angles of reinforcement.

Lightweight products

Because of their strength and stiffness characteristics, composite structures can be designed and fabricated to weigh only 65 to 75 percent of the weight of comparable aluminum structures for the same design specification. This weight saving can have very significant advantages in the

Composite structures weigh only 70 percent of comparable aluminum structures.

automotive and aerospace industries. For example, nearly 90 percent of the *Voyager* aircraft that flew nonstop round the world without refueling was fabricated from composites with graphite fibers.

Superior fatigue properties

Composite parts are very durable in vibrational environments.

The fatigue properties of epoxy materials reinforced with carbon fiber and also aramid-reinforced epoxy materials can be approximately 60 percent of their ultimate tensile strengths. This is much superior to the aluminum alloys and the steels, and it has been responsible for their use in helicopter rotors.

Superior damping

Polymeric composites have better damping characteristics than the steels.

The energy-dissipation properties, or damping properties, of polymeric composites are superior to the commercial metals. Consequently the transmission of vibrational loads is reduced and noise radiation is minimized. This property has been exploited in the design of the interior of public transport vehicles and in airframe structures.

Superior corrosion resistance

Composites have high resistance to corrosion.

The corrosion resistance of most composites is excellent compared to the steels. Consequently maintenance is minimized and the service life of the product is extended. Glass-reinforced polymeric materials offer significant advantages in the boat-building industry. The primary advantage is resistance to harsh wet environments, but the seamless construction also reduces assembly operations and minimizes leaks.

Minimal material wastage

Polymeric composites are manufactured by efficient processes with minimal waste of material.

There is less material waste when composite parts are manufactured because polymeric parts are generally fabricated sequentially, one layer upon a previous layer, using a mold. This procedure contrasts with many manufacturing processes for the commercial metals where parts are machined from stock or cut from sheet material. These latter processes generate a great deal of scrap. Of course this scrap can be recycled if the company adopts a green design approach to the product realization process (see Chapter 6).

Superior impact properties

The impact resistance and damage tolerance of some composite materials, such as the aramid-epoxies, are much better than the commercial metals.

Design cascading consequences

The effect of a composite part upon the overall performance of a product is often overlooked in design review meetings because attention often focuses only upon the cost of the composite part compared to the cost of the part made from a monolithic material. This should be avoided because the appropriate criterion for evaluating the viability of a composite design is the return on the investment for the complete system—that is, for the product.

Evaluate the consequences of designing a polymeric composite part carefully. Don't focus only on the cost of the part; are there other subtle advantages?

Cascading manufacturing benefits arise from the reduced number of joining steps when one composite part replaces several smaller parts made from traditional materials. This reduces manufacturing requirements, inventory, assembly, inspection, and such things as the machining of bolt holes. Fewer joints can improve dimensional accuracy, thereby reducing rejection and rework rates for the overall system. Thus the design of parts in composite materials precipitates cascading effects at both the product performance level and at the manufacturing level. These cascading advantages in both performance and cost must be considered when cost/benefit payoffs are evaluated.

In one industrial case study published in the marketing literature distributed by Hercules Aerospace Company, a picker stick for a Northrop Sensamatic textile loom was designed in a Magnamite graphite epoxy material and was manufactured by pultrusion. This part was more expensive than the traditional wooden picker stick, but the polymeric part permitted the loom to be more productive. This was

The more expensive composite picker stick for the textile loom enabled the machine to achieve higher productivity because of:

- higher operating speed
- longer life
- less downtime for maintenance

accomplished because the operating speed of the loom was increased by 10 percent and the longevity of the part was increased by 600 percent. Thus there were less downtimes for maintenance, and the initial higher cost of the composite part was offset by the higher productivity of the loom. This scenario is typical of many design situations.

Tribological properties

The wear and frictional properties of carbon-fiber-based composites such as those reinforced with nylon are excellent. They can be used in bearing applications with steel shafts without lubrication and they exhibit performances comparable to lubricated steel systems. These designs can reduce maintenance costs in many systems.

Transportation

The lightweight characteristics of composite parts make handling easy during manufacture, and they frequently reduce freight costs.

Thermal properties

Aramid fibers have a negative coefficient of thermal expansion.

Thermal stresses in a composite part can be minimized by an appropriate mechanical design and by choosing a material with an appropriate macrostructure. This can be accomplished by exploiting the negative coefficient of expansion of aramid fibers in conjunction with positive expansion coefficients of the other common polymeric fibers, namely by hybridization. Macrostructural design is discussed further in Section 18.4.

18.3 Disadvantages of Composites

Polymeric composite materials technologies are somewhat immature relative to those for commercial metals. Consequently there are a number of impediments to the growth of this field and to the development of commercial

- Cost
- Complex modelling
- Lack of high-productivity manufacturing processes
- Lack of concurrent design and manufacturing protocols
- Different types of manufacturing equipment

Table 18.3

Some disadvantages of fibrous composite materials

and industrial products outside the aerospace industry. Some of these are summarized in Table 18.3

Cost

One of the greatest disadvantages of composite materials is the cost of these materials in comparison to the cost of commercial metals. Costs depend on the materials, attributes of the part, and the production volume. Naturally there are situations where the parts consolidation advantages offered by these polymeric materials can offset the higher cost of the material. Furthermore the composite part may enhance the performance of a product so much that the higher cost of the part is offset by the better performance of the product.

Cost depends on production volume, materials, and design functionality.

Composite parts are often more expensive than metal parts but they may have significant advantages.

Better performance can be measured by the increased operating speed of production machinery or the superior performance of a composite tennis racket, for example, where cost may not be so important. Freight containers are now being manufactured from fiber-reinforced plastics rather than metal despite the higher initial cost of the composite product. This cost is offset by the longer life of the new lightweight composite designs.

Complex modeling

Mathematical models for predicting the behavior of composite materials are generally more complex than the models for monolithic materials. The development of these

The mathematical modelling of materials with macrostructure is naturally more challenging than modelling the commercial metals.

models or the use of software containing models of materials with macrostructure requires personnel with a strong background in applied mathematics and in broad aspects of design-for-manufacturability of polymeric materials. There is an absence of design guidelines and of rules for design-for-manufacturability. Consequently engineers with no background in these materials have quite a challenge when they are required to design a composite part.

Lack of high-productivity processes

Many products for commercial and industrial marketplaces must be manufactured quickly to satisfy the demands of a mass production environment. This is generally not a disadvantage for aerospace and military products where only a few items are required each year. However, despite efforts to improve production rates, the speed of manufacturing composite parts for ordinary commerce and industry is relatively slow compared to production rates for metallic parts. These composite-based products usually have lower costs and lower performance requirements than aerospace and military applications. A flavor of the primary parameters for creating composite parts in these marketplaces is presented in Figure 18.5 where the three parameters of cost, performance, and production rate are compared.

Rates of producing composite parts are slower than for conventional materials.

Lack of concurrent design and manufacture protocols

The creation of cost-effective composite parts mandates a multidisciplinary, concurrent engineering approach. Since this approach often requires supplier involvement, precautions may be needed to protect proprietary information. The approach must be a design and manufacturing protocol that integrates the attributes of the part with the attributes of a manufacturing process used to produce the part. Since design procedures involve creativity, analysis, and evaluation, it must be an iterative procedure. A methodology for these tasks is presented in Section 18.5 of this chapter.

Production of a composite part mandates the integration of both design and manufacturing.

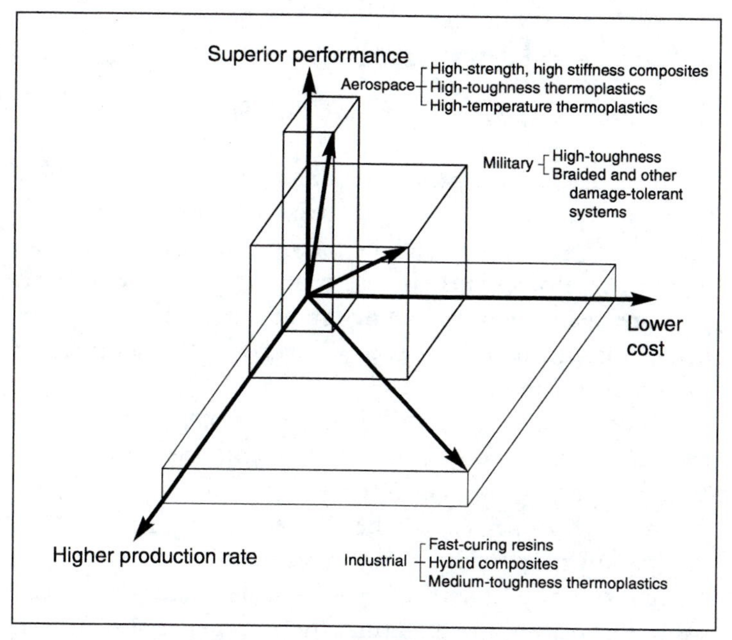

Figure 18.5

The cost, performance, and production characteristics of composite parts for aerospace, military, and commercial applications.

Different manufacturing equipment

The manufacturing equipment needed for the production of composite parts is quite different from conventional equipment used to make metal parts such as machine tools and welding equipment. Polymeric parts are often made by using high temperature ovens, filament-winding machines, preheated dies, and autoclaves. Thus new skills must be developed by an organization switching from conventional materials to polymeric composites. Furthermore, the manufacturing processes are complex and are sometimes difficult to control.

Composite parts are manufactured by processes quite different from traditional industrial processes.

18.4 The Design of Fibrous Composite Materials

The properties of a composite part depend upon:

- Properties of the matrix
- Properties of the reinforcement
- Distribution of the reinforcement
- Orientation of the reinforcement

The mechanical properties of a composite part depend upon the properties of the matrix, the properties of the reinforcing materials, and the distribution and orientation of these reinforcements. These properties of the composite part are determined by the design and manufacturing processes, but ultimately they are governed by the manufacturing process alone.

This process is responsible, for example, for the cure of the part as well as for the distribution and orientation of the constituents that are collectively responsible for the part's mechanical properties. If the desired strength and energy dissipation properties are not developed correctly because the continuous fibers in a part are not oriented as they should be during manufacture, then properties used by the design team in their calculations will be different from those exhibited by the actual part. The part may not satisfy the design specification.

Alternatively in an injection molding operation involving a viscous resin containing chopped fibers, knit lines that form at the confluence of two fluid streams in the mold cavity may cause a line of weakness in the part. If that possibility is overlooked in the design phase, part failure is possible. Such a situation contrasts with the processing of homogeneous and isotropic materials for which the mechanical properties are largely independent of the manufacturing method.

The properties of metal parts are generally not affected by the manufacturing process.

It is therefore important to recognize that the manufacturing processes used to create a composite part are also responsible for some of the properties of the material as well as for the creation of the part. This is depicted in Figure 18.6. This is not true for monolithic materials.

Consider, for example, a piece of low carbon steel at room temperature. This material is an isotropic linear elastic material at low strains, with a predictable density, energy dissipation characteristics, thermal properties, electrical prop-

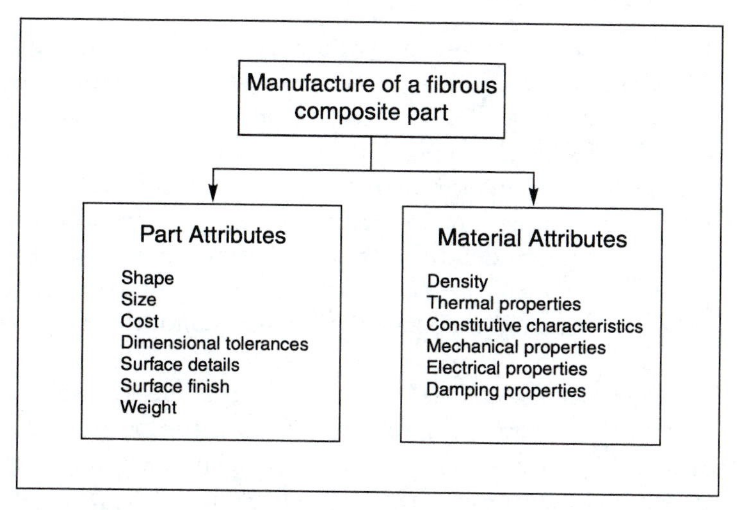

Figure 18.6

The manufacturing process for a fibrous composite part is responsible for both the part attributes and the properties of the composite material.

erties etc.. Furthermore these attributes are largely independent of the process employed to manufacture parts from the metal and there is no freedom to radically change these properties. Therefore strength of materials calculations to predict deflections and stresses, as well as the vibrational response, can be undertaken using only a few material parameters such as Young's modulus, Poisson's ratio, and the specific gravity.

Polymeric composites however have a broader range of design parameters for creating the material's macrostructure, and consequently the potential exists for creating a material with a broader range of desirable properties. This is both good and bad. It is good because there is considerable freedom to create a set of material properties that satisfy a large number of diverse specifications for materials with unique properties. However, it is also bad because the mathematical analyses associated with the design process are much more complicated than with monolithic materials, and the manufacturing processes are also more complex.

The properties of fibrous composite materials depend upon a large number of variables. These can be tailored to design a material with unique properties.

The mathematical modeling of composite materials is more complex than that of monolithic materials.

Thus, for example, composite materials can exhibit elastic, viscoelastic, or viscoplastic behavior. They can ex-

Macroscopic design freedom permits parts with different properties to be created:

- Elastic
- Viscoelastic
- Isotropic
- Orthotropic

hibit isotropic, orthotropic, or anisotropic properties; and furthermore, these properties can be varied spatially throughout the material. Special thermal and electrical properties can also be developed in a part to comply with a design specification. Thus the design team has considerable freedom to manipulate many design parameters. However the mathematical models for creating a composite with specific properties, or for predicting the behavior of composite parts, are much more complex than for monolithic materials.

The mass, stiffness, and energy dissipation properties govern the dynamical response of a part.

The mechanical, electrical, and thermal properties that govern the dynamic response of a fibrous composite material for a prescribed set of time-dependent excitations are governed by the design parameters listed in Table 18.4. The selection of these parameters governs the properties of the composite material and hence the performance of the part. For example, these parameters govern the mass, stiffness, and energy dissipation properties of a composite part, and

Fiber properties: graphites, carbons, glasses, aramids, etc.

Form of the fibrous reinforcement: mats, rovings, braids, woven rovings etc.

Matrix properties: epoxy resins, polyester resins, polyimide resins, polyether etherketones, etc.

Fiber length: continuous fibers, chopped fibers of various lengths, etc.

Fiber volume fraction: the fraction of fibrous reinforcement in unit volume of the material

Fiber orientation: the alignment of the longitudinal axis of the fiber relative to a prescribed axis

Stacking sequence: a description of the anatomy of a composite laminate that describes the orientation of the plies and their sequence in the laminate

Ply thickness: the thickness of prepreg tape, woven fabric, a single pass in a filament winding operation, etc.

Table 18.4

Parameters for designing fibrous composite materials.

hence they govern the dynamical response for a given time-dependent mechanical loading.

These fibrous forms, matrices, and fibers permit a variety of material properties to be created, and these capabilities can be exploited by the engineering profession in the design of innovative materials for new products.

Fibrous reinforcements assume different forms in different manufacturing processes, and these various forms should be considered as parameters for designing a composite material. Each manufacturing process imposes constraints on the selection of the form of fibrous reinforcement that can be used to create a composite part with the desired macroscopic properties. Various forms of reinforcement are listed in Table 18.5.

Each manufacturing process imposes constraints upon the types of reinforcements.

The design of a part to be made from a composite material requires consideration of many design parameters. Whereas the methodologies for designing composite-based components with simple shapes have been well developed,

Fibers: artifacts with one dimension many times greater than their diameters, which are typically of the order of 10 mm

Filament: a single fiber

Rovings: collections of continuous filaments

Yarns: collected filaments with an applied twist

Woven roving: a heavy fabric made by weaving rovings or yarn bundles in an interlacing manner

Braid: interlaced yarn or roving woven in a tubular configuration

Mat: a blanket of loose fibrous material, typically comprising swirled filaments and chopped fibers of random orientation

Tapes: collections of parallel filaments which are mixed with a binder to form flat ribbons much longer than their width, which is much greater than their thickness

Chopped staple: material obtained by cutting continuous filaments to lengths of approximately 10 mm to 50 mm

Table 18.5

Forms of fibrous reinforcement.

they do not explicitly incorporate the constraints and characteristics imposed by the fabrication processes used to make composite components. Thus many companies design composite parts in a way similar to the traditional design of metal components—that is, the part is designed first and then the manufacturing process is chosen.

Consider for example the proposition that a connecting rod for an internal combustion engine might be made either from a traditional metal or from an advanced composite material. A photograph of a typical connecting rod is presented in Figure 18.7.

Figure 18.7

An automotive connecting rod fabricated in steel.

Given the specifications, the design process for a metal rod conveniently decomposes into the distinct phases shown in Figure 18.8: kinematic analysis, material selection, machine dynamics, and stress analysis. Stress analysis determines the final details of the form-design of the part, and generally decisions about the appropriate manufacturing processes are made in isolation from the design process. The choice of the manufacturing process—forging or casting, for instance—affects the part attributes, as discussed earlier under design-for-manufacture, but these manufacturing processes can also influence Young's modulus and Poisson's ratio for the material because of grain structure imparted by each process. Properties that are affected by a manufacturing process can be incorporated into the design specification and hence influence the design of a part's shape.

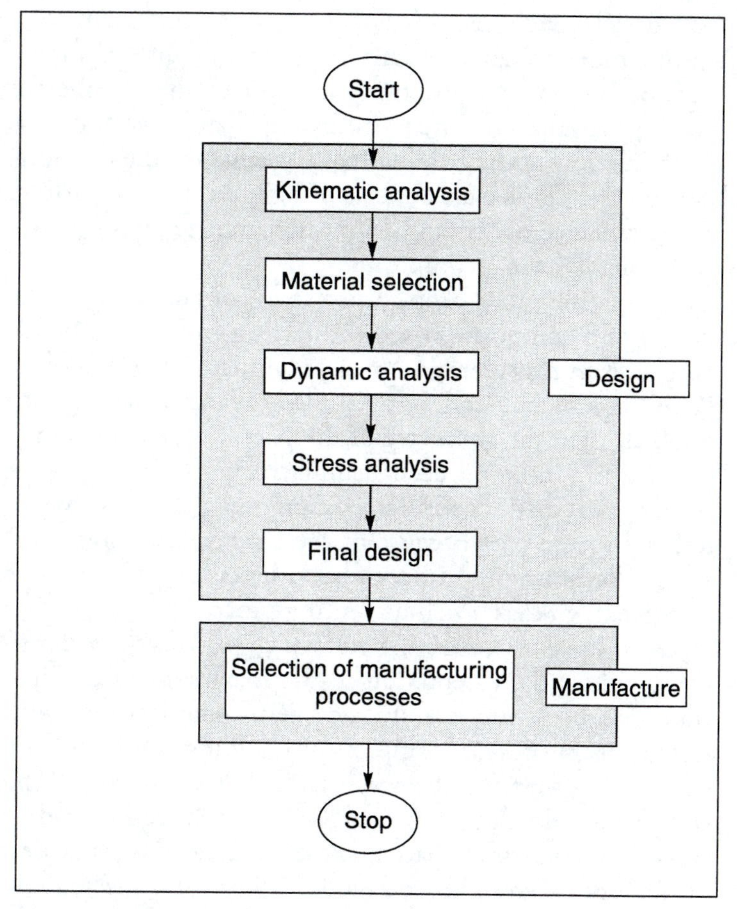

Figure 18.8

Traditional sequential product development cycle for an automotive connecting rod.

The design and fabrication of a connecting-rod in an advanced composite material has several advantages over a traditional metal design. These include lower weight, increased specific stiffnesses and strengths, and superior damping characteristics which not only result in superior vibrational response characteristics but also in significantly decreased noise from the engine. Note that these advantages for an individual component also improve the overall system design. Such design-cascading effects may permit the elimination of a balance shaft in the engine, the simplification of engine mounts, the reduction of the number of bearings required for the transmission shafts, and of course the reduction of the bearing requirements for the connecting rod and crankshaft because of reduced inertial loading. These enhancements can be almost guaranteed if the piston is also made from a composite.

Design-cascading consequences of a composite connecting rod:

- Elimination of balance shaft
- Simplified engine mounts
- Less bearings on transmission shafts
- Quieter engine
- Lower vibrational loads
- Cheaper bearings for connecting rods

In contrast to a drop-forged steel connecting-rod, the design of a composite-based connecting-rod cannot be divorced from consideration of the manufacturing processes. If the connecting-rod is pultruded, filament wound, or braided, then the geometry of the part will be constrained by each process, and each manufacturing process will, in turn, impart distinct stiffnesses, damping, and mass characteristics to the part because of the fiber volume fractions, spatial distribution of fibers, fiber orientations, and lay-ups associated with each manufacturing process.

Each manufacturing process imparts distinct fiber orientations and fiber volume fractions while constraining stacking sequences, fiber lengths, matrix selection, and form of reinforcement.

A composite connecting-rod which permits the fibers to have preferential orientation in the maximum load directions has been successfully designed, manufactured, and tested. In this design, compression molding is used for the bearing housing for the small-end of the rod, while the compressive loading imposed on the connecting rod is carried by a pultruded part, and the tensile loading is carried by a filament-wound part. Figure 18.9 presents a schematic diagram of the design. Innovative designs of this kind can capitalize on coherent design-for-manufacture methodologies in order to satisfy the insatiable demand for high-performance low-cost components in the marketplace.

However, there is a dearth of process-driven design-for-manufacturability methodologies for designing and fabricating cost-effective, high-strength, lightweight components for the variety of composite-based systems encountered in the defense, manufacturing, and aerospace industries.

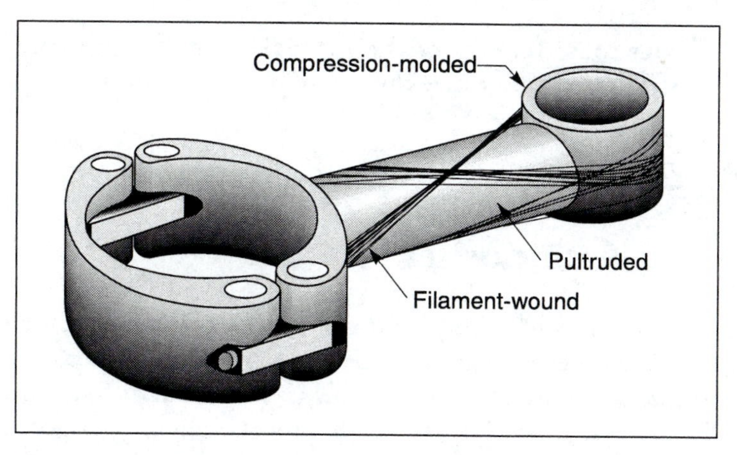

Figure 18.9

Schematic of an automotive connecting rod fabricated from a composite material.

In contrast to the classical design-for-manufacturability approaches for monolithic materials, viable design-for-manufacturability methodologies for composite materials not only incorporate the manufacturing processes but they are in fact exclusively motivated by the manufacturing processes. These methodologies will manifest themselves at the macromechanical level as desired material properties and part attributes are established, while at the subassembly level, these methodologies render obsolete some conventional design-for-assembly methodologies. Parts which are currently fabricated as metal subassemblies can be made at a reduced cost as a single part in a polymeric composite material. Injection molding is often used to do this. See Figure 18.10.

DFM methodologies for composite parts are driven by the manufacturing processes.

Figure 18.10

A single injection molded automotive part replaces a multipart assembly of steel components.

For example, in the traditional approach a typical subassembly begins with the fabrication of parts with accurately machined costly interfaces in order to facilitate the subsequent assembly of these parts. This is labor-intensive and may account for up to 50 percent of the total cost of the subassembly. Design-for-manufacturability methodologies for composite parts and subassemblies can significantly reduce the costs of machining, assembly, and labor by permitting the fabrication of a complete subassembly as a

Composite parts can be designed to replace more expensive assemblies of metallic parts.

fibrous composite material by a single manufacturing process. These processes are discussed in the next section.

18.5 Processing of Polymeric Composites

Automated processes include:

- Braiding
- Filament winding
- Pultrusion

After the desired attributes of the fibers and matrix for a part have been identified, the appropriate manufacturing technique must be selected. This is often undertaken in an iterative manner because each process has implicit constraints and attributes that can affect the placement of fibers and the shape of the part—and consequently the mechanical, thermal, and electrical properties of the part. The methods of combining reinforcements with a matrix have evolved from the traditional manual lay-up of impregnated laminates to automated high-volume processes such as filament winding, pultrusion, and braiding to which have been melded improved quality control, reproducibility, and precision. Some of these processes are listed in Table 18.6. The following discussion begins with a consideration of their effect on the design of composite structures as a background for a broader discussion of design-for-manufacture of composite parts.

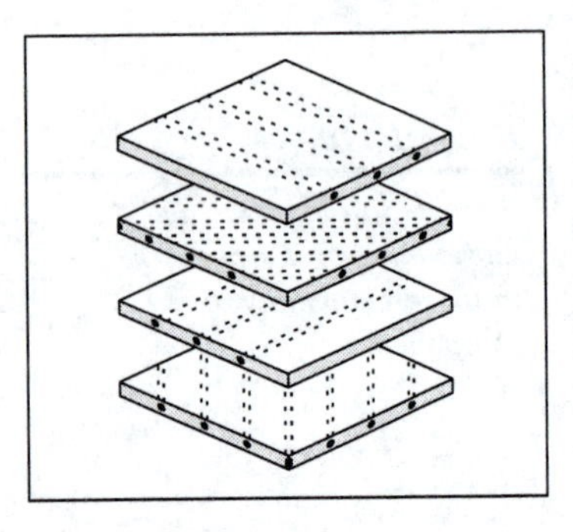

Figure 18.11

An exploded view of a laminate illustrating the orientation of the continuous fibers of each ply and the stacking sequence.

Wet hand lay-up

Hand lay-up is the simplest fabrication technique. It involves placing dry reinforcing fabric into an open mold before application of the resin. The structure is built up by hand one layer upon another until the desired thickness is obtained. This structure is termed a *laminate.* See Figure 18.11. The approach is illustrated in Figure 18.12. To strengthen the part, air pockets trapped in the layers of fabric normally are removed with rollers applied by hand. The fabric mats are usually preimpregnated with resin, but resin can be applied with a brush. This technique requires little equipment and offers great flexibility. The most com-

Hand lay-up
Vacuum-bag molding
Autoclave molding
Spray-up
Injection molding
Compression molding
Resin transfer molding
Filament winding
Pultrusion
Braiding

Table 18.6

The primary processes for manufacturing fibrous composite parts.

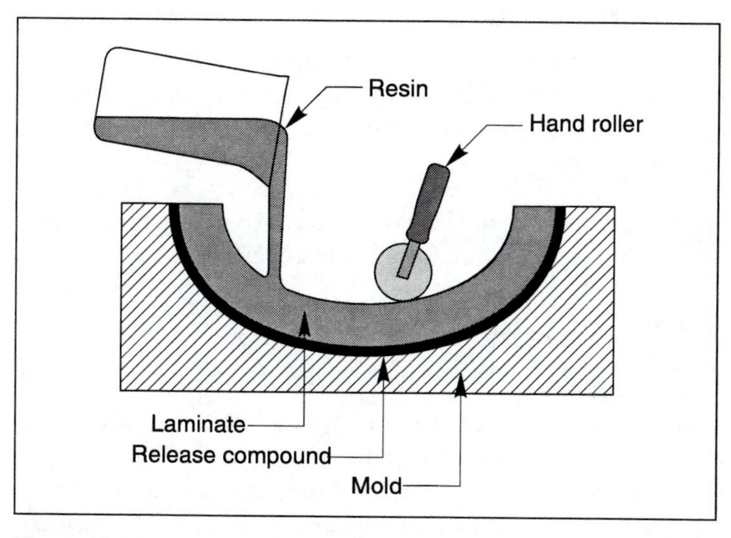

Figure 18.12

The wet hand lay-up technique.

mon matrix resins used in this process are the epoxies and polyesters that cure at room temperatures, and the most common reinforcement is a fabric of glass fibers often called "glass cloth".

Design advantages

- Tooling can be readily changed if a part is redesigned.
- Semiskilled labor is required.
- Investment in equipment is low.

Design disadvantages

- It is difficult to compact hand-laid-up structures adequately during manufacture. Therefore they are weaker than most other composites because of voids and because they have a high resin content. Resins are considerably weaker than reinforcements.
- The mechanical properties of composites made with tightly woven fabrics and bonded by high-viscosity resins are low because the fabric often is not adequately saturated by the matrix before it is cured.
- The macrostructure varies considerably throughout each part and also from part to part. This variation mandates larger safety factors in the design calculations.
- Only the face of the part adjacent to the surface of the mold is typically of good surface quality.

Vacuum-bag molding

Vacuum-bag molding is an extension of hand lay-up; it produces parts of higher quality.

Vacuum-bag molding represents an extension of the hand lay-up process. Its principal advantage is that it can produce higher quality parts than the wet hand lay-up technique. In vacuum-bag molding the resin and fibers are laid up in the mold either by hand or by spray lay-up, after which the assembly is covered with a flexible film whose edges can be sealed. A vacuum is drawn to compress the assembly uniformly with atmospheric pressure. This compression eliminates excess resin, voids of trapped air, and it also develops superior fiber-matrix adhesion and a higher reinforcement concentration. Consequently a part of superior mechanical properties and quality is produced. The method is shown schematically in Figure 18.13.

The vacuum is responsible for atmospheric pressure compressing the laminate to:

- eliminate voids
- eliminate excess resin
- ensure better fiber–resin adhesion
- ensure a higher concentration of reinforcement

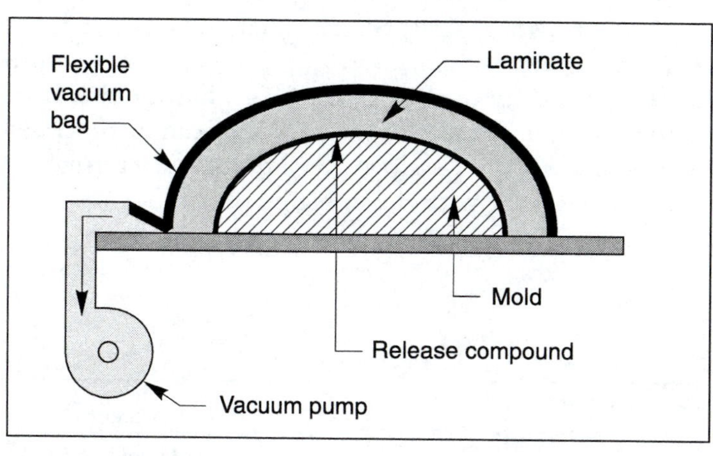

Figure 18.13

Vacuum bag molding.

Design advantages

- Mechanical properties are superior to parts made by wet lay-up.

Design disadvantages

- Complex shapes with sharp curves in two directions are difficult to bag because the film wrinkles and cannot conform smoothly with the surface contours.

Autoclave molding

Autoclave equipment is typically used in conjunction with hand lay-up, preimpregnated fabrics, and vacuum bagging to manufacture high-quality composite parts that are void-free and that have up to 65 percent reinforcement.

Autoclave modeling involves:

- Prepregs
- Vacuum bagging
- Application of heat and pressure

Very high quality parts can be fabricated by autoclave molding.

These attributes are achieved because of several characteristics of autoclave molding. First, the technique usually employs preimpregnated fabrics ("prepregs") that are unidirectional tapes or fabrics impregnated with a partially cured resin of highly controlled fiber-resin content. Second, the technique uses vacuum bagging to achieve a high fiber–resin ratio. Third, autoclave equipment applies heat as well as pressure to the part permitting more exotic resins to be used. The higher pressure generated by the equipment ensures superior compactness. Autoclave molding can produce high-quality parts (Figure 18.14), but the size of the autoclave limits the size of the parts that can be processed. The equipment also is expensive. This manufacturing process is shown schematically in Figure 18.15.

Figure 18.14

An aerospace autoclave part.

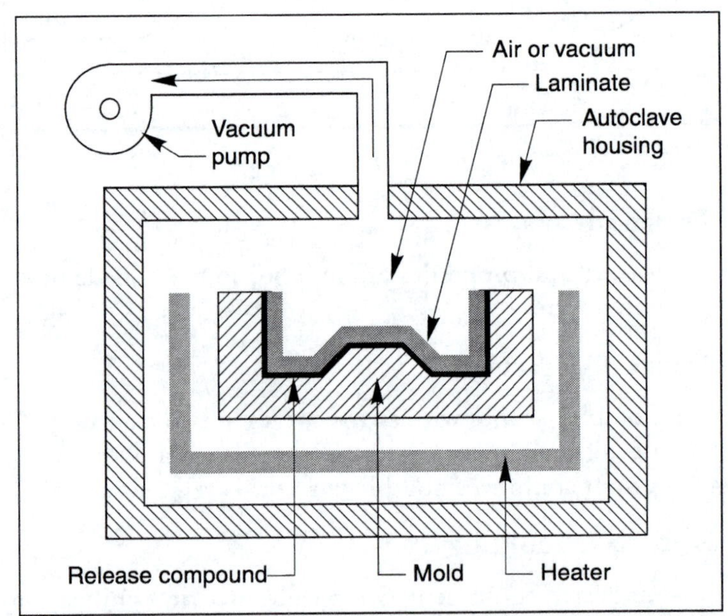

Figure 18.15

Autoclave molding.

Design advantages

- A superior part can be produced because the resin-fiber distribution and ratio are controlled more accurately than for a part made by wet lay-up.
- The parts have superior part definition and consolidation as well as a higher fiber content than parts made by wet lay-up or vacuum bagging.

Design disadvantages

- There is only one surface with accurate dimensions and contour.
- The method is slow relative to automated methods.

Spray-up

Spray-up is similar to wet hand lay-up except that the method of applying the reinforcement and the resin is different. This manufacturing process involves feeding a continuous strand of reinforcement roving through a cutter which generates a chopped fiber roving. This chopped roving is sprayed directly into an open mold along with a resin stream, as shown in Figure 18.16. Then the mixture is rolled to remove entrapped air and to ensure a good coating of the fibers with the resin.

The spray-up process uses chopped fibers (typically glass fiber and polyester resin).

This is an inexpensive process because roving is the cheapest form of reinforcement and spray-up equipment is inexpensive. The most common resin systems are the low-heat-cure and the room-temperature-curing polyesters. The spray-up process has the advantage of requiring less skilled labor than hand lay-up, and like hand lay-up there is no inherent limitation on the size of the part. Much higher production rates than the hand lay-up method can be achieved, and very large, complex structures can be made. However, the mechanical properties of parts are inferior to those produced by the hand lay-up of continuous fibers because a structure fabricated with chopped fibers is inherently weaker than one fabricated with continuous fibers.

Design advantages

- No limitation on the size of the part.
- Higher production rates than hand lay-up.
- Lower manufacturing costs than hand lay-up.

Design disadvantages

- Because only one surface of the part is in contact with the mold, only one surface has an accurate contour and dimensions.
- Because the process uses chopped fibers, the resulting material properties are inferior to those of a similar part manufactured with continuous fibers.
- Parts have low levels of reinforcement because of inadequate compaction during manufacture.
- The process requires more skilled operators than those needed for hand lay-up.
- Only creates thin planar parts.

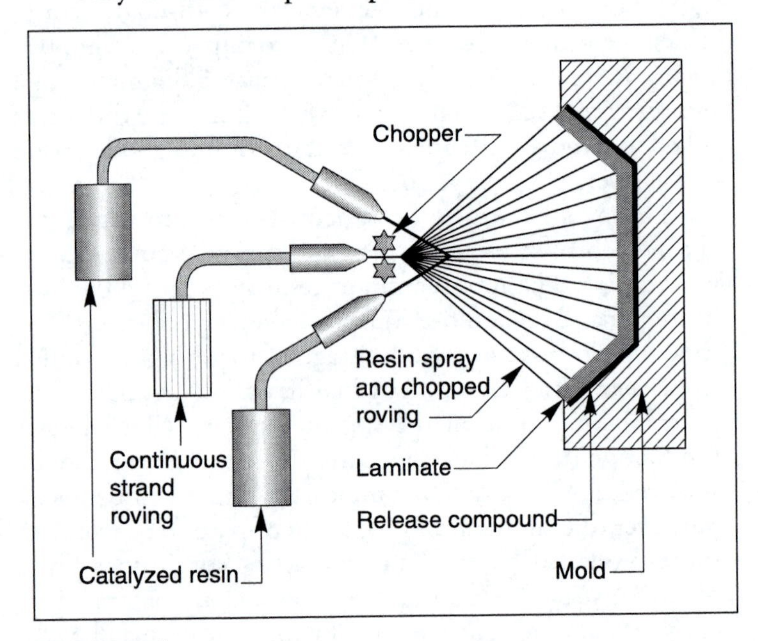

Figure 18.16

Spray-up.

Injection molding

Injection molding is a high volume method for manufacturing thermoplastic and thermoset parts by injecting or forcing a fluid plastic into a hot, closed mold. The mold is cooled, the plastic solidifies, and the part is ejected from the mold. A schematic of the process is in Figure 18.17.

Injection molding is the most important method of manufacturing thermoplastic parts.

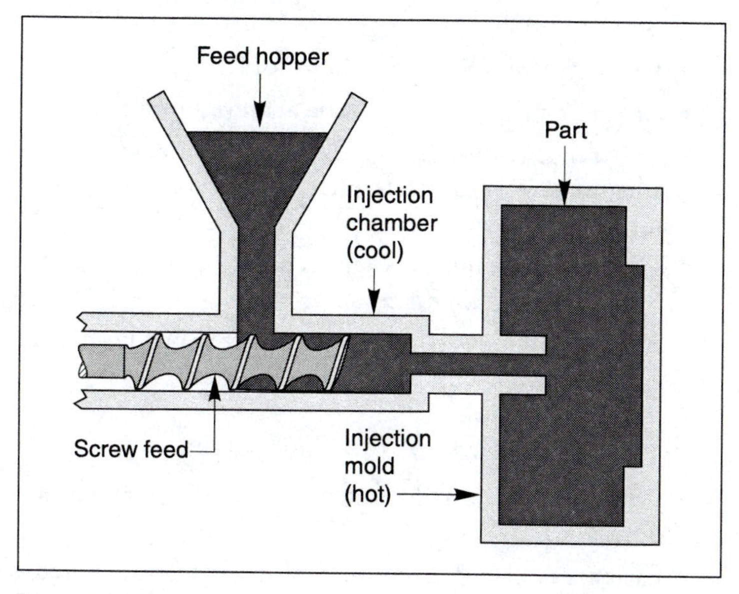

Figure 18.17

Injection molding.

Injection molding is the most important of the traditional methods for making thermoplastic parts. It is typically initiated with whiskers or blends of chopped reinforcement up to 25 mm long and with fiber concentrations up to 40 percent by weight. With this process a compromise must generally be achieved between the mechanical properties of the part which depend upon the length of the fibers and the processing constraints such as temperature and injection characteristics that can degrade fibers. Furthermore, the size of the gate or passage into the mold cavity can impose a restriction on the fiber concentration. Fluid dynamics of molds with multiple gating systems

Injection molding requires a compromise between the part properties and the manufacturing conditions.

Knit lines create weaknesses.

requires careful mold design to avoid the lines of weakness called *knit lines* where two or more fluid streams meet. Fluid dynamics also affects the orientation of the reinforcing fibers in knit lines. It is generally desirable to have a random orientation of fibers throughout the part to insure uniform mechanical properties.

Design advantages

- This high volume process has short mold cycles.
- Excellent surface detail can be achieved on parts.
- Parts range from a few grams up to approximately 2 kilograms. Typical parts include children's toys and automobile bumpers.
- The closed die ensures an accurate part with good consolidation properties.
- Parts of complex shape can be molded. Furthermore parts can incorporate holes and inserts.

Design disadvantages

- Parts with nominally isotropic properties are produced.
- Parts are weaker than comparable parts manufactured with continuous fibers.
- The process requires more equipment than hand lay-up.
- Molds are expensive to design and manufacture; therefore product design changes should be avoided.

Compression molding

BMC: bulk molding compound
SMC: sheet molding compound

This process is initiated by a weighed charge of bulk molding compound (BMC), or alternatively a charge of sheet-molding compound (SMC), that is placed in an open mold which is then closed. Heat and pressure are applied for a specific time to cure and shape the part. The process is schematically presented in Figure 18.18. Compression molding is generally used to make parts with a glass fiber reinforcement and either an epoxy or polyester matrix. Usually

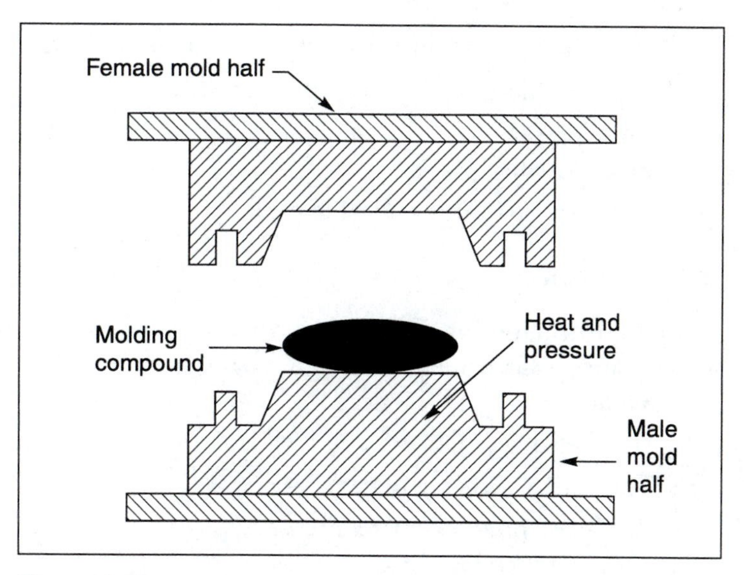

Figure 18.18

Compression molding.

these materials are in an arbitrarily shaped viscous bulk form or a sheet form. A bulk molding compound is a mixture of randomly oriented chopped fibrous reinforcement, resin, fillers, and an initiator. The fiber length usually is less than 25 mm and the fiber content is typically about 40 percent. Fillers are added to minimize the cost of the resin. A sheet molding compound can take several forms including a mixture of chopped or swirling continuous fibers, random chopped fibers, or unidirectional continuous fibers with a filler resin–initiator mixture that is sandwiched between two polyethylene films. These films are removed before the manufacturing process begins. Compression molding has been used in the auto industry to manufacture the entire cabs of trucks and the hoods and tailgates of minivans, for example.

Bulk molding compound has fiber lengths less than 25 mm and a fiber content about 40%.

Sheet molding compounds assume several forms:

- Chopped fibers
- Swirling continuous flows
- Random chopped fibers
- Unidirectional continuous flows

Design advantages

- Part shapes are complex; attachments and inserts are readily accommodated.
- Parts with high tolerances can be produced.

- Parts with high-quality surface finishes can be molded.
- High production rates can be achieved.
- The closed die ensures good interior and exterior surface quality.
- Parts require minimal edge trimming.

Design disadvantages

- Molds and tooling are expensive to design and manufacture. Consequently design changes should be avoided.
- The shelf-lives of BMC and SMC are limited.
- Equipment costs are high.
- SMC glass-polyester material costs approximately four times as much as steel.
- The primary problems with compression molded parts are related to the flow of the heterogeneous viscous fluid in the mold. These problems include folds and kinks in fibers, warpage, weld lines, delamination, and the incorrect orientation of fibers.

Resin transfer molding

Low viscosity resin impregnates the reinforcement in a heated mold.

Large parts of high strength can be manufactured by resin transfer molding.

The manufacture of a part by resin transfer molding (RTM) involves the initial placement of a dry preform or fiber mat in the desired orientation in a mold that is then closed and subsequently filled with a low-viscosity resin that impregnates the reinforcement. The mold is then heated to cure the reactive resin system. A solid composite structure forms and is then removed from the mold. Resin transfer molding, therefore, has several similarities with compression molding. Glass, carbon, and aramid are typical reinforcing fibers. The fibers must be able to retain their shape during injection of the resin. Very large parts with high reinforcement-to-resin ratios can be made by this technique. Typical parts include domestic shower enclosures and cabinets for industrial products.

Design advantages

- This is a high volume manufacturing process. It is employed by the automotive industry to produce body panels for vehicles.
- Parts are typically characterized by good surface qualities on both inner and outer surfaces.
- Large parts of complex shape can be molded with excellent structural properties.
- Special reinforcing features and inserts can be incorporated into the part.
- Only low-skilled operators are required.
- As part of a DFM initiative, the process is often used to produce large parts that were traditionally made as an assemblage of many smaller parts.

Resin transfer molding can produce large parts of complex shape.

Design disadvantages

- The design and manufacture of the molds is expensive; therefore product redesigns should be avoided.
- It is difficult to control the distribution of resin throughout the part. Consequently both the internal and the external edges are resin rich.
- The properties of RTM parts are comparable with parts manufactured by compression molding but they are not as good as parts made by pultrusion, autoclave molding, or filament winding.
- There is a significant financial investment in equipment, but this process utilizes lower pressures than those required in compression molding.

Filament winding

This manufacturing process consists of feeding continuous-strand reinforcement filament, or roving, through a resin bath prior to winding it on a rotating mandrel. Alternatively, preimpregnated roving can be used. The part is fabricated one layer at a time until the desired thickness and

Filament winding creates parts one layer at a time.

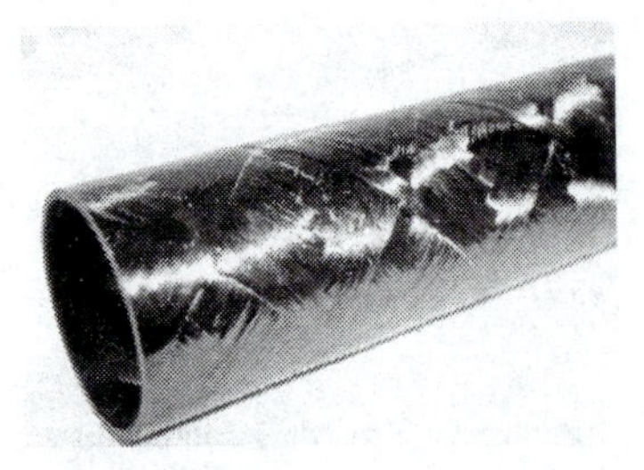

Figure 18.19

A Filament-wound tube.

pattern of continuous fibers have been created. It is possible to vary the wind angle, the resin content, and the tension of the winding in each layer of reinforcement. After the predetermined roving patterns have been laid down by the special-purpose winding machinery, the mandrel and part are cured before the molded part is removed from the mandrel. This technique is frequently restricted to the creation of parts with surfaces of revolution. See Figure 18.19. A schematic of the process is in Figure 18.20.

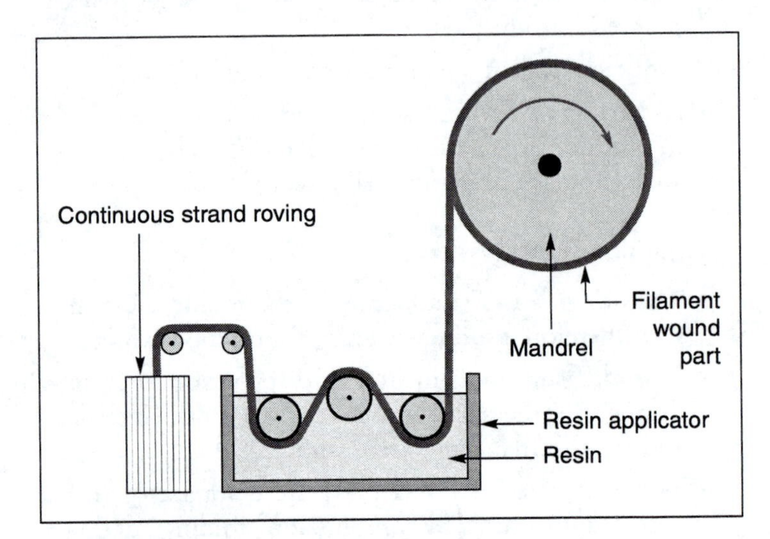

Figure 18.20

Filament winding.

Filament winding can be used to produce large cylindrical parts.

The most common resins used for filament winding are the epoxies and the polyesters. The most common reinforcements are glasses, carbons, and aramids which are generally used in the form of rovings. Parts 25 mm to 6 meters in diameter have been wound for pressure vessels, ducts, and storage tanks. The size and complexity of these cylindrical parts are dictated by the characteristics of the winding machine, such as the size of the mandrel, and the winding patterns that can be developed—by numerical control, for example.

Design advantages

- Parts can vary widely in diameter and length.
- Parts with different properties in different directions can be readily wound.
- High quality parts can be fabricated, especially if the part is cured in an autoclave after winding.
- There is minimal material wastage during this process.
- To facilitate the fabrication of noncircular parts, a flexible mandrel can be used and the part is formed by distorting the mandrel after it is wound.
- Filament winding is a high-volume manufacturing process.
- The process permits large numbers of identical parts to be manufactured quickly at low cost.
- Parts with a high fiber volume can be wound using continuous fibers that create a high strength part.

Design disadvantages

- Some desired patterns of rovings can be difficult to wind.
- The process requires expensive equipment.
- The shape of the component must facilitate removal of the mandrel.
- Parts with concave shapes cannot be filament wound without complicated post-winding activities before the part is cured.
- It is difficult to change the fiber direction while laying down one layer.
- Only long slender hollow parts are made.
- Mandrels can be expensive and complicated.
- The external surface of the part is not as refined as the inner surface adjacent to the mandrel.

Figure 18.21

Pultrusion.

Pultrusion

This automated manufacturing process involves pulling a continuous strand, or roving, through a resin bath before the impregnated reinforcement is drawn through a die which controls both the cross-sectional shape of the subsequent stock and also the ratio of resin to reinforcement. Figure 18.21 presents a schematic of the process. Generally the die is heated to initiate cure, and the final cure is completed in an oven through which the stock is drawn. The continuous reinforcement can be any commercial fiber, and a part can be 80 percent roving by weight. The matrix is typically a thermosetting resin such as polyester.

Pultrusions are made from continuous fibers up to 80% impregnated by a thermosetting resin.

Hollow parts or those with a solid, constant cross-section (pipes and beams) are manufactured by pultrusion (see Figure 18.22), as well as are automobile leaf springs, floors, stringers for aircraft, ladder rails, fishing rods, electrical insulator rods, and panels for truck trailers.

Pultrusion is used to produce long slender parts of constant cross section.

Figure 18.22

Pultruded parts. (Courtesy of DFI Pultruded Composites, Inc.)

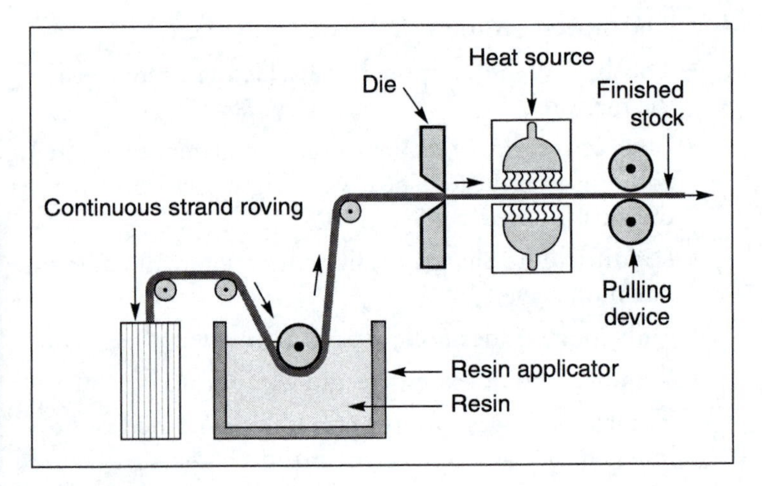

Figure 18.21

Pultrusion

Design advantages

- A high volume continuous process.
- The process leaves less scrap than the hand lay-up method.
- The process accommodates a wide variety of fibrous forms, such as rovings, fabric, and mat.
- Parts can be fabricated in glasses, aramids and carbons.
- Pultruded parts can vary in size from 25 mm to 5 meters in diameter, and part thicknesses can be up to 75 mm for parts fabricated with roving.
- Parts of any transportable length can be manufactured.
- Pultrusions with complex thin-walled geometries can be produced.

Design disadvantages

- Pultrusions have a constant cross-sectional shape.
- Parts fabricated from only unidirectional fibers exhibit excellent longitudinal characteristics but poor transverse characteristics.
- The design and manufacture of dies is expensive.
- The process requires expensive equipment.

Braiding

The braiding of fibrous composite parts is a mechanized intertwining of two or more systems of yarns to form an integrated structure. While the process has some similarities with knitting and weaving, the method of creating the fabric is different. Continuous fibers are braided over a rotating and removable mandrel. The mandrel is fed through the center of the braiding machine upon which are hung numerous carriers from which the fibers are braided at a controlled angle. These carriers work in pairs, somewhat akin to folk dancers around a maypole, to perform the over-and-under braiding sequence. This process permits com-

A braid is an integrated structure created by the intertwining of two or more yarns.

Braids can be manufactured with different fiber orientations, shapes, fiber volume fractions, and hybrid fiber combinations.

Fibers:

- Glasses
- Graphites
- Aramids

Resins:

- Polyesters
- Epoxies

posite parts to be manufactured with a wide variety of fiber orientations, fiber volume fractions, and part shapes. The braided structure (Figure 18.23) is then impregnated with resin, but more economical prepregs can be used. Braiding is shown schematically in Figure 18.24. Continuous reinforcing fibers can be glasses, aramids, or graphites. Hybrids can be developed from different combinations of fibrous materials. The matrix is typically a polyester or an epoxy resin. Two dimensional braiding is capable of manufacturing prismatic tubelike parts but three dimensional braiding can produce both hollow and solid prismatic parts of complex shape. Braided structures possess superior damage resistance compared to parts fabricated by other processes, and they have excellent torsional stability. These characteristics are exploited in aircraft propellers, drive shafts, and sporting goods (hockey sticks, squash rackets, fishing rods, and ski poles) as well as pressure vessels, artificial limbs, boat masts, and tables for X-ray machines.

Figure 18.23

Braids of different sizes, patterns, and materials.

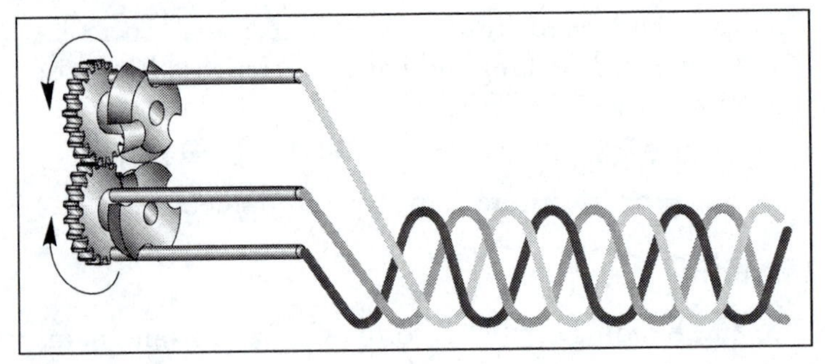

Figure 18.24

Braiding.

Design advantages

- Braiding is a low cost and potentially continuous process.
- Parts can contain multidirectional reinforcements.
- The tooling for braiding is relatively cheap.

- The process is a fast method of making fibrous composite parts when the fiber volume fraction and the orientation of the fibers is not critical.
- Braided parts can assume more complex shapes than filament-wound parts.
- Braided parts have higher resistance to damage and higher structural integrity than other composite structures because of the interlacing of the yarns in the braiding sequence.
- Three-dimensionally braided parts can be used to avoid the often unacceptably poor interlaminar properties of laminates and the intralaminar properties of filament-wound parts.

Braided parts can have hollow or solid cross sections.

Braided parts are tough, with good damage resistance.

Design disadvantages

- The shape of the part affects the ability to lay down the reinforcement where it is required.
- Fiber volume fractions are generally low. Consequently the resulting part does not possess the same quality of mechanical properties as parts manufactured by other methods such as filament winding.
- There are difficulties in debulking the parts to control resin-rich areas and remove voids in the braided structure.
- The mechanical properties of the part are governed by the class of braiding machine, the number of yarn carriers, and the quantity of yarn that can be spooled.

18.6 DFM Methodology for Parts Made from Fibrous Composite Materials

A rigorous design-for-manufacture methodology for composite parts requires an iterative approach that uses the basic strategy for solving open-ended design problems at all

DFM methodology for composite parts involves the iteration of:

- Creativity
- Analysis
- Evaluation

The design of a polymeric fibrous material is dependent upon both the design protocols and the manufacturing process.

the primary stages: namely the cycling of synthesis, analysis, and evaluation. Furthermore, the methodology must include design of the material, which is governed by both design protocols and the manufacturing processes, and also the geometrical attributes of the part.

This design-for-manufacture methodology for composites should include an iterative procedure involving the design specification for the composite part, the materials synthesis process, creation of the appropriate part attributes, identification of a suitable manufacturing process, and a computational analysis capability for evaluation of proposed designs. The methodology is presented in Figure 18.25. Parameters governing the design of fibrous composites are listed in Table 18.4; they include the fiber and matrix properties and the fiber orientation and distribution. Properties of the part are its geometrical shape, surface finish, tolerances, etc., that were discussed in Chapter Twelve. Characteristics of the manufacturing processes include the advantages and disadvantages of each method of processing fibrous composite parts as discussed in Section 18.5.

DFM methodology:

- Design specification
- Macroscopic material design
- Manufacturing process attributes
- Part attributes
- Part performance

The methodology begins with a part or subassembly specification that typically includes constraints on the static and dynamic deflections, mass, and inertial characteristics, loading, impact resistance, fatigue life, environmental factors such as temperature and humidity, and of course cost, quantity, and production rate. This specification provides the basis for an iterative design-for-manufacture methodology that simultaneously considers interdisciplinary attributes such as the form of the part, manufacturing processes, and material formation. Furthermore the design specification provides the basis for evaluation of the product.

A methodology for creating a part can begin with the process for creating the material. It should focus on the specification of a matrix and fibers and their distribution and orientation throughout the part. These attributes depend on characteristics of the part and of the manufacturing processes to be used. Part attributes are typically the size, shape, surface details, and surface finish, while attributes of

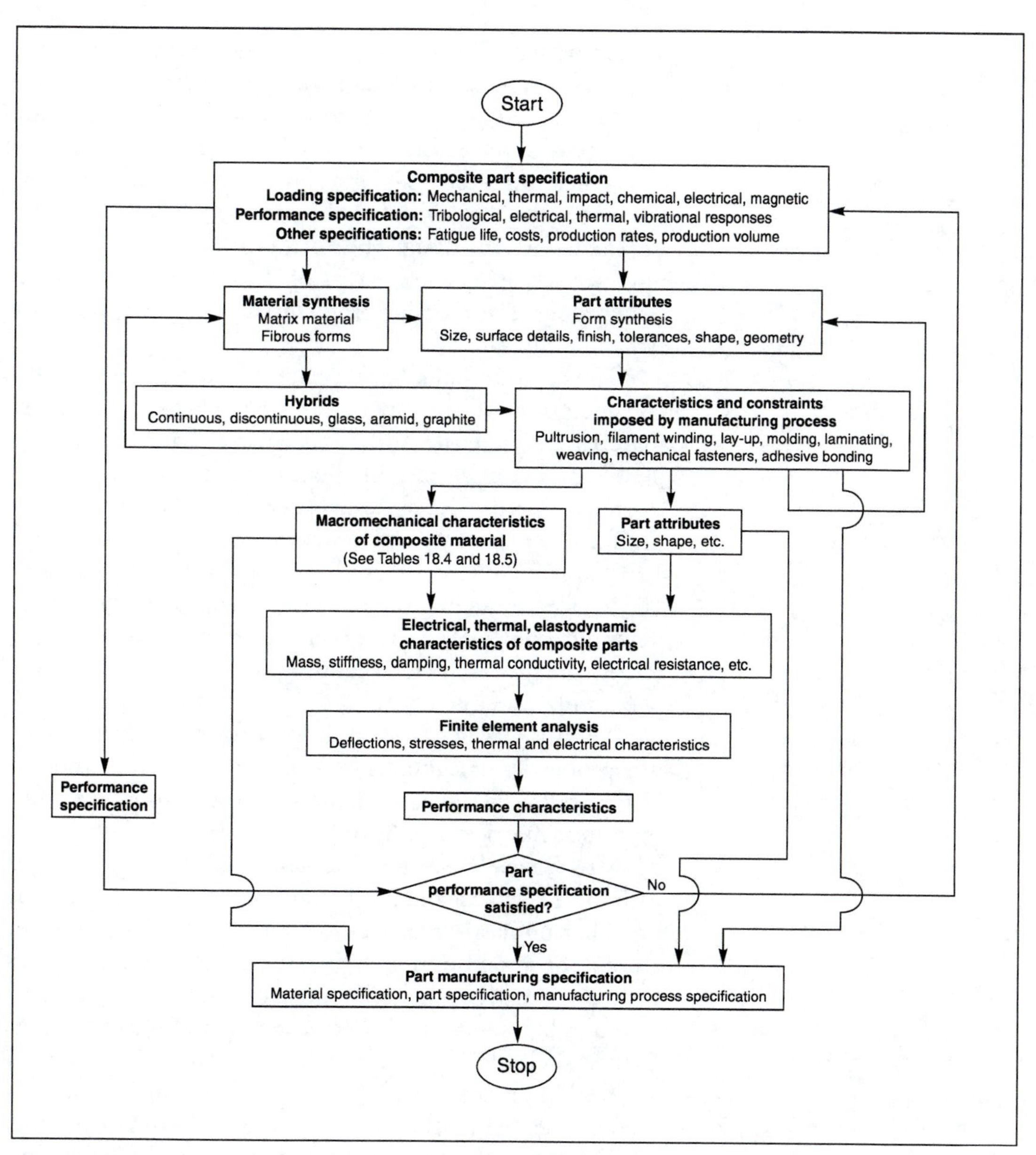

Figure 18.25

A design-for-manufacture methodology for parts fabricated in fibrous polymeric composite materials.

the manufacturing process include fiber forms, production rate, size, shape, surface details, and dimensional tolerances.

An iterative process is used to design the part, its material, and the manufacturing process. Once this has been completed, then the basis exists for analyzing and evaluating the part and its performance characteristics and for comparing them with the design specification. Thus the methodology invokes the essential steps of the classical approach to solving open-ended design problems: namely creativity, analysis, and evaluation.

The creation of an appropriate macrostructure involves consideration of the fibers and the matrix. The matrix material can be either a thermoset or a thermoplastic, depending on temperature and manufacturing process considerations. Thermosetting resins such as the epoxy resins are limited to a maximum service temperature of approximately 120°C for highly loaded applications and to approximately 105°C for toughened epoxy resin systems. However, the maximum hot/wet inservice temperature of thermosetting resins can be extended to approximately 260°C by using polyamide systems.

Thermoplastic resins have significantly superior hot/wet properties and impact resistance compared to thermosetting resins. Because of their high strains to failure, thermoplastic matrices are the only ones currently available that allow the new class of intermediate-modulus high-strength carbon fibers to use their full-strain potential in the composite. Thermoplastic matrices do not absorb any significant amount of water but they are vulnerable to organic solvents. Furthermore, the shelf life of thermoplastic matrices is unlimited, in contrast to thermosetting resins which have a limited shelf life.

Shelf-life:

Thermoplastics:	unlimited
Thermosets:	limited

Carbon fibers:
- Strong but expensive

Glass fibers:
- Cheap, good impact resistance

Aramid fibers:
- Damage tolerant but more expensive than glass

Carbon, aramid, and glass fibers are readily available on the commercial market. Carbon fibers are commonly used for high-strength applications and are characterized by longitudinal Young's moduli of 450 GPa and tensile strengths of 4000 MPa. Glass fibers are much cheaper, with typical Young's moduli of 70 GPa and tensile strengths of 3450 MPa and are more suitable for impact-loading situations.

Furthermore, glass fibers have good handleability, ease of processing, and environmental durability. However, the sensitivity of the glass fiber surface to attack by moisture and their stress rupture under long-term loading (at stress levels exceeding the threshold) are important design considerations. Aramids exhibit typical Young's moduli of 130 GPa and tensile strengths of 3400 MPa. However, these aromatic polyamide fibers display far lower strengths in comparison to carbon fibers, they have a tendency to absorb moisture, and they adhere relatively poorly to matrix resin systems. In spite of these drawbacks, aramid fibers are surprisingly damage tolerant and are frequently used in filament-wound rocket motor cases, pressure vessels, and secondary structures on fixed wing commercial aircraft and helicopters.

High-stiffness components which must operate in environments with a likelihood high-impact loads can be designed in materials which combine the high stiffness properties of graphite fibers with the high impact resistance properties of glass fibers. Graphite/glass materials are examples of hybrid material systems that capitalize on the ability of composites to be tailored to provide desired properties. The freedom offered by this material synthesis process can be further exploited in situations where the primary requirements are high strength and low cost. Thus by specifying the more expensive high-strength graphite fibers in the highly stressed regions of a part while specifying cheaper glass fibers in the lower-stressed regions, the designer can control the material cost of the part while satisfying the high strength constraint.

Hybrid materials:

- A nominally homogeneous mixture of graphite and glass fibers provides high strength and impact resistance.
- A high-strength design subject to a cost constraint can be obtained with high-strength fibers in the highly stressed regions and a cheaper low-strength fiber elsewhere.

The fibers employed in the manufacture of a composite part may be either continuous or discontinuous. Continuous reinforcing fibers are commercially available in forms ranging from monofilaments to filament fiber bundles and from unidirectional ribbons to single layer fabrics and multilayer fabric mats. The reinforcing fibers and matrix resins can be combined into many different intermediary products that are designed for specific fabrication processes. Combinations of fiber ribbons or woven fabrics with the resin system are called *prepregs*. They have very precisely con-

Composite parts can be fabricated from continuous or discontinuous fibers. The decision influences the mechanical properties and the manufacturing processes.

Prepegs:

- Precise fiber-resin ratio
- Quality tack and drape
- Control of resin flow during cure
- Create a better part

Continuous fibers are needed for:

- Filament winding
- Pultrusion
- 3-D weaving

trolled fiber/resin ratios, highly controlled tack and drape for thermoset matrices, controlled resin flow during the cure process, and in some fabrication processes allow better control of fiber angle and placement to facilitate the fabrication of composite parts to stringent tolerances.

The choice of discontinuous fibers for a part constrains the manufacturing process to molding, whereas the choice of continuous fibers is a necessary condition for fabricating the part by filament-winding, pultrusion, or 3-D weaving and lay-up techniques. The constraints imposed upon the selection of a manufacturing process after specification of a continuous or a discontinuous fiber are shown in Figure 18.26. In a composite reinforced with continuous fibers, the fibers provide virtually all the load-carrying capacity, and they redistribute the load if several fibers fail. This is like a green-stick fracture and is much more desirable than a catastrophic failure of the structure. Of course a structure made with discontinuous fibers would generally be much weaker than an equivalent structure made with continuous fibers and hence is more likely to fail catastrophically .

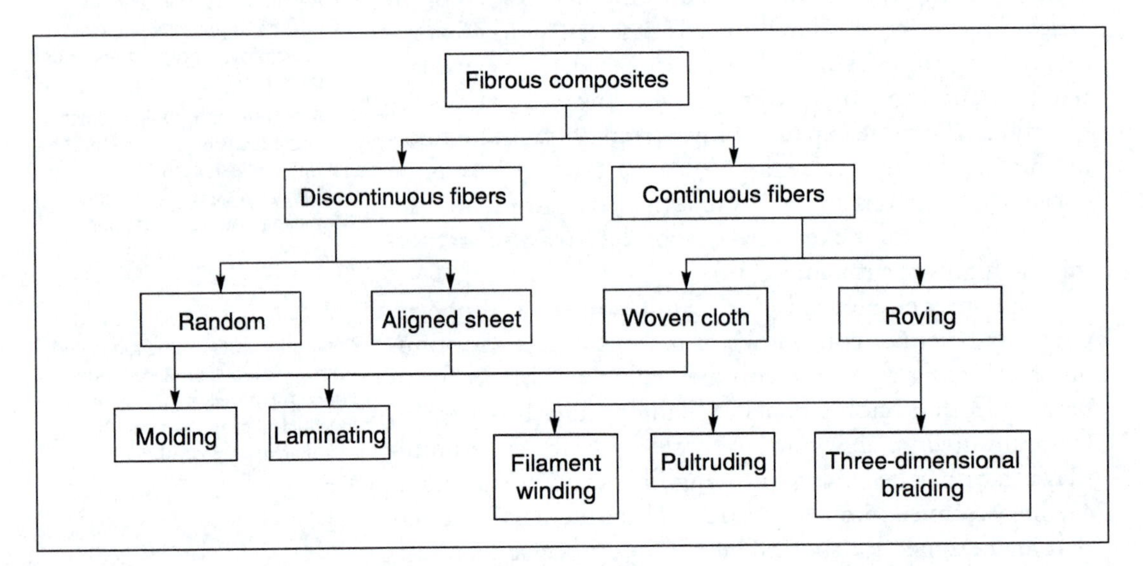

Figure 18.26

Effect of continuous and discontinuous fiber selection on the choice of manufacturing process.

The selection of a manufacturing process is governed by the shapes that process can form.. For example, filament winding is ideally suited to making pipelike parts using continuous fibers, while injection molding can create parts of complex shape in a closed mold using chopped fibers. More complex shapes can be made by sophisticated filament winding techniques. Since the internal surface on a filament-wound part conforms to the winding mandrel, the dimensional tolerances on this surface are quite accurate. However, the external shell is usually rough, which may be undesirable. This can be overcome by several approaches such as adding an external sacrificial fiberglass wrap that can be ground down to the final dimensions. Similarly, pultrusion is cost-effective when constant cross-sectional profiles are sought. Hand lay-up is efficient and cost effective for low-volume custom-designed parts with intricate geometries. Considerations of this kind are necessary to create a viable part.

Each manufacturing process imposes constraints on the size and shape of the part.

It is evident that the choice of continuous versus discontinuous fibers in the material-design phase significantly affects the choice of a manufacturing process, as shown in Figure 18.26. The manufacturing process in turn constrains the geometry of the part. Furthermore each manufacturing process results in a distinct fiber volume fraction as well as a distinct fiber orientation and fiber distribution in the part. This imparts anisotropies, such as aeolotropies and orthotropies, which produce certain mechanical characteristics such as stiffness, damping, mass, etc. For example, in the case of filament winding techniques alone, different combinations of hoop, longitudinal, and helical winding patterns result in parts with different stiffnesses, damping capacities, and mass characteristics. These in turn govern the overall vibrational performance of the part under service conditions.

Each manufacturing process imposes constraints on the distribution of fibers in the part and this in turn governs the mechanical properties.

Thus there is an iterative cycle of selecting a material with the desired properties, the design of a part with the desired characteristics, and the selection of a manufacturing process that will produce an acceptable version of the part. Its attributes are then responsible for the performance of the

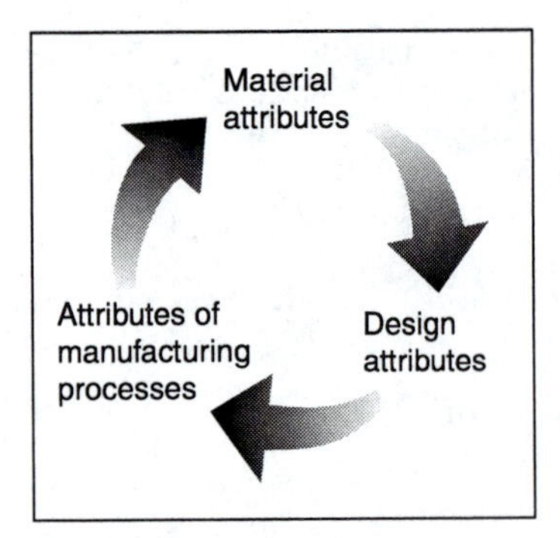

Figure 18.27

The iterative DFM methodology for polymeric composites.

Finite element methods can be used to analyze the behavior of composite parts subjected to complicated loads.

part under the conditions that the part must withstand. This iterative cycling is illustrated in Figure 18.27.

The elastodynamic behavior or indeed other performance characteristics such as thermal or electrical responses of composite parts can be simulated by finite element analyses of a part created by this iterative design methodology. These analytical methods are mature and are capable of accurately predicting the response of a part with a complex shape and macrostructural properties as it is subjected to complex loading. This is accomplished by breaking down the part domain into small elements of different shapes and sizes prior to generating a set of nonhomogeneous algebraic equations to yield a solution at discrete points throughout the part. These numerical solutions must then be evaluated by comparing them with the performance specification for the part. Typically several different parts will be created and characterized in several different materials by several different sets of parts attributes, and by several different manufacturing processes. The performances of the various designs can be simulated and subsequently evaluated relative to the original design specification by fabrication and physical testing in the laboratory. The most appropriate design is subjected to further synthesis, analysis, and evaluation. Ultimately, the methodology provides the specification of the composite material, the attributes of the part, and the specification for the manufacturing process.

The DFM methodology provides the:

- material specification
- attributes of the part
- specification of manufacturing process

The macrostructure of a material designed by this methodology, the attributes of the shape and surface properties of the part, the characteristics imposed by the manufacturing process, and the costs are intertwined. However, the large number of design parameters associated with composites can facilitate the development of cost-effective, high-performance composite parts. This can only be obtained by an iterative approach that includes many manufacturing considerations.

Summary

Composite materials are a combination of two or more constituents differing in form or composition at the macroscopic level. They are nontraditional materials that can be made with unusual properties to satisfy a prescribed need. This is accomplished by selecting an appropriate form and distribution of ingredients in the composite and by the careful selection of the manufacturing process. This manufacturing process is responsible not only for creating the part but also for creating the properties of the material. The advantages and disadvantages of composite materials and of manufacturing processes have been discussed. Polymeric fibrous composite parts can be developed by iteratively and interactively considering the design-for-manufacture rules presented in Table 18.7.

- Design specification of the part
- Geometrical attributes of the part
- Macromechanical design of the material
- Hybrid materials
- Attributes of the manufacturing processes

Table 18.7

DFM rules for polymeric composite parts.

Key Concepts

Aramids. High-strength, high-modulus fibers used to reinforce composite parts. Examples include the Kevlars and Nomex products.

Braiding. A method for manufacturing fibrous polymeric parts by weaving fibers into a tubular shape rather than a flat fabric.

Carbon fiber. An important high-performance fiber used to provide high strength and high stiffness in composites.

Composite material. A solid material made up of two or more materials called reinforcing elements and matrix binders, sometimes with fillers. The constituents do not dissolve or merge but retain their identities. Normally the constituents can be physically identified and they exhibit an interface.

Compression molding. A process for molding composite parts that is characterized by an open mold for accommodating the material before closure and curing by the application of heat.

Cure. To change irreversibly the physical properties of a thermoset material by a chemical reaction.

Fiber. A general term for a piece of filamentary material with a length that is typically at least 100 times greater than its diameter. The diameter is typically of the order of 0.10 to 0.13 mm.

Filament winding. A process for making fibrous composite parts by winding continuous reinforcing fibers around a rotating removable mandrel.

Fiberglass reinforcement. A strengthening material used in thermosets and thermoplastics. Available in a wide variety of fibrous forms.

Hand lay-up. A fabrication process for composite materials involving the placement in a mold of successive layers of reinforcement which have either been preimpregnated with a binder or which will be coated with one. The material is then cured to create the final part.

Hybrids. A composite material consisting of two or more material systems. Hybrid designs can include different classes of fibers such as carbon and glass or they can be a composite containing a mixture of continuous and discontinuous fabrics.

Laminate. A fibrous composite material made up of plies or layers which are generally bonded together by the application of pressure and heat.

Macrostructure. The general form and distribution of constituents, impurities, etc., in a metal, alloy, or composite material. The macrostructural characteristics are revealed by examination with the unaided eye or at low magnification.

Matrix. The component of a composite in which reinforcing fibers are embedded.

Metal-matrix composite. Composite materials made of carbon or ceramic fibers embedded in a metal matrix.

Microstructure. The structure of metals, alloys, and composite materials as revealed under the microscope.

Ply. A fabric of reinforcing fibers employed in the manufacture of composite materials.

Prepreg. A resin-impregnated cloth of accurate filament-resin ratio that can be stored until used to manufacture a composite part.

Pultrusion. A continuous process for manufacturing composite parts that have a constant cross-section and contain continuous fibers.

Reaction injection molding. A molding process for manufacturing composite parts in which two or more liquid polymers are mixed prior to entering a closed mold where they react and solidify.

Reinforcement. A material added to a matrix binder to create a composite material. Typical reinforcements are fibers and whiskers.

Resin. A polymer used to bind together the reinforcing material of a fibrous composite.

Resin transfer molding. A molding process used to manufacture composite parts in which resin is added to a closed mold containing reinforcing fibers. The finished part is released after it has cured.

Spray-up. A method for fabricating parts from chopped fibers by using a spray gun that mixes chopped fibers and resin prior to depositing the mixture into a mold where it subsequently hardens.

Thermoplastic. A plastic material capable of being softened repeatedly by heat and hardened by cooling.

Thermoset. A plastic material that can be irreversibly cured to a solid state.

Vacuum-bag molding. A process in which a flexible film is placed over the lay-up in a mold, after which the film edges are sealed and a vacuum is drawn to remove air and compress the assembly into a high quality part.

Review Questions

1. Define a composite material and discuss some of its common ingredients.
2. What are the principal kinds of composite materials?
3. What are the roles of reinforcements and of the matrix in a composite material?

4. Discuss the different classes of reinforcement and their use in composite materials.
5. List some advantages of composite materials.
6. List some disadvantages of composite materials.
7. What is a hybrid composite? Why are hybrids useful?
8. What role can composites play in parts consolidation?
9. List five applications of fibrous composite materials.
10. List and discuss the important design parameters in the creation of a composite material.
11. List and discuss the important forms of fibrous reinforcement.
12. Why are the design and manufacture of fibrous composites intertwined?
13. Describe the advantages and disadvantages of each of the processes listed in Table 18.6.

Problems

1. The introduction of composite parts requires complex trade-offs between conflicting constraints and design cascading consequences. Green design issues, cost, and performance are three important factors in the materials selection process. Discuss these issues using automotive drive shaft data presented in the next paragraph.

   Consider four design candidates for an automobile drive shaft:

   - Design 1 is the conventional design fabricated in steel. It is the reference design relative to which all others are measured.
   - Design 2 is fabricated in aluminum and weighs only 0.8 of Design 1 but costs 1.1 times Design 1.

- Design 3 is fabricated in glass-fiber-reinforced plastic and weighs 0.92 of Design 1 but costs 1.5 times Design 1.
- Design 4 is fabricated in carbon-fiber-reinforced plastic and weighs 0.52 of Design 1 but costs 1.7 times Design 1.

Design 4 is a one-piece design but all others are two-piece designs containing a midspan universal joint. Evaluate these design candidates in the context of both design and manufacturing issues such as functionality, process selection, weight, performance, recyclability, corrosion, service life, and cost.

A one piece lightweight part was fabricated as a fiberglass-graphite hybrid pultruded over an aluminum tube by Morrison Molded Fiber Glass Company (MMFG). This drive shaft was introduced on the General Motors 1988 model GMT-400 pickup truck. Comment on this design relative to your analysis and evaluation of the previous design scenario.

2. During the past hundred years or so several new designs for canoe paddles made from a variety of materials have been proposed.

   a. Develop a design specification for a canoe paddle. Which issues are important and what types of loads are imposed upon the product?

   b. Propose several conceptual designs involving both metals and polymeric fibrous materials. Subsequently, evaluate them. Specify the manufacturing processes to be used to produce these paddles.

3. Baseball bats have one axis of symmetry and they can be purchased in several forms in sporting goods stores, including those made from fibrous polymeric composites. Visit a sporting goods store and analyze the different bats on sale there.

   a. Develop a design specification for a baseball bat. What are the desirable material properties?

   b. Generate several conceptual designs, including some fabricated from composite materials.

   c. Analyze and evaluate each concept and develop an optimal design.

4. Use the approach of Question 3 to develop optimal designs for fishing poles, golf clubs, and tennis rackets.

5. New materials have changed the bow and arrow over the years, as is evident from a visit to a sporting goods store. Early bows were made from naturally shaped woods including the English yew purportedly used by Robin Hood in Sherwood Forest. Leonardo da Vinci proposed a laminate design for the bow, and others have been built in metal and composite materials with complex shapes. Arrows have undergone a similar metamorphosis.

   a. Develop a design specification for a bow and arrow.

   b. Develop several conceptual designs for a bow and for an arrow. Include composite materials.

   c. Analyze and evaluate the concepts and propose an optimal design.

6. A planar slider crank mechanism has a connecting rod which vibrates in the plane of the mechanism. If the link is assumed to be modeled as a straight, slender, flexible beam subjected to inertial bending loading, select an appropriate material from which to manufacture the articulating member. Comment on the design criteria used in this materials selection exercise. Comment on the disadvantages of the proposed design and material selected.

7. A planar four-bar linkage is often used as the basic kinematic chain for robotic systems. The linkage is often situated on a rotating pedestal which provides the

three-dimensional motion and the coupler link is extended beyond one of the revolute joints of the classical four-bar mechanism. At the extremity of this extension is the end-effector or robot hand. Imagine this machine modelled as a combination of hollow box-shaped members and assume that the dominant criterion for evaluating its performance is small transverse deflections in the plane of the articulating linkage and also a rapidly attenuating transient response upon completing a maneuver. Discus the material selection process and the fiber orientations if the articulating members were fabricated as laminates of continuous fibers.

Further Reading

Argarwal, B. D. and L. J. Broutman. *Analysis and Performance of Fiber Composites.* New York: John Wiley, 1980.

Ashbee, K. *Fundamental Principles of Fiber Reinforced Composites.* Lancaster, PA: Technomic, 1989.

Broutman, L. J. and R. H. Krock, eds. *Composite Materials, Volumes 1-8.* New York: Academic Press, 1975.

Christensen, R. M. *Mechanics of Composite Materials.* New York: John Wiley, 1979.

Dostal, C. A., ed. *Engineered Materials Handbook, Volume One: Composites.* Materials Park, OH: ASM International, 1987.

Gandhi, M. V. and B. S. Thompson. *Smart Materials and Structures.* London: Chapman & Hall, 1992.

Harris, B. *Engineering Composite Materials.* London: The Institute of Metals, 1986.

Langley, M., ed. *Carbon Fibres in Engineering.* London: McGraw-Hill, 1973.

Jones, R. M. *Mechanics of Composite Materials.* Washington DC: Scripta Book Company, 1975.

Lubin, G., ed. *Handbook of Composites.* New York: Van Nostrand-Reinhold, 1982.

Milewski, J. V. and H. S. Katz, eds. *Handbook of Reinforcements for Plastics.* New York: Van Nostrand-Reinhold, 1987.

Philips, L. N., ed. *Design with Advanced Composite Materials.* London: The Design Council, 1989.

Piggot, M. R. *Load Bearing Fibre Composites.* Oxford, UK: Pergamon, 1980.

Rosato, D. V., D. P. DiMattia, and D. V. Rosato. *Designing with Plastics and Composites: A Handbook.* New York: Van Nostrand-Reinhold, 1991.

Schwartz, M. M. *Composite Materials Handbook.* New York: McGraw-Hill, 1984.

Strong, A. B. *Fundamentals of Composites Manufacturing: Materials, Methods, and Applications.* Dearborn, MI: Society of Manufacturing Engineers, 1989.

Tsai, S. W. and H. T. Hahn. *Introduction to Composite Materials.* Westport, CT: Technomic, 1980.

Vincent, J. F. V. *Structural Biomaterials.* London: Macmillan Press, 1982.

Whitney, J. M., I. M. Daniel, and R. B. Pipes. *Experimental Mechanics of Fiber Reinforced Composite Materials.* Brookfield Center, CT: Society for Experimental Stress Analysis Monograph No. 4, 1982.

Epilogue

Never give in, never give in, never, never, never, never—*in nothing, great or small, large or petty—never give in except to convictions of honour and good sense.*

Sir Winston S. Churchill
(1874 - 1965)

From a speech given October 29, 1941
at Harrow School, Harrow-on-the-Hill
England

Disce Ut Semper Victurus

Vive Ut Cras Moituris

Learn as if you were to live forever.
Live as if you would die tomorrow.

Isidore Archbishop of Seville
(c 570-636)

An Epilogue: The Designer's Toolkit

Creative people such as painters, sculptors, carpenters, and die-makers employ tools in order to function innovatively in their respective crafts. Thus a sculptor uses mallets and chisels to fashion stone while a carpenter uses saws and planes to fashion wood.

The engineering designer is also a creative individual—a craftsman—who uses a tool kit. This text and the resource literature listed in it are some of the essential tools required by engineering designers as they ply their trade.

These tools enable the designer to define problems, create diverse solutions, critically evaluate solutions, optimize solutions, and to arrive at cost-effective solutions through the use of various design strategies. It is important that these tools do not collect dust on the bookshelf of the engineer's library. Instead, they should be honed, sharpened, and refined in practice for the betterment of both humanity and the nation. The development of creative solutions to engineering design problems is a challenging activity that requires fortitude, strength, conviction, and tenacity. Good Luck!

Part Three

Appendices

Appendix A

Communication

Appendix B

Design-Build-Test Projects

Appendix C

Glossary

Appendix D

Bibliography

Appendix A: Communication

Thinking cannot be clear till it has had expression. We must write, or speak, or act our own thoughts or they will remain in a half torpid form.

Henry W. Beecher
(1813–1887)

Communication

An essential ingredient of any successful enterprise is the flow of information and ideas among individuals and departments. This flow of information is called communication. Communication is concerned with personal interactions such as listening, reading, writing reports, and the preparation of technical drawings and graphics. Thus communication is an essential ingredient of the engineering profession.

Contents

A.0 Introduction to Communication

Since most communication activities involve solving open-ended problems, they can be addressed by the iterative use of creativity, analysis, and evaluation first presented in Chapter One, Figure 1.1. These communication skills are developed by repeated practice and refinement.

In Chapter One we learned that communication is an essential weapon in the arsenal of a successful engineer. If an engineer cannot communicate effectively—especially to decision makers in a company—then brilliant technical ideas and innovative design concepts may be lost. Good communication relates to all stages of the product realization process, but it is often difficult to achieve because engineers are often more interested in tinkering with a piece of equipment or undertaking mathematical analyses than in communication.

In this chapter we shall discuss three kinds of communication: writing, public speaking, and some unspoken aspects of public appearances. Common to all three is the need to think and reason. Clear communication requires a clear understanding of the subject and the refinement of ideas based upon that understanding. The final form of the communication requires the author to think, evaluate ideas, and learn from them.

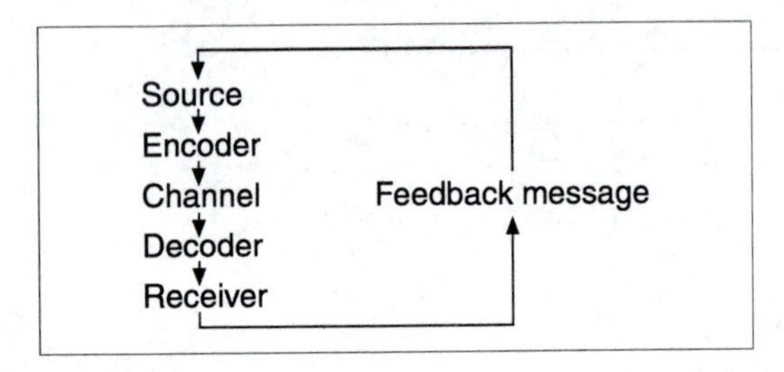

Figure A.1

A schematic model of communication.

A.1 Some Basics

Communication is important in engineering because, as you learned in Chapter 5, the product realization process is a team activity. Rarely will an individual in industry be solely responsible for designing a product. The engineer must interact effectively with a wide variety of people in different disciplines and be able to influence others through communication. You will need to communicate effectively with your subordinates, peers, and management.

It has been said that "Communication takes place when one mind so acts upon its environment that another mind is influenced, and in that other mind an experiment occurs which is like the experience in the first mind, and is caused in part by that experience." Several different communication processes exist, but they all stimulate the human senses. They are modeled in Figure A.1. Thus, communication consists of six elements: a source; an encoder; a message; a channel; a decoder; and a receiver.

Communication begins with a *source.* Information is assembled by one or more individuals to be communicated to one or more people. This information is then *en-*

coded into a form of message that depends upon the medium to be employed for transmission: the *channel.* To communicate you must identify the audience and the means of reaching them. Will you telephone, write a letter using paper or e-mail, send a memo, generate a sketch or an engineering drawing, or write a formal report? Are you to speak to a group or simply exchange ideas with a coworker? Which is most appropriate? Hence there are numerous channels for transmitting information. Upon stimulating the human senses, the message is *decoded* before it is processed by a *receiver* such as an individual or a group of people, and *feedback* is generated. This feedback to the source of the communication can indicate whether the intended message has been received and understood. Appropriate modifications can then be made if necessary to clarify the message.

A.2 The Design Notebook

Communication derives from some source, as shown in Figure A.1. In engineering design the source for the written word or the spoken word is often the design notebook. It chronicles the product realization process and contains essential information about a product from early marketing studies and the product design specification to manufacturing data. Each member of a design team probably will own several of these books. They are receptors for all information about a product and its evolution.

A design notebook should be a resource from which a set of events can be recalled accurately several years later. It must contain the original entries describing events, assumptions, decisions, and conclusions along with relevant theoretical and experimental data. Table A.1 lists some characteristics of a design notebook. This notebook is of crucial importance in the context of intellectual properties: namely ideas and how they are represented. It is particularly relevant to publications, computer programs, and inventions that are legally protected by copyrights, patents, and trademarks.

Clearly, design notebooks are very valuable tools and resources in the product evolution process and are essential in all engineering design activities.

A.3 The Written Word

Writing should be clear and unambiguous. Furthermore, its style should hold the reader's attention. This can only be accomplished by practice and then more practice—writing and then rewriting. Writing and graphical communication skills can be enhanced by reading quality literature such as the *National Geographic Magazine* or *Scientific American.* Reading high-quality publications can help your design activities in numerous ways.

The writing of technical literature, such as memoranda, reports, and proposals is an iterative process whose primary attributes are presented in Table A.2. Six steps of the process are now discussed in more detail.

1. It should be durable, hardbound, (not loose-leaf), and page-numbered sequentially. All entries should be written in permanent ink and dated. Pages should not be torn out of the notebook.
2. The project number or project title should be recorded to readily identify the project.
3. It should record all original work pertaining to a design project. Thus it may contain a mixture of useful and not-so-useful information. All information *must* be recorded because the future is always uncertain and—in a new context—information that appears useless today may be valuable tomorrow.
4. It should contain all of the plans, calculations, observations, experimental procedures, and data from experiments and prototype testing in such detail that the work could be repeated later by someone else. All assumptions, mathematical models, decisions, and evaluative thinking—both good and bad—should be recorded. This information should be signed, dated, and witnessed. The witness must be an individual who understands the technical aspects of your work.
5. It should be a diary of events; consequently important dates, meetings, and attendees should be recorded. Relevant telephone and fax numbers, E-mail addresses and other information should be recorded. Summaries of important telephone conversations and other communications should be written as they occur.
6. It should contain an index of important information as the project evolves. This index can be established at the beginning of the notebook so that it evolves with the project. The retrieval of information can be simplified by an organizational structure of headings and cross references throughout the notebook. Use the pages liberally. Do not try to cover every square millimeter with writing, but cross out and initial blank spaces to prevent later insertion of information. Clarity is important if information is to be retrieved quickly.
7. Mistakes should be crossed out in ink, initialed, and dated. The corrected version should appear on subsequent pages.

Table A.1

Characteristics of a design notebook.

Define the Purpose of the Document

Before you begin to write, be absolutely clear about the purpose of the document. Define the problem! Write a sentence defining the objective of the manuscript in the least number of words. Do you want to impart knowledge? Are you asking the recipient to do something? Are you soliciting a response? Clarify your purpose before you formulate a plan to achieve your objectives.

Assemble Information

Writing begins with a plan. Plan a communication project as you would any other. Use a Gantt chart or some other time-management scheme. Understand the specifics of the assignment, the characteristics of your readership, any constraints on mathematical explanations, and the types of words used in the document. Are there constraints on the number of pages, page format, number and sizes of figures, tables and graphs, etc? You must know the deadlines and constraints on your time to avoid being rushed during writing.

Assemble information you think may be necessary. You may need your design notebook and related literature. Start putting ideas down on paper. Don't worry about organization yet—just write down your thoughts and notes. Discouragement is quite natural at this stage of many writing tasks. You may feel overwhelmed.

Define the purpose of the document
Assemble information
Outline the document
Write the first daft
Edit the document
Rewrite the document

Table A.2

How to write a document.

Outline the Document

Out of the chaos and clutter of assembling information you will begin to see the big ideas and the less important ones—the first glimmer of an outline. An outline, as it grows and takes shape—organizes your thoughts and directs your writing so that it becomes a logical progression of ideas. The various sections become a series of headings and subheadings. This skeleton provides the basis for your subsequent writing, and it may be used as a guide for readers.

Most documents are based upon the three-part model of an introduction, a main body, and a conclusion. Thus it embodies the old adage of ...

1. Introduction: tell 'em what you're gonna tell 'em
2. Body: tell 'em
3. Conclusion: tell 'em what you've told 'em.

The introduction contains a preview of what you are writing about. It introduces the body of the document so that the background and ramifications are clear to readers who are not experts in the field. The introduction should strive to ignite an interest in the main body of the document and should encourage the reader to continue.

The body of the manuscript is its longest part. It contains facts, ideas, logical arguments, analyses, supportive themes, details, and illustrative examples. Finally the conclusion restates the primary information and summarizes the document.

Write the First Draft

The secret of first drafts is to simply record your ideas on paper or into your computer. Use your outline and expand it—even if only in the form of sequential lists. Initially write the text to present the logical flow of ideas. Don't wait until the perfect sentence materializes. Don't worry about typos, spelling, or syntax. You can address these issues later. Neither should you feel compelled to begin with the introduction and write the manuscript from beginning to end. It is often easier to write the introduction after the main body of your document has been written. Provide ample space for the inevitable corrections and modifications. If you are using a pencil and paper, leave a blank line between each line of writing. If you are using a word processor, print your document double-spaced.

Edit the Manuscript

Editing is an analytical process involving a critical review of the first and subsequent drafts of the manuscript. It is an iterative process. This is best undertaken at least a day after the first draft was written so that the arguments and logic used to develop the draft are forgotten and you can scrutinize your work with a fresh, critical eye. The text must be read with great attention to details. Reading aloud may help you identify the "feel" of sentences and sections that need to be reworked. If you use a word processor, you may find it helpful to print a paper copy now and then to develop a different feel for the work.

Editing is of two kinds: content editing and copy editing. When you edit the content of a document your primary concern is to develop a manuscript with the best organization and the best combination of words to express your ideas clearly. Copy editing is a later editing stage in which the nearly-finished document is meticulously examined for misspellings, grammatical errors, superfluous words, and other details. We will discuss it more fully in a section on the details of writing. At this stage the objective is clear writing. However, this objective cannot be achieved the first time; good writing requires rewriting.

Rewrite the Document

Writing is an open-ended problem—a design problem. It can be addressed by the methodology presented in Section 1.1 involving creativity (a draft document), analysis and evaluation (editing), and further creativity (rewriting). This is an iterative process. Most of the time you spend creating a document should be spent rewriting it. Rewriting can consume eighty percent of your writing time, and the document can be rewritten many, many times if it is complex and important.

The Details of Writing

The previous section took a global look at writing a document. We now focus on the details. Good technical writing requires accurate facts, sound ideas, and a logical organization. But the little things set professional work apart from the unsuccessful beginner. Several rules may help you write a good manuscript. They are summarized in Table A.3 and a discussion of each one follows.

Checkpoints
Watch for spelling, punctuation, and grammatical errors
Give your report a descriptive title
Start with the big picture—an overview of the report
Using headings and subheadings to give structure
Keep paragraphs short and logically related
Use short, variable-length sentences
Use the correct tense
Avoid personal pronouns

Table A.3

Checkpoints as you write a technical report.

- **The devil is in the details**

Spelling must be correct. Look up words you're not sure of, and if possible, use a spelling checker on a computer. Similarly, there is no excuse for poor punctuation in technical literature. Modern punctuation is quite simple. Read—and reread—a good up-to-date guide. Mistakes in the little details can suggest to the reader that you may be quite careless or inaccurate in your technical work.

The tools for this phase of writing are punctuation, spelling, word selection, sentence structure and the flow of ideas from paragraph to paragraph. Use a dictionary and a thesaurus. While text can be edited on the screen of a computer, some people like to print a hard copy of the manuscript periodically and make corrections in the margins with a colored pen.

- **Create a descriptive title**

The title of a report should describe its contents. Thus "Design Report" is an unacceptable title; it is not descriptive enough. A more appropriate one might be "The Design, Fabrication, and Testing of a Birthday-Candle-Powered Vehicle." A delicate compromise must be sought between brevity and detail. There are no hard-and-fast rules but an upper limit on the length of the title might be twelve or fifteen words.

- **Start with the big picture**

The early pages of the document should introduce its primary topic. If the targeted audience contains nonexperts, the early pages should give them an overview and some background information. Discuss the details later.

- **Provide structure for the document**

The document should have a clear structure defined by headings and subheadings. This naturally follows if a good outline has been developed. This structure can be further clarified by the selection of appropriate fonts and type sizes. Headings and subheadings serve to divide a document into major elements in a logical way, and they clarify themes.

- **Paragraph format**

Keep paragraphs short and logically connected, with ideas flowing smoothly through them. Each paragraph should begin with a topic sentence that expresses the primary idea of that paragraph. Other sentences amplify the primary idea.

- **Use short variable-length sentences**

Information transfer and the maintenance of interest by the reader can best be achieved by using short sentences of variable length and structure because short sentences help keep the structure simple, whereas sentences longer than about 40 words, like this one that has 70, are likely to involve more complicated grammatical construction, contain separate ideas that can be better stated by several sentences, and can make documents difficult to read. A balance of short and longer sentences often helps ideas flow smoothly.

- **Use the correct tense**

Use the past tense to describe work done in the past. Use the present tense to describe facts, theories, and results that are as valid now as they were in the past.

- **No personal pronouns**

Technical reports in the engineering profession do not contain personal pronouns. Thus they are devoid of words like "we," "I," "my," etc. Be careful that other pronouns don't create ambiguous language.

Writing a technical report

An engineer must write many kinds of documents—memos, letters, industrial reports, papers for publication in academic journals, etc. One of the most important of these documents is a technical report. The ability to produce a high-quality report is an important skill for any engineer. Indeed these documents are frequently the only permanent record of the evolution of a product other than the design notebooks. Writing a good technical report also is important in design classes. The form of a technical report is suggested in Table A.4. This format is only a guide because many organizations have a preferred format often described in a company style guide.

Format
Title page
Summary
Contents list
Nomenclature
Introduction
Theoretical developments
Experimental investigation
Results
Discussion of results
Conclusions
Acknowledgements
References
Appendices
Tables
Figures

Table A.4

Format for a technical report.

The *title page* contains a concise descriptive statement of the subject of the report. It should tell the reader in less than about a dozen words what is the primary thesis of the manuscript. There should be no doubt. In addition, the title page should contain the name of the author or group of individuals who wrote the report and their affiliation.

This first page is followed by the *summary* or the *abstract*. It should be no longer than a single page of text that concisely describes the contents of the document. The purpose of the abstract is to provide a clear indication of the scope and outcome of the investigation, and the reader should then have sufficient information to decide whether to read the complete document. It is sometimes also used to provide key words and phrases for storage and retrieval by the abstracting services.

The *contents list* provides the skeleton of the document by listing the headings and subheadings of the various sections and their page numbers. Some contents lists also identify appendices, tables, and figures.

The *nomenclature* section lists the symbols and their interpretations used in the document. This is particularly common in mathematical documents. The list should be ordered alphabetically, beginning with uppercase letters and then lowercase letters. Greek symbols should be followed by subscripts and superscripts.

The *introduction* puts the report into perspective with other common information. It provides the reader with background information and explains how the report relates to other reports, incidents, and technical fields.

The main body of the paper can contain several parts, depending upon the type of report. These typically include theoretical developments, experimental investigations, results, and a discussion of results. The *theoretical developments* section contains a description of the system being analyzed, the assumptions of any mathematical models, the justification for these assumptions, and explanations of the computer programs developed or used to generate simulations. The *experimental investigation* section explains the development and operation of equipment used to investigate the behavior of a physical system. Enough detail should be included to permit someone else to create the same system and verify results at a later date. This section also includes an explanation of the experimental procedures used to operate the physical system and to generate experimental data.

Experimental data are presented in the *results* section of the report. Invariably a combination of experimental and theoretical results are presented, along with error bands on the data. Results can be presented as tabulated data, graphs, or figures. Sometimes voluminous results are best presented in appendices or other latter portions of the report. Data must be clearly identified by table and figure numbers, and in the main text the reader should be advised where they can be found.

Results are evaluated and interpreted in the *discussion of results* section of the report. There results are analyzed to develop a specific theme or recommendation. Theoretical and experimental results should be compared and arguments proposed to explain why there are differences between them. Discuss how analytical and computational models could be improved to enhance the predictive capabilities of the mathematical tools. Statistical arguments and experimental errors are also developed in this section.

The *conclusion* section contains the essence of the results and their ramifications. The conclusions are directly related to the title of the report, and this section summarizes findings presented earlier in the document. No new information appears in this section.

Acknowledgements formally recognizes the contributions made by individuals other than the authors. It can also be used to recognize the support staff and others who made contributions. This section often identifies organizations funding the work upon which the report is based.

References to the work of others typically appear in two forms in a technical report or a paper. Publications are cited within the main body of the report and also as a list of references at the end of the report. In the main body of the report the reader can be referred to prior published literature in order to put the current text into perspective relative to the existing literature, to add clarification, and to provide contrast. Various methods are employed to do this; they depend on the policy of a company or the policy of the professional journal in which the paper is to be published. For example, square brackets containing one or more numbers at the end of a sentence may be used to refer the reader to numbered publications in the *references* section. Other publications use an author's name followed by a superscripted number to designate the reference. Some cite the author's name and the year of publication within curved brackets. If an author has several cited papers published in the same year, then the lower case letter *a* can be appended to the year of the first paper and a lowercase *b* can be appended to the year for the second publication. If a publication with three or more authors is referred to in the main body of a report, only the last name of the first author is generally used along with the abbreviation, *et al* (Latin *et alii,* meaning *and others*). All authors' names are listed in the *references* section.

Examples of these three methods are:

> "A major consequence of the lack of tools to assist with design for assembly is that those who wish to improve the design process by using concurrent design must do so totally manually [4–6]."

> "The volumetric flow of fluid through a porous bed was modelled by Darcy[3] in 1856."

> "Weibull (1939) demonstrated a statistical analysis that was particularly effective in describing experimental fatigue data."

At the end of the technical report, a *references* section contains a list of the original sources of cited material. In the absence of a style set by others, references should be listed in alphabetical order dependent upon the last name of the first author of each paper. (Some style manuals require references to be listed in the order of their citation in the report.) The surnames of the authors should be presented along with their initials. Reference to books and industrial or government technical reports should include the authors and title of the publication, the publisher, the city and year in which the report was published, and any necessary page numbers. Professional papers appear in magazinelike publications called *journals* that appear at regular intervals. References to journal articles should include the names of the authors, the title of the article, the name of the journal (sometimes in a standard abbreviation set by a profession society), and the volume, page numbers, and year of the particular issue.

Several formats are in general use to define references. Journals published by the American Society of Mechanical Engineers (ASME) require an unabbreviated format unlike many other journals. Their format for books, a journal article, a thesis, and a conference paper are illustrated as follows. (A slightly different but common format is used for references in this book.)

Wilson, C. E. and Sadler, J. P., 1993, *Kinematics and Dynamics of Machinery*, Harper Collins, New York.

Stoll, H. W., 1986, "Design for manufacture: An overview," *ASME Applied Mechanics Reviews*, Vol. 39, No. 9, 1356-1364.

Tung, C. Y., 1982, " Evaporative Heat Transfer in the Contact Line of a Mixture,"

Ph.D. Thesis, Rensselear Polytechnic Institute, Troy, NY.

Takagi, T., 1992, "A Concept of Intelligent Materials and the Current Activities of Intelligent Materials in Japan," *Proceedings of the First International Conference on Smart Structures and Materials*, Glasgow, Scotland, B. Culshaw, P. T. Gardiner and A. McDonach, eds., copublished by the Institute of Physics publishing and EOS/SPIE, pp. 13–19.

Appendices are bodies of text, data, mathematical proofs, etc., that support the main body of the report but which, because of their detail, length, or complexity, would detract from the flow of the report.

Short *Tables* of data can be included in the main text but extensive sets of tabulated data extending over several pages are best presented in appendices. They always should have sequential table numbers and a title. The table number should be used for reference in the text. If a table is presented within the main text, it should not occupty more than one page.

Figures (line drawings, graphs, and photographs) should be numbered consecutively and should be identified with a title.

A.4 The Spoken Word

Most people spend a considerable proportion of their day talking informally. While this is a common occurrence, many individuals do not relish speaking to a large group. They have stage fright. Nevertheless public speaking is a frequent requirement for a professional design engineer. Designers must present information at regular program meetings or before an audience that may include senior management or clients. The objective might be to explain the status of a project. Because this activity is commonplace in industry, it is an important part of the undergraduate design curriculum.

In this section we suggest some guidelines that will help you prepare and deliver an oral presentation. Much of what has been said about report writing applies equally to preparing to talk publicly. You must know your audience, know your subject, and organize your message as you would if it were written. Let's reflect upon these topics again and think about what else you should do when you must *speak* to your audience. Table A.5 lists some important points.

- Know the purpose of your presentation
- Analyze your audience
- Write down ideas to crystalize your thinking
- Decide on your presentation format
- Repeat important points more than in a written presentation
- Use slides or other forms of illustration
- Use show-and-tell hardward where possible
- Practice your talk
- Anticipate questions and prepare answers
- Think about unspoken messages you convey
- Check out the room and its facilities ahead of time

Table A.5

Checkpoints during speech preparation.

• Determine the purpose of your presentation

Why are you speaking to your audience? You need to define the problem just as when you write a document. What are you trying to accomplish? Are you trying to motivate a group to do something? Are you trying to educate people? The result of this analysis will determine your format once you have analyzed your audience.

• Analyze your audience

Tailor your speech to fit the audience; you must analyze your audience to understand their needs. Typical attributes include their education and knowledge of the topic of your presentation, their relationship to the work you are presenting, and the size of the group. These attributes govern the vocabulary that you can use, the complexity of the ideas you can present, and the kinds of facts you need to introduce.

Thus if you are going to talk to company managers about a product, you might need to focus on development problems, production alternatives, process costs, etc. However, if you are to speak to a group of technicians about the same product, then other topics become more important. These might include maintenance protocols, operating instructions, ergonomics, safety issues, service manuals, and reliability.

What is the attitude of the audience toward the topic of your presentation? Are they inclined to agree, disagree, or be indifferent? If they are likely to belong to the latter category, you may need a stimulating or possibly humorous argument.

The size of your audience affects the formality of your presentation. Speaking to half a dozen engineers gathered around a conference table requires a different approach than addressing an audience of five hundred from a lectern on a stage. But don't let the size of the audience determine the seriousness of your presentation. Your career is just as likely to rise or fall from a presentation in a conference room as from one in an auditorium.

• Write down the contents

You should carefully plan your oral presentation. Write down your ideas much as you would prepare to write a document. This will crystallize your thinking. Henry Beecher's statement on the cover page of this appendix eloquently captures this activity.

Many of the concepts in Section A.3 on The Written Word are applicable to planning and preparing for a speech. You will need an introduction, a main body, and a conclusion. Ideas should flow logically, and you should consider what diagrams and data you will need to present. Should you bring parts, physical models, or engineering hardware as illustrations? Will you need a projector for slides or computer images?

• Presentation format

A speech can be delivered in many ways. You could read from a typed document, you could speak from notes, you could speak in an impromptu fashion, or you could memorize your speech. Whichever method you choose, it must fit the occasion and you must be comfortable with it.

If you decide to read your presentation, then you need not concern yourself with forgetting something, and with practice you can ensure that you can deliver the desired information in the allotted time. However, reading a manuscript rarely leads to a successful speech unless you are trained and experienced in that art—not the usual case for a young engineer. If you decide to read your speech you are likely to speak in a monotone that quickly loses listeners and you will probably never look at the audience. A poorly read speech suggests to the listener that the speaker doesn't know the subject well enough to talk about it in an organized way. Nevertheless, if you do decide to read your presentation, you should type the document in a large-type and double-spaced format so that it can be easily read. In addition, you need to remember to look up at your audience often.

If you decide on an impromptu format for your speech, you will require minimal preparation time. You fly by the seat of your pants. This approach requires intimate familiarity with your subject and permits you to deliver a spontaneous presentation that can be adjusted quickly to the mood and response of the audience. However, you may suffer from an unorganized delivery and consequently may fail to deliver the desired information.

If you decide to speak from notes—often called extemporaneous speaking—you must outline your ideas on a sheet of paper and use it to guide your presentation. This is the most commonly successful type of delivery. It allows spontaneity yet maintains the structure of a written outline. Of course a speech delivered from notes requires a comprehensive knowledge of the subject and a thorough rehearsal.

If you decide to memorize your speech you will need to compose it and store it on paper before committing it to memory. This can involve many extra hours of work. Furthermore, you might forget important points during your delivery, or you might become so engrossed with the recollection process that your presentation becomes unnatural. This approach is only for a trained speaker in a special situation.

- Repetition

Oral and written communication have much in common. One is the structure of the presentation. You should introduce the topic before the main body of the talk and then conclude with summarizing remarks. (Recall a previous section on outlining a document.) There is, therefore, a repetitive element to the presentation. If repetition is overdone it can become boring, but complex technical information presented orally may need more repetition than written material. It may help to use a slide projector or projected computer screen to keep information before the audience and to guide their thinking. You can use a chalk board in the same way.

- Illustrations

Illustrations, diagrams, photographs, charts, films, and engineering drawings are part of most presentations. They often are more effective than speech in emphasizing important points. Recall the old Chinese proverb that "A picture is worth a thousand words." Furthermore, engineers are adept at thinking visually.

Audio-visual aids can add variety to your presentation. They can illustrate, clarify, or dramatize an idea. When preparing these visual aids be sure they are not cluttered. Use no more than about four bullet headings per slide to provide you with prompts for your presentation and a focal point for the audience. Ensure that the text is easily read by individuals at the back of the room. Use color whenever you can.

Once you have placed information on a projection screen, don't simply read the prose; explain and interpret it. This is your queue, your prompt, your note card. Finally, allow your audience at least a minute to absorb each image.

If you write on a chalk board, you must remember to keep turning to your audience to maintain eye contact. Do not stand between a projection screen and your audience and do not talk to the screen instead of to your listeners. Talk to the audience!

- Physical models

Oral presentations are enhanced not only by illustrations but also by show-and-tell activities with physical models. Engineers like to see pieces of engineering hardware, parts, and equipment. Real objects improve understanding and help hold your audience's attention.

Information is collected by the brain through various senses. One of these is touch. If you can bring equipment or parts to your presentation, then do so. If the group is small, pass the parts around or invite people to examine large heavy hardware at the end of your talk.

- Practice your talk

From sports to speeches, practice makes perfect. Adopt the iterative approach presented in Chapter One. After you create your speech, practice in front of a mirror, record the presentation on video or audio tape, or collect a group of friends to listen and criticize your presentation. Analyze the speech, evaluate it, and refine it. Repeat that cycle as often as you can. Repetitive practice builds the confidence that helps overcome any nervousness. Talk, talk, talk, and talk again. Between each practice presentation critically evaluate your performance and then refine it again with more practice. Rehearsals are essential to good public speaking.

It is especially important to check the timing of your talk because invariably you will only have a specified block of time in which to deliver your presentation. If you are required to talk for fifteen minutes then a half hour talk is too long!

Before you start speaking, breathe deeply and look out over your audience. If you have memorized the first few sentences, then your talk will be off to a good start and the subsequent topics will follow naturally. A good start is always a confidence booster. Your opening remarks are important because an audience can be won over or lost in the first five minutes. One of your goals should be to convey enthusiasm. If you sound interested and display enthusiasm then your audience will be more inclined to listen attentively and more likely to learn. Try to talk naturally by varying the cadence, the rate at which you speak, the power of delivery, the emphasis of certain words or phrases, and the tone of your voice. Use pauses where appropriate, just as you would in a normal one-on-one conversation with another person.

Articulate your words carefully and ensure that words are not slurred together. Don't use colloquial or slang expressions in a formal presentation. Avoid the annoying use of filler words such as "like", "okay", "uh", "you know", "er" and "sort of". Obviously you will need to project your voice enough that everyone can hear you. This depends on the size of the room and the background noise level. Will you be using a microphone? Then just speak in a normal voice.

- Questions

Most oral presentations include the opportunity for the audience to ask questions during your delivery or at the end. This is quite normal because the purpose of the presentation is to impart knowledge to your audience and to stimulate a desire to learn more. Therefore as you prepare your talk, anticipate the questions you may be asked and prepare your responses.

If the question-and-answer period is at the end of your presentation, you will need to wait for several seconds at the conclusion of your delivery so that your listeners can collect their thoughts and formulate their questions. If the room is large, repeat the question so everyone can hear it and you can verify that you have understood it. This will also give you time to formulate your response. If you are unable to answer the question, then say so and—if it seems appropriate—offer to try to find the answer and subsequently contact the questioner.

- Body language

During any interaction between people, information is transmitted by words, appearance, and gestures. When you deliver a talk, you need to dress appropriately for the situation. If you are to give a progress report to a dozen coworkers in the company conference room, that might mean your normal work dress. If you are to present a paper at a professional society meeting, you should not appear to have just returned from a photo shoot for a fashion magazine but you should dress conservatively and professionally.

Eye contact with your audience holds attention and induces a feeling of one-on-one conversation. Your body language should suggest that you are knowledgeable about the subject, excited by it, and want to share your excitement with everyone in the room. Express this excitement through the tone and volume of your voice. Avoid a dull monotone. Stand up straight, don't slouch, smile when appropriate, and try to look comfortable and relaxed even if you are tense with stage fright. Deep breathing helps relieve nervousness.

- Venue details

Despite perfect preparation, details associated with the place for your presentation can sometimes interfere with a good speech. Is the room arranged for the type of talk that you have prepared? Is there a cord long enough to reach an electrical outlet to power your projector? Is there a spare bulb? Has the easel that you requested been delivered to support your posterboards? Is there chalk or a marker for the board? How are the lights dimmed when you use a projection screen? These are a few of the questions that need to be answered before your presentation begins. You may want to arrive early at the room so that you can check these details and prepare for any inadequacies.

Summary

You have learned in Part A of this appendix just how important good communication skills are in engineering practice. These skills include writing and speaking as well as certain unspoken kinds of communication.

In a design environment all communication media use the design notebook as a source of the information to be communicated. Once you have decided to communicate information the next step is to decide upon the communication medium. Writing and speaking have much in common. Both need a well-defined purpose, careful planning, repeated rewriting or rehearsal, and much evaluation and refining. Tables A.1 through A.5 summarize some of the important ideas.

Further Reading

The Chicago Manual of Style. 14th ed. Chicago: The University of Chicago Press, 1993.

Barass, R. *Scientists must Write*. New York: Chapman and Hall, 1993.

Beer, D. F. *Writing and Speaking in the Technology Professions*. New York: IEEE Press, 1992.

Blake, G. and R. W. Bly. *Elements of Technical Writing*. New York: Macmillan, 1993.

Ebel, H. F., C. Bliefert, and W. E. Russey. *The Art of Scientific Writing*. VCH, 1987.

Harbinger, S. A., A. B. Whitmer, and R. Price. *English for Engineers*. New York: McGraw-Hill, 1951.

Harris, C. E., M. S. Pritchard, and M. J. Rabins. *Engineering Ethics*. Belmont, CA: Wadsworth Publishing Company, 1995.

Michaelson, H. B. *How to Write and Publish Engineering Papers and Reports*. Philadelphia, PA: ISI Press, 1982.

Minto, B. *The Pyramid Principle: Logic in Writing and Thinking*. London: Pitman, 1991.

Rosenstein, A. B., R. R. Rathbone, and W. F. Schneerer. *Engineering Communications*. Englewood Cliffs, NJ: Prentice-Hall, 1964

Rubens, P. *Science and Technical Writing*. New York: Henry Holt and Company, 1992

Strunk, W., and E. B. White, *The Elements of Style*, 3rd ed. New York, MacMillan, 1979

Appendix B: Student Design-Build-Test Projects

Contents

Student Design-Build-Test Projects

This section of the appendix describes eight shoe-box size design-build-test projects suitable for undergraduate students. Inexpensive projects like these are frequently used to encourage students to develop skills for solving open-ended design problems. As described earlier in this text, such skills are quite diverse and are relevant to industrial practice. The projects require the students

- to plan and manage their activities for several weeks
- to define the problem
- to develop a design specification
- to create several potential solutions
- to develop mathematical models
- to evaluate and test their conceptual designs
- to undertake economic analyses and to work within a budget
- to design and manufacture the product
- to develop oral and written communication skills

Plans for a variety of student projects are described on the following pages. Depending upon the class objectives and student experience, they can be tailored to a specific situation. Some projects limit the objects which can be used to create the device while others do not. Some of these projects involve vehicles because experience suggests that students find these projects to be the most exciting and satisfying.

A Rube Goldberg Device

Objectives

Design and construct a Rube Goldberg device to separate three kinds of objects into their respective groups within two minutes.

Rules

- The device should embody the theme of the famous cartoonist who depicted extremely complicated, intricate, and outlandish devices to perform simple tasks. Typical events might include a series of dominos that are flipped, a rolling object, or a swinging pendulum.
- This particular Rube Goldberg device shall utilize five different sources of energy. Typical energy sources include electrical, mechanical, chemical, thermal, potential, electrochemical, etc.
- One of the objectives of this competition is for the device to complete as many separate events as possible while separating the three classes of objects.
- The device shall be started by pouring into a hopper a mixture of 10 paper clips, 5 glass marbles and 12 toothpicks.
- The device must separate these three kinds of objects into three separate containers labelled toothpicks, marbles and paper clips
- Once the device has started, it must not receive further information from the contestants; it must operate autonomously.
- The device must emit an acoustical signal one minute after starting the series of separation processes and again after two minutes. This signal shall be of no more than 10 seconds duration and no less than 3 seconds duration. The final signal must be sounded by the device as the final event is completed. Only timers designed by the contestants are admissible; commercial timers are not permitted.
- The device must fit inside a space of one cubic meter. All energy sources for activating the device must be located within this cube and all events must occur within this space.

Criteria for determining the winner

Designs should be evaluated by assigning points to each event, to the number of different energy sources used, and to the successful completion of the assigned task. The Rube Goldberg device with the largest number of points is declared the winner.

Mousetrap Vehicle

Objective

Design and build a vehicle powered solely by a conventional unmodified commercial mousetrap.

Rules

- The vehicle shall travel 40 meters down a flat corridor in a straight line and come to a complete stop in the shortest possible time.
- The finish line shall be straddled by a 30 cm by 30 cm square zone. Part of the mouse-trap must be located within that zone when the vehicle has come to a complete stop.
- All parts of the vehicle must travel forward over the race track. Launch ramps are not permitted.
- Although components can be fixed to the mousetrap it must not be modified; the coil spring and the release mechanism must not be changed in any way.
- The vehicle must not be pushed to initiate motion. Motion shall be initiated by activating the standard commercial spring-release lever of the mouse trap with a pencil or similar rod.

Discussion

There are many variations of this particular competition, such as introducing different terrains over which the vehicles must operate, or requiring the vehicles to travel the furthest distance along a building corridor.

Criteria for determining the winner

The vehicle to complete the course in the fastest time shall be declared the winner. Each vehicle should have a reference mark for evaluating ties. If two vehicles have the same time then the vehicle with this reference mark closest to the center of the finishing zone shall be the winner. If a vehicle does not stop within the finishing zone, an additional time penalty shall be imposed.

Egg Transporter

Objective

Design and build a transport system for moving a raw grade-A chicken egg as quickly as possible a distance of 5 to 10 meters from a start position. The egg must stop as close as possible to a rigid terminal barrier without cracking.

Rules

- The energy for propelling the vehicle shall come from a mass of 2 kilograms falling from a height of no more than one meter above the floor.
- The instructor can specify the length of the test track on the day of the race or include it earlier as part of the assignment. If the former approach is adopted then the students must have test-data ready on race day and a method of changing the performance of their vehicle quickly in response to the specification given by the Race Director.
- The entire vehicle launch system must fit inside a cube with a side of length one meter.
- The egg transport system must be at the floor level at the beginning of the test run.
- The system cannot be modified in any way after the falling mass has been initiated. Push or pull starts will not be permitted.
- The egg transport system must have a rigid structure behind the egg and the egg must rest against it; the egg mount cannot be padded.
- The raw egg shall be mounted at the front of the vehicle so that the pointed end of the egg protrudes beyond the front by at least 2 cm. The egg must be mounted 5-15 cm above the floor so that it will make contact with the terminal barrier first in the event of a collision. Tape shall be used to secure the egg to the vehicle. However, no tape shall cover the impact region at the pointed end of the egg.
- To start the test run, all parts of the launch system must be behind the start line and the pointed end of the egg must be on the start line.
- Event timing shall begin when the falling mass is set in motion and shall end when the vehicle has stopped. The stopping distance is defined as the distance from the terminal barrier to the nearest point of the egg.
- Each group shall be allotted five minutes to complete the test run.
- A cracked egg is defined as one which generates a damp spot on a paper towel.

Criteria for determining the winner

The performance of the transporter is defined by the following formula:

Index = 3 x (running time in seconds) + (stopping distance in cm).

The lowest score wins.

A team with a cracked egg shall be ranked behind all others.

Somersaulting Vehicle

Objective

Design and construct a device that repeatedly cycles through a vertical jump, a flip, and a landing on its base.

Rules

- The device shall be powered by two rubber bands of length 3.5 inches and width 0.125 inches. These bands are sold as Number 33 rubber bands.
- The design team shall define the base of the device before the competition begins.
- The device must come to rest before the subsequent flip is initiated.
- The device must fit inside a 6 x 6 x 12 inch box.

Criterion for determining the winner

The winner is the device that repeats the cycle of jumping vertically, flipping, and landing on its base the most times in one minute.

This contest is based upon the ASME 1992-1993 National Student Design Contest.

Flywheel-Powered Gondola

Objective

Design and build a cable car powered by one or more rotating flywheels and suspended from a length of taut music wire. The angle of the wire is constant relative to the horizontal but it gradually increases during a series of competitive heats.

Rules

- The diameter of the wire is 0.032 inch and the length 20 feet.
- A vehicle successfully completes a heat of the competition if it travels 15 feet along the wire.
- The gondola shall receive motive power only from one or more flywheels.
- A physical mark on each gondola shall be used to determine the distance travelled along the wire.
- The gondola shall weigh no more than 12 ounces and it must fit inside a cube of side 12 inches.
- The only energy imparted to the gondola shall be from a human hand uncoiling a cord 24 inches long and wrapped round a sheave, shaft, or reel. This cord must spool off the shaft completely on the first pull.

Criteria for determining the winner

The best gondola design shall be determined by requiring each gondola to participate in a series of contests. Each contest shall be characterized by a different angle of the wire relative to the horizontal plane. Upon completing each heat, the slope of the wire shall be increased by 10 degrees from zero degrees to a maximum of 90 degrees until the winner is determined. The winner shall be the cable car that climbs the wire inclined at an angle that no other design can climb. In the event of a tie, the shortest time to cover a prescribed distance can be used to differentiate between two contestants. Alternatively the furthest distance travelled up the wire can be used.

Propeller-Driven Vehicle

Objectives

Design and construct a propeller-powered land vehicle that shall travel along a straight course for the greatest distance.

Rules

- The vehicle shall remain in contact with the floor at all times.
- External guides are not permitted on the test track although passive guides are permitted at the start of the test run. These guides may touch the sides of the vehicle before the vehicle enters the test course. They cannot be longer than 3 feet and they cannot cross the start line onto the track.
- Vehicles shall be simply released to start the contest; pushing the vehicle is not permitted.
- The only admissible power source is six No. 32 rubber bands approximately 0.125 inch wide by 3 inches long by 0.03125 inches thick. Rubber band driven propellers are the only propulsion method allowed. However, the vehicle can feature single or multiple propellers and a single rubber band, or a combination of several rubber bands. All rubber bands must be attached to the vehicle for the duration of the test run. The rubber bands must not drive the wheels.
- The design team can employ a device for restraining the rotation of the propeller(s) after the rubber band(s) has/have been wound or stretched. This restraining device must not impart energy to the vehicle.
- There is no size limitation on the design of the vehicle.
- It must have three or four wheels.
- Vehicles cannot be assembled from commercial kits; propellers must be crafted by the students. Off-the-shelf items such as ball bearings, extruded tubes, adhesives, wooden sections etc. may be used.

Criterion for determining the winner

The winner shall be the vehicle that travels the furthest distance along the corridor.

Egg-Drop Device

Objective

Design and construct a protective crate to protect a raw egg from breaking when dropped in a free fall from a height specified in the competition rules.

Rules

- The free fall distance is usually between 2 to 7 meters.
- The objective of the contest is to design and fabricate a device that controls the descent of the egg so that it lands at the center of a designated target area without breaking.
- Contestants shall be provided with one raw egg and they must design their protective crate using only the following items:

  Five rubber bands.
  100 cm of masking tape of any width.
  One sheet of standard paper used in a copying machine, size 8.5" by 11".
  Fifteen conventional drinking straws.
  Fifteen tongue depressors or popsicle sticks.
  100 cm of string.

Criteria for determining the winner

The winner shall be the egg-drop device that lands closest to the center of the target area and remains there without breaking the egg. The egg is deemed to have broken if there is an obvious crack or there is enough liquid on the surface of the shell after impact to dampen a domestic paper towel.

An Automatic Ball Sorter

Objective

Design and build a system for automatically separating a mixture of 50 glass marbles and 50 steel balls of equal size.

Rules

- The mixture of 50 glass marbles and 50 steel balls should be of equal size. Samples must be made available to contestants before the day of the competition.
- The mixture will be supplied to the contestants on the day of the competition and the same mixture shall be used by each team.
- The mixture will be supplied in a plastic bucket which may be emptied once into another receptacle, for example, a tray, hopper, or funnel. After this, the balls may not be manipulated further except by the machine.
- The machine may be powered by mechanical and/or electrical means, such as springs and rubber bands, or electric relays, solenoids, and electric motors, and it can feature a variety of sensors, switches and traps.
- The device must operate autonomously. There must be no interaction with the device once the mixture has been poured out of the bucket. The separation process must be self-starting.
- The machine must fit inside a cube of side 1 meter.

Criteria for determining the winner

The winner is the device that achieves the separation in the shortest time. A 5 second penalty should be added to the score for each missorted ball.

Appendix C: Glossary

A

Adaptive design. Either the product resulting from a conceptual design activity involving the adaption of an existing design or the activity itself. The solution is generally achieved without changing the solution principle although it usually requires new parts and assemblies.

Aesthetics. Those attributes of a product that are perceptible through the senses. Aesthetics is concerned with the color, surface texture, and visual form of products, in addition to the consequences of sensory stimulation aroused by smell, sight, taste, movement, and touch, for example. Aesthetics is particularly important in the development of successful consumer products.

Allowance. The tightest fit between two parts. Describes the minimum clearance between or maximum interference between two parts.

Alloy. Material resulting from the addition of one or more other elements to a metal. The motivation is to modify the mechanical and/or physical properties of the original metal.

Alloy steel. A steel which contains small amounts of other elements that modify its mechanical and/or physical properties. These elements include chromium, cobalt, manganese, nickel, phosphorous, tungsten and vanadium.

Ambient environmental conditions. Temperature, pressure, humidity, chemicals, and air quality and their impact on a product or production process.

Amorphous material. Non-crystalline matter; a glass or plastic; not a solid.

Analogy. A method for creatively solving design problems by recognizing some similarity between two things that are otherwise not alike. Helps solve a design problem by adaptation or modification of ideas from other situations.

Analysis. The separation or decomposition of a whole entity into its individual parts for examination. This examination centers on characteristics, function, interrelationships, and behavior.

Anisotropic. A property that depends upon the direction of the axis along which the property is tested. Not isotropic.

Annealing. The process of heating a material which has a stressed structural state to a temperature that will alleviate this stress, before cooling it slowly to ambient temperature. This procedure is employed to enhance the mechanical properties, enhance machinability or to facilitate cold working.

Anodizing. Development of an oxide coating on the surface of a metal.

ANSI. An abbreviation for the American National Standards Institute. This governing body sets standards for conventional engineering products, attributes and technical drawings. There are ANSI standards for interference fits, bolts, taper pins, and key ways for example.

Anthropometry. The measurement of the size and proportions of the human body.

Arc welding. A technique for fusion welding that employs an electric arc as a heat source.

Archiving. The process of storing and retrieving engineering documentation. This pertains to both paper documents and computer stored information.

Artistic thinking. This type of thinking involves randomness, imagination, visualization, synthesis, nonjudgemental scrutiny, irregularity and sensuality. It is the antithesis of deductive thinking and occurs in the right side of the brain. Sometimes called associative thinking, creative thinking, or lateral thinking.

Assembly. The fitting together of discrete parts to make a more complex whole. This can be accomplished by deformation, phase change, addition of material, interference or interlocking.

Assembly drawing. A drawing showing how each part of an assembly is assembled to complete the whole product. This drawing consists of all the parts shown in their operating positions and a bill of materials or parts list.

Associative thinking. Thinking associated with randomness, imagination, visualization, synthesis, nonjudgmental scrutiny, irregularity, and sensuality. It is the antithesis of deductive thinking and occurs in the right side of the brain. It is also called creative thinking, divergent thinking, or lateral thinking.

Autoclave. A closed pressure vessel for conducting a chemical reaction or other operation under pressure and heat.

B

Basic design. A simple sketch to establish the required principle of operation of a potential product. This concept provides a framework for the final product. Generally several basic designs are developed in response to a problem statement. Subsequently they are evaluated and the best concept is transformed into plans for manufacturing the final product. The same as the conceptual design phase.

Benchmarking. A process of comparing a product with competitors' products to measure its value, quality, etc.

Bill of materials. A list of parts or materials used in the manufacture or assembly of a product. This list usually appears on the working drawing of the product.

Billet. A piece of metal with a square cross-section area of approximately 36 square inches that is stock for a manufacturing process. The billet is generally manufactured by forging or rolling.

Biomimetics. A strategy for design synthesis that is based on a direct analogy between a biological system and an engineering system. Thus a biological system is mimicked in engineering practice.

Boring. A metal cutting process to enlarge a hole that has already been drilled. Undertaken with a single-point cutting tool on a lathe or a specialized machine tool called a boring machine. The accuracy of the cut is dependent upon the rigidity of the boring bar spindle and the cutting tool rather than the guiding effect of the drilled hole. A bored hole is accurate and has a good surface finish.

Blank. The initial workpiece for the shaping of a single component or semi-finished part.

Braiding. A manufacturing technique used to produce parts from composite fibrous materials by weaving fibers into a tubular shape rather than a flat fabric.

Brainstorming. A method for creative thinking in which a group of individuals with diverse backgrounds attempts to solve a specified problem by recording all the ideas spontaneously suggested by the group members. The ideas are documented and evaluated before the procedure is initiated again.

Brass. An alloy of copper with zinc.

Brazing. The joining of metallic parts by using a brass filler metal. The melting point of this filler is lower than the parent metals, hence the parts do not melt during the joining.

Brittle material. A material which exhibits minimal plastic deformation before fracture.

Bronze. An alloy of copper with tin. Sometimes this term is used to describe other copper based alloys such as manganese bronze and aluminum bronze.

Burr. The rough sharp ribbon or ridge of metal formed at the intersection of two surfaces after one face has been machined.

Butt joint. A joint in which two parts are joined edge to edge.

Butt weld. A weld between two members which are aligned edge to edge and joined by a weld of substantially the same sectional area and shape as that of the butting edge of the smaller of the two members.

C

CAD. An acronym for computer-aided design or computer-aided drafting. The phrase can refer to the activity of using a computer-based system to support engineering graphics or it can refer to the physical system itself.

CAE. An acronym for computer-aided engineering which is the use of a computer to perform a broad range of tasks needed in engineering a product. These tasks include material selection, design-for-manufacturability protocols, modal analysis, and finite element analysis.

CAM. An acronym for computer-aided manufacturing. This approach uses computers to control production processes. Examples include numerically controlled (NC) machine tools and robots for welding and assembly.

Carbon fiber. An important high performance fiber used in composite parts. It is lightweight, strong, and stiff.

Carbon steel. A type of steel (an alloy of iron and carbon) whose properties depend upon the carbon content rather than other alloying elements.

Case hardening. Processes employed to change the composition of surface of ferrous materials to create a hardened surface zone.

Cast iron. Alloys of iron and carbon containing, as cast, more than 1.7 percent carbon. Generally the carbon content is between 2.4 and 4.0 percent. Silicon, sulphur, phosphorus, and manganese are generally present.

Casting. The manufacturing process of pouring a liquid metal into a mold and allowing it to cool and solidify into a desired shape.

Checklists. This method of creative thinking involves reviewing a list of questions that can trigger associative thinking and the stimulation of new ideas. This approach is particularly useful in variant design and adaptive design. Typical questions are: Magnify? Combine? Substitute? Reverse? Rearrange? Combine? Modify? Adapt? Minify?

Chips. Pieces of metal prized away from a workpiece by a hard cutting tool to create a new surface. These chips can be continuous or discontinuous.

Chuck. A device for holding the workpiece or a toll. For example chucks grasp workpieces in a lathe and a drill bit in a drill press.

CIM. An acronym for computer-integrated manufacturing. This approach links and embeds all manufacturing activities in a single computer system. These activities include accounting, marketing, personnel and other regular business functions.

Closed-ended problems. Problems which are well defined with clearly prescribed methods of solution using analytical methods.

CNC. An acronym for computer numerical control. This approach permits the programmable automation of a machine tool using a computer.

Combinations of elements. This method of creativity involves the recognition of a new relationship among a set of conventional machine elements. A prerequisite is a large knowledge base of off-the-shelf elements that are subjected to continual regrouping until a potential solution is discovered.

Commercial scrap. Recyclable material derived from the activities of businesses such as airlines, restaurants, sports complexes, and stores.

Communication. The transmission of information.

Composite material. A solid material made up of two or more materials at the macroscopic level, such as a fibrous component and a resin matrix.

Compression molding. A technique for manufacturing parts from fibrous composite materials and thermoset plastics in which the material is placed in an open mold cavity before closing the mold and applying heat and pressure to shape the part.

Conceptual design. A freehand sketch to establish the operating principle, vision, or skeleton of a product. Generally a number of different conceptual designs are created during the conceptual design phase of the design process. The same as a basic design.

Conceptual design phase. The very earliest phase of the design process where different concepts are created in response to the design specification. These conceptual designs are evaluated relative to the design specification to identify the most promising for further refinement, analysis, and optimization.

This iterative process culminates in the product design phase of the design process. The conceptual design phase is the same as the basic design phase.

Concurrent design. An integrated design process which simultaneously addresses the interests of product design, process planning, manufacturing, marketing, service, and recycling.

Concurrent engineering (CE). An integrated approach to the development of a new product in which multidisciplinary teams work in parallel from product benchmarking and the development of design specification through conceptualization to the introduction of the product into the marketplace and beyond. Concurrent engineering is the same as simultaneous engineering, unified life-cycle engineering, and integrated product development.

Concurrent product development. An integrated approach to the development of a new product in which multi-disciplinary teams work in parallel from conceptualization through to the introduction of the product in the marketplace. Sometimes called concurrent engineering, unified life-cycle engineering, or integrated product development.

Consumer product. The Consumer Product Safety Act states that a consumer product is any article, or component thereof, produced or distributed for sale to a consumer for personal use, consumption, or enjoyment in or around a permanent or temporary household or residence, a school, in recreation, or otherwise.

Continuous improvement. A procedure in which products and manufacturing processes are improved continuously over a period of years through experience and analysis. The goal is the reduction of costs.

Contour turning. This process occurs when the path of a lathe tool is of curvilinear form relative to the rotational axis of the workpiece. A curved profile is generated on the part.

Convergent thinking. This type of thinking is associated with logic, analysis, judgement and evaluation. It is the antithesis of creative thinking and it occurs in the left brain. Sometimes called vertical thinking, realistic thinking and deductive thinking.

Cope. The top flask or segment of the body of a sand mold for sand casting.

Core. Part of a sand mold that is used to generate a hollow space inside a casting.

Corrosion. The gradual decomposition of the surface of a metal due to chemical or electro-chemical attack. Oxidation is probably the most common form.

Cost analysis. An economic analysis undertaken to determine the cost of a product. There are two types: The first focusses on the resources required to undertake a design project; the second involves the total cost of materials and manufacturing of a product.

Creative thinking. Thinking associated with randomness, imagination, visualization, synthesis, nonjudgmental scrutiny, irregularity, and sensuality. It is the antithesis of deductive thinking and occurs in the right hemisphere of the brain. It is also called associative thinking, divergent thinking, or lateral thinking.

Creep. Plastic continuous deformation of a material over time when subjected to a constant stress. It is dependent upon the material, the initial stress level and the temperature of the material.

Critical path analysis (CPA). Sometimes called the critical path method(CPM). A project planning and sequencing method using a logic network. The method identifies the critical sequence of activities that will influence the completion date of the project.

Cultural impediments. These impediments to creative thinking are imposed on the individual by societal values. Traditions are one type of cultural impediment.

Cure. To change irreversibly the physical properties of a thermoset by chemical reaction.

Cutting tool. A wedge-shaped piece of material used to cut a workpiece. The cutting tool is harder than the workpiece and is shaped to cut. Tools are usually made from high speed steels or cemented carbides.

D

Damping capacity. The ability of a material to absorb dynamic strain caused by shock loading or mechanical vibrations.

Decision matrix. A matrix of rows and columns used to evaluate conceptual designs relative to a number of design criteria. The characteristics of each conceptual design are quantified by one of several approaches and the best design is frequently compared with the ideal design.

Deductive thinking. This thinking is associated with logic, analysis, judgment, and evaluation. It is also called convergent thinking, realistic thinking, or vertical thinking and occurs in the left hemisphere of the brain. It is the antithesis of creative thinking.

Delamination. The physical separation or loss of bond between two plies of a composite laminate.

Delivery date. The date when the product is to be delivered to the customer. The date influences the time and resources to be expended on the design process and also the manufacturing processes used.

Depth of cut. This is the normal distance between the surface being cut and the new surface generated by the cutting operation.

Design. Throughout this text, except for portions of Chapter Two, design refers to the design task in engineering. Design involves the application of diverse knowledge to the development of drawings and documentation that describe how a product is to be manufactured.

Also see:

- Adaptive design
- Basic design
- Conceptual design
- Detail design
- Embodiment design
- Engineering design
- Form design
- Original design
- Product design
- Variant design

Design cycle. An iterative approach to solving open-ended design problems using creativity, analysis, and evaluation.

Designer. An individual who practices the arts, sciences, and technologies inherent in the design process. This individual is a generalist who exploits the characteristics of the right brain and the left brain to perform both creative and analytical tasks in the evolution of a product.

Design-for-assembly (DFA). A systematic approach to ensuring that a product can be easily assembled from discrete parts. The central aim of DFA is to reduce the number of parts and to ensure that they are easy to assemble.

Design-for-disassembly. A methodology for designing products that can be disassembled easily, typically at the end of their useful life, so that parts and materials can be recycled as part of materials management.

Design-for-the-environment (DFE). A design process in which environmental concerns such as energy conservation, materials management, and waste prevention are the primary foci. Concerns include reduction, recycling, and remanufacturing. Also called "green design". Products created according to this paradigm are called "green products."

Design-for-manufacture (DFM). Design-for-manufacture, or design-for-manufacturability, is the practice of designing a product with an emphasis on ease of manufacture. DFM reduces development costs and the time needed for product development. Hence it shortens the time to market of the product and ensures cost-effective manufacture.

Design-for-X (DFX). A group of approaches focused upon only one aspect of the design process rather than the function of the design. This single aspect of the design process is designated by the "X". Typical groups of approaches include design-for-assembly, design-for-cost, design-for-quality, and design-for-manufacture.

Design life. The time during which the product completely satisfies the design specification. It is typically measured in years or hours. Design life is a consequence of the reliability of the product.

Design methodology. A procedure for solving design problems by defining the creative and analytical aspects of the design process using a series of steps involving feedback, updating, and refinement.

Design notebook. An account of all the information describing the evolution of a product from the initial market survey to manufacture. This book should contain enough information to ensure that events, reasons for decisions, and justification of assumptions can be accurately recalled at any time.

Design parameter. A design requirement that must be satisfied by the product. A set of these design parameters constitutes the design specification.

Design parameter importance factor. A numerical scheme imposed on the design parameters to quantify their relative importance. This clarifies the design specification.

Design review. A periodic meeting of a design team and appropriate management personnel which focuses on monitoring and reviewing the progress of the design project. The design is often compared with the product design specification and an agreed-upon project schedule to ascertain the state of the project.

Design specification. A comprehensive statement of the requirements that must be satisfied by the product. The design specification is a problem definition that controls the design process from the initial conception until delivery to the customer. It states the design requirements, not how they are to be satisfied. Sometimes a design specification is called a "statement of requirements" or a "design brief."

Detail design. The outcome of the detail design phase. It specifies relevant manufacturing details such as material selection, production processes, shape, surface finish, hardness, and tolerances.

Detail design phase. The detail design phase is one of the final stages of transforming a conceptual design into an entity that can be manufactured. Specifications are developed for subassemblies and individual parts that provide the basis for the detail design tasks of analysis and synthesis. The detail design phase is the same as the final stages of the product design phase.

Detail drawing. A multidimensional drawing of a single part, typically using ANSI standard conventions, to describe the part's shape, material, size and surface finish. This drawing should be sufficiently descriptive to ensure that the part can be manufactured with the information supplied.

Deviation. The difference between the size of a manufactured part and the specified size.

Die casting. A casting process in which liquid metal under pressure is injected into a cooled metal mold.

Divergent thinking. Thinking associated with randomness, imagination, visualization, synthesis, nonjudgmental scrutiny, irregularity, and sensuality. It is the antithesis of deductive thinking and occurs in the right hemisphere of the brain. It is also called associative thinking, creative thinking, or lateral thinking.

Downgate. Any channel which conveys metal downwards into the structure of the sand mold, used in the casting process, before it enters the mold cavity.

Draft. A small taper introduced to those walls of a pattern that are parallel to the direction of pattern removal in the sand casting process. Also introduced into the shape of die cavities in the closed die forging process.

Drag. The bottom flask or segment of the body of a sand mold for sand casting.

Drilling. A process to produce or enlarge round holes with a drill bit. Drilling can be done with a drill press or other machine tools such as a lathe and tailstock combination.

Drop forging. A forging operation in which a piece of metal is formed to the desired shape by two metal die halves under a drop hammer.

Ductile material. A material that can exhibit considerable plastic deformation before breaking.

Duplication. A method of developing conceptual designs by either directly duplicating someone's concept with modifications to avoid patent infringements or else by recognizing the association between the problem to be solved and a problem with a known solution.

E

Economic analysis. A financial analysis, typically involving manufacturing, materials, and design attributes, to determine the cost of a product or process.

Embodiment design. The embodiment design phase is the first step in transforming a conceptual design into an entity that can be manufactured. It is the development of a tangible, visible form for a concept. It is situated between the conceptual design phase and the detail design phase. Embodiment design is the step in which the original qualitative concept is first quantified through appropriate analysis. Embodiment design includes an appropriate layout, materials selection, preliminary manufacturing considerations, economics, and form design considerations justified by calculations.

Emotional impediments. Impediments to creative thinking associated with the psychological safety of the individual—that is, with the desire of the individual for security, his or her fear of making a mistake, and any lack of the patience needed for embryonic ideas to mature.

Empathy. A method for creative thinking in which the individual imagines that he or she is a part of the situation for which a solution is sought.

Energy consumption. A design parameter that requires the identification and isolation of all sources of energy consumption and of energy losses. It is important for the improvement of product efficiency.

Energy utilization. The conversion of energy in a product. This attribute is the biggest cause of pollution and global warming, while directly affecting the operating cost. It should be minimized as part of a green design strategy.

Engineering. The application of scientific principles to practical ends such as the design, construction, and operation of efficient and economical structures, equipment, and systems.

Engineering analysis. Procedures that permit the performance, characteristics, quality, and cost of a part, product, or process to be ascertained by using mathematical models incorporating appropriate assumptions and approximations. Engineering analysis is the reverse of engineering design.

Engineering design. The technical element in the product realization process that uses the knowledge and techniques of engineering, science, aesthetics, economics, and psychology to establish specifications for products and their production processes; the process by which engineering descriptions and specifications are formulated to ensure that a product will possess the desired behavior, performance, quality, and cost. (Courtesy of *Improving Engineering Design,* The National Research Council, National Academy Press, 1991)

Engineering design practices. The assemblage of empirical, analytical, heuristic, and computational methods used by engineering designers. Examples include finite element analysis, design-for-manufacture, and Taguchi methods.

Engineering graphics. A method of communicating engineering concepts using geometrical and spatial representations. The representations are produced according to standards and conventions so that the ideas can be understood by any individual conversant with these protocols.

Engineering science. The traditional idealized analytical engineering disciplines including thermodynamics, strength of materials, kinematics, controls, etc.

Environmental friendliness. Concerns the impact of a product on the environment. It is assuming greater significance as societal values change more towards protecting ecosystems and preserving natural resources.

Environmental impediments. Impediments to creative thinking imposed upon individuals by the environment in which they work. They include the physical surroundings and relationships with colleagues and other workers.

Ergonomics. Originally, the ability of people to perform work. (Derived from the Greek *ergon* (work) and *nomos* (law)). Now more broadly interpreted as design focused on the interface between the product and people. Considerations include convenience and comfort. In the U.S.A. it is also called human factors engineering or human engineering.

Evaluation. The appraising of conceptual design candidates to determine which best satisfies the product design specification.

Exploded assembly drawing. Usually a three dimensional assembly drawing in which the parts of the assembly are exploded apart and spread over the drawing from their functional locations to provide a clearer understanding of the individual parts and features. This type of drawing is used extensively when technical illustrations are required in maintenance handbooks and parts catalogs.

Extrusion. A manufacturing process which produces material of long length and uniform cross section from a metal billet by causing the billet to flow under pressure through a die.

F

Face turning. Sometimes called facing, this process occurs when a lathe tool is fed at right angles to the longitudinal axis of the rotating workpiece. It generates a planar surface on the end of the workpiece or on a shoulder.

Family of parts. A set of similar parts that can be grouped together. Thus the ANSI slotted 100-degree flat countersunk head machine screws are a family of parts.

Fastener. A mechanical device that constrains the motion of two or more parts. This constraining action includes both the rigid joining of parts and the flexible joining that permits the certain relative motions. Some fasteners facilitate dis-assembly (a bolt and nut) while others do not (a rivet).

Fatigue. The failure of the mechanical properties of a part due to the repeated generation of alternating or cyclic stresses of an intensity below the tensile strength.

Feed. This machining term is defined as the distance that the tool advances along or into the workpiece each time the tool point passes a specific point in its travel over the surface.

Ferrous materials. Materials with iron as the major constituent. Pure iron is a relatively soft material and of little commercial value. However, iron alloys have major significance in engineering and offer a broad range of properties depending upon their alloying elements.

Fiber. A filamentary material with a length typically at least 100 times greater than the diameter. The diameter is of the order of 0.10 mm.

Filament winding. A process for fabricating fibrous composite parts by winding continuous reinforcing fibers around a rotating removable mandrel.

Filler rod. Filler metal in the form of a rod or wire that is added during welding to provide a joint with the desired properties.

Fillet. A rounded interior corner of a casting or forging.

Fillet weld. Any weld of triangular cross section that is not classified as a butt weld.

Final report. A detailed report at the end of a design project. Using both text and graphics, it contains in-depth accounts of decisions made and options studied.

Finite Element Modeling and Analysis. A computational method used extensively to solve field problems during the analysis phase of the design process. In the modeling phase the object is divided into finite elements connected to adjacent elements at points called nodes. During analysis, the model is subjected to forces and boundary conditions at the nodes. This numerical method is often used to perform stress analyses.

Fixturing. A manufacturing procedure involving fixtures. A fixture is a production tool that permits parts to be accurately located.

Flash. A thin web of metal formed at the sides of a forging where a small portion of the metal is forced out of the die cavity and collects between the two die halves in a gutter.

Flask. A hollow frame, open at the top and bottom, used as an enclosure for retaining the sand around the pattern during the construction of the sand mold for the casting process.

Flux. A compound which, when subjected to heat, combines with an oxide to render it controllable or to create a fluid. Used in some welding processes to control adverse effects of oxygen in the weld pool.

Forging. A manufacturing process in which metal is shaped by impact or the steady compression of the material. Usually the metal is first heated to increase deformability.

Form design. The phase that determines the shape of individual components and also the selection of the material from which they are to be manufactured. These tasks involve both analysis and creativity using information from materials science and manufacturing along with appropriate analytical and computational techniques. Form design is one of the activities of the product design phase.

Foundry. A facility where metal is melted and castings are produced.

Function. A parameter that defines what a product must do. There is usually a performance specification associated with this attribute.

G

Gantt chart. A horizontal bar chart for planning and scheduling a project. The chart indicates those tasks that are prerequisites for subsequent tasks and those tasks that can be worked on concurrently.

Gas welding. A process in which parts are melted by heat from burning gases.

Global warming. The increase of the earth's temperature because atmospheric pollution prevents heat from being radiated into space.

Governmental regulations. Legal constraints on products and their manufacture.

Grain. The natural formation of crystals in a metal when growth of the crystals has occurred simultaneously from numerous sites.

Grain flow. The elongated arrangement of the grains in the macrostructure of a metal. This is usually caused by mechanical working, such as during forging.

Gray cast iron. A cast iron in which most of the carbon in the metal exists as graphite. The face of a fracture is gray.

Green design. A design process that considers energy conservation, materials management, waste prevention, disposal, recycling, etc. Good practice results in environmentally-friendly products with good functionality, life, and performance. It is sometimes called design-for-the-environment.

Green products. Products developed by green design; products that are environmentally-friendly.

Greenhouse effect. Pollution in the earth's atmosphere prevents heat from being radiated back into space. This is responsible for an increase in the earth's temperature and changes in ecosystems. The phenomenon is analogous to a horticultural greenhouse.

Grinding. The interaction of a workpiece with a rotating abrasive wheel. Each grain acts as a cutting tool. Precision grinding is used to generate parts with high quality surfaces in both conventional metals and very hard materials. Grinding can be used to generate flat surfaces and surfaces of revolution.

Group technology. The identification of parts with similar design attributes and manufacturing characteristics so that parts families can be established.

H

Hand lay-up. A fabrication technique for manufacturing parts using composite materials by placing successive layers of reinforcement in a mold. The composite material is either preimpregnated with a binder or is subsequently coated. Subsequently the material is cured to yield the final part.

Hardness. The resistance to abrasion, indentation and cutting. A surface characteristic of a material.

Hardening. The heat treatment of a steel to increase its hardness by first quenching and then tempering.

Heat affected zone. A region of parent metal that is metallurgically affected by the heat of a welding process but it is neither plastically deformed or melted.

Heat treatment. A process used to develop certain desired properties in a metal by subjecting it to one or more thermal cycles in the solid state.

Heuristics. A group of ad hoc methods or rules of thumb procedures based upon empirical experiential information.

Historical evolution. A process for generating ideas for conceptual designs based on the premise that old solutions to the same class of problems were probably constrained by scientific and technological limitations that have now been resolved. Consequently new solutions can be created by infusing new technologies into old designs.

Human factors analysis. An evaluation process to determine whether a product serves the desired needs of the customer. This evaluation considers aesthetic, emotional, quality, safety, physical and psychological attributes.

Human factors engineering (HFE). Same as ergonomics. Design focused on the interface between the product and people.

Hypothesize-and-test. A process for creating new designs based on the formulation of an idea that is subsequently tested with a prototype of the product. Sometimes testing can be done by computational simulation.

I

Ideation. This is the conceptual design phase of the product realization process where ideas are formed.

Incremental improvement. A series of small modifications introduced on a continuing basis to reduce product cost and improve quality through experience accumulated as a product is used.

Industrial design. A discipline concerned with the aesthetics and human factors attributes of a product. Industrial designers need a broad background in engineering, fine arts, and some of the medical disciplines. They are skilled in the creation of consumer products where color, shape, style, and ergonomic aspects are important.

Industrial product. Products used to manufacture other products. They include equipment, machinery, accessories, and tools often purchased directly from the manufacturer.

Industrial scrap. Recyclable materials generated by manufacturing processes or from industrial products that are no longer functional.

Information resources. These sources include textbooks, technical journals, trade publications, vendor catalogues, the internet and www, consultants, professional codes and practices, and the patent gazette. The ability to identify appropriate sources of information is important.

Information searches. The generation of knowledge by the retrieval and study of material from libraries, reference books, industry, government, and universities.

Innovation. The introduction of novel ideas into products, organizations, or systems.

Intellectual impediments. Impediments to creative thinking associated with the inability of an individual to reason, understand, or perceive relationships and differences. They include inappropriate solution strategies, limited educational background, and the inability to communicate effectively.

Integrated product development (IPD). The development of a product in which multi-disciplinary teams work together through the entire product realisation process. Sometimes called concurrent engineering, or unified life-cycle engineering.

Interference fit. This type of fit between two parts unifies them to create a single part. This is accomplished because the interface region is created with the male dimension larger than the female dimension, and force is required to complete the assembly process.

Invention. The creation of a new device, process, or product that has never existed before.

Interlaminar. A term used in the field of composite materials to describe the interface between adjacent plies of a composite structure. Typical characteristics include fracture, shear stresses, or voids.

Intralaminar. A term used in the field of composite materials to describe a characteristic of a single lamina without reference to any adjacent lamina. The characteristic is confined to that lamina and includes fracture, shear stresses, or voids.

Investment casting. Sometimes called the "lost wax" process, this process involves the creation of a mold with a relatively thin uniform skin on a pattern. This pattern is subsequently removed, generally by melting the wax pattern, to generate a finished shell impression.

J

Jig. A device or structure used during a machining operation to hold and support the workpiece while guiding the cutting tool.

Joining. The unification of two or more parts to create a single object or an assembly. This can be accomplished in many ways including adhesives, mechanical fasteners and plastic deformation to create either temporary or permanent joints.

Just-in-time (JIT). A method of significantly reducing inventory costs by manufacturing or delivering parts and assemblies just before they are needed.

L

Landfill. A depression or excavation in the ground to hold waste material that will subsequently be covered with a layer of soil.

Laminate. A fibrous composite material made up of layers or plies that are generally bonded together by the application of pressure and heat.

Lapping. A microfinishing process in which a master form called a "lap" and a fine abrasive are moved slowly and continuously over the surface of the workpiece with light pressure. The resulting surface finish and dimensional accuracy of the surface are superior to those attained by grinding.

Lap weld. Any welded joint in which two overlapping plates or flanges are united.

Lateral thinking. This type of thinking involves randomness, imagination, visualization, synthesis, nonjudgemental scrutiny, irregularity and sensuality. It is the antithesis of deductive thinking and occurs in the right brain. Sometimes called creative thinking or artistic thinking.

Lathe. A machine tool used to produce surfaces of revolution. The workpiece is rotated about an axis while the cutting tool moves past it.

Life cycle. The lifetime of a product that includes product specification, conceptual design, product design, extraction of raw materials, manufacture, service, and disposal, including remanufacturing, recycling, and reusing.

Life-cycle cost. The total cost of a product throughout its life, from the initial concept through manufacture, use, and disposal.

Life-cycle phases. Sometimes called design phases, these stages of a product's life are planning and design specification, conceptual design, product design, manufacture; service, and product retirement.

Limits. The largest and smallest sizes shown on a drawing by a toleranced dimension. The maximum size is the upper limit and the minimum size is the lower limit.

M

Machinability. The relative ease with which a material can be machined. Material with high machinability permits parts to be manufactured at an acceptable quality at low cost with minimal difficulty.

Machine tool. A powered device used to produce a part of specified size and shape by removal of material as chips. The tool and workpiece must be rigidly supported while a mechanism moves the tool relative to the workpiece.

Machining allowance. An allowance incorporated in the shape of castings and forgings so that the original surface can be machined to generate a desired surface quality.

Macrostructure. The general form and distribution of constituents, impurities, etc., in a metal, alloy, or composite material that can be seen with the unaided eye.

Maintenance protocols. Statements that define tasks that must be performed to ensure that a product continues to perform according to the design specification.

Malleable irons. These materials are manufactured by the prolonged annealing of white iron. The resulting materials have higher ductility than gray cast irons and higher tensile strengths.

Management. The art, act, or manner of directing or controlling the activities of a group of individuals undertaking a task.

Manufacturing process. The process for producing component parts, assemblies, and complete products. It includes fabrication, assembly, testing, storage, and distribution.

Market. The target group that has a demand for a product and also the ability to purchase it.

Marketing. The conception, pricing, distribution, and promotion of ideas, goods, and services to create exchanges that satisfy individual and organizational objectives. This activity includes product, price, place, and promotion. The department in a company that convinces customers to purchase company products or services. This department will collect information on potential customers and use this data to design products to more closely satisfy customer needs.

Market investigation phase. The phase preceding the specification phase that focuses on understanding the context in which a proposed product is to function. It generates and analyzes information pertaining to customers, such as the price range that customers are willing to pay, where they are likely to purchase products, and what type of message will most likely influence them. It also is concerned with industry codes and standards, market forecasts, product evolution and status, and competitive products.

Material control. The management of materials flow throughout the production of a product. JIT is a subset of this approach which recognizes when materials are needed and coordinates their delivery.

Materials handling. The movement of parts, assemblies and raw material in a production operation or plant.

Materials selection. The selection of a material for a part through the interaction of part shape, manufacturing processes, economics, material properties and characteristics, and design functionality.

Materials management. A part of green design concerned with minimizing the generation of waste throughout a product's service life and during the extraction of raw materials, in addition to maximizing the recovery, recycling, and remanufacturing of materials.

Mathematical model. An abstract mathematical representation of a physical system used to predict its behavior in practice.

Matrix. A polymeric or metal material in which the reinforcing fibers of a composite material are embedded.

Matrix charts. Two or three dimensional grid representations of variables using orthogonal scales. The interior grid cells contain words or symbols representing the relationships. These are frequently qualitative.

Metal inert-gas welding (MIG). Electric arc welding with a consumable metallic electrode surrounded by a stream of inert gas. The arc is struck between the electrode and the workpiece and the melting electrode provides the filler for the joint.

Metal-matrix composite. Composite materials comprising carbon or ceramic fibers embedded in a metal matrix.

Microstructure. The structure of metals, alloys, and composite materials as revealed by examination under the microscope.

Mild steel. A low-carbon steel containing between 0.12 and 0.25 percent carbon.

Milling. A metal-shaping process in which a workpiece is moved past a rotating cutter to form surfaces of various shapes.

Milling cutter. A tool with multiple cutting faces on the periphery used by milling machines. They rotate relative to the workpiece as the cutter is fed through the workpiece.

Modeling. A mathematical, often computational, model representing a conceptual or more mature design. This data base is often used to communicate ideas, provide the basis for production planning and the graphical representation enhances visualization.

Modular design. A design philosophy whereby common interfaces provide for easy assembly, easy replacement of parts, easy servicing, easy upgrades and therefore reduced costs.

Mold. Any receptacle into which molten metal is introduced for the purpose of solidification into a desired shape.

Monolithic material. A material with a single homogeneous macrostructure. Examples include the commercial metals.

Morphological synthesis. A systematic method of prospecting for creative solutions to design problems by establishing a morphological matrix. Each side of each cell in the typically three-dimensional matrix is associated with a potential solution. The various permutations of these solutions permit a large variety of solutions to be developed systematically.

Multidisciplinary design activities. Design activities that involve diverse disciplines in both the arts and sciences.

Multifunctional part. A part that has more than one function. This approach reduces costs because of savings in manufacture inventory and assembly.

Multipoint cutting tool. Cutting tools with at least two cutting edges that sequentially interact with the workpiece. They are capable of high rates of material removal. A face milling cutter and the common twist drill are examples of multipoint cutting tools.

Municipal solid waste. Waste material generated in a municipality by both residential and commercial activities.

N

Numerical control (NC). A type of programmable automation of a machine tool using coordinates, letters, and symbols to describe the part geometry.

Noise. Loud, discordant, or disagreeable sounds generated by a product. Noise is increasingly regarded as an undesirable characteristic. If people must operate near the product, noise is subject to governmental regulations.

Non-ferrous materials. Materials that are not based upon iron. Examples include magnesium, aluminum and nickel.

O

Open-ended problems. Ill-defined problems that have no definitive description. They are not clearly understood, are vague, and have no single solution.

Operating costs. The cost of operating a product within its performance specification. Energy consumption, maintenance, reliability, and service life are primary cost factors.

Original design. Either the product of a conceptual design activity or the activity itself, involving the creation of a radically new product that likely will be the first of a new generation of products. This type of conceptual design is generally associated with the highest level of creative design activity.

Original equipment manufacturer (OEM). A manufacturer of industrial products that are sold to a second manufacturer as a component of a more complex product. The second manufacturer installs the component in the final product.

Orthotropic. An elastic material having three mutually perpendicular planes of symmetry.

Oxidation. Chemical combination of an element with oxygen to form an oxide. This process can be fast as in combustion or slow as in atmospheric corrosion.

Oxyacetylene welding. Gas welding in which the heat is derived from acetylene gas burning with oxygen.

P

Packaging and transportation. Design parameters concerned with ensuring that the customer receives the product in the same condition that it passed the final inspection of the quality control process.

Parent metal. In welding, this is the metal of the parts to be welded together—not the filler metal.

Part. A piece or component. Typically a metallic or plastic object that is an item in a product, such as a casting or a setscrew.

Part count. The number of parts in an assembly.

Parting line. A line around a forged part that denotes the interface or join between the two die halves. It is usually visible as a slight but distinct ridge on the surface of the forging.

Patent. A legal document granted by the government to an inventor, a company or a group of individuals which excludes others from making, using, selling a product for a limited time.

Pattern. A modified replica of the exterior shape of a metal casting. Patterns usually are made of wood and impart their shape to the sand cavity of the mold into which molten metal is poured.

Perceptual impediments. Impediments to creative thinking associated with the inability to correctly define a design problem or to recognize the information needed to define the problem. They are connected with the stereotyping of ideas or the inability to consider a problem from several different perspectives.

Planing machine. A machine tool used to perform linear cutting operations using a single-point cutting tool. The tool and the workpiece move relative to each other in a translational motion. The workpiece is clamped to a long horizontal reciprocating machine table while the tool is attached to a substantially rigid stationary structure.

Planning. The identification and sequencing of the tasks in a project.

Plastics. Materials manufactured from synthetic or natural resins. These materials can be shaped and they include polystyrene, polycarbonate, polyamides, polypropylene, polyethylenes, and epoxy.

Plastic deformation. The deformation which remains in a body after the forces causing this deformed state have been removed.

Ply. A fabric of reinforcing fibers employed in the manufacture of composite materials.

Porosity. A nonhomogeneous structure caused by voids created during the casting process.

Powder metallurgy. The process of manufacturing parts with complex solid shapes and/or parts with unique properties and microstructures from metallic powders mixed sometimes with non-metallic additives.

Prepreg. A resin-impregnated cloth in a flat form that can be stored until being used to manufacture a composite part.

Primary forming processes. Manufacturing processes in which the cast structure of a metal is destroyed by successive deformation steps. The resulting product is a semifinished product because it is subjected to further manufacturing processes in order to generate a part with the desired shape, tolerances, and surface qualities. Typical primary forming processes are extrusion, forging, rolling, and drawing.

Process planning. The activity responsible for the conversion of design data into instructions for manufacturing a product. It is the function within a manufacturing facility that establishes the processes and the process parameters to be used in to convert a piece part from its initial form into its predetermined final form described on a engineering drawing.

Processible material. A material that can be easily shaped during manufacture.

Product. Something produced as the outcome of a design process.

Product cost. The total cost of a product in the marketplace. Typical components include the cost of materials, manufacture, transportation, packaging, and financial overhead.

Product delivery process (PDP). The development of a design specification through conceptualization to the introduction of the product into the marketplace and beyond.

Product design. The detailed plans and documentation that enable a product to be made in such a way that it economically satisfies the design specification. The result of the product design phase of the product realization process.

Product design phase. The phase of the design process initiated by a concept and terminated with the design of a product that is manufacturable through the specification of materials, tolerances, surface finishes, dimensions, heat treatment, bought-out parts, etc.

Product development cycle. The total time to conceptualize a new product and bring it to market.

Product development teams. Groups of individuals with multidisciplinary skills, typically from different departments in a company and also outside vendors. Typically these teams include accountants, production personnel, manufacturing engineers and product designers.

Product liability. The aspect of design concerned with legislation pertaining to the liability of a company and its the individual designers responsible for developing a product when injury or death occurs as a result of human interaction with a product. The responsibilities of the design team in this regard mandate meticulous engineering practices and the development of comprehensive documentation.

Product life cycle. The period of time that includes product specification, conceptual design, product design, extraction of raw materials, manufacture, service, and disposal, including remanufacturing, recycling, and reusing.

Product realization process (PRP). The process by which new or improved products are conceived, designed, manufactured, brought to market, and subsequently supported. It includes determining customers' needs, translating these needs into engineering specifications, designing the product, determining the appropriate production processes, deciding upon appropriate support processes, and coordinating all of these activities.

Product retirement. When the product no longer complies with the design specification it is retired. This decision triggers several options including disassembly, remanufacture, recycling, reuse of components and disposal.

Product testing. The procedure for experimentally determining the product performance relative to the performance originally written in the product design specification. Often undertaken in a test laboratory using prototypes.

Product upgrade. The extension of the service life of a product by replacement of obsolete parts to enable the product to comply with or exceed the original performance specification rather than be scrapped.

Production. The manufacture of the product, the creation of an artifact. Design activities are limited to correcting unforeseen problems.

Production process. This can mean the action of converting raw materials into the finished product. A very broad term including all of the ancillary activities as well as a machine tool cutting metal. Alternatively the term means the manufacturing process where the part is being shaped.

Production volume. The number of items, products, assemblies or parts manufactured, typically relative to time. A larger production volume usually results in a reduced part price.

Program evaluation and review technique (PERT). A logic network for planning and sequencing a project.

Progress report. An in-depth account of a program containing both text and graphics. They are created periodically to review the status of the program and document the decisions made.

Prototype. A physical model of a product that will enable the experimental evaluation of a design and its performance characteristics prior to production of a product.

Pultrusion. A continuous process for manufacturing composite parts with a constant cross-sectional shape.

Q

Quality. The totality of characteristics of a product or service that bear on its ability to satisfy stated or implied needs. Quality is closely related to cost; consequently quality involves the ability of a product to satisfy the customers requirements at a price that is acceptable to the customer.

Quality control. A discipline dedicated to determining whether products comply with the specifications.

Quality engineering. An engineering philosophy that increases the competitiveness of new products while simultaneously increasing their quality and decreasing their cost.

Quality function deployment (QFD). A systematic process for translating customer requirements into appropriate technical specifications during all stages of the product realization process. The interdisciplinary QFD team identifies product attributes desired by the customer before benchmarking these relative to the competition and developing attributes that offer a competitive advantage.

Quality loss function. A function employed in the Taguchi methods of quality engineering and defined as the cost of the variation of the characteristics of a product relative to some target quantity due to product tolerances, environmental conditions, and operating costs, etc. This financial loss is sometimes referred to as the quality loss.

Quality rating. A factor employed to evaluate the quality of a conceptual design by comparing it with an ideal design. It is the ratio of the number of points assigned to the particular conceptual design divided by the number of points assigned to the ideal design. This approach can also be applied to subassemblies and it enables areas of weakness to be identified.

Quantity. The number of products requested by the customer. It dictates the manufacturing processes to be employed in production and the cost of the product.

R

Rapid prototyping. The creation of physical models from a CAD database by, for example, stereolithography or fused vapor deposition modeling.

Reaction injection molding. A molding process for manufacturing composite parts in which two or more liquid polymers are mixed prior to entering a closed mold where they react and solidify.

Realistic thinking. This type of thinking is associated with logic, analysis, judgement and evaluation. It is the antithesis of creative thinking and it occurs in the left brain. Sometimes called convergent thinking, vertical thinking and deductive thinking.

Reaming. The sizing of a drilled hole by slightly enlarging it to a good surface finish by a long cylindrical fluted tool called a reamer.

Recyclability. The relative ease of reintroducing into the raw materials supply a recovered material that would otherwise enter the waste stream.

Recycled materials. This green design philosophy is concerned with the reassimilation of a material into the raw material stream in a manner such that it may be used for the purpose identical or similar to that of its first use.

Recycled parts. The remanufacturing of used parts to a condition where they perform as well as a new part.

Recycling. The diversion from the waste stream of scrap material that can be used to manufacture new products. This reduces the volume of virgin materials consumed in the manufacture of new products.

Reduce, reuse, and recycle. An environmentally-sensitive approach to the development of products from conception to disposal and beyond into the next generation of products.

Reinforcement. A material added to a matrix binder to form a composite material with the desired properties. Typical reinforcements are fibers or whiskers.

Reliability. Product reliability is the probability that a product will perform its function when subjected to service conditions for a time period all specified by the customer. Reliability is a function of the product realisation process.

Remanufacturing. The restoration of used products to a condition which, if not precisely as-new, has performance characteristics that approximate new. Remanufacturing involves the disassembly, cleaning, and refurbishment of usable parts in addition to the provision of new parts where necessary, followed by reassembly and testing.

Residual stresses. Stresses in a casting, forging, or rolled section caused by uneven cooling rates.

Resin. A polymeric material used to bind together the reinforcing material of a composite.

Resin-transfer molding (RTM). A molding process for manufacturing composite parts in which resin is transferred into a closed mold in which the reinforcing fibers have been placed. The part is released after it has cured.

Reuse. The reutilization of a product such as a refillable bottle. It is typically part of an environmentally sensitive protocol.

Reverse engineering. The process of disassembling and analyzing a product to determine how it was designed and manufactured. Enhancements are then suggested or ideas incorporated in other products.

Riser. A vertical cavity connecting the mold cavity to the upper surface of the sand in the cope of the sand casting process. It can permit gases to escape and provide a reservoir of molten metal during solidification.

Robot. A computer-controlled device used to perform many repetitive manufacturing tasks, such as painting, pick-and-place, and assembly.

Robust design. An approach focused on the reduction of quality loss in products and processes. The features and function of the product are determined before systematically studying all of its components in order to ascertain those which affect manufacturability or reliability. Subsequently, the tolerances imposed upon the parts are determined in order to provide the greatest freedom to manufacture and operate the product. This approach is closely related to both cost and reliability.

S

Safety. The design parameter concerned with the health of human beings in the neighborhood of products, but it can also pertain to equipment and property.

Sanitary landfill. A depression in the ground containing an impervious liner to minimize water and soil pollution where waste material is dumped prior to being buried beneath a layer of soil.

Sand casting. The technique of manufacturing a casting in a sand-filled mold.

Scheduling. The assignment of the tasks in a project to specific calendar dates.

Scrap. A by-product of the material-purification process, the material-conversion process, or the residue at the end of the product service life.

Secondary forming processes. Manufacturing processes in which semifinished products are transformed into finished products by developing parts of greater accuracy and surface finish. Typical secondary processes are the sheet-metal-working processes, machining, and finishing operations.

Semifinished product. A product, typically manufactured by one of the primary manufacturing processes, that is subsequently reworked to produce the final product. Examples include plates, sheet material, bar stock, rolled sections, and extrusions which may be employed as the stock material for a milling operation or the parts for a weldment.

Sequential engineering. The classical method of developing products involving a number of discrete isolated departments in a company. The product is developed in a chronological sequence of events from the initial market research to the final sale to the customer. The sequential evolution of a product from the market research group, through research, design, development, and manufacturing to sales. Sequential engineering is sometimes called sequential product development.

Sequential product development. The classical method of developing products involving a number of discrete isolated steps. The product is developed in a chronological sequence of events from the initial market research to the final sale to the customer. Sequential product development is sometimes called sequential engineering.

Service life. The duration of time that the product satisfies the design specification while operating under normal service conditions.

Servicing. An activity that supports the installation, training, maintenance, and repair of a product for the customer.

Set-up time. The time expended to adjust and configure a machine tool so that it can perform a prescribed set of operations. It involves the installation of cutting tools, the selection of cutting speeds and feed rates, and the adjustment of stops to limit travel.

Shape. The form of an object created by the internal spatial relationships governing the arrangement of edges and surfaces.

Shaping machine. A machine tool used to produce flat surfaces. A reciprocating motion occurs between a single-point cutting tool and a stationary workpiece. The tool generates a chip during its forward stroke and, as the tool returns, the workpiece is moved laterally for the next cutting stroke.

Sheet. A flat-finished part of very large surface area and relatively small uniform thickness. Metal sheet is manufactured by rolling or forging. The thickness is usually less than 6 mm or one-quarter of an inch.

Shelf life. The time that a product can be stored prior to use. It may be unimportant for some products, but other products must be stored under prescribed conditions and for no more than a specified time because of product degradation, etc.

Shrinkage cavity. A void in a casting that is formed in the last region of the casting to solidify. It is created by surface tension phenomena which cause the liquid metal to be attracted to the previously solidified metal. Shrinkage cavities cause weakness in the casting.

Simultaneous engineering (SE). The same as concurrent engineering. An integrated approach to the development of a new product in which multidisciplinary teams work in parallel for the entire product realisation process.

Simultaneous product development. The same as concurrent engineering. An integrated approach to the development of a new product in which multidisciplinary teams work in parallel for the entire product realisation process.

Single-point cutting tool. A tool with only one cutting edge, such as a lathe tool.

Six-sigma method. A statistical method for ensuring that failure will occur less than three times in a million opportunities. It quantifies the degree of deviation from a desired performance of parts, products and processes.

Size. The spatial dimensions of an object. It can also be used in a relative sense in the comparison of objects. This design parameter should be minimized because of the cost of materials and other cascading consequences.

Solid modeling. The representation of solid objects using a digital computer.

Specification phase. The phase of the design process preceding the conceptual design phase and in which the design specification for a product is generated. The design specification is a comprehensive statement of the product requirements.

Spheroidal-graphite irons. A class of cast irons in which the graphite is present in finely dispersed spheroids. These materials are strong and more ductile than the gray irons.

Spot welding. Resistance welding in which a weld is produced at a spot between two lapped thin sheets held by electrodes. Pressure is exerted by the pair of electrodes. The resistance at the interface creates heat and unification occurs when a potential difference is applied.

Stainless steel. One type of steel alloy that contains high percentages of chromium to provide a high resistance to corrosion.

Standards. Rules that facilitate clear communication of technical ideas governing how parts are manufactured and how they are represented on technical drawings. There are several standards organizations in the world including the American National Standards Institute (ANSI), the International Standards Organization (ISO), Japanese Standards (JIS), British Standards (BS), German standards (VDI), Department of Defense (DOD) and the U.S. Military (MIL).

Statistical process control (SPC). A process focused on product quality that permits a manufacturer to calculate the number of parts to be sampled and inspected in order to ensure that desired quality standards are being satisfied. From the analysis of a small group of parts, the manufacturer can estimate the probability that the rest of the group does or does not meet the specifications.

Steel. An alloy of iron and carbon in which the carbon is present as a carbide. Steel may also contain other elements in controlled quantities. The metal can be hot- or cold-worked.

Stereolithography. A method of rapid prototyping. A computer model of the part is thinly sliced before it is constructed. A pair of light beams move their focal point around in a bath of photosensitive polymer to create one layer of solid polymer at a time. Once one layer has solidified the next layer is deposited upon it to gradually build up the part.

Straddle milling. The simultaneous milling of two sides of a workpiece by side milling cutters.

Strain hardening. The increase in hardness and strength developed during cold working, sometimes called work hardening, where there is plastic deformation below the recrystallization temperature.

Stress relieving. Stresses are relieved by heating the metal to an appropriate temperature before permitting it to cool slowly.

Sub-assembly. A rational assembly of parts, that typically are a functioning entity, which is only part of a larger assembly. These smaller assemblies work in harmony together to form the complete system.

Surface finish. Associated with the quality of a part and measured by microscopic characteristics of the surface such as waviness or roughness. Typical measures include the centerline average, the height of the waviness, the roughness width, and the waviness width. Each manufacturing process has a range of surface finishes.

Synectics. A technique of defining and attempting to solve problems that utilizes creative thinking. It involves the exchange of ideas among a small group of individuals who have diverse areas of specialization.

Systematic design procedure. The creation of a product and the development of plans to manufacture it requires the definition of the problem, the creation of numerous potential solutions, the evaluation of these concepts, and the transformation of this vision into a product.

T

Taguchi methods. A variety of methods that statistically determine the required quantitative features of a design or a manufacturing process to render it robust against uncertainties, variations, and disturbances in the governing parameters. The objective of these methods is to deliver high quality products at low cost.

Tap. A tool used to cut threads in a hole of diameter nominally equal to the crest diameter of the threads.

Taper turning. This machining process occurs when a single-point lathe tool moves in a straight line inclined to the longitudinal axis of the rotating workpiece. A conical face is created.

Team. A group of individuals who work together toward some objective.

Technical drawing. A specialized type of graphics used to communicate technical information. Includes 3-D computer modeling, drafting and the illustrating of a technical device.

Technical impediments. Impediments to creative solutions to engineering problems from inappropriate computational tools, technological limitations, materials limitations, and immature manufacturing processes.

Tempering. The heating of a previously quenched material to develop an increase in ductility.

Thermoplastic. A plastic material that is capable of being repeatedly softened by heating and hardened by cooling.

Thermoset. A plastic material which can be irreversibly cured to a solid state.

Tolerance. The guaranteed maximum deviation from the specified value of a component characteristic under stated conditions. An example is the set of limits on the dimensions of a part to ensure that it will fulfill the part specification. Tolerances are necessary because no manufacturing process can routinely make a part to exact dimensions. Tolerances should be set as large as possible within the specification because close tolerances are synonymous with much higher costs. Tolerances are typically imposed upon geometric properties such as perpendicularity, parallelism, concentricity, flatness, and straightness.

Tool steel. Steel capable of being hardened and tempered which, along with other properties, makes it suitable as a tool material.

Total quality management (TQM). A management philosophy focused on product quality and increasing the value of the product for the customer. TQM requires continual attention to quality at every phase of the product realization process by every member of the organization.

Toughness. The capacity of a material to resist loadings that generate high strain rates.

Tungsten inert-gas welding (TIG). Electric arc welding with a non-consumable tungsten electrode surrounded by a stream of inert gas. The arc is struck between the electrode and the workpiece and the filler metal for the joint is provided by a filler rod. There is independent control of the heat input to the joint and the amount of filler material.

Turning. A machining process using a lathe to create cylindrical or irregularly shaped internal and external surfaces of revolution. This is accomplished by rotating or "turning" the workpiece as it interacts with a cutting tool.

Twist drill. A tool with two cutting edges and two helical flutes along the length of the drill. This tool is used to produce holes in a part by rotating the drill about its longitudinal axis and forcing the tool against the workpiece in a drill press.

U

Unified life-cycle engineering. An integrated approach to the development of a new product in which multi-disciplinary teams work in parallel from conceptualization through to the introduction of the product in the marketplace. Sometimes called concurrent engineering, concurrent product development, or integrated product development.

V

Value engineering. The systematic application of techniques which identify function, prior to establishing a value for that function, and subsequently providing that function at the lowest cost. This approach provides a tool for ascertaining the financial consequences of design decisions.

Variant design. Either the product resulting from a conceptual design activity or the activity itself. It involves a variation of the size or configuration of an existing product. The function and the solution principle remain unchanged.

Vendor. A company selling parts or assemblies used in a product. Sometimes called a supplier.

Vertical thinking. This type of thinking is associated with logic, analysis, judgement and evaluation. It is the antithesis of creative thinking and it occurs in the left brain. Sometimes called convergent thinking, realistic thinking and deductive thinking.

Virgin material. Material that has not been subjected to any industrial processing, such as bauxite ore for the manufacture of aluminum, petroleum for the manufacture of plastics, or wood for the manufacture of paper.

Void. A pocket of gas trapped in a laminate subsequently cured, or a gas pocket trapped in a casting.

W

Waste. Solid, liquid, and gaseous materials with low economic worth that are disposed of by depositing them in landfills, discharging them into treatment facilities, or venting them to the atmosphere. Waste is those products or residuals that cannot be used economically.

Waste stream. The confluence, unification, and subsequent flow through time of the total waste material generated by the numerous sources in a region, a community, a facility, or a residence.

Weight. A design parameter closely associated with the material selection process. It should be minimized because additional weight increases the product cost for manufacture, transportation, and energy consumption.

Weld. A union between pieces of metal at faces rendered plastic or in a liquid state by the application of heat, pressure, or both. Filler metal may be added to improve this union.

Weldment. A structure fabricated from two or more parts by one or more welding processes.

Weld zone. The region of a welded joint comprising the sum of the heat-affected-zone and the weld-metal zone.

White irons. A class of irons containing hard cementite but no graphite. These ferrous materials are difficult to machine because they are hard and brittle. They are primarily used when a wear resistance surface is needed.

Wireframe model. A computer model of a part or assemblage using only edges and vertices.

Working drawing. Sometimes called a blueprint, this type of drawing specifies the manufacture and a set of these drawings can specify the assembly of the design. Generally these drawings include specifications and written instructions.

Workpiece. The object, part, or work that interacts with a tool and subsequently becomes a finished or a semifinished part.

Appendix D: Bibliography

Adams, J. L. *Conceptual Blockbusting*. Reading, MA: Addison-Wesley, 1974.

Akao, Y., ed. *Quality Function Deployment*. Cambridge, MA: Productivity Press, 1990.

Alger, J. R. M. and C. V. Hayes. *Creative Synthesis in Design*. Englewood Cliffs, NJ: Prentice-Hall, 1964.

Allen, C. W., ed. *Simultaneous Engineering*. Dearborn, MI: Society of Manufacturing Engineers, 1990.

Allen, M. S., *Morphological Creativity*. Englewood Cliffs, NJ: Prentice-Hall, 1962.

Andreasen, M. M., S. Kahler, T. Lund, and K. G. Swift. *Design for Assembly*. Kempston, UK: IFS Publications Ltd., 1988.

Argarwal, B. D. and L. J. Broutman. *Analysis and Performance of Fiber Composites*. New York: John Wiley, 1980.

Ashbee, K. *Fundamental Principles of Fiber Reinforced Composites*. Lancaster, PA: Technomic, 1989.

Ashby, M. F., *Materials Selection in Mechanical Design*. Oxford, UK: Pergamon Press, 1992.

Asimow, M. *Introduction to Design Methods*. Englewood Cliffs, NJ: Prentice-Hall, 1982.

ASM Engineered Materials Reference Book. Metals Park, OH: ASM International, 1989.

ASM Metals Handbook. Metals Park, OH: ASM International, 1990.

Avallone, E. A. and G. Baumeister, eds. *Mark's Standard Handbook of Mechanical Engineering*. New York: Reinhold, 1988.

Bakerjian, R., ed. *Tool and Manufacturing Engineers Handbook: Volume 6, Design for Manufacturability*. Dearborn, MI: Society of Manufacturing Engineers, 1992.

Barass, R. *Scientists must Write*. New York: Chapman and Hall, 1993.

Beer, D. F. *Writing and Speaking in the Technology Professions*. New York: IEEE Press, 1992.

Blake, G. and R. W. Bly. *Elements of Technical Writing*. New York: Macmillan, 1993.

Boothroyd, G. and P. Dewhurst. *Product Design for Assembly*, Boothroyd and Dewhurst, Inc., Wakefield, RI, 1987.

Boothroyd, G., Dewhurst, P., and Knight, W. *Product Design for Manufacture and Assembly*, Marcel Dekker, Inc., New York, 1994.

Bralla, J. G., ed. *Handbook of Product Design for Manufacturing*. New York: McGraw-Hill, 1986.

Broutman, L. J. and R. H. Krock, eds. *Composite Materials, Volumes 1-8*. New York: Academic Press, 1975.

Buhl, H. R. *Creative Engineering Design*. Ames, Iowa: The Iowa State University Press, 1968.

Burall, P. *Green Design*. London: The Design Council, 1991.

Burgess, J. H. *Designing for Humans: The Human Factor in Engineering*. Princeton, NJ: Petrocelli Books, 1986.

Burke, James. *Connections*. Boston: Little, Brown & Company, 1978.

Burstal, A. F. *A History of Mechanical Engineering*. London: Faber and Faber, 1963.

Chaplin, C. R. *Creativity in Engineering Design: the Educational Function*. The Royal Academy of Engineering, London, 1989.

Chironis, N. P. *Mechanisms and Mechanical Devices Sourcebook*. New York: McGraw-Hill, 1991.

Chow, W., *Cost Reduction in Product Design*. New York: Van Nostrand Reinhold, 1978.

Christensen, R. M. *Mechanics of Composite* Materials. New York: John Wiley, 1979.

Cohen, L., *Quality Function Deployment*. New York: Addison-Wesley, 1995.

Corbett, J., M. Dooner, J. Meleka, and C. Pym. *Design for Manufacture*. Reading, MA: Addison-Wesley, 1991.

Cornish, E. H. *Materials and the Designer.* Cambridge, UK: Cambridge University Press, 1987.

Cox, C. M. *The Early Mental Traits of Three Hundred Geniuses.* Stanford, CA: Stanford University Press, 1926.

Crane, F. A. A. and J. A. Charles. *Selection and Use of Engineering Materials.* London, UK: Butterworths, 1984.

Cross, N. *Engineering Design Methods.* New York: John Wiley, 1989.

DeBono, E. *Lateral Thinking.* New York, Harper & Row, 1970.

Dertouzos, M. L., R. K. Lester, and R. M. Solow. *Made in America.* Cambridge MA: MIT Press, 1989.

Dieter, G. E. Engineering Design. *A Materials and Processing Approach.* New York: McGraw Hill, 1983.

Dixon, J. R. *Design Engineering: Inventiveness, Analysis, and Decision Making.* New York: McGraw-Hill, 1966.

Dostal, C. A., ed. *Engineered Materials Handbook, Volume One: Composites.* Materials Park, OH: ASM International, 1987.

Doyle, L. E., C. A. Keyser, J. L. Leach, G. F. Schrader, and M. B. Singer. *Manufacturing Processes and Materials for Engineers.* Englewood Cliffs, NJ: Prentice-Hall, 1985.

Ealey, L. A. Quality by Design: *Taguchi Methods and U.S. Industry.* Dearborn, MI: ASI Press, 1988.

Ebel, H. F., C. Bliefert, and W. E. Russey. *The Art of Scientific Writing.* VCH, 1987.

Eckold, G. Design and Manufacture of Composite Structures. New York: McGraw-Hill, 1994.

Eder, W. E. *A European Outlook on Engineering Design in Education*, Journal of Engineering Education, 82, (2), 118-122.

El Wakil, S. D. *Processes and Design for Manufacturing*. Englewood Cliffs, NJ: Prentice-Hall, 1989.

Engineers Guide to Composite Materials. Metals Park, OH: ASM International, 1987.

Farag, M. M. *Selection of Materials and Manufacturing Processes for Engineering Design*. Hemel Hempstead, UK: Prentice Hall International, 1989.

Fleming, Q. W., J. W. Bronn, and G. C. Humphreys. *Project and Production Scheduling*. Chicago: Probus Publishing Company, 1987.

French, M. J. *Invention and Evolution-Design in Nature and Engineering*. Cambridge: Cambridge University Press, 1988.

French, M. J. *Conceptual Design for Engineers*. The Design Council: London, 1985.

French, M. J. *Form, Structure, and Mechanism*. Springer Verlag: New York, 1992.

Fritz, R. *Creativity*. New York: Fawcett, 1991.

Funk, E. R., and L. J. Rieber. *Handbook of Welding*. Boston, MA: Breton Publishers, 1985.

Galluzzi, P., ed. *Leonardo da Vinci: Engineer and Architect*. Montreal: The Montreal Museum of Fine Art, 1987.

Gandhi, M. V. and B. S. Thompson. *Smart Materials and Structures*. London: Chapman & Hall, 1992.

Gibbs-Smith, C., and G. Rees. *The Inventions of Leonardo da Vinci*. Oxford: Phaidon Press Limited, 1978.

Golley, J. *Whittle: The True Story*. Washington, D.C.: Smithsonian Institution Press, 1987.

Gombrich, E. H. *The Story of Art*. London: Phidon Press, 1967.

Gordon, W. J. J. *Synectics, the Development of Creative Capacity*. New York: Harper Brothers, 1961.

Grayson, C. J. and C. O'Dell. *American Business a Two-minute Warning: Ten Changes Managers Must Make to Survive Into the 21st Century*. New York: Free Press, Macmillan, 1988.

Gray, T. G. F., J. Spence, and T. H. North. *Rational Welding Design*, Butterworths, London 1975.

Green Products by Design: Choices for a Cleaner Environment. Washington, DC: U.S. Congress, Office of Technology Assessment, OTA-E-541, 1992.

Harbinger, S. A., A. B. Whitmer, and R. Price. *English for Engineers*. New York: McGraw-Hill, 1951.

Harris, B. *Engineering Composite Materials*. London: The Institute of Metals, 1986.

Harris, C. E., M. S. Pritchard, and M. J. Rabins. *Engineering Ethics*. Belmont, CA: Wadsworth Publishing Company, 1995.

Hartley, J. R. *Concurrent Engineering*. Cambridge, MA: Productivity Press, 1992.

Henstock, M. E. *Design for Recyclability*. London: The Institute of Metals, 1988.

Hine, C. R. *Machine Tools and Processes for Engineers*. New York: McGraw-Hill, 1971.

Hollins, W. and S. Pugh. *Successful Product Design*. London: Butterworths, 1990.

Hubka, V. and W. E. Ernst. *Theory of Technical Systems: A Total Concept Theory for Engineering Design*. New York: Springer Verlag, 1988.

Hunter, T. A. *Engineering Design for Safety*. New York: McGraw-Hill, 1992.

Improving Engineering Design. National Research Council, National Academy Press, Washington DC, 1991.

Jones, F. D. *Ingenious Mechanisms for Designers and Inventors, Vols. 1, 2, and 3*. New York: Industrial Press, 1930, 1936, 1951.

Jones, J. C. *Design Methods*. New York: Van Nostrand Reinhold, 1992.

Jones, R. M. *Mechanics of Composite Materials.* Washington DC: Scripta Book Company, 1975.

Kerzner, H. *Product Management: A Systems Approach to Planning, Scheduling, and Controlling.* New York: Van Nostrand Reinhold Company, 1979.

Kim, S. H. *Essence of Creativity.* New York: Oxford University Press, 1990.

Laithwaite, E. *An Inventor in the Garden of Eden.* New York: Cambridge University Press, 1994.

Langley, M., ed. *Carbon Fibres in Engineering.* London: McGraw-Hill, 1973.

Lindberg, R. A. and N. R. Braton. *Welding and other Joining Processes.* Boston: Allyn and Bacon, 1976.

Love, S. F. *Planning and Creating Successful Engineered Designs.* New York: Van Nostrand Reinhold, 1980.

Lubin, G., ed. *Handbook of Composites.* New York: Van Nostrand-Reinhold, 1982.

Lumsdaine, E. and M. Lumsdaine. *Creative Problem Solving: Thinking Skills for a Changing World.* New York: McGraw-Hill, 1995.

Matousek, R. *Engineering Design: A Systematic Approach.* London: Blackie and Son, 1969.

Mayer, R. E. *Thinking, Problem Solving, Cognition.* New York: W. H. Freeman and Company, 1983.

McCormick, E. J., and M. S. Sanders. *Human Factors in Engineering and Design.* New York: McGraw-Hill, 1982.

Michaelson, H. B. *How to Write and Publish Engineering Papers and Reports.* Philadelphia, PA: ISI Press, 1982.

Michaels, J. V. and W. P. Wood. *Design to Cost.* New York: Wiley, 1989.

Milewski, J. V. and H. S. Katz, eds. *Handbook of Reinforcements for Plastics.* New York: VanNostrand-Reinhold, 1987.

Miller, L. C. G. *Concurrent Engineering Design*. Dearborn, MI: Society of Manufacturing Engineers, 1993.

Minto, B. *The Pyramid Principle: Logic in Writing and Thinking*. London: Pitman, 1991.

Mumford, L. *Technics and Civilizations*. New York: Harcourt, 1934.

Newell, J. A. and H. L. Horton. *Ingenious Mechanisms for Designers and Inventors, Vol. 4*. New York: Industrial Press, 1967.

Niebel, B. W., A. B. Draper, and R. A. Wysk. *Modern Manufacturing Process Engineering*. New York: McGraw-Hill, 1989.

Orstein, R. and R. F. Thompson. *The Amazing Brain*. London: The Hogarth Press, 1985.

Osborn, A. F. *Applied Imagination. Principles and Procedures of Creative Problem Solving*. New York: Scribner and Sons, 1953.

Pacey, A. *Technology in World Civilization-A Thousand Year History*. Cambridge, MA: MIT Press, 1990.

Pahl, G., and W. Beitz. *Engineering Design*. English edition edited by K. Wallace. London: The Design Council, 1984.

Parnes, S. J. and H. F. Harding. *A Source Book for Creative Thinking*. New York: Scribner's and Sons, 1962.

Parsons, W. B. *Engineers and Engineering in the Renaissance*. Baltimore: Williams and Wilkins, 1939.

Philips, L. N., ed. *Design with Advanced Composite Materials*. London: The Design Council, 1989.

Piggot, M. R. *Load Bearing Fibre Composites*. Oxford, UK: Pergamon, 1980.

Plastics Recycling: A Strategic Vision Beyond Beverage Bottles. Washington, DC: Plastics Recycling Foundation Inc., 1992.

Poincare, H. *Science and Method*. London: Nelson, 1914.

Pugh, S. *Total Design: Integrated Methods for Successful Product Engineering*. Reading, MA: Addison-Wesley, 1991.

Ray, M. S. *Elements of Engineering Design*. Englewood Cliffs, NJ: Prentice-Hall, 1985.

Redford, A. and J. Chal. *Design for Assembly: Principles and Practice*. New York: McGraw-Hill, 1991.

Reuleaux, F. *Der Construkteur*. Braunschweig: Vieweg, 1882.

Reuleaux, F., and C. L. Moll. *Construktionlehre für den Maschinenbau*. Braunschweig: Vieweg, 1862.

Roberts, R. M. *Serendipity: Accidental Discoveries in Science*. New York: John Wiley, 1989.

Romans, D. and E. N. Simons. *Welding Processes and Technology*. London: Pitman, 1974.

Rosato, D. V., D. P. DiMattia, and D. V. Rosato. *Designing with Plastics and Composites: A Handbook.* New York: Van Nostrand-Reinhold, 1991.

Rosenstein, A. B., R. R. Rathbone, and W. F. Schneerer. *Engineering Communications*. Englewood Cliffs, NJ: Prentice-Hall, 1964

Rubens, P. *Science and Technical Writing*. New York: Henry Holt and Company, 1992

Schey, J. A. *Introduction to Manufacturing Processes*. New York: McGraw-Hill, 1987.

Schwartz, M. M. *Composite Materials Handbook*. New York: McGraw-Hill, 1984.

Sheppard, A. Aesthetics: *An Introduction to the Philosophy of Art*. Oxford: Oxford University Press, 1987

Singer, C. and T. I. Williams, eds. *A History of Technology. Vols. I, II, III, IV, V*. New York: Oxford University Press, 1954-1984

Souder, W. E. *Management Decision Methods for Managers of Engineering and Research*. New York: Van Nostrand Reinhold, 1980.

Stoll, H. W. *Design for Manufacture: an Overview*, ASME Applied Mechanics Reviews, 39 (9, September 1986): 1356-1364.

Strong, A. B. *Fundamentals of Composites Manufacturing: Materials, Methods, and Applications*. Dearborn, MI: Society of Manufacturing Engineers, 1989.

Strunk, W., and E. B. White, *The Elements of Style*, 3rd ed. New York, MacMillan, 1979

Suh, N. P. *The Principles of Design*. Oxford: Oxford University Press, 1990.

The Chicago Manual of Style. 14th ed. Chicago: The University of Chicago Press, 1993.

"The application of Design-to-Cost at Rolls Royce", NATO/AGARD Lecture Series 107, May 1980.

The Encyclopaedia Britannica, 15th ed. Chicago: Encyclopaedia Britannica, Inc., 1992.

Trucks, H. E. *Designing for Economical Production*. Dearborn, MI: Society of Manufacturing Engineers, 1987.

Tsai, S. W. and H. T. Hahn. *Introduction to Composite Materials*. Westport, CT: Technomic, 1980.

Turner, K., ed. *Sustainable Environmental Management*. London: Bellhaven Press, 1988.

Ullman, D. G. *The Mechanical Design Process*. New York: McGraw-Hill, 1992.

Urban, G. L. and J. R. Hauser. *Design and Marketing of New Products*. Englewood Cliffs, N. J.: Prentice Hall, 1993.

Van Gundy, A. B. *Techniques of Structured Problem Solving*. New York: Van Nostrand-Reinhold, 1988.

Van Weenen, J. C. *Waste Prevention: Theory and Practice*. The Hague: CIP-Gegevens Koninklijke Bibliotheek, 1990.

Vincent, J. F. V. *Structural Biomaterials*. London: Macmillan Press, 1982.

Weisberg, R. W. *Creativity: Beyond the Myth of Genius*. New York: W. H. Freeman and Company, 1993.

Whitehouse, G. E. *Systems Analysis and Design Using Network Techniques*. Englewood Cliffs, NJ: Prentice Hall, 1973.

Whitney, J. M., I. M. Daniel, and R. B. Pipes. *Experimental Mechanics of Fiber Reinforced Composite Materials.* Brookfield Center, CT: Society for Experimental Stress Analysis Monograph No. 4, 1982.

Womack, J. P., D. T. Jones, and D. Roos. *The Machine That Changed the World.* New York: Harper Collins, 1991.

Yaeger, J., D. Rutan, and P. Patton. *Voyager.* New York: Alfred A. Knopf, 1987.

Yankee, H. W. *Manufacturing Processes.* Englewood Cliffs, NJ: Prentice-Hall, 1979.

Index

Symbols

A

B

D

F

N

Q

R

S

T

U

V

W

X

Y

Z

no entries